PHOTON 2000

Related Titles from AIP Conference Proceedings

PHOTON 2000

International Conference on
the Structure and Interactions of the Photon

Ambleside, England   26–31 August 2000

EDITOR

A. J. Finch

Lancaster University, United Kingdom

Melville, New York, 2001
AIP CONFERENCE PROCEEDINGS ■ VOLUME 571

Editor:

A. J. Finch
Department of Physics
Lancaster University
Lancaster LA1 4YB
UNITED KINGDOM

E-mail: a.finch@lancaster.ac.uk

L.C. Catalog Card No. 2001091923
ISBN 0-7354-0010-5
ISSN 0094-243X
Printed in the United States of America

CONTENTS

Preface. ix
Committees. xi

PHOTON STRUCTURE FUNCTION

Structure Functions for the Virtual and Real Photons 3
    M. Krawczyk
A Model of F_2^γ 18
    B. Badełek, M. Krawczyk, and J. Kweiciński, and A. M. Staśto
A High-Q^2 Measurement of the Photon Structure Function F_2^γ
at LEP2 25
    R. J. Taylor
Measurement of the Low-x Behavior of the Photon Structure
Function F_2^γ 31
    E. W. Clay
First Measurement of the Photon Structure Function $F_{2,c}^\gamma$ 37
    R. Nisius (for the OPAL Collaboration)
Polarized and Unpolarized Structures of the Virtual Photon 43
    K. Sasaki and T. Uematsu
Partonic Structure of γ_L^* in Hard Collisions. 49
    J. Chýla
QCD Analysis of $F_2^\gamma(x,Q^2)$: An Unconventional View 55
    J. Chýla

DIFFRACTION

Introduction to Diffractive Photoprocesses 63
    G. Shaw
The Scattering of Photons on Proton and Photon Targets 76
    E. Gotsman, E. Levin, U. Maor, and E. Naftali
The BFKL Equation beyond Leading Order 82
    D. Ross
Photoproduction at HERA with a Leading Proton 88
    H. Mahlke-Krüger (for the H1 Collaboration)
Rapidity Gaps between Jets at Large Rapidity Separations in
Photoproduction at HERA 96
    A. Wyatt (for the H1 Collaboration)
Inclusive Diffraction at HERA 102
    C. L. Johnson
Jet Production in Diffractive ep Scattering. 108
    B. List (for the H1 and ZEUS Collaborations)
Deeply Virtual Compton Scattering at HERA 117
    R. Stamen (for the H1 and ZEUS Collaborations)

Proton-Dissociative Diffractive Photoproduction of Vector Mesons at Large $|t|$ at HERA 125
  K. Klimek (for the ZEUS Collaboration)
Total Photonic and Hadronic Cross Sections 131
  R. M. Godbole, A. Grau, and G. Pancheri
Double Tag Events in Two-Photon Collisions at LEP 140
  M. Wadhwa
Measurement of the Cross Section for the Process ee→ee$\gamma^*\gamma^*-$eeX at $\sqrt{s_{ee}}=189-202$ GeV 147
  M. Przybycień
Total Cross Sections in $\gamma\gamma$ Collisions 154
  M. N. Kienzle-Focacci

INCLUSIVES

Introduction to High-p_T Inclusives 163
  M. Wing
Inclusive π^0 and K_S^0 in $\gamma\gamma$ Reactions at L3 172
  P. Achard (for the L3 Collaboration)
Di-Jet Production in Photon-Photon Collisions at $\sqrt{s_{ee}}$ from 189 to 202 GeV 179
  B. Surrow and T. Wengler
Measurements of Di-Jet Cross Sections in $\gamma\gamma\rightarrow$Hadrons with ALEPH 185
  P. Hodgson, M. Lehto (for the ALEPH Collaboration),
  and B. Pötter
Dijet Cross Sections in Photoproduction and Photon Structure 191
  S. Maxfield (for the H1 Collaboration)
Total Cross Sections and Event Properties from Real to Virtual Photons 197
  C. Friberg
Jet Substructure at HERA 203
  C. Glasman (for the ZEUS Collaboration)
The Structure of the Virtual Photon 209
  C. Glasman (for the ZEUS Collaboration)
Jets in DIS and the Virtual Photon Structure 216
  S. Maxfield (for the H1 Collaboration)
Prompt-Photon Production at HERA 225
  J. Terrón (for the ZEUS Collaboration)
Measurement of Dijet Cross Sections with Leading Neutrons in Photoproduction at HERA 231
  A. Bunyatyan (for the H1 and ZEUS Collaborations)

HEAVY FLAVOUR PRODUCTION

NLO Calculations for Heavy Flavour Production in Two-Photon
Collisions 239
    E. Laenen, S. Frixione, and M. Krämer
Charm Production in Two-Photon Collisions Measured by the
ALEPH Detector at LEP II Energies. 245
    U. Sieler
Inclusive D-Meson and Λ_c Production in Two-Photon Collisions
at LEP 252
    M. Chapkin, V. Obraztsov, and A. Sokolov
Charm and Bottom Production in Two-Photon Collisions at LEP
with the L3 Detector 266
    S. Saremi
Charm and Bottom Production in Two-Photon Collisions with OPAL. 276
    Á. Csilling
Open Charm and Beauty Production at HERA. 283
    K. Daum
Recent Results on Charm Photoproduction 292
    D. Hochman (for the ZEUS Collaboration)
Exclusive Photoproduction of J/ψ Mesons at HERA. 299
    A. A. Savin
Exclusive Electroproduction of Charmonium at HERA 305
    T. Abe (for the H1 and ZEUS Collaborations)

EXCLUSIVES

Two-Photon Exclusive Processes in QCD 315
    S. J. Brodsky
Interference of the Two-Photon and Bremsstrahlung Production
for the $\pi^+\pi^-$ System at B- and ϕ-Factories 326
    I. F. Ginzburg, A. Schiller, and V. G. Serbo
$\gamma^*\gamma$ to $\pi\pi$ at Large Q^2 333
    M. Diehl, T. Gousset, and B. Pire
Cross-Section Measurement of τ Pairs in Two-Photon Collisions
with the L3 Detector at LEP 2 339
    D. Haas
Two-Photon Annihilation into Pion Pairs 345
    C. Vogt
Measurement of the Mass, Width, and Two-Photon Width of the
$\eta_c(1S)$ 351
    H. P. Paar
Study of the Two-Photons Decay of Charmonium States Formed
in $\bar{p}p$ Annihilations 358
    W. Baldini (for the E835 Collaboration)

$K_s^0 K_s^0$ Final State and Glueball Searches and $\Lambda\bar{\Lambda}$ Production
in Two-Photon Collisions in L3 at LEP 366
    S. Braccini
Search for the Glueball Candidates $f_0(1500)$ and $f_J(1710)$
in $\gamma\gamma$ Collisions in ALEPH 374
    R. W. L. Jones (for the ALEPH Collaboration)
Resonance Production in Two-Photon Collisions at LEP
with the L3 Detector 380
    V. P. Andreev
Is the $\sigma(600)$ a Glueball? Two-Photon Reactions Can Tell Us. 388
    M. R. Pennington
Exclusive Electroproduction of ρ^0 Mesons with the ZEUS Detector
at HERA 394
    S. Kananov (for the ZEUS Collaboration)

FUTURE LINEAR COLLIDER

Introduction and Recent Developments in $\gamma\gamma,\gamma e$ Colliders 403
    V. Telnov
Why Photon Colliders Are Necessary in Future Collider Program. 409
    I. F. Ginzburg
Two-Photon Physics at Future Linear Colliders 416
    A. De Roeck
Charm Quark Production at Linear e^+e^- and Photon Colliders 423
    P. Jankowski
Progress towards a $\gamma\gamma{\to}4$ Leptons Monte Carlo. 430
    C. Carimalo, W. Da Silva, and F. Kapusta

RELATED PROCESSES

Results from the OPAL Experiment Using Photonic Final States. 439
    A. Macchiolo
Vector Properties of the Central Production Pomeron 448
    A. Singovski (for the WA102 Collaboration)

SUMMARY

Summary of Photon 2000 457
    M. Krämer and S. Söldner-Rembold

List of Participants. 479
Conference Photos 489
Author Index. 491

PREFACE

Photon 2000 was an International Conference on the Structure and Interactions of the Photon. It was staged by the Lancaster University Physics Department in the grounds of St. Martins College, Ambleside in the highly popular tourist destination of Cumbria, England. Cumbria is famous for its beautiful scenery of lakes and mountains, and its associations with poets and authors such as William Wordsworth, Beatrix Potter, and Arthur Ransome. Lancaster University Physics Dept. has held a number of conferences at St. Martins College (Ambleside) and once again we were not disappointed with the quality of the catering and helpfulness of the staff in ensuring the smooth running of the conference.

The conference is the latest in a series stretching back to the first Workshop on Photon Photon collisions in 1973. In 1995 we welcomed the photoproduction community from HERA and the series became known as PHOTON-x. In view of the highly focussed subject matter of the conference it has become traditional at this series for all talks to be plenary in nature. This requires considerable discipline from the speakers and chairmen to ensure the conference keeps to the timetable. We want to take this opportunity to thank all the participants for ensuring that the meeting was never more than 10 minutes behind schedule. The `traffic light' system provided by our technician Mr. S. Holt also helped in that respect and we recommend use of such a facility to other conference organisers.

Although this conference was unusually held only 15 months since the previous one in the series, the active nature of the field is clear in the number of new results presented. This reflects the large amount of high quality data available at LEPII and HERA, and the corresponding number of Ph.D. students working in the field. Topics which have come to the fore in recent months include the use of double tagged events to study the nature of the Pomeron, and studies of virtual photons at both LEP and HERA. The conference program was compiled by one, two or three convenors for each topic, typically one theorist and one experimentalist, in consultation with the programme committee. We would like to take this opportunity of thanking all of them for their hard work in producing a well rounded programme covering all the topical subjects.

After two and a half days of intense talks it was time for some light relief, and for most of the participants this involved a cruise on Lake Windermere followed by a visit to The Aquarium of the Lakes, where we learned something of the natural history of the Lake District. A few of the hardier delegates tackled a walk on the fells led by Terry Sloan, and were rewarded with sunshine overhead, even if their feet were in bog. In the evening we were all entertained by a display of traditional English folk dancing by the Crook Morris dance team, before the Conference Dinner at Low Wood Hotel.

Finally we would like to thank Ms. Heather Matthews for her assistance before during and after the conference, without whom it would not have been such a success.

Alexander Finch

Supported by the Particle Physics and Astronomy Research Council and the Institute of Physics

PHOTON STRUCTURE FUNCTION

All the fifty years of conscious brooding
have brought me no closer to the answer to the question,
"What are light quanta?"
Of course today every rascal thinks he knows the answer,
but he is deluding himself.
A. Einstein, 1951

Structure Functions for the Virtual and Real Photons

Maria Krawczyk[1]

Institute of Theoretical Physics, Warsaw University, Warsaw, Poland

Abstract. Development of concepts related to the photon and its high energy hadronic interaction is briefly reviewed. A photon considered as an ideal probe of hadron structure, paradoxically is also considered as an ideal target to test the perturbative QCD. The present status of the theoretical and experimental analysis of the structure functions for a virtual and real photon is presented.

100 YEARS OF LIGHT QUANTA

The photon is the best known boson and one of the oldest elementary particles. This year is special - we celebrate 100 years of light quanta; below I list the crucial dates in the development of notion of the photon [1]:

- 1900 - Planck's hypothesis of quantum of electromagnetic energy, $E = h\nu$
- 1905 - Einstein's hypothesis of quantum of light (γ), $E = h\nu = pc$
- 1915 - Millikan's experiment: photo-emission from metal
- 1922 - Compton experiment: $\gamma e \rightarrow \gamma e$

In the next years (1925-7) the Quantum Electrodynamics (QED) was invented by Born, Heisenberg, Jordan, Dirac and others [2], with photon playing the role of a gauge particle of electromagnetic interaction. Few years later (1931) Wigner gave the complete group theoretical description of angular-momentum states, according to which photon means helicity states of spin 1 massless particle [3]. The current name: the photon was given in 1926 by the chemist G.N. Lewis [4].

---

[1] Partly supported by KBN Grant No 2P03B0511(2000)

CP571, *PHOTON 2000*, edited by A. J. Finch
© 2001 American Institute of Physics 0-7354-0010-5/01/$18.00

Photon-hadron interaction

The photon properties, as follows from Quantum Electrodynamics, are well known: the photon is a massless, chargeless object with a pointlike coupling to the charged fundamental particles. As such, it is an ideal tool for probing structure of more complicated objects, for example hadrons, acting as a microscope with the resolution given by its wavelength.

However, in the high energy photon-hadron interaction there are phenomena which can (should ?) be interpreted in terms of " hadronic (partonic) structure" of the *photon*. This way one can effectively describe leading contributions to the rates of certain processes. Some early ideas and facts are recalled below:
- 1960-72 - observation of hadronic properties of the photon in soft processes, the $\rho(\omega, \phi)$-photon analogy, Vector Dominance Model (VDM) (also GVDM) [5]
- 1969-71 - an importance of $\gamma\gamma \to hadrons$ processes in e^+e^- collisions [6]
- 1970-74 - a deep inelastic scattering on the real photon [7]a) and Parton Model predictions for structure functions $F_{1,2(L)}^\gamma$, and F_3^γ, g_1^γ [7]b)
- 1977-80 - asymptotic (point-like) solution in QCD (LO [8] and NLO [9] results); structure functions of a real photon as a unique test of QCD
- 1979-84 - singularities in the asymptotic solutions for a real photon at small x; negative F_2^γ in the NLO QCD analysis [9–12]
- 1981-84 - hadronic contribution to F_2^γ as a cure of the problem at small x [11,12]
- 1981-84 - structure functions of virtual photons (singularity free LO and NLO predictions) - a unique test of QCD [10,12]
- 1989-93 - relation of spin dependent structure functions of photons to QED and QCD anomaly [13].

Starting from 1981 structure functions of unpolarized real and virtual photons, F_2^γ and $F_{eff}^{\gamma*}$, respectively, are being measured. Ten years later the first data on the production of large p_T particles and jets in the photon-induced processes, so called resolved photon processes, have appeared for both real and virtual photons. Review of data can be found in [14], [15].

The photon A.D.2000

The above short outline of historical development of basic concepts related to the hadronic "structure" of the photon shows how in the past high expectations were followed by deep defeats. Even nowadays the situation is far from being clear. Although for many hadronic processes involving photons there exist already NLO QCD calculations, a proper way of describing the photon interaction is still a subject of ongoing discussions, *e.g.* how to count the order of the perturbation [16]. Still, after so many years of photon physics there is a lot of confusion, even terminology seems to be inadequate and generates additional problems, *e.g.* [17]. The fact that a photon has a double face - being a probe and a target, sometimes in the same process, obviously does not help.

The main source of data on the strong (hadronic,partonic) properties of the real photon comes from $DIS_{e\gamma}$ experiments in the e^+e^- collisions. Recent results on the structure function F_2^γ, based on few years runs' at LEP1 and TRISTAN at the CM energy ~ 90 and 60 GeV, respectively, are now available. There are also new data taken at LEP at higher energies. Altogether the existing data cover a wide range of the (average) Q^2 from 0.2 to ~ 706 GeV^2 [18]. The range of the (center of bin) x_{Bj} variable extends from ~ 0.001 [19] to 0.98. Recently the first dedicated measurement of $F_{2,c}^\gamma$ has been performed at LEP2, see [20]. In addition there exist data of the leptonic structure functions, see $e.g.$ [14].

The complementary data on the photon structure are coming from measurements of the resolved real photon processes, $i.e.$ production of large p_T jets (also individual hadrons, photons, heavy quarks), in $\gamma\gamma$ collisions at e^+e^- machines and in the photoproduction at the $e^\pm p$ collider HERA ($\sqrt{s} \sim 300$ - 320 GeV), see [21].

The determination of the hadronic structure function F_2^γ for a real photon in e^+e^- collision relies on the unfolding, therefore precision of these data depends crucially on the accurate description of the hadronic final state by the Monte Carlo models. The improvement in the unfolding in DIS -type experiments has been obtained recently, still the dependence on the chosen MC used in the analysis cannot be avoided [22], see also [23]a-b). In the modeling of the final state, one includes [24,25] various initial "states" of the target photon: the photon as a $q\bar{q}$ pair and the photon as a ρ meson, with a relevant structure. The virtual photon (a probe) can interact by its constituents as well, see $e.g.$ [26]. So, in the DIS-type analyses also processes with a resolved γ (γ^*) are being involved.

During the last years data on the "structure of the virtual photon" have appeared, mostly from the resolved virtual photon processes at the ep collider HERA (with the virtuality of the photon from 0.1 to 85 GeV^2), see [21]. Fifteen years after the first measurement of the $DIS_{e\gamma^*}$ events by PLUTO collaboration [27], a new measurement has just been performed at LEP (L3 Coll. [28]). Extraction of the effective structure function or of effective parton densities in a virtual photon from the e^+e^- and ep data is a difficult task, especially if interference terms are large as observed in the OPAL experiment for double-tag leptonic events [29,14].

Basic QCD predictions for processes with real and virtual photon are definitely in agreement with the data. However the discrepancies between the data and predictions of Monte Carlo models for various distributions in the photon-induced processes are observed both in ep and e^+e^- collision. Implementation of the modified transverse momenta distribution of the partons in the γ and taking into account multiple parton interaction in the MC programs [23]c) help to describe the data. Nevertheless, the existing data, also these on heavy quark production [30], seem to give us a message that the partonic content of the photon is not properly described by existing parametrizations.

Taking all these facts into account I think it is sensible to start from scratch with a basic introduction to the concept of photon structure functions. I use as a model the "leptonic structure" of a real photon. Next I discuss the e^+e^- environment of the DIS-type experiments for photons, where our basic knowledge comes from. At this stage the leptonic structure of the virtual photon can be introduced. Then the same steps will lead us to the concept of hadronic structure functions of the photon.

THE "STRUCTURE" OF THE LIGHT QUANTA

In the quantum field theory, a photon, as any elementary particle, can fluctuate into various states consisting of leptons, quarks, $W^{\pm}$ bosons, hadrons. *"...through an interaction with a Coulomb field the photon could materialize as a pair of electrons, $\gamma \to e^+e^-$. Although not usually thought of in these terms, this phenomenon was the earliest manifestation of photon structure"* [5]b).

Leptonic structure of the photon

the target = γ

One can test leptonic structure of photon in the deep inelastic scattering, $e\gamma \to e + leptons$. Let us take a (unpolarized) real photon ($p^2 = 0$) as a target and assume that the probe, highly virtual photon γ^*, with large virtuality $-q^2 = Q^2$, couples directly to the electric charge of the fundamental particles. To describe this process one can introduce the corresponding structure functions, as for the proton case. The contribution of the lowest order QED process, $\gamma^*(q)\gamma(p) \to l^+l^-$, Fig.1(left), to $F_2^\gamma|_l$ is given by (the *box* contribution) [6]:

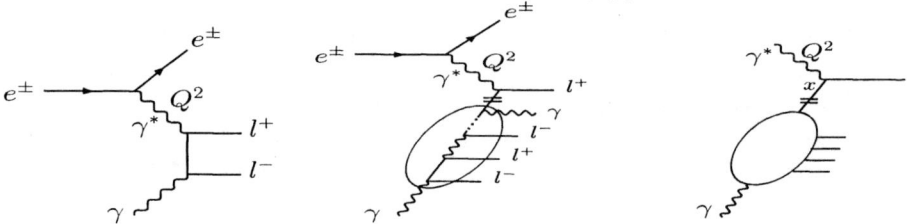

FIGURE 1. The lowest QED process for production of l^+l^- final state (left); the leptonic content of the photon in the LLA (center); a general representation of the $\mathrm{DIS}_{e\gamma}$ with a point-like coupling of the probe to the elementary constituent of the photon target with $x = x_{Bj}$ (right).

$$F_2^\gamma|_l = \frac{\alpha}{\pi}Q_l^4 x_{Bj}[(-1 + 8x_{Bj}(1 - x_{Bj}) - x_{Bj}(1 - x_{Bj})\frac{4m_l^2}{Q^2})\beta$$

$$+[x_{Bj}^2 + (1 - x_{Bj})^2 + x_{Bj}(1 - 3x_{Bj})\frac{4m_l^2}{Q^2} - x_{Bj}^2\frac{8m_l^2}{Q^2}]\ln\frac{1 + \beta}{1 - \beta}], \qquad (1)$$

where $x_{Bj} = Q^2/2pq$, m_l - lepton mass, Q_i - electric charge, $Q_l = 1$, and β is the lepton velocity in the $\gamma^*\gamma$ CM system. For large invariant mass of the l^+l^- system, $W \gg m_l$, x_{Bj} not too close to 0 and 1, and neglecting terms $\sim m_l^2/Q^2$ one gets (keeping the leading logarithmic (LL) term)

$$F_2^\gamma|_l = \frac{\alpha}{\pi} Q_l^4 x_{Bj}[x_{Bj}^2 + (1 - x_{Bj})^2] \ln \frac{Q^2}{m_l^2}. \tag{2}$$

The obtained $F_2^\gamma|_l$ is solely due to the QED interaction. It is large at large x_{Bj} and it has the characteristic logarithmic rise with Q^2 from the *collinear* configuration in the $\gamma \to l^+l^-$ splitting. In principle one can also introduce a *leptonic density* in the real photon: in the LLA one gets $l^+(x_{Bj}, Q^2) = \frac{\alpha}{2\pi} Q_l^2 x_{Bj}[x_{Bj}^2 + (1 - x_{Bj})^2] \ln \frac{Q^2}{m_l^2}$ (the same for l^-). Here $x_{Bj} = x$, where x - the part of the four-momentum of the initial photon taken by its leptonic constituent (Fig.1 (right)).

Leptonic structure of the real photon $F_2^\gamma|_l$ can be measured at future colliders, where beams of energetic real photons can be obtained in the backward Compton scattering (Photon Colliders) [31]. Nowdays the leptonic structure functions of a real photon are measured in various experiments at e^+e^- colliders. Here (quasi) real photons with a Weizsäcker-Williams energy spectrum play a role of a target. The $F_2^\gamma|_l$ data for $l = \mu$ together with QED predictions are summarized in Fig.2,

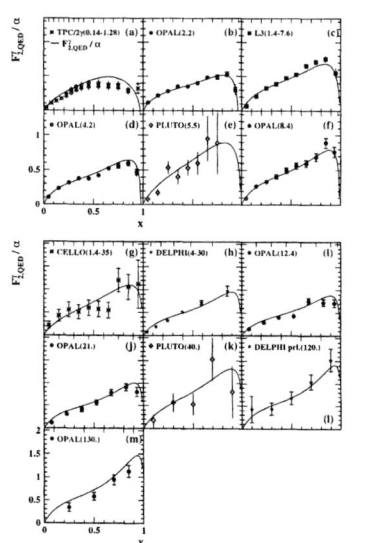
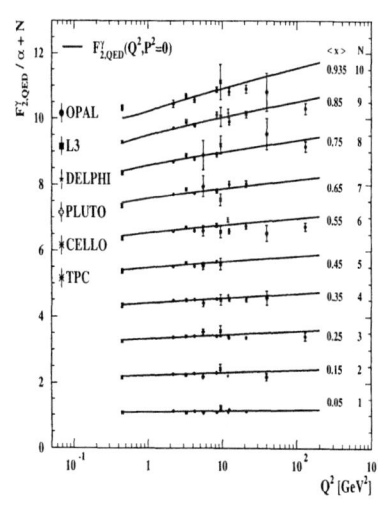

FIGURE 2. Summary of existing $F_2^\gamma|_l/\alpha$ data, for $\mu^+\mu^-$ final state, shown with QED predictions for $P^2=0$, (left) - as a function of x_{Bj} for broad $< Q^2 >$ range, (right) - as a function of Q^2 for fixed x_{Bj} bins, from [14].

from [14].

The basic features of the lowest QED order result (2) will be modified only softly by higher QED corrections. Formally leading logarithmic QED corrections, powers of $\alpha \log Q^2/m_e^2$, can be summed up using the *evolution equation* in Q^2 [32], with an *inhomogenous term* due to the $\log Q^2$- dependence present already in the lowest order QED prediction (2). The resulting *collinear QED cascade* included in the LLA in the leptonic density $l^+(x, Q^2)$ is represented in Fig.1(center). Here, starting from the first splitting of the initial photon all emission processes up to the interaction with a probe are based on point-like couplings.

the target = γ^*

Leptonic structure functions can also be introduced for a virtual photon, *i.e.* with $|p^2| = P^2 \neq 0$, to be measured in the lepton beams collision. Let us discuss the production of an *arbitrary* state X in the process $e(p_1)e(p_2) \rightarrow e(p_1')e(p_2')X$, via the $\gamma^*(q_1)\gamma^*(q_2)$ collision ($q_1 = p_1 - p_1'$, $q_2 = p_2 - p_2'$), Fig.3 (left). The corresponding cross section for the unpolarized lepton beams, assuming $|q_{1,2}^2| \gg m_e^2$, and typical conditions in present experiments, is given by (see [6], also [14])

$$E_1' E_2' \frac{d\sigma(ee \rightarrow eeX)}{d^3p_1' d^3p_2'} = L_{TT}(\sigma_{eff} + \frac{1}{2}\tau_{TT}\cos 2\bar{\phi} - 4\tau_{TL}\cos \bar{\phi}), \tag{3}$$

where helicity states of photons are denoted by T - transverse (+ or -) and L - longitudinal (0). An effective cross section is defined as $\sigma_{eff} = \sigma_{TT} + \sigma_{LT} + \sigma_{TL} + \sigma_{LT}$, where $\sigma_{TT,TL,LT,LL}$ (the first subscript is for the photon with q_1) denote the corresponding cross sections, $\tau_{TT,TL}$ - the interference terms. The $\bar{\phi}$ is the angle between two scattering planes of the scattered electrons in the $\gamma^*\gamma^*$ CM system.

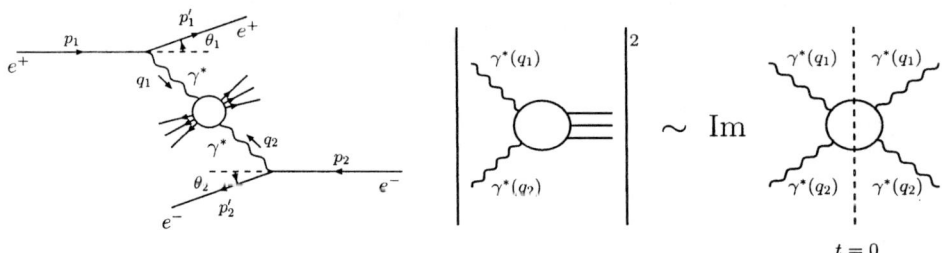

FIGURE 3. The $\gamma^*\gamma^* \rightarrow X$ process in e^+e^- collision (left). The cross section for the $\gamma^*\gamma^* \rightarrow X$ scattering and its relation to the imaginary part of the forward $\gamma^*\gamma^* \rightarrow \gamma^*\gamma^*$ amplitude (right).

One can relate these quantities to the imaginary part of the forward helicity amplitudes $W_{\lambda_1'\lambda_2',\lambda_1\lambda_2}$ for the process $\gamma^*(\lambda_1)\gamma^*(\lambda_2) \rightarrow \gamma^*(\lambda_1')\gamma^*(\lambda_2')$ [6], see Fig.3 (right). Amplitudes with $\lambda_{1(2)} = \lambda_{1(2)}'$ are related to corresponding cross sections; the interference terms correspond to helicity-flip forward amplitudes: $\tau_{TT} = AW_{++,--}$ and $\tau_{TL} = A(W_{++,00} - W_{0+,-0})/2$, where $2\,A = ((q_1 q_2)^2 - q_1^2 q_2^2)^{1/2}$.

To measure structure functions of virtual photon γ^* the double-tag events with $|q_1^2| \gg |q_2^2| \neq 0$ are used. Below I will concentrate on $\mathrm{DIS}_{e\gamma^*}$ events with $l^+ l^-$ final state, where $Q^2 \gg P^2 \gg \mu^2$ (using a standard notation for a probe $q \equiv q_1$ and $Q^2 = -q^2$, and for a target $p \equiv q_2$ and $P^2 = -p^2$), with μ - a characteristic scale for studied phenomena (here m_l). The structure functions for a polarization-averaged virtual photon target, $F_1^{\gamma^*}, F_2^{\gamma^*}$ and $F_L^{\gamma^*} = F_2^{\gamma^*} - 2xF_1^{\gamma^*}$ can be introduced, $e.g.$

$$F_2^{\gamma^*} = \frac{Q^2}{4\pi^2\alpha} \frac{A}{2} (\sigma_{TT} + \sigma_{LT} - \frac{1}{2}\sigma_{LL} - \frac{1}{2}\sigma_{TL}). \tag{4}$$

If, after integration of the differential cross section (3) over $\bar{\phi}$, the contributions of the terms τ_{TT} and τ_{TL} vanish, one can relate the measured cross section to an effective structure function for a virtual photon, $F_{eff}^{\gamma^*} \sim \sigma_{eff}$ [6]. This is not the case in the recent OPAL measurement of the μ-pair production [29,14], as can be seen in Fig.4. Here the Monte Carlo predictions based on the QED calculations, also for the options with τ_{TT}=0 and $\tau_{TT} = \tau_{TL}$=0, to test relevance of the interference terms, are displayed.

Large (negative) intereference terms found for $x > 0.1$ in double-tag leptonic events at LEP1, as apparent from Fig.4, make the extracting of the corresponding leptonic structure function for γ^* unfeasible, and also shed a light on potential problems in extracting the hadronic structure function for γ^*.

Note that for $P^2 \to 0$ only one interference term, τ_{TT}, remains in cross section (3). It is called also the F_3^γ structure function [7]b). In the lowest order QED it has

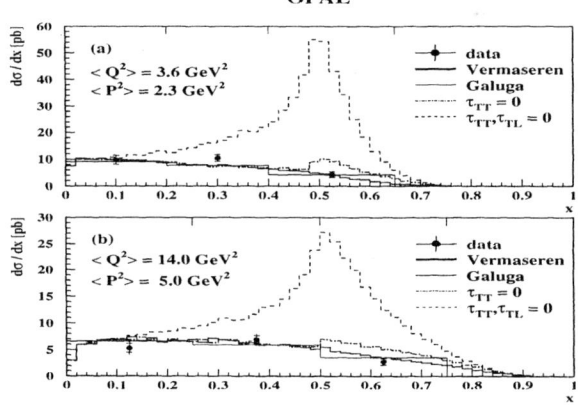

FIGURE 4. Differential cross section for $\mu^+\mu^-$ production in $\gamma^*(Q^2)\gamma^*(P^2)$ collision at LEP1 (OPAL Coll.). The thick line is a QED prediction of Vermaseren Monte Carlo; predictions represented by the solid line and dot-dashed (dashed) line were obtained by using GALUGA MC program for all contributions and assuming $\tau_{TT} = 0$ ($\tau_{TT} = \tau_{TL} = 0$), from [29].

the scaling property, $F_3^\gamma|_l = -\alpha/\pi Q_l^4 x_{Bj}^3$, so it should not influence significantly the extraction of F_2^γ for a real photon at large Q^2.

Hadronic structure function of the photon

the target = γ

In principle one can introduce also "partonic structure of the photon" since the photon could materialize itself as a pair of quarks. The splitting $\gamma \to q\bar{q}$ leads to the corresponding photon structure functions already in the lowest order QED, equivalent here to the Parton Model (PM).

The Parton Model prediction for the deep inelastic scattering $e\gamma \to e + hadrons$ is based on the process $\gamma^*\gamma \to q\bar{q}$. Prediction for the (hadronic) $F_2^\gamma|_q$ is given by the formula as for a leptonic final state (eqs. 1 and 2), with some modifications: $m_l \to m_q$, $Q_l \to Q_q$ and the color factor, $N_c = 3$, has to appear. For a final state with heavy quarks this is the modification (QPM formulae), for light quarks one usually uses the massless approximation, with the QCD parameter, Λ_{QCD}, as an argument in the leading logarithm. So, we have in the Parton Model (in LLA) the following expressions for hadronic F_2^γ and for the (light) *quarks densities*:

$$F_2^\gamma = \sum_{q,\bar{q}} Q_q^2 x_{Bj} q^\gamma(x_{Bj}, Q^2), \tag{5}$$

$$q^\gamma(x_{Bj}, Q^2) = \frac{\alpha}{2\pi} Q_q^2 N_c [x_{Bj}^2 + (1 - x_{Bj})^2] \ln \frac{Q^2}{\Lambda_{QCD}^2}. \tag{6}$$

As previously for leptons, F_2^γ has been calculated within the QED - it is proportional to α! Both the x_{Bj} and the Q^2 dependence are obtained, both are the same as for leptonic final state: large F_2^γ value at large x_{Bj}, a logarithmic rise with Q^2 (here called a *scaling violation*).

The leading logarithmic QCD corrections introduce logarithmic (Q^2) modifications of the basic predictions of the Parton Model (5),(6), and include also a *gluonic content* of the photon. These corrections can be summmed up by solving the corresponding inhomogeneous evolution equations. By solving them without any input (boundary condition), assuming only that the *particular* solution of the equation has the Q^2 dependence as in PM (eqs. 5-6) one obtains the so called *asymptotic solution* [8]. It corresponds to the collinear configuration of successive emissions of quarks and gluons (as in Fig.1 (center)), all of them based on the point-like couplings of QED (the first and the last one), and of QCD.

However, the asymptotic, purly perturbative solutions suffer from *power singularities*. For example the moments $(f^n(Q^2) = \int dx x^{n-1} f(x, Q^2))$ of the non-singlet structure function are given by $f^n(Q^2)|_{asym} \sim \log Q^2/(1 - d_n)$, where d_n are proportional to the nth -moment of the PM term: $[x^2 + (1 - x)^2]$. Simple poles occur for $d_n = 1$, leading after inverting the moments to the following small x behavior

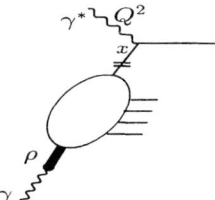

FIGURE 5. The deep inelastic scattering on the photon in the ρ state.

of F_2^γ: $(1/x)^{n=1.596}$ (LLA) and $(1/x)^{n=2}$ (NLLA) [9–12]. The singularities become increasingly severe with higher order of QCD calculation [12]. Already in the NLLA they lead to the negative value of F_2^γ [9]b).

To cure the problem of singularities in the (asymptotic) structure functions for a real photon, one should include in the calculation also the hadron-like (*non-perturbative (NP)*) contribution [11,12]. The hadronic properties of the photon are apparent in the *soft* photon-hadron interaction, where the similarity between photon and vector mesons ρ, ω, ϕ interaction is observed (VMD model [5]). This NP component can be included *e.g.* in a boundary condition at Q_0^2 scale, [11]c,

$$f^n(Q^2) = \frac{4\pi}{\alpha_s(Q^2)}[1 - (\frac{\alpha_s(Q^2)}{\alpha_s(Q_0^2)})^{1-d_n}]\frac{\tilde{a}_n}{1-d_n} + [\frac{\alpha_s(Q^2)}{\alpha_s(Q_0^2)}]^{-d_n}f_n(Q_0^2), \qquad (7)$$

where the input at scale Q_0^2 (even $f^n(Q_0^2)$=0!) regularizes the bad behaviour present for $d_n \to 1$, since $\frac{1}{\epsilon}(1 - w^\epsilon) \to -\log w$ for $\epsilon \to 0$. By doing this we get rid of the power singularities for the real photon structure functions, at the same time we lose an "absolute" predictivity of QCD for this quantity.

Equation (7) shows why it is customary to treat the structure functions or the quark densities in the photon as being $\sim \alpha/\alpha_s$ although the primary $\log Q^2$ dependence present in the Parton Model (egs. 5-6), which remains also after QCD corrections, has nothing to do with α_s. This way of counting changes organization of the perturbation expansion in the QCD calculations, see [16].

In some approaches one treats the photon in the hadronic mode almost as an independent object. Probing the "structure" of the photon in, say, state of ρ can be performed in an analogous way as testing the structure of other hadrons, *e.g.* in the deep inelastic scattering [25], see Fig.5. The general behaviour of the partonic content of the γ in the ρ mode is known - the scaling property in the PM and the logarithmic scaling violation due to the QCD corrections. The corresponding DGLAP evolution equation are homogeneous as for the proton, and an input at some scale is needed to solve the equation, etc...

Various parton parametrizations were constructed for a real photon in the past (there are about 20 of them, see compilations in [14], [15]b). The earlier ones were based on a simple Parton Model formula (for quarks) or the asymptotic solutions.

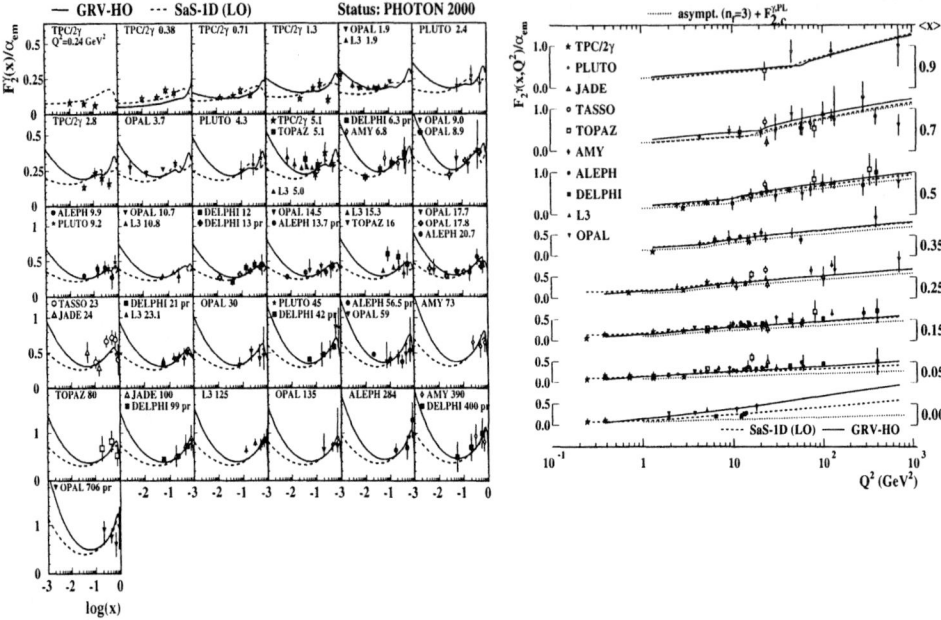

FIGURE 6. A compilation of the photon structure function F_2^γ/α data as a function of x_{Bj} in bins of Q^2 (left) and as a function of Q^2 for $< x_{Bj} >$ bins (right) compared to the GRV NLO (solid line) and SaS1D (LO) (dashed line) parametrizations of parton distributions in the photon, from [33].

The later parametrizations were based on approaches incorporating hadronic-like (NP) contributions at some stage. Recently parametrizations for a real photon are obtained from parametrizations for virtual photon for $P^2 \to 0$ [25], [34]b).

The compilations of the all existing data for the structure function for the real photon [33] are presented in Fig.6 in comparison with two parton parametrizations, obtained in two different approaches to the hadron-like contributions, GRV [34]a) and SaS [25]. The general behaviour of the F_2^γ measured for hadronic and leptonic final states is very similar, compare Fig.6 and Fig.2.

An additional information on the "structure" of the photon is coming from the production of heavy quarks in photon-induced processes. The recent DIS-type measurement at LEP led to the extraction, for the first time, of the charm contribution to F_2^γ, $F_{2,c}^\gamma$, see [20]. The QCD description of heavy quark production in processes induced by photons is not satisfactory [30], however similar problem with proper description of a heavy quark production exists also in pure hadronic processes.

$$\text{the target} = \gamma^*$$

The structure function of the virtual photon can be calculated in the Parton Model (QED!) from the $\gamma^*\gamma^* \to q\bar{q}$ process. For $Q^2 \gg P^2 \gg m_q^2$ ($x_{Bj} = Q^2/2p \cdot q$)

one obtains

$$F_2^{\gamma^*}(x_{Bj}, Q^2, P^2) = N_c \sum_{q,\bar{q}} \frac{\alpha}{\pi} Q_i^4 x_{Bj} [[x_{Bj}^2 + (1 - x_{Bj})^2] \ln \frac{Q^2}{P^2 x_{Bj}^2} + 6 x_{Bj}(1 - x_{Bj}) - 2], \quad (8)$$

to be compared with the eq. (1). The corresponding PM quark density in the virtual photon defined in the LL approximation has the form:

$$q^{\gamma}(x_{Bj}, Q^2, P^2) = \frac{\alpha}{2\pi} Q_q^2 N_c [x_{Bj}^2 + (1 - x_{Bj})^2] \ln \frac{Q^2}{P^2}. \quad (9)$$

The QCD evolution equations in Q^2 for the virtual photon are analogous to those for the real photon with the inhomogeneous term given by the corresponding PM expression. In the case of the virtual photon one can solve the evolution equation without the initial conditions. Assuming that for $Q^2 \gg P^2 \gg \Lambda_{QCD}^2$ the nonperturbative effects are absent (see ref. [10]) one obtains for moments of the non-singlet structure function

$$f^n(Q^2) = \frac{4\pi}{\alpha_s(Q^2)} [1 - (\frac{\alpha_s(Q^2)}{\alpha_s(P^2)})^{1-d_n}] \frac{\tilde{a}_n}{1 - d_n} \sim \log \frac{Q^2}{P^2}, \quad (10)$$

So, without additional experimental or model assumption the definite, *singularity free* (asymptotic) predictions can be derived for both the x and the Q^2 dependence - a unique situation in QCD. Note however that in all recent analyses nonperturbative component in $F_2^{\gamma^*}$ is introduced [34]b), [25].

Measurements of the structure functions of γ^* can be performed in e^+e^- collision, as discussed for a leptonic final state. The $DIS_{e\gamma^*}$ events with hadronic final state were studied experimentally by PLUTO Coll. [27], new data have appeared from LEP (L3 Coll. [28]). In Fig.7 the results from both experiments, corresponding to effective structure functions (cross sections), are presented in comparison with predictions of the QPM, soft VDM, GRS [34]b) models, see however [35].

There are already few parton parametrizations for a virtual photon (see collections in [14], [15]b)), they are valid for $0 \le P^2$ and become the corresponding parametrizations for the real photon in the limit $P^2 \to 0$. All these parametrizations deal with the *transversely* polarized *virtual* photon, with one exception of the Chýla parametrization [36] for a *longitudinal* virtual photon, see also [37].

New insight into the photon structure may come from spin-dependent structure functions [38], not measured so far. Especially the structure function g_1^{γ} is of great importance, since its first moments (*a sum rule for the "spin " of the photon*) involve strong and electromagnetic anomalies, and it is deeply connected with the chiral properties of QCD [13]. It maybe studied at future Linear Colliders [39].

CONCLUSION AND OUTLOOK

A photon, considered as an ideal probe of hadron structure, paradoxically is also considered as an ideal target to test the perturbative QCD. Both a real and a virtual photon may reveal "inner structure" in the interaction with other particles.

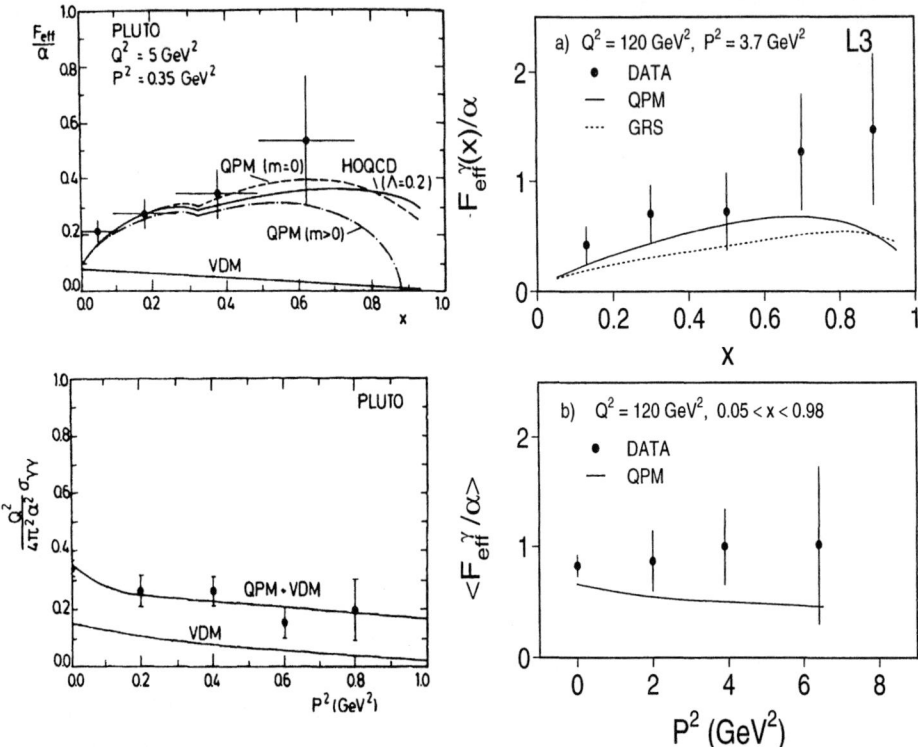

FIGURE 7. The data for the F_{eff}^{γ}/α and comparison with various predictions for the virtual photon. PLUTO Coll. data for $< Q^2 >= 5$ GeV^2, and $< P^2 >= 0.35$ GeV^2 [27], and L3 Coll. data for $< Q^2 >=120$ GeV^2 and $< P^2 >=3.7$ GeV^2 [28]. The x-dependence (upper panels) and P^2-dependence (lower panels) are presented.

An apparent hadronic structure of the photon is clearly seen in the data. However, there is no full agreement between the standard QCD predictions and experiment for various distributions for hadronic processes induced both by real and virtual photons. During last years we learned that an improvement of the description of the data can be obtained by introducing extra p_t for constituents in the target photon, and/or the contributions due to multiple interaction. Still there are problems in describing x_γ distribution from the dijet events, prompt photon events in the photoproduction at HERA, heavy quark production in the e^+e^- and ep processes.

For a virtual proton even more basic questions are open. How important are interference terms, both for $\gamma^*\gamma^*$ and γ^*p processes? Do we see more than just the PM content of the virtual photon in present data? If yes, do we need a "structure" of the longitudinal virtual photon?

We have a lot to improve in our description of the photon-hadron interaction.

ACKNOWLEDGMENTS

I would like to thank organizers of this interesting conference for fruitful atmosphere. I am grateful to F.Kapusta, R. Nisus, T. Sjöstrand, J. Chýla, S. Söldner-Rembold, D. Miller, I. Ginzburg, U. Jezuita-Dabrowska, P. Jankowski and Magda Staszel for many illuminating discussions. I appreciate the collaboration with U. Jezuita-Dabrowska and P. Jankowski on this contribution.

REFERENCES

1. Crease, R.P., and Mann, C., *The Second Creation*, Collier Books Maximillan Publishing Company, New York, 1986
2. Schweber, S.S., *QED and the Man Who Made it: Dyson, Feynman, Schwinger, and Tomonaga*, Princeton Univ. Press, 1997.
3. Kim, Y.S., "Wigner's Photons", hep-ph/9802011.
4. Lewis, G.N., *Nature* **118** no 2981, 874 (1926).
5. a) Sakurai, J.J., *Ann. Phys.* **11**, 1 (1960); Stodolsky, L., *Phys. Rev.* B **134**, 1099 (1964); Sakurai, J.J., and Schildknecht, D.,*Phys. Lett.* B**40**, 121 (1972); b) Bauer, T.H., et al., *Rev. of Modern Phys.* **50**, 262 (1978).
6. Artega-Romero, N., et al., *Compt. Rend* **269B**, 153,1129 (1969), Jaccarini, A., et al., *Il Nouvo Cimento* **IV**, 933 (1970); Arteaga-Romero, N., et al., *Phys. Rev.* D**3** 1569 (1971); Balakin,V.E., et al., *JETP Lett.* **11**, 388 (1970); Brodsky S.J., et al., *Phys. Rev. Lett.* **25**, 972 (1970), *ibid* **27**, 280 (1971); *Phys. Rev.* D**4**, 1532 (1971) Budnev, V.M., and Ginzburg, I.F., *Phys. Lett.* B**37**, 320 (1971); Budnev, V.M., et al., *Phys. Rep.* C**15**, 181 (1975); Brown R.W., and Muzinich I.J., *Phys. Rev.* D**4** 1496 (1971); Carlson C.E., and Tung Wu-Ki, *Phys. Rev.* D**4**, 2873 (1971).
7. a) Walsh, T.F., *Phys. Lett.* B**36**, 121 (1971); b) Walsh, T.F., and Zerwas, P., *Phys. Lett.* B**44**, 195 (1973) ; Zerwas P., *Phys. Rev.* D**10**, 1485 (1974); Kingsley, R.L., *Nucl. Phys.* B**60**, 45 (1973); Chernyak, V.L., and Serbo, V.G., *Nucl. Phys.* B**71**, 395 (1974); Ahmed M.A., and Ross G.G., *Phys. Lett.* B**59**, 369 (1975); Worden, R.P., *Phys. Lett.* B**51**, 57 (1974).
8. a) Witten, E., *Nucl. Phys.* B**120**, 189 (1977); b) Llewellyn-Smith, C.H., *Phys. Lett.* B**79**, 83 (1978); DeWitt, R.J., et al., *Phys. Rev.* D**19**, 2046 (1979); Brodsky, S.J., et al., *Phys. Rev. Lett.* **41**, 672 (1978); *Phys. Rev.* D**19** , 1418 (1979); Frazer, W.R., and Gunion, J.F., *Phys. Rev.* D**20** (1979) 147; Sasaki, K., *Phys. Rev.* D**22**, 2143 (1980); Delduc, F., et al., *Nucl. Phys.* B**174**, 147, 157 (1980); Peterson, C., et al., *Nucl. Phys.* B**174**, 424 (1980);*ibid* **229**, 301 (1983); Frazer, W.R., Rossi, G., *Phys. Rev.* D**25**, 843 (1982);
9. a) Bardeen,W.A., and Buras,A.J., *Phys. Rev.* D**20**, 166 (1979); *Phys. Rev.* D**20**, 2041 (1980); b) Duke, D.W., and Owens, J.F., *Phys. Rev.* D**22**, 2280 (1980).
10. Hill, C.T., and Ross, G.G., *Nucl. Phys.* B**148**, 373 (1979); Uematsu, T., Walsh, T.F., *Phys. Lett.* B**101**, 263 (1981); *Nucl. Phys.* B**199**, 93 (1982).
11. a) Bardeen, W.A., Int. Conf. Photon-Lepton (1981), ed. W. Pfeil, Bonn, p. 432; b)

Antoniadis, I., and Grunberg G., *Nucl. Phys.* **B213**, 445 (1983); c) Glück, M., and Reya, E., *Phys. Rev.* **D28**, 2749 (1983),Glück, M., et al. *Phys. Rev.* **D30** (1984).

12. Rossi, G., Ph.D.Thesis (1983), UCSD-10p10-227, *Phys. Lett.* **B130**, 105 (1983), *Phys. Rev.* **D29**, 852 (1984)

13. Gorsky, A.S., et al., *Phys. Lett.* **B227**, 474 (1989); Gorsky, A.S., and Ioffe, B.L., *Particle World* **4**, 114 (1990); Efremov, A.V., and Teryaev, O.V., *Phys. Lett.* **B240**, 200 (1990); Bass, S.D.,*Int. J. of Mod. Phys.* **A7**, 6039 (1992); Bass, S., et al., *Phys. Lett.* **B437**, 417 (1998); Narison, S., et al., *Nucl. Phys.* **B381**, 69 (1993), Shore, G., and Veneziano, G., *Mod. Phys. Lett.* **A8**, 373 (1993).

14. Nisius, R., *Phys. Rep.* **332** 166 (2000).

15. a)Erdmann, M., *The Partonic Structure of the Photon*, Springer Tracts in Modern Physics **138**, Springer 1997; b) Krawczyk, M. et al, hep-ph/0011083, *Phys. Rep.*

16. Chýla, J.,*JHEP* 04, 007 (2000), these proceedings; also Krawczyk, M., *Acta Phys. Polon.* **B21**, 999 (1990), Krawczyk,M., and Zembrzuski A., hep-ph/9810253.

17. Chýla J., "Hard collision of photons: Plea for common language", Nov. 2000.

18. Taylor, R., these proceedings.

19. Clay, R., these proceedings.

20. R. Nisius, these proceedings.

21. "Pt-inclusives session"-these proceedings.

22. The LEP Working group for Two-Photon Physics, CERN-EP-2000-109 (2000).

23. a) D.J. Miller, Proc. of the Workshop on Two-Photon Physics at LEP and HERA, Lund, May 1994, eds. G. Jarlskog, L. Jönsson, p. 4, *J. Phys.* **G24** (1998) 317; b) M. N. Kienzle-Focacci,Proc. of the Int. Conf. on the Structure and Interactions of the Photon and the 12th Int. Workshop on Photon-Photon Collisions, Freiburg, Germany, May 1999, ed. S. Söldner-Rembold, *Nucl. Phys.* B (Proc. Suppl.) **82**, 447 (2000); A. J. Finch, *ibid* p. 156. c) Butterworth J.M., and Taylor R.J., *ibid* p. 112.

24. Field, J.H., et al, *Phys. Lett.* **B181**, 362 (1986), *Z. Phys.* **C36**, 121 (1987);Kapusta, F., *Z. Phys.* **C42**, 225 (1989).

25. Schuler,G.A., Sjöstrand, T.,*Z. Phys.* **C68**,607 (1995);*Phys. Lett.* **B376**, 193 (1996).

26. Kapusta, F., Proc. of the 9th Int. Workshop on Photon-Photon Collisions, eds. Caldwell D.O., and Paar H.P., World Scientific 1992, p. 142.

27. Barger, C., et al *Phys. Lett.* **B142**, 119 (1984).

28. Acciarri, M., *Phys. Lett.* **B483**, 373 (2000).

29. Abbiendi, G., et al., *Eur. Phys. J.* **C11**, 409 (1999).

30. "Heavy Flavour session" - these proceedings.

31. Ginzburg,I.,et al.,*Nucl.Instr. & Meth* **205**, 47 (1983);Telnov,V., *ibid* **A 294**, 72 (1990);*ibid* **A 355**, 3 (1995); R.Brinkmann et al., *ibid* **A 406**, 13 (1998); also Telnov, V.,these proceedings.

32. Gribov, V., and Lipatov, L., *Sov. J. Nucl. Phys.* **15**, 438 (1972).

33. Söldner-Rembold, S., http://s.home.cern.ch/s/soldner/www.

34. a) Glück, M., Reya, E., Vogt, A., *Phys. Rev.* **D46**, 1973 (1992), *ibid* **45**, 3986 (1992);b) Glück, M., Reya E., and Stratman, M., *Phys. Rev.* **D51**,3220 (1995); Drees M., and Godbole R. M., *Phys. Rev.* **D50**, 3124 (1994); Gorski, A.S., et al., *Z. Phys.* **C44)**, 529 (1989), Ioffe B.L., Oganesian A., *Z. Phys.* **C44**, 529 (1989); Glück M., Reya E., and Schienbein M., *Phys. Rev.* **D60**,054019 (1999).

35. Glück, M.,et al., *Has the QCD RG-Improved Parton Content of Virtual Photons been Observed?* hep-ph/0009348.
36. Chýla J., hep-ph/0006232 and this proceedings.
37. Friberg,C.,and Sjöstrand, T., hep-ph/0009003.
38. Uematsu T., these proceedings.
39. Stratman, M., Proc. of the Photon 99, Freiburg, Germany, May 1999, ed. S. Söldner-Rembold, *Nucl. Phys.* B (Proc. Suppl.) **82**, 400 (2000).

A Model of F_2^γ

B. Badełek[*,1], M. Krawczyk[#,1], J. Kwieciński[†,1], and A.M. Staśto[†,1]

*Department of Physics, Uppsala University, P.O.Box 530, 751 21 Uppsala, Sweden and
Institute of Experimental Physics, Warsaw University, 00-681 Warsaw, Poland
Institute of Theoretical Physics, Warsaw University, 00-681 Warsaw, Poland
†Department of Theoretical Physics, H. Niewodniczanski Institute of Nuclear Physics,
31-342 Cracow, Poland*

Abstract. A model of the real photon structure function F_2^γ which can be used in the low Q^2, low x region is formulated. It includes both the VMD and the QCD components, the latter based on arbitrary parton distributions in the photon and suitably extrapolated to the low Q^2 region.

INTRODUCTION

The deep inelastic electron–photon scattering, $\text{DIS}_{e\gamma}$,

$$e \, \gamma \to \gamma^* \gamma \to hadrons, \tag{1}$$

studied in high energy $e^+ e^-$ collisions with tagged electron is an analogue of the deep inelastic lepton–nucleon scattering [1]. Here the probe – a virtual photon of four momentum q ($q^2 = -Q^2 < 0$), tests a structure of the target particle, the real photon of four momentum p ($p^2 = 0$). The corresponding spin averaged cross section for process (1) it is related to the imaginary part of the elastic amplitude for $\gamma^* \gamma \to \gamma^* \gamma$ (Fig.1). It can be parametrized by the photon structure functions, e.g. $F_1^\gamma(x, Q^2)$ and $F_2^\gamma(x, Q^2)$, and interpreted in the Parton Model and QCD for very high Q^2 [2,3]. The status of the measurements of the structure functions for the real photon is presented in recent review articles [4].

Here we present a model [6] of the photon structure function F_2^γ which includes both the VMD contribution [5] and the QCD term, suitably extrapolated to the low Q^2 region. This approach is based on the extension of a similar representation of the nucleon structure function [7,8] to the case of the photon. Our framework permits also to describe $\gamma^* \gamma$ and the $\gamma\gamma$ total cross sections as functions of energy. It may be used to describe the region of very high energies which can become accessible in future linear colliders, $e^+ e^-$ LC and Photon Colliders [14,15].

---

[1] Supported by KBN Grant No 2P03B0511 (2000)

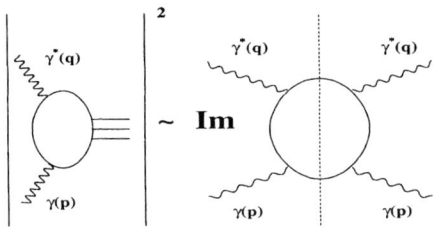

FIGURE 1. The cross section for the $\gamma^*\gamma \to hadrons$ scattering and its relation to the imaginary part of the forward $\gamma^*\gamma \to \gamma^*\gamma$ amplitude.

Possible parametrisations of the photon structure function which extend to the low Q^2 region have also been discussed in literature, see [9,10]. The energy dependence of $\sigma_{\gamma\gamma}$ is also described by other models, see [10,11].

DISPERSIVE RELATION FOR $\gamma^*\gamma$ SCATTERING

The QCD describes the photon structure functions in the large Q^2 region [3,2]. However, in the low Q^2 region, below 1 GeV^2, one expects that the VMD mechanism [5] is important. By the Vector Meson Dominance mechanism in this case we understand the model in which the virtual photon of virtuality Q^2 fluctuates into vector mesons which next undergo interaction with the (real) photon of virtuality $p^2 \simeq 0$, see Fig.2 (left). In order to be able to describe the photon structure

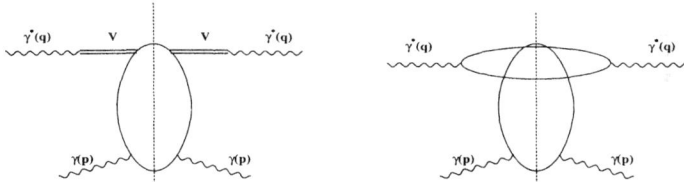

FIGURE 2. Left: Diagrammatic representation of the Vector Meson Dominance model in the $\gamma^*\gamma$ scattering. Right: The arbitrary mass state for the virtual photon in $\gamma^*\gamma$ scattering.

function for arbitrary values of Q^2 it would be very useful to have a unified scheme which contains at each Q^2 both the VMD contribution and the contribution, which corresponds to the direct γ^* coupling to quarks from the photon target, with possible higher order QCD corrections, i.e. the QCD term. The unified scheme can be obtained by utilising the dispersive representation in Q^2. The $\gamma^*\gamma$ collision can be viewed as the interaction of a real photon target with a photon with virtuality Q^2 which fluctuates onto general hadronic state, cf. Fig.2 (right). One can separate regions of integration over low- and high values of virtualities of the hadronic system Q'^2.

For low values of Q'^2, $Q'^2 < Q_0^2$, one uses the Vector Meson Dominance model [5], where the virtual photon forms a vector meson. The resulting VMD contribution to the photon structure function F_2^γ, taking into account only diagonal transitions between states having the same masses, is given by:

$$F_2^{\text{VMD}}(x, Q^2) = \frac{Q^2}{4\pi} \sum_V \frac{M_V^4 \, \sigma_{V\gamma}(W^2)}{\gamma_V^2 (Q^2 + M_V^2)^2}, \tag{2}$$

where the soft cross section for $V\gamma$ interaction is described by the Regge formula [6]. In eq.2 we consider mesons with masses $M_V^2 < Q_0^2$ only.

The contribution coming from the region of high values of Q'^2 ($Q'^2 > Q_0^2$) can be related to the parametrization of the photon structure function valid in the large Q^2 domain, $F_2^{QCD}(x, Q^2)$. It defines the partonic contribution F_2^{partons} to the structure function F_2^γ, extended to arbitrary low values of Q^2. For convenience, we adopt the approximation used in Ref. [7], namely:

$$F_2^{\text{partons}}(x, Q^2) = \frac{Q^2}{Q^2 + Q_0^2} F_2^{QCD}(\bar{x}, Q^2 + Q_0^2), \quad \bar{x} = \frac{Q^2 + Q_0^2}{W^2 + Q^2 + Q_0^2}. \tag{3}$$

Modifications of the QCD contribution: replacement of the parameter x by $\bar{x}$, shift of the scale $Q^2 \to Q^2 + Q_0^2$ and the factor $Q^2/(Q^2 + Q_0^2)$ instead of 1, introduce power corrections which vanish as $1/Q^2$ and are negligible at large Q^2. The magnitude of Q_0^2 is set to 1.2 GeV^2 as in the case of the proton [7].

MODEL OF F_2^γ AND OF THE TOTAL PHOTON–PHOTON CROSS SECTIONS

Our model of the photon structure function $F_2^\gamma(x, Q^2)$ is based on the following decomposition:

$$F_2^\gamma(x, Q^2) = F_2^{\text{VMD}}(x, Q^2) + F_2^{\text{partons}}(x, Q^2), \tag{4}$$

where F_2^{VMD} and $F_2^{\text{partons}}(x, Q^2)$ are defined by equations (2) and (3). A total $\gamma^*\gamma$ cross-section in the high energy limit is given by

$$\sigma_{\gamma^*\gamma}(W, Q^2) = \frac{4\pi^2 \alpha}{Q^2} F_2^\gamma(x, Q^2), \tag{5}$$

with $x = Q^2/(Q^2 + W^2)$. The $Q^2 = 0$ (for fixed W) limit of eq. (5) gives the total cross-section $\sigma_{\gamma\gamma}(W^2)$ for the interaction of two real photons. From (4), (2) and (3) we obtain the following expression for this cross-section at high energy:

$$\sigma_{\gamma\gamma}(W) = \alpha\pi \sum_{V=\rho,\omega,\phi} \frac{\sigma_{V\gamma}(W^2)}{\gamma_V^2} + \frac{4\pi^2\alpha}{Q_0^2} F_2^{QCD}(Q_0^2/W^2, Q_0^2). \tag{6}$$

At large Q^2 the structure function given by eq. (4) becomes equal to the QCD contribution, $F_2^{QCD}(x, Q^2)$, in a standard form, as given by a chosen parton parametrization for γ. The VMD component gives the power correction term which vanishes as $1/Q^2$ for large Q^2, $Q_0^2 = 1.2$ GeV^2.

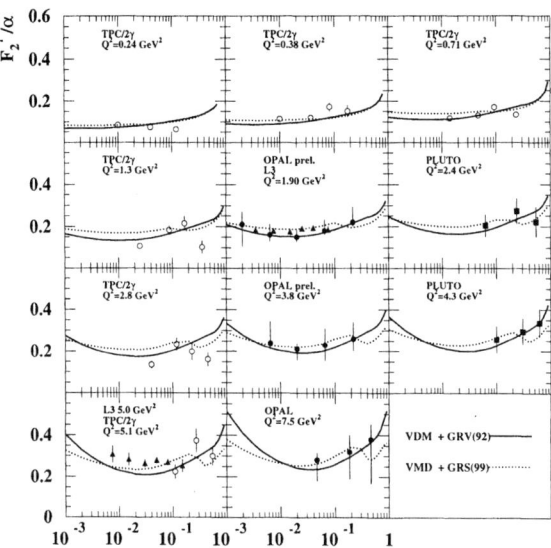

FIGURE 3. Comparison of our predictions for F_2^γ/α as a function of x for different values of Q^2 in the low Q^2 region, with experimental data. The solid (dotted) line were obtained using the LO parametrisations of F_2^{QCD}/α: GRV and GRS$'$, respectively.

COMPARISON WITH DATA AND CONCLUSION

In the quantitative analysis of the photon structure function and of the total cross sections we have used the estimation of VMD term as described in [6]. The form of structure function F_2^{QCD} was taken from two LO parton parametrizations, parametrization GRV [12], and based on updated data parametrization GRS' [13], with a number of active flavours equal four.

Results of the real photon structure function $F_2^\gamma(x, Q^2)$ are presented in Fig.3 as the function of x in the region of small Q^2. Our prediction reproduces reasonably well the data [18] independently of the parametrization (GRV or GRS') of F_2^{QCD} used in the model. Results for the two-photon cross sections, $\sigma_{\gamma\gamma}(W)$ are given in Fig.4 (left) in comparison with corresponding data [19].The calculated total $\gamma\gamma$ cross-section exhibits an approximate power-law increase with increasing energy W, i.e. $\sigma_{\gamma\gamma}(W) \sim (W^2)^{\lambda_{eff}}$ with λ_{eff} slowly increasing with energy: $\lambda_{eff} \sim 0.1 - 0.12$ for 30 GeV $< W < 10^3$ GeV (for the "GRV" curve). The VDM part is described by the soft Pomeron contribution with $\lambda = 0.0808$.

In Fig. 4 (right) the decomposition of the cross section for $\gamma\gamma$ is presented to show the role of the QCD term in rise of the cross section at very large energies. The large difference in the predictions of the $F_2^{partons}$ obtained using GRV and GRS parametrizations is seen in the medium W region. This is responsible for the corresponding difference in the cross sections $\sigma_{\gamma\gamma}(W)$ presented in Fig. 4 (left).

21

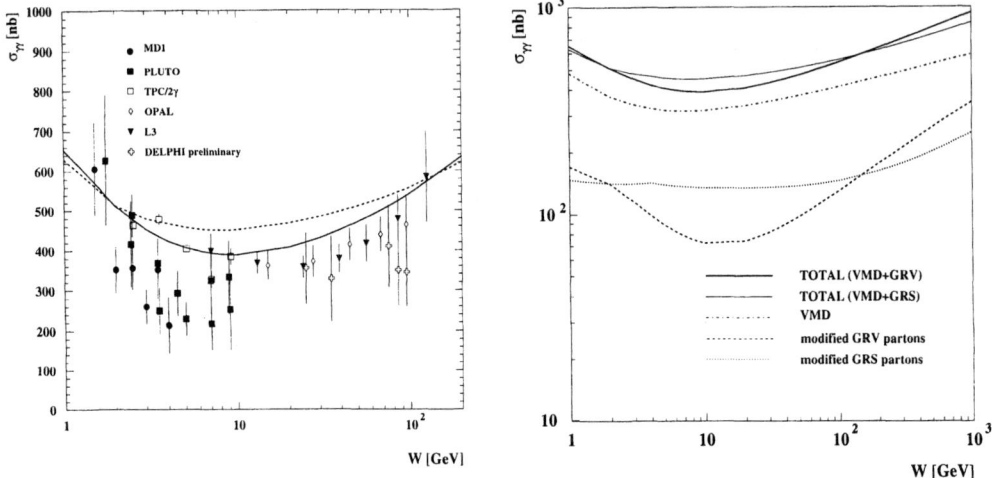

FIGURE 4. Left: Comparison of our predictions for $\sigma_{\gamma\gamma}(W)$ with experimental results. The curves correspond to different form of F_2^{QCD}: GRV (solid line) and GRS$'$ (dashed line). Right: Decomposition of the $\sigma_{\gamma\gamma}(W)$ into contributions: VDM and QCD, as follows from GRV and GRS parametrizations.

To conclude, we have presented an extension of the representation developed for the nucleon structure function F_2^p for arbitrary values of Q^2, [7], onto the structure function of the real photon. This representation includes both the VMD contribution and the QCD component, obtained from the QCD parton parametrizations for the photon, suitably extrapolated to the region of low Q^2. In the $Q^2=0$ limit the model gives predictions for the total cross section $\sigma_{\gamma\gamma}$ for the interaction of two real photons. We showed that our framework is fairly successful in describing the experimental data on $\sigma_{\gamma\gamma}(W)$ (and also on $\sigma_{\gamma^*\gamma}(W, Q^2)$) and on the photon structure function F_2^γ at low Q^2. We also showed that one can naturally explain the fact that the increase of the total $\gamma\gamma$ cross-section with increasing CM energy W is stronger than that implied by soft Pomeron exchange and it is driven by the partonic contribution.

ACKNOWLEDGMENTS

One of us (MK) is grateful to Organizers of this excellent Conference for an fruitful and stimulating atmosphere. She is also grateful for a financial support.

REFERENCES

1. Brodsky, S.J., et al, *Phys. Rev. Lett.* **27**, 280 (1971); Carlson, C.E. and Tung, W. *Phys. Rev.* **D4**, 287 (1971);
2. Walsh, T.F., *Phys. Lett.* **B36**, 121 (1971); Peterson,C., Walsh, T.F., and Zerwas, P.M., *Nucl.Phys.* **B174**, 424 (1980); Zerwas, P. M., *Phys.Rev.* **D10**, 1485 (1974).
3. Witten, E., *Nucl.Phys.* **B120**, 189 (1977).
4. Krawczyk, M., Staszel, M., Zembrzuski, A., IFT-99-15 (hep-ph/001183), *Physics Rep.* to appear; Erdman, M., *Springer Tracts in Modern Physics* **138** (1997); Nisius, R.,*Physics Rep.* **332**, 166 (2000).
5. Sakurai, J., J., *Ann. Phys.* **11**, 1 (1960), Sakurai, J., J., and Schildknecht, D.,*Phys. Lett.* **B40**, 121 (1972); Bauer,T.H., et al.,*Rev. Mod. Phys.* **50** (1978) 261.
6. Badełek, B., Krawczyk,M., Kwieciński,J., Staśto, A.M., *Phys. Rev.*, 074021 **D62** (2000).
7. Kwieciński,J., and Badełek, B., *Z.Phys.* **C43**, 251 (1989); Badełek, B., and Kwieciński, J., *Phys. Lett.* **B295**, 263 (1992).
8. Martin, A.D., Ryskin, M.G., and Staśto, A.M., *Eur. Phys. J.* **C7**, 643 (1999); *Nucl. Phys. Proc. Suppl.* **74**, 121 (1999).
9. Gotsman, E., Levy, A., and Maor, U.,*Z. Phys.* **C40**, 117 (1988); Donnachie, A., Dosch, H. G. and Rueter, M., *Phys.Rev.* **D59**, 074011 (1999).
10. Schuler, G. A., *Comput. Phys. Commun.* **108**, 279 (1998);
11. Donnachie, A., Dosch, H. G., and Rueter, M., *Phys.Rev.* **D59**, 074011 (1999) and *Eur. Phys. J.* **C13**, 141 (2000); Forshaw, J. R., and Storrow, J.K., *Phys. Rev.* **D46**, 4955 (1992); *Phys. Lett.* **B278**, 193 (1992); Schuler, G.A., and Sjöstrand, T., *Z.Phys.* **C73**, 677 (1997); Corsetti, A., Godbole, R. M., and Pancheri,G. *Phys. Lett.* **B435**, 441 (1998) and IISC-CTS-8-99 (hep-ph/9912395); Donnachie, A., and Soldner-Rembold, S.,FREIBURG-EHEP-2000-01 (hep-ph/0001035); Gotsman, E., et al., TAUP-2618-00 (hep-ph/0001080).
12. Glück, M., Reya, E., and Vogt, A. *Phys.Rev.* **D46**, 1973 (1992).
13. Glück, M., Reya, E., and Schienbein, I., *Phys.Rev.* **D60**, 054019 (1999).
14. Cooper-Sarkar, A.M., Devenish, R.C.E., and De Roeck, A., *Int.J.Mod.Phys.* **A13**, 3385 (1998).
15. Vogt, A., Proc. of the Int. Conference on the Structure and Interactions of the Photon, Freiburg, 1999, *Nucl. Phys. Proc. Suppl.***82**, 394 (2000); Accomando, E., et al., *Phys. Rep.* **299**, 21 (1998).
16. Schuler, G.A., and Sjöstrand, T., *Z. Phys.* **C68**, 607 (1995).
17. Donnachie, A., and Landshoff,P.V., *Phys. Lett.* **B296**, 227 (1992).
18. PLUTO Coll.; Berger, Ch., et al., *Phys. Lett.* **B142**, 111 (1984); *Nucl. Phys.* **B281**, 365 (1987); TPC/2γ Coll.; Aihara, H., et.al., *Z. Phys.* **C34**,1 (1987); L3 Coll.; Acciarri, M., et al., *Phys. Lett.* **B 436**, 403 (1998); OPAL Coll., *Physics Note* PN389 (1999); OPAL Coll., Ackerstaff, K., et al., *Z.Phys.* **C74**, 33 (1997)
19. PLUTO Coll.; Berger, Ch., et al., *Phys. Lett.* **B149**, 421 (1984); *Z. Phys.* **C26**, 353 (1984), TPC/2γ Coll.; Aihara, H., et al., *Phys. Rev.* **D41**, 2667 (1990); MD-1 Coll.; Baru, S.E., et al., *Z. Phys.* **C53**, 219 (1992); L3 Coll.; L3 Note 2570, submitted to the XXXth Int. Conference on High Energy Physics, Osaka, Japan, July 27-August

2, 2000; OPAL Coll.; Abbiendi, G., et al., *Eur.Phys.J.* **C14**, 199 (2000); DELPHI Coll.; Zimin, N., Proc. of the Int. Conference on the Structure and Interactions of the Photon, Friburg, 1999, *Nucl. Phys.* **B (Proc. Suppl.) 82**, 139 (2000).

A high-Q^2 measurement of the photon structure function F_2^γ at LEP2

Russell J. Taylor

Department of Physics and Astronomy, UCL,
Gower Street, London WC1E 6BT, United Kingdom

Abstract. The photon structure function $F_2^\gamma(x, Q^2)$ has been measured at $\langle Q^2 \rangle$ of 706 GeV^2 using a sample of two-photon events with a scattered electron observed in the OPAL electromagnetic endcap calorimeter. The data were taken during the years 1997-1999, when LEP operated at e^+e^- centre-of-mass energies ranging from 183 to 202 GeV, and correspond to an integrated luminosity of 424 pb^{-1}. This analysis represents the highest $\langle Q^2 \rangle$ measurement of F_2^γ made to date.

INTRODUCTION

We present a measurement of the hadronic photon structure function $F_2^\gamma(x, Q^2)$ at a higher value of the average momentum transfer squared, $\langle Q^2 \rangle$, than has ever previously been reported. The measurement of F_2^γ is interesting because of its potential to test perturbative QCD [1,2]. In the high-Q^2 domain the perturbatively calculable point-like contribution to F_2^γ, which rises logarithmically with Q^2, dominates over the non-perturbative hadron-like part.

The structure function F_2^γ has been measured at $\langle Q^2 \rangle$ of 706 GeV^2 using a sample of single-tagged two-photon events recorded by the OPAL detector between 1997 and 1999. These events (also referred to as $\gamma^\star\gamma$ events) can be regarded as deep inelastic scattering of an electron on a quasi-real photon, and the flux of quasi-real photons can be calculated using the equivalent photon approximation [3].

To study $F_2^\gamma(x, Q^2)$ the distribution of events in x and Q^2 is needed. These variables are related to experimentally measurable quantities by

$$Q^2 = 2\, E_{\rm b}\, E_{\rm tag}\, (1 - \cos\theta_{\rm tag}) \tag{1}$$

and

$$x = \frac{Q^2}{Q^2 + W^2 + P^2}, \tag{2}$$

where $E_{\rm b}$ is the energy of the beam electron, $E_{\rm tag}$ and $\theta_{\rm tag}$ are the energy and polar angle of the deeply inelastically scattered electron, W^2 is the invariant mass

CP571, *PHOTON 2000*, edited by A. J. Finch
© 2001 American Institute of Physics 0-7354-0010-5/01/$18.00

squared of the hadronic final state and $P^2 = -p^2$, where p is the four-momentum of the quasi real target photon. The requirement that the associated electron is not visible in the detector ensures that $P^2 \ll Q^2$, so P^2 can be neglected when calculating x from Equation 2.

DATA SELECTION

This analysis uses data from the 1997 to 1999 LEP runs, with e^+e^- centre-of-mass energies ranging from 183 to 202 GeV. The total integrated e^+e^- luminosity is 424 pb^{-1}. Candidate $\gamma^\star\gamma \rightarrow$ hadrons events are required to satisfy the following selection criteria, in addition to several technical cuts to ensure good detector status and track quality.

1. A tagged electron is required; that is, a cluster in the OPAL electromagnetic endcap calorimeter with an energy of at least $0.6E_b$ and a polar angle θ in the range 230–500 mrad with respect to either beam direction.

2. The energy of the most energetic electromagnetic cluster in the hemisphere opposite to that which contains the tagged electron must be less than $0.25E_b$.

3. The number of tracks originating from the hadronic final state must be at least 3.

4. The visible invariant mass W_{vis} of the hadronic system is required to be in the range 2.5 GeV $\leq W_{\text{vis}} \leq$ 50 GeV.

5. The number of objects (tracks plus unassociated clusters), belonging to the hadronic final state must be at least 9.

6. The energy deposited in a cone of 200 mrad half-angle about the direction of the tag, excluding the tag itself, must not be more than 2 GeV.

Cuts 1–4 select a sample of candidate single-tag hadronic two-photon events, with double-tag events excluded by cut 2. Events with leptonic final states are rejected by cuts 3 and 5. The invariant mass cuts have two functions. The lower limit removes the low-W region which is dominated by resonance production and is very difficult to model accurately. The upper limit rejects background events from hadronic decays of Z^0 bosons, as does cut 6.

A total of 348 events pass these cuts, with the data covering the range 270 GeV^2 < Q^2 < 2200 GeV^2. There is a two sigma difference between the number of events selected in the 1997/8 data and that recorded in 1999, with the 1999 data lying below the Monte Carlo expectation, particularly at low W_{vis}. No explanation has been found for this. The larger samples of $\gamma^\star\gamma$ events with the electron tagged in subdetectors at lower polar angles are consistent between the two periods, suggesting that the observed difference could well be purely statistical. However, as a precaution, the difference is included in the systematic error in this preliminary analysis.

26

The OPAL LEP1 analysis of F_2^γ using tags in the same subdetector [4] found the trigger efficiency to be 100%. The present analysis uses a tighter set of cuts, thus no inefficiency is to be expected, and a trigger efficiency of 100% is assumed.

MONTE CARLO MODELLING AND BACKGROUND

Monte Carlo programs are used to simulate $\gamma^\star\gamma$ events and to provide background estimates. The Monte Carlo generator used to simulate signal $\gamma^\star\gamma$ multiperipheral events is HERWIG 5.9+k_t(dyn) [5]. The GRV LO [6] parameterisation of F_2^γ was used as the input structure function.

The dominant background comes from the reaction $Z^0/\gamma^\star \to$ hadrons. Also significant are non-multiperipheral four-fermion processes with $e^+e^-q\bar{q}$ final states and the QED process $\gamma^\star\gamma \to \tau^+\tau^-$. Less severe sources of background are estimated to account for around 1% of the data sample.

Figure 1 shows comparisons between data and Monte Carlo distributions. Figure 1(b) shows the polar angle of the tagged electron. It can be seen that the Monte Carlo is somewhat higher than the data in the polar range 260–350 mrad. Turning to variables describing the hadronic final state, it can be seen that the number of charged tracks is reasonably well described, Figure 1(c), but that the Monte Carlo lies above the data at low W_{vis} in Figure 1(d) - which correlates with high x.

DETERMINATION OF F_2^γ

The perennial problem in measurements of F_2^γ is that, because the $\gamma^\star\gamma$ centre-of-mass system does not coincide with the laboratory system, the hadronic final state, which must be measured to determine W, is only partially observed in the detector. This leads to a dependence of the F_2^γ measurement on the Monte Carlo modelling, which is needed for the unfolding process used to relate the visible distributions to the underlying x distribution.

In the high-Q^2 measurement presented here, however, the situation is not as serious as at lower values of Q^2. Because of the larger tagging angle, the hadronic final state has much more transverse momentum and as a consequence is better contained in the detector. Figure 2 shows the correlation between the measured invariant mass W_{vis} and the generated W as given by HERWIG 5.9+k_t(dyn). It can be seen that the correlation is maintained throughout. This is in contrast with the situation observed in the lower Q^2 analysis [7] where the correlation deteriorates at high W. As a consequence of this the result can be expected to have a smaller dependence on the Monte Carlo modelling of the hadronic final state.

After subtraction of background, the data are unfolded on a linear scale in x in the range $0.1 \le x \le 0.98$ using the GURU program [8]. Each data point is corrected for radiative effects and bin-centre corrections are applied. In Figure 3 the data are compared to several theoretical calculations. The leading order

OPAL Preliminary

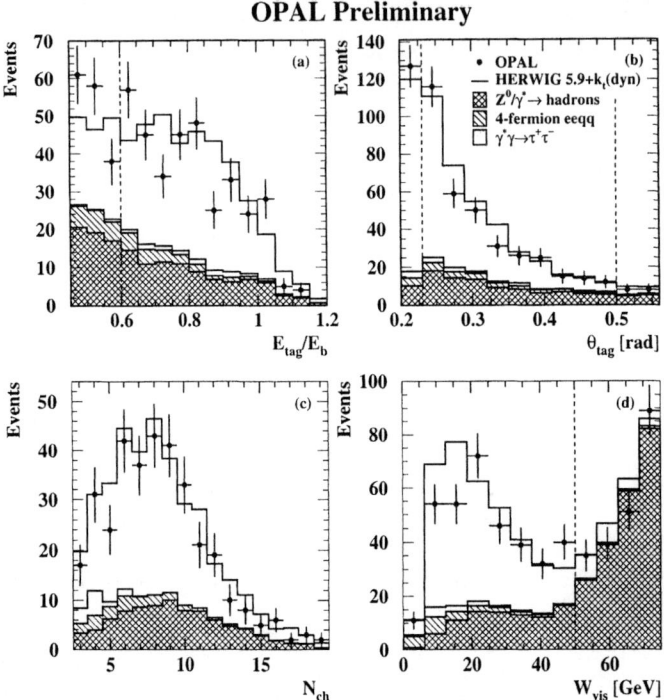

FIGURE 1. Comparison of data distributions with the Monte Carlo prediction. The open histogram is the sum of the HERWIG 5.9+k_t(dyn) prediction and the contributions of the major background sources (shown as shaded histograms). All selection cuts have been applied, except for any cut on the variable in the plot (indicated as dashed lines). The distributions shown are: (a) E_{tag}/E_b, the energy of the tagged electron as a fraction of the beam energy, (b) θ_{tag}, the polar angle of the tagged electron, (c) N_{ch}, the number of tracks originating from the hadronic final state, and (d) W_{vis}, the measured invariant mass of the hadronic final state.

parameterisations of F_2^γ from GRV, SaS1d [9] and WHIT1 [10], which all include a contribution from massive charm quarks, are described in detail in reference [2]. The naive quark-parton model (QPM) simulates only the point-like component of F_2^γ, and is calculated for four active flavours with masses of 0.325 GeV for light quarks and 1.5 GeV for charm quarks. It can be seen that in this high-Q^2 regime the differences between the models are relatively small, particularly in the central x-region. The differences between the QPM and the other models are much smaller than at lower Q^2, where the photon has been shown to have a significant hadron-like component [7]. All the predictions are compatible with the data in three of the x bins, but overshoot the data in one bin.

28

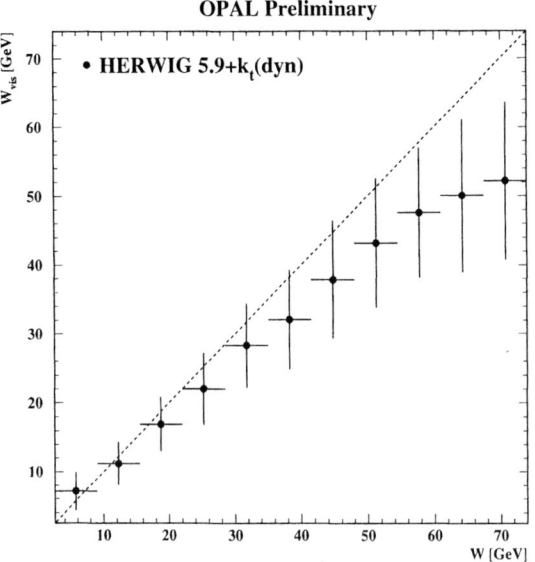

FIGURE 2. The correlation between the generated hadronic invariant mass and the measured value, as given by HERWIG 5.9+k_t(dyn). The vertical error bars represent the spread within each bin. The dashed line corresponds to perfect correlation.

CONCLUSIONS

The photon structure function F_2^γ has been measured using deep inelastic electron-photon scattering events recorded by the OPAL detector during the years 1997–1999. The $\langle Q^2 \rangle$ value of 706 GeV^2 represents the highest measured thus far. F_2^γ has now been measured by OPAL at $\langle Q^2 \rangle$ values ranging from 1.9–706 GeV^2.

REFERENCES

1. E. Witten, Nucl. Phys. **B120** (1977) 189;
   A. Cordier and P.M. Zerwas, in ECFA Workshop on LEP200, Vol. 1, CERN 87-08, ECFA 87/108, edited by A. Böhm and W. Hoogland, p. 242.
2. R. Nisius, Phys. Rep. **332** (2000) 165.
3. C.F. von Weizsäcker, Z. Phys. **88** (1934) 612;
   E.J. Williams, Phys. Rev. **45** (1934) 729;
   V.M. Budnev et al., Phys. Rep. **15** (1975) 181.
4. OPAL Collaboration, K. Ackerstaff et al., Z. Phys. **C74** (1997) 33.
5. G. Marchesini et al., Comp. Phys. Comm. **79** (1992) 465.

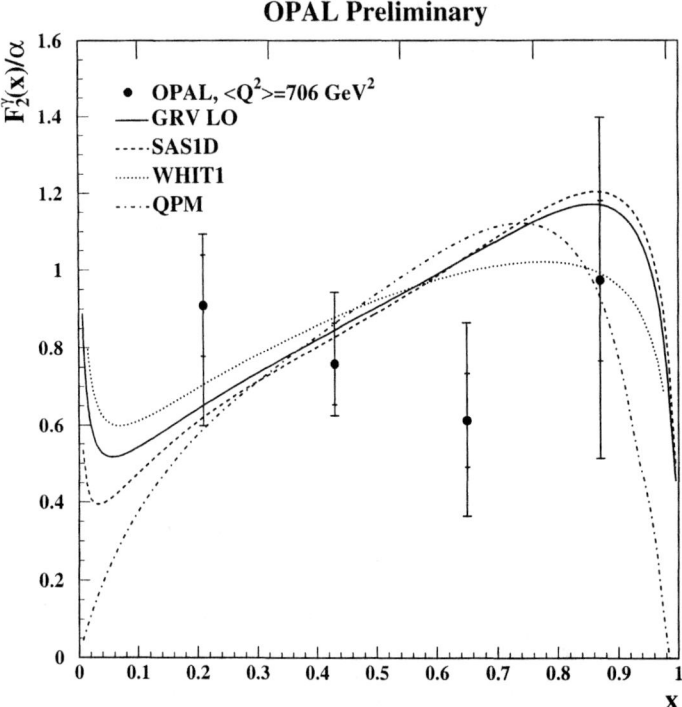

FIGURE 3. The measured F_2^γ at $\langle Q^2 \rangle = 706$ GeV^2. The inner bars indicate the statistical error and the full bars the total error. The bin boundaries are indicated by the vertical lines at the top of the figure. The curves show the predictions of the GRV LO, SaS1d, WHIT1 and QPM structure functions, all for the $\langle Q^2 \rangle$ value of the sample.

The LEP Working Group for Two-Photon Physics, ALEPH, L3 and OPAL Collaborations, *Comparison of Deep Inelastic Electron-Photon Scattering Data with the Herwig and Phojet Monte Carlo Models*, CERN-EP/2000-109, Submitted to Eur. Phys. J. **C**.

6. M. Glück, E. Reya and A. Vogt, Phys. Rev. **D45** (1992) 3986;
   M. Glück, E. Reya and A. Vogt, Phys. Rev. **D46** (1992) 1973.

7. E.W. Clay, these proceedings;
   OPAL Collaboration, G. Abbiendi *et al.*, *Measurement of the low-x behaviour of the photon structure function F_2^γ*, CERN-EP/2000-082, Accepted by Eur. Phys. J. **C**.

8. A. Höcker, V. Kartvelishvili, Nucl. Instr. and Meth. **A372** (1996) 469.

9. G. A. Schuler and T. Sjöstrand, Z. Phys. **C68** (1995) 607.

10. K. Hagiwara *et al.*, Phys. Rev. **D51** (1995) 3197.

Measurement of the Low-x Behaviour of the Photon Structure Function F_2^γ

Edmund W. Clay

Department of Physics and Astronomy, UCL, Gower Street, London WC1E 6BT, United Kingdom.

Abstract. The photon structure function $F_2^\gamma(x, Q^2)$ has been measured using data taken by the OPAL detector at e^+e^- centre-of-mass energies of 91 GeV, 183 GeV and 189 GeV, in Q^2 ranges of 1.5–30.0 GeV^2 (LEP1), and 7.0–30.0 GeV^2 (LEP2), probing lower values of x than ever before. Since previous OPAL analyses new Monte Carlo models and new methods such as multi-variable unfolding have been introduced, reducing significantly the model dependent systematic errors in the measurement.

INTRODUCTION

The measurement of the hadronic photon structure function $F_2^\gamma(x, Q^2)$ is a classic test of QCD predictions [1]. The structure function of the photon differs from that of the proton because the photon can couple directly to quark charges, as well as fluctuate into a hadronic state. The value of $F_2^{\text{proton}}(x, Q^2)$ exhibits a clear rise towards low values of Bjorken x [2,3], consistent with general QCD expectations [4]. A similar rise is predicted for the photon structure function. The photon structure function is studied at LEP using samples of events of the type $e^+e^- \rightarrow e^+e^- + \text{hadrons}$. The analysis presented here uses single-tagged events (from here on referred to as $\gamma^*\gamma$ events), in which only one of the scattered beam electrons[1] is observed in the detector. These events can be regarded as deep inelastic scattering of an electron off a quasi-real target photon.

In contrast to measurements of the proton structure function, here the energy of the target particle is not known. In consequence, the kinematics cannot be fully determined without measuring the hadronic final state, which is only partially observed in the detector. This leads to a dependence of the F_2^γ measurement on Monte Carlo modelling of the hadronic final state.

The analysis presented here uses OPAL data collected during the years 1993–1995, 1997 and 1998, at e^+e^- centre-of-mass energies of 91 GeV (LEP1), and 183 GeV and 189 GeV (LEP2). A more detailed account of the analysis is given in [5].

---

[1] For conciseness, positrons are also referred to as electrons.

CP571, *PHOTON 2000*, edited by A. J. Finch

KINEMATICS AND DATA SELECTION

To measure $F_2^\gamma(x, Q^2)$, the distribution of events in x and Q^2 is required. These variables are related to the experimentally measurable quantities

$$Q^2 = 2\, E_b\, E_{tag}\, (1 - \cos\theta_{tag}), \qquad x = \frac{Q^2}{Q^2 + W^2} \tag{1}$$

where E_b is the energy of the beam electrons, E_{tag} and θ_{tag} are the energy and polar angle of the deeply inelastically scattered (or 'tagged') electron, W^2 is the invariant mass squared of the hadronic final state and $P^2 = -p^2$ is the negative value of the virtuality squared of the quasi-real photon.

Electrons are tagged in the SW (27/33–55 mrad at LEP1/LEP2) and FD (60–120 mrad) subdetectors of OPAL. The SW subdetector is used with both LEP1 and LEP2 data, while FD is only used at LEP1, to provide a sample in the same range of Q^2 as the LEP2 SW sample. Each sample is split into two bins of Q^2, giving six Q^2 regions in total.

Events are selected by applying cuts on the scattered electrons and on the hadronic final state. A tagged electron within the clear acceptance of SW or FD is selected by requiring $E_{tag} \geq 0.75\, E_b$ at LEP1 or $E_{tag} \geq 0.775\, E_b$ at LEP2. The highest energy cluster in the hemisphere opposite the tagged electron must have an energy $E_a \leq 0.25\, E_b$. The number of tracks in the event passing quality cuts and originating from the hadronic final state, N_{ch}, must be at least three, of which at least two tracks must not be identified as electrons. Finally, the visible invariant mass W_{vis} of the hadronic system is required to be in the range 2.5 GeV $\leq W_{vis} \leq$ 40/60 GeV at LEP1/LEP2. The numbers of events in each sample passing the cuts, the integrated luminosities and the Q^2 ranges are listed in Table 1. The trigger efficiencies were evaluated from the data using sets of separate triggers and found to be larger than 99% for all of the samples.

$\langle Q^2 \rangle$ [GeV^2]	sample	luminosity [pb^{-1}]	events	Q^2 range [GeV^2]
1.9	LEP1 SW	74.6	4356	1.5–2.5
3.7			4010	2.5–6.0
8.9	LEP1 FD	97.8	1909	6.0–12.0
17.5			1578	12.0–30.0
10.7	LEP2 SW	222.9	4593	7.0–13.0
17.8			5495	13.0–30.0

TABLE 1. The integrated luminosity, number of selected events, and Q^2 range for each data sample. The error on the luminosity is negligible.

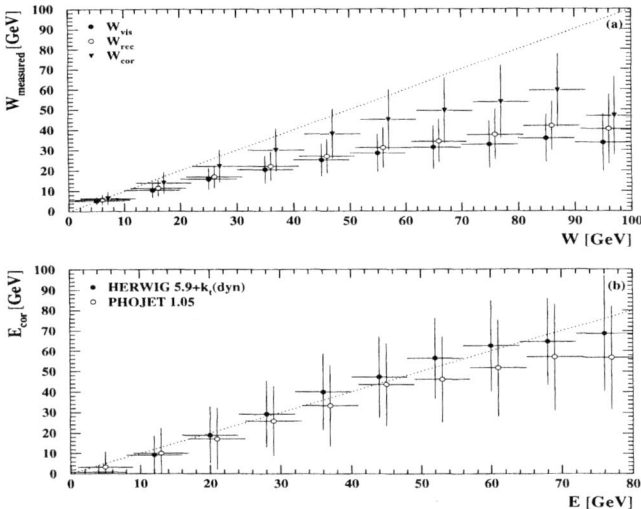

FIGURE 1. (a) The correlation between the generated and measured invariant mass, W, for the HERWIG 5.9+k_t(dyn) LEP2 SW Monte Carlo sample, using W_{fd}, W_{rec} and W_{cor}. (b) The corrected energy E_{cor}, in the forward region against the total generated energy E, deposited in that region, for the same sample as in plot (a).

MONTE CARLO MODELS

It has been seen in previous studies [6] that Monte Carlo modelling of the hadronic final state is a large source of systematic error in the F_2^γ measurement. While there is still no $\gamma^\star\gamma$ Monte Carlo generator which describes all the features of the observed hadronic final state, improved Monte Carlo programs have become available since the previous OPAL measurements.

The Monte Carlo generators used to simulate signal events are HERWIG 5.9+k_t(dyn) [7] and PHOJET 1.05 [8].

The dominant background comes from the reaction $e^+e^- \rightarrow e^+e^-\tau^+\tau^-$ proceeding via the multiperipheral process, with $e^+e^- \rightarrow e^+e^-e^+e^-$ giving a smaller contribution. The process $Z^0 \rightarrow$ hadrons also contributes significantly at LEP1. Other sources of background were found to be negligible.

DETERMINATION OF F_2^γ

The visible hadronic mass squared W_{vis}^2 is evaluated by summing over all tracks and clusters in the hadronic final state. Because hadronic showers are not well contained in the electromagnetic calorimeters, and some energy is lost in the beampipe, the measured energy in the forward region is less than the true energy. It is possible to improve the reconstruction of W_{vis} by including kinematic information from the tagged electron [9]. The (reconstructed) variable formed in this way is called W_{rec}. The value of W_{rec} is still generally smaller than the true value. This is mainly due to energy losses in the forward calorimeters. In an attempt to make the energy response of the detector more uniform a new (corrected) variable is formed: W_{cor}, in which the contribution from the forward region has been scaled by a factor of 2.5. Figure 1a shows the correlation between the generated W and the three measured quantities W_{vis}, W_{rec} and W_{cor}. In general, W_{cor} is still lower than W, mainly because of energy lost in the beampipe. Figure 1b shows the correlation between the generated energy in the forward region E and the scaled observed energy in that region, E_{cor}.

The main unfolding program used for this analysis was GURU [10], which can be modified to perform unfolding in two dimensions. As with the attempts to improve the W reconstruction described above, the motivation is to reduce the dependence of the unfolding on a particular Monte Carlo model.

The second unfolding variable was $E_{\text{T}}^{\text{out}}/E_{\text{tot}}$, the transverse hadronic energy out of the plane containing the beam line and the tagged electron, divided by the total observed energy. This variable was chosen because of its sensitivity to the angular distribution of the hadrons in the final state.

RESULTS

The photon structure function F_2^{γ} is measured by unfolding each data sample in bins of $\log(x)$. Each OPAL data sample is divided into two ranges of Q^2. The ranges correspond to $\langle Q^2 \rangle$ values of 1.9 and 3.7 GeV^2 for the LEP1 SW sample, and 8.9 (10.7) and 17.5 (17.8) GeV^2 for the LEP1 FD (LEP2 SW) sample.

The central value of F_2^{γ} in each x bin is the average of the data unfolded with HERWIG 5.9+k_{t}(dyn) and PHOJET, using two dimensional unfolding with x_{cor} as the first variable and $E_{\text{T}}^{\text{out}}/E_{\text{tot}}$ as the second variable.

In Figures 2 and 3 the results are compared to measurements of F_2^{γ} from other experiments, which are found to be generally consistent with the new OPAL measurements.

CONCLUSIONS

The photon structure function F_2^{γ} has been measured as a function of x to the lowest attainable x values, in six ranges of Q^2 (including two overlapping pairs) corresponding to average Q^2 values of 1.9, 3.7 GeV^2 for LEP1 SW, and 8.9 (10.7), 17.5 (17.8) GeV^2 for LEP1 FD (LEP2 SW).

The contribution to the systematic error from Monte Carlo modelling has been reduced compared to previous OPAL analyses by using new Monte Carlo models, improved methods for the reconstruction of the invariant mass of the hadronic final state, and two dimensional unfolding.

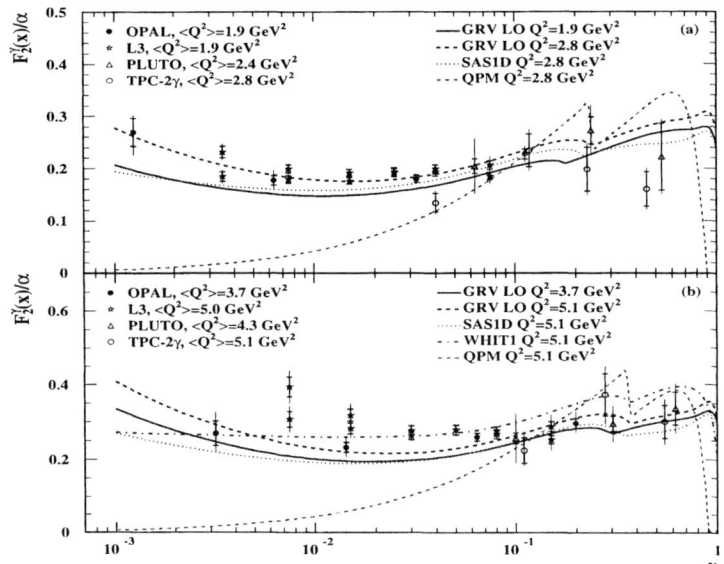

FIGURE 2. The measurement of F_2^γ/α using the LEP1 SW sample, for $\langle Q^2 \rangle$ values of (a) 1.9 and (b) 3.7 GeV^2. Also shown is a selection of results from other experiments. For each point, the inner error bars show the statistical error and the full error bars show the total error.

Monte Carlo modelling of the final state is still a significant source of systematic error, but it no longer dominates all other sources.

The GRV LO [11] and SaS1D [12] parameterisations are generally consistent with the OPAL data in all the accessible x and Q^2 regions. In contrast, the naive quark-parton model is not able to describe the data for $x < 0.1$. These results show that the photon must contain a significant hadron-like component at low x.

REFERENCES

1. E. Witten, Nucl. Phys. **B120** (1977) 189;
   A. Cordier and P. M. Zerwas, *ECFA Workshop on LEP200*, Vol. 1, CERN 87-08, ECFA 87/108, eds A. Böhm and W. Hoogland (1987) 242;
   C. Berger and W. Wagner, Phys. Rep. **146** (1987) 1;
   R. Nisius, Phys. Rep. **332** (2000) 165.
2. H1 Collaboration, S. Aid *et al.*, Nucl. Phys. **B470** (1996) 3;
   H1 Collaboration, C. Adloff *et al.*, Nucl. Phys. **B497** (1997) 3.
3. ZEUS Collaboration, M. Derrick *et al.*, Z. Phys. **C72** (1996) 399;
   ZEUS Collaboration, J. Breitweg *et al.*, Eur. Phys. J. **C7** (1999) 609.
4. De Rujula *et al.*, Phys. Rev. **D10** (1974) 1649.

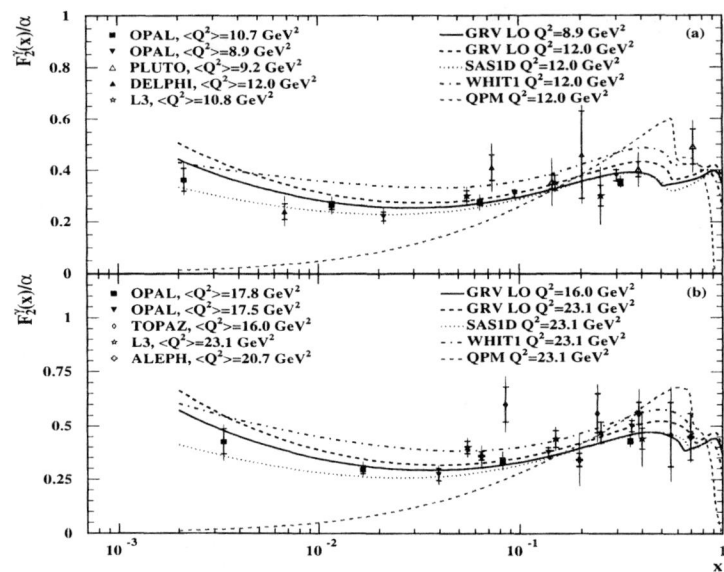

FIGURE 3. The measurement of F_2^γ/α using the LEP1 FD and LEP2 SW samples for $\langle Q^2 \rangle$ values of (a) 8.9 (10.7) and (b) 17.5 (17.8) GeV^2 for LEP1 (LEP2). Also shown is a selection of results from other experiments. For each point, the inner error bars show the statistical error and the full error bars show the total error.

5. OPAL Collaboration, G. Abbiendi *et al.*, CERN-EP-2000-091, HEP-EX/0007018, Submitted to Eur. Phys. J. C.

6. OPAL Collaboration, R. Akers *et al.*, Z. Phys. **C61** (1994) 199;
   OPAL Collaboration, K. Ackerstaff *et al.*, Z. Phys **C74** (1997) 33;
   OPAL Collaboration, K. Ackerstaff *et al.*, Phys. Lett. **B411** (1997) 387;
   OPAL Collaboration, K. Ackerstaff *et al.*, Phys. Lett. **B412** (1997) 225.

7. J. A. Lauber, L. Lönnblad and M. H. Seymour, *Photon '97, 10–15 May 1997*, eds A. Buijs, F.C. Erné, World Scientific (1997) 52;
   S. Cartwright, M.H. Seymour et al., J. Phys. **G24** (1998) 457;
   The LEP Working Group for Two-Photon Physics, ALEPH, L3 and OPAL Collaborations, CERN-EP-2000-109, Submitted to Eur. Phys. J. C.

8. R. Engel, Z. Phys. **C66** (1995) 203;
   R. Engel, J. Ranft and S. Roesler, Phys. Rev. **D52** (1995) 1459;
   R. Engel and J. Ranft, Phys. Rev. **D54** (1996) 4244.

9. L. Lönnblad *et al.*, CERN 96-01, *Physics at LEP2*, eds G. Altarelli, T. Sjöstrand and F. Zwirner (1996) Vol. 2, p. 201.

10. A. Höcker, V. Kartvelishvili, Nucl. Instr. and Meth. **A372** (1996) 469.

11. M. Glück, E. Reya and A. Vogt, Phys. Rev. **D45** (1992) 3986;
    M. Glück, E. Reya and A. Vogt, Phys. Rev. **D46** (1992) 1973.

12. G. A. Schuler and T. Sjöstrand, Z. Phys. **C68** (1995) 607.

First Measurement of the Photon Structure Function $F_{2,c}^\gamma$

Richard Nisius (OPAL Collaboration)

CERN, CH-1211 Genève, Switzerland, Richard.Nisius@cern.ch

Abstract. The first measurement of $F_{2,c}^\gamma$ is presented. At low x the measurement indicates a non-zero hadron-like component to $F_{2,c}^\gamma$. At large x the measurement constitutes a test of perturbative QCD at next-to-leading order, with only m_c and α_s as free parameters, with a precision of $\mathcal{O}(40\%)$.

INTRODUCTION

For about 20 years measurements of photon structure functions give deep insight into the rich structure of a fundamental gauge boson, the photon. A recent review on this subject can be found in [1]. Here, the discussion is restricted to the measurement of $F_{2,c}^\gamma$, which recently has been achieved for the first time. Only the main features of the analysis are given, the experimental details can be found in [2].

The differential cross-section for deep inelastic electron-photon scattering, shown in Figure 1, is given by

$$\frac{\mathrm{d}^2\sigma}{\mathrm{d}x\,\mathrm{d}Q^2} = \frac{2\pi\alpha^2}{x\,Q^4}\left[\left(1 + (1-y)^2\right) F_2^\gamma(x, Q^2) - y^2 F_L^\gamma(x, Q^2)\right].\tag{1}$$

Here Q^2 is the absolute value of the four momentum squared of the exchanged virtual photon, $\gamma^\star$, x and y are the usual dimensionless variables of deep inelastic scattering and α is the fine structure constant. In experimental analyses y^2 is usually small. Consequently, the term proportional to the longitudinal structure function F_L^γ can be neglected and the differential cross-section is directly proportional to F_2^γ. For light quarks F_2^γ is related to the sum over the quark parton distribution functions q^γ of the quasi-real photon, γ, via

$$F_2^\gamma(x, Q^2) = x \sum_{q=u,d,s} e_q^2\left[q^\gamma(x, Q^2) + \bar{q}^\gamma(x, Q^2)\right].\tag{2}$$

Due to the large scale established by their masses, the contribution to F_2^γ from heavy quarks can be calculated in perturbative QCD. At present collider energies

CP571, *PHOTON 2000*, edited by A. J. Finch
© 2001 American Institute of Physics 0-7354-0010-5/01/$18.00

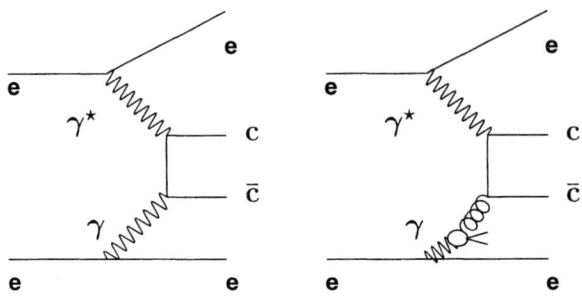

FIGURE 1. Examples of leading order diagrams contributing to (left) the point-like, and (right) the hadron-like part of $F_{2,c}^{\gamma}$.

only the contribution of charm quarks $F_{2,c}^{\gamma}$ is important. Like the structure function for light quarks, $F_{2,c}^{\gamma}$ receives contributions from the point-like and the hadron-like component of the photon shown in Figure 1.

Because of the charge of the charm quarks their contribution to F_2^{γ} is large and the importance increases for increasing values of Q^2, as can be seen from Figure 2, which shows the contributions from light quarks and from charm quarks separately, as predicted by the GRV parametrisations [3]. Charm quarks can only

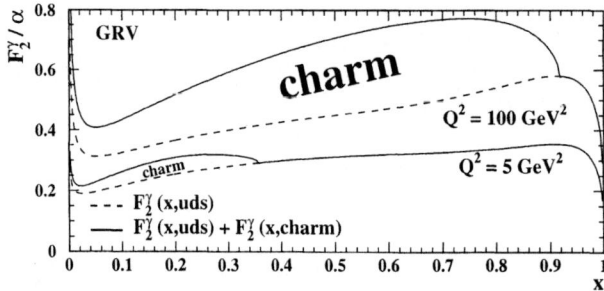

FIGURE 2. The structure function F_2^{γ} for u, d, s quarks alone and for u, d, s, c quarks, as a function of x and for two different values of Q^2.

be produced if the photon-photon invariant mass W is at least twice the mass of the charm quarks m_c. Using $x - Q^2/(Q^2 + W^2)$ this leads to the varying production threshold in x as a function of Q^2 seen in Figure 2.

Close to the production threshold, the point-like contribution to $F_{2,c}^{\gamma}$ is accurately approximated by the prediction of the lowest order Bethe-Heitler formula. For quasi-real photons also the next-to-leading order (NLO) predictions have been calculated in [4]. For the hadron-like contribution the photon-quark coupling must be replaced by the gluon-quark coupling, and the Bethe-Heitler formula has to be integrated over the allowed range in fractional momentum of the gluon using a parametrisation of the gluon distribution function of the photon, see e.g. [1].

The predicted behaviour of the point-like and hadron-like component of $F_{2,c}^{\gamma}$

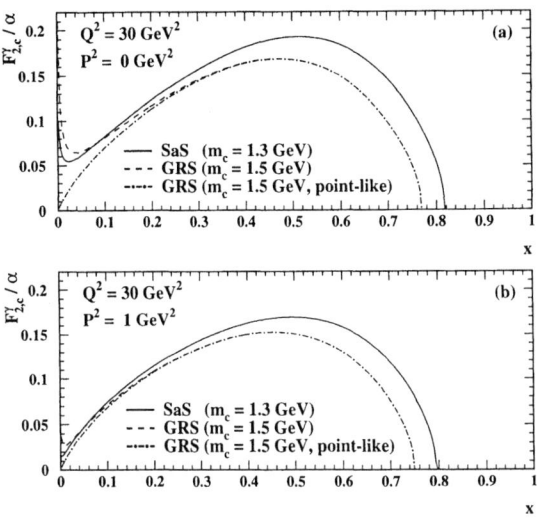

FIGURE 3. The predictions of the SaS1D (full) and the GRS (dash) parametrisations for $Q^2 = 30$ GeV^2 and for (a) $P^2 = 0$ and (b) $P^2 = 1$ GeV^2. In addition, the point-like contribution for $m_c = 1.5$ GeV(dot-dash) is shown.

for different values of m_c is shown in Figure 3, using the SaS1D [5] and GRS [6] parametrisations. For $x > 0.1$ the structure function is saturated by the point-like component which is only slowly suppressed for increasing virtualities P^2 of the quasi-real photon. In contrast, the hadron-like contribution dominates at small values of x and decreases much faster for increasing P^2. Finally, lowering the mass of the charm quarks leads to a higher threshold in x.

Given this predicted behaviour, the region of $x > 0.1$ can be used to test a purely perturbative NLO QCD prediction with only m_c and α_s as free parameters, and the low x behaviour mainly probes the gluon distribution function of the photon.

MONTE CARLO MODELS

The LO Monte Carlo generators HERWIG 5.9 [7] and Vermaseren [8] are used, both with $m_c = 1.5$ GeV. In HERWIG charm production is modelled using matrix elements for massless charm quarks, together with the GRV parametrisation for the parton distributions of the photon, again for massless charm quarks. The effect of the charm quark mass is only accounted for by not simulating events with $W < 2m_c$. Due to the massless approach used in HERWIG and the crude treatment at threshold, the predicted charm production cross-section is likely to be too large. The Vermaseren generator is based on the Quark Parton Model (QPM) and consequently does not take into account the hadron-like component of the photon structure. However, the complete dependence of the cross-section on the different photon helicities is modelled.

THE MEASUREMENT OF $F_{2,c}^{\gamma}$

The measurement of $F_{2,c}^{\gamma}$ proceeds along the same lines as the measurement of F_2^{γ} with the addition of the identification of the charm quarks via the reconstruction of D* mesons.

Events are selected with an energy of the scattered electron above 50 GeV, measured in the angular ranges (a) $33 - 55$ mrad or (b) $60 - 120$ mrad from either beam direction, thereby covering the approximate range in Q^2 of $5 - 100$ GeV^2. The visible hadronic mass W_{vis} is required to be below 60 GeV. Charm quarks are identified via D* $\rightarrow$ D$^0\pi$, followed by D$^0 \rightarrow K\pi$ or D$^0 \rightarrow K\pi\pi$. Using $f(\text{c} \rightarrow D^\star) = 0.235 \pm 0.011$ and the branching ratios of the D^0 decay modes of 0.02630 ± 0.00082 and 0.0519 ± 0.0029, this analysis covers only about 4% of all events containing a pair of charm quarks. For a clear acceptance the D* mesons are further required to fulfill, $|\eta^{D^\star}| < 1.5$ and $p_T^{D^\star} > 1$ or 3 GeV for (a) or (b). Together with a typical selection efficiency of about 25% only about 1% of all c$\bar{\text{c}}$ events are positively identified.

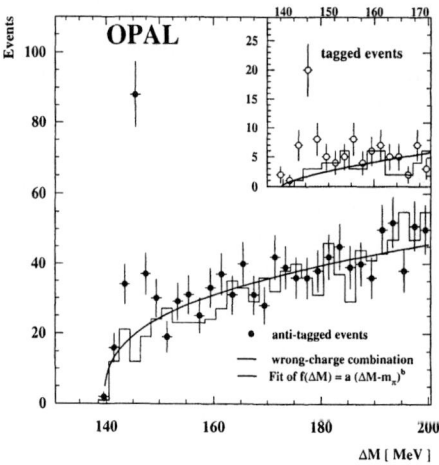

FIGURE 4. Mass difference ΔM for the anti-tagged and tagged sample.

Figure 4 shows the distribution of the difference between the D* and the D^0 candidate mass. In both samples a clear peak is visible around $\Delta M = 145.4$ MeV. Subtracting the background, obtained from a fit to the upper sideband of the signal, $29.8 \pm 5.9(\text{stat})$ D* mesons are found in the peak region.

Figure 5 shows the distributions of W_{vis} and of the measured Q^2 in comparison to the predictions of the HERWIG and Vermaseren Monte Carlo generators normalised

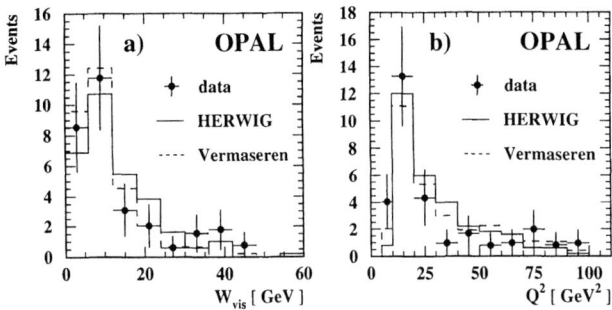

FIGURE 5. The distributions of a) the visible invariant mass W_{vis}, and b) the negative four-momentum squared Q^2.

to the number of data events. Both Monte Carlo generators give a good description of the shape of the data distributions.

The cross-section for $D^\star$ production is determined in the well-measured kinematic range described above. Based on this and the extrapolation factors obtained from the Monte Carlo models the full cross-section $\sigma(e^+e^- \to e^+e^-c\bar{c})$ and $F^\gamma_{2,c}$ are evaluated in two bins of x with $0.0014 < x < 0.1$ and $0.1 < x < 0.87$.

For $x > 0.1$ the predictions of the Vermaseren and HERWIG Monte Carlo models are very similar. In contrast, for $x < 0.1$, there are two main differences. Firstly the selection efficiency for events with $D^\star$ mesons fulfilling the kinematical requirements is different, and secondly, and even more important, the predicted cross-section within the invisible phase-space is largely different for the two models, resulting in different extrapolation factors, 12.9/5.1 for HERWIG/Vermaseren.

Since the hadron-like contribution is neglected in the QPM, the Vermaseren cross-section is much smaller than the LO and the NLO cross-section for $x < 0.1$. In contrast, mainly due to the massless approach taken, the prediction from HERWIG is higher than the cross-section from the LO and the NLO calculation. Therefore it is likely that the correct cross-section, and therefore the correct extrapolation factor, lies within the range of the two Monte Carlo predictions.

The measured $F^\gamma_{2,c}$ for $\langle Q^2 \rangle = 20$ GeV^2 is shown in Figure 6. The central values are obtained by averaging the results using the HERWIG and Vermaseren Monte Carlo models, and half the difference is taken as extrapolation error, which dominates the uncertainty for $x < 0.1$. The NLO prediction is based on $m_c = 1.5$ GeV, the renormalisation and factorisation scales are chosen to be $\mu_R = \mu_F = Q$, and the hadron-like contribution to $F^\gamma_{2,c}$ uses the GRV parametrisation. The NLO corrections are small for the whole x range. The band for the NLO calculation is evaluated by varying m_c between 1.3 and 1.7 GeV and using $Q/2 \le \mu_R = \mu_F \le 2\,Q$.

For $x > 0.1$ the error of the measured cross-section is dominated by the statistical uncertainty, and the NLO calculation with only m_c and α_s as free parameters is in good agreement with the data. In contrast, for $x < 0.1$, the result suffers from the strong model dependence discussed above. Despite this uncertainty the corrected

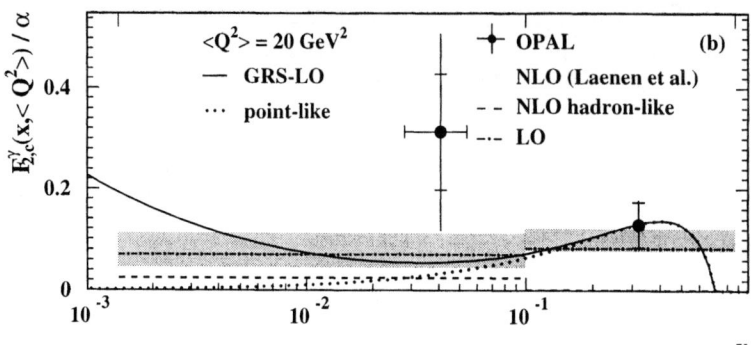

FIGURE 6. $F_{2,c}^\gamma$ compared to several predictions explained in the text.

data suggest a cross-section which is above the purely point-like component, i.e. the hadron-like component of $F_{2,c}^\gamma$ is non-zero.

Conclusion and Outlook

In conclusion, for $x > 0.1$, the purely perturbative NLO calculation is in good agreement with the measurement and for $x < 0.1$, the measurement suggests a non-zero hadron-like component of $F_{2,c}^\gamma$.

By using the massive matrix elements available in HERWIG6.1 and the full integrated luminosity of more than 500 pb^{-1} of the LEP2 programme, the measurement is likely to be improved considerably, both concerning the statistical and the systematic error.

Acknowledgement:

I wish to thank the organisers of this interesting conference, and especially Alex Finch. They created a fruitful atmosphere throughout the meeting and made attending the meeting a very nice experience.

REFERENCES

1. R. Nisius, Phys. Rep. **332**, 165 (2000).
2. OPAL Collab., G. Abbiendi et al., CERN-EP/99-157, accepted by Eur. Phys. J. **C**.
3. M. Glück, E. Reya, and A. Vogt, Phys. Rev. **D45**, 3986 and **D46**, 1973 (1992).
4. E. Laenen et al., Phys. Rev. **D49**, 5753 (1994).
5. G.A. Schuler and T. Sjöstrand, Z. Phys. **C68**, 607 (1995).
6. M. Glück, E. Reya, and M. Stratmann, Phys. Rev. **D51**, 3220 (1995).
7. G. Marchesini et al., Comp. Phys. Comm. **67**, 465 (1992).
8. R. Bhattacharya, G. Grammer Jr., and J. Smith, Phys. Rev. **D15**, 3267 (1977);
   J. Smith, J.A.M. Vermaseren, and G. Grammer Jr., Phys. Rev. **D15**, 3280 (1977).

Polarized and Unpolarized Structures of the Virtual Photon

Ken Sasaki* and Tsuneo Uematsu[†]

*Department of Physics, Faculty of Engineering, Yokohama National University,
Yokohama 240-8501, Japan
[†]Department of Fundamental Sciences, FIHS, Kyoto University, Kyoto 606-8501, Japan

Abstract. We discuss the structure functions and the parton distributions in the virtual photon target, both polarized and unpolarized, beyond the leading order in QCD. We study the factorization-scheme dependence of the parton distributions.

INTRODUCTION

As Maria Krawczyk remarked in her introductory talk on structure functions [1] , the virtual photon structure provides a unique test of QCD. In this talk I would like to discuss the polarized and unpolarized virtual photon structures. But because of the limitation of the allocated time, I will mainly focus my talk on the polarized virtual photon structure.

Recently there has been growing interest in the polarized photon structure functions. Especially, the 1st moment of a photon structure function g_1^γ has attracted much attention in connection with its relevance for the axial anomaly, which has also played an important role in the QCD analysis of the nucleon spin structure functions. Now the information on the spin structure of the photon will be obtained from the resolved photon process in polarized electron and proton collision in the polarized version of the ep collider. More directly, the spin-dependent structure function of the photon can be measured by the polarized e^+e^- collision in the future linear colliders.

Here we investigate two-photon process (Figure 1) with the kinematical region where the mass squared of the probe photon (Q^2) is much larger than that of the target photon (P^2) which is in turn much bigger than the Λ^2, the QCD scale parameter squared. The advantage for studying the virtual photon target is that we can calculate whole structure functions up to next-leading-order (NLO), in contrast to the real photon target where there remain uncalculable non-perturbative pieces. This is true for summing up the QCD logarithmic terms due to twist-2 operators corresponding to the QCD parton picture. Here we neglect all the power corrections arising from the higher-twist effects and target mass effects of the form $(P^2/Q^2)^k$

CP571, *PHOTON 2000*, edited by A. J. Finch

$(k = 1, 2, \cdots)$. Some non-perturbative effects like gluon condensations reside in the higher-twist effects. Our aim here is to study the polarized virtual photon structure function $g_1^\gamma(x, Q^2, P^2)$ at the same level of unpolarized structure function $F_2^\gamma(x, Q^2, P^2)$ (For the experimental status, see [2]). We can also investigate the parton distributions inside the polarized virtual photon. As we will see, the spin structure of the polarized virtual photon would offer a good testing ground for factorization scheme dependence of the parton distribution functions.

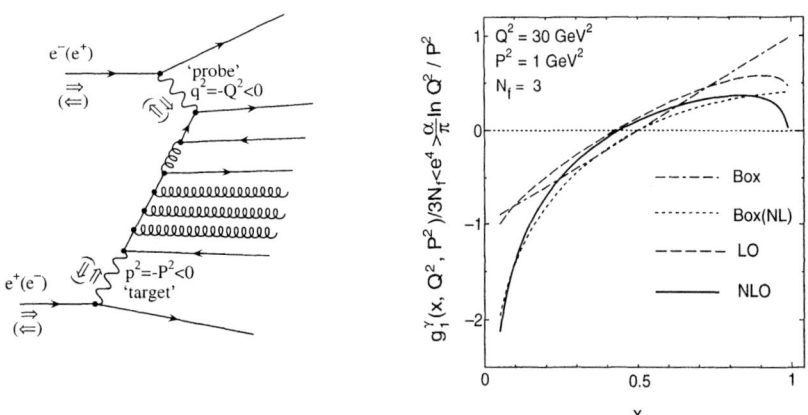

FIGURE 1. Two-photon process in polarized e^+e^- collision for $\Lambda^2 \ll P^2 \ll Q^2$ and the polarized photon structure function $g_1^\gamma(x, Q^2, P^2)$ for $Q^2 = 30$ GeV^2 and $P^2 = 1$ GeV^2 with $N_f = 3$. LO, NLO and Box (NL) denote QCD LO, NLO and Box-diagram (non-leading) results.

PQCD CALCULATION

We can apply the same framework used in the analysis of nucleon spin structure functions, namely the operator product expansion (OPE) supplemented by the renormalization group (RG) method or equivalently DGLAP type parton evolution equations. The NLO calculation has become possible since the two-loop anomalous dimensions of the quark and gluon operators in OPE or equivalently two-loop parton splitting functions were calculated by two groups [3,4] The n-th moment of $g_1^\gamma(x, Q^2, P^2)$ for the kinematical region:

$$\Lambda^2 \ll P^2 \ll Q^2$$

is given by

$$\int_0^1 dx\, x^{n-1} g_1^\gamma(x, Q^2, P^2) = \frac{\alpha}{4\pi} \frac{1}{2\beta_0} \left[\sum_{i=+,-,NS} L_i^n \frac{4\pi}{\alpha_s(Q^2)} \left\{ 1 - \left(\frac{\alpha_s(Q^2)}{\alpha_s(P^2)} \right)^{\lambda_i^n/2\beta_0+1} \right\} \right.$$

$$+ \sum_{i=+,-,NS} \mathcal{A}_i^n \left\{ 1 - \left(\frac{\alpha_s(Q^2)}{\alpha_s(P^2)} \right)^{\lambda_i^n/2\beta_0} \right\} + \sum_{i=+,-,NS} \mathcal{B}_i^n \left\{ 1 - \left(\frac{\alpha_s(Q^2)}{\alpha_s(P^2)} \right)^{\lambda_i^n/2\beta_0+1} \right\}$$

$$\left. + \mathcal{C}^n + \mathcal{O}(\alpha_s) \right] \qquad (1)$$

44

where L_i^n, $\mathcal{A}_i^n$, $\mathcal{B}_i^n$ and $\mathcal{C}^n$ are computed from the 1- and 2-loop anomalous dimensions as well as from 1-loop coefficient functions. λ_i^n ($i = +, -, NS$) denote the eigenvalues of 1-loop anomalous dimension matrices. $\alpha_s(Q^2)$ is the QCD running coupling constant. In Figure 1 we have shown the $g_1^\gamma(x, Q^2, P^2)$ evaluated from (1) by inverse Mellin transform for $Q^2 = 30$ GeV^2 and $P^2 = 1$ GeV^2 with $N_f = 3$ [10]. Note that the same formula with different coefficients, L_i^n, $\mathcal{A}_i^n$, $\mathcal{B}_i^n$, $\mathcal{C}^n$ and λ_i^n ($i = +, -, NS$) holds for the unpolarized structure function $F_2^\gamma(x, Q^2, P^2)$ [5].

SUM RULE

For a real photon target ($P^2 = 0$), Bass, Brodsky and Schmidt have shown that the 1st moment of $g_1^\gamma(x, Q^2)$ vanishes to all orders of $\alpha_s(Q^2)$ in QCD [6]:

$$\int_0^1 dx\, g_1^\gamma(x, Q^2) = 0. \tag{2}$$

Now the question is what about the $n = 1$ moment of the virtual photon case. Here we note that the eigenvalues of one-loop anomalous dimension matrix are $\lambda_+^{n=1} = 0, \lambda_-^{n=1} = -2\beta_0$. Taking $n \to 1$ limit of (1) the first three terms vanish. Denoting e_i, the i-th quark charge and N_f, the number of active flavors, we have

$$\int_0^1 dx\, g_1^\gamma(x, Q^2, P^2) = -\frac{3\alpha}{\pi} \sum_{i=1}^{N_f} e_i^4 + \mathcal{O}(\alpha_s) \tag{3}$$

We can go a step further to $\mathcal{O}(\alpha_s)$ QCD corrections which turn out to be [7]:

$$\int_0^1 dx\, g_1^\gamma(x, Q^2, P^2) = -\frac{3\alpha}{\pi} \left[\sum_{i=1}^{N_f} e_i^4 \left(1 - \frac{\alpha_s(Q^2)}{\pi} \right) - \frac{2}{\beta_0} (\sum_{i=1}^{N_f} e_i^2)^2 \left(\frac{\alpha_s(P^2)}{\pi} - \frac{\alpha_s(Q^2)}{\pi} \right) \right]$$
$$+ \mathcal{O}(\alpha_s^2). \tag{4}$$

This result coincides with the one obtained by Narison, Shore and Veneziano [8], apart from the overall sign for the definition of g_1^γ.

PARTON DISTRIBUTIONS

Spin-dependent parton distributions

Factorization theorem tells us that the physically observable quantities like cross sections or structure functions can be factored into the long-distance part (distribution function) and short-distance part (coefficient function). Thus the polarized photon structure function can be written schematically as

$$g_1^\gamma = \Delta \vec{q}^\gamma \otimes \Delta \vec{C}^\gamma \tag{5}$$

45

where spin-dependent parton distributions $\Delta \vec{q}$:

$$\Delta \vec{q}^{\gamma}(x, Q^2, P^2) = (\Delta q_S^{\gamma}, \Delta G^{\gamma}, \Delta q_{NS}^{\gamma}, \Delta \Gamma^{\gamma}) \tag{6}$$

are polarized flavor-singlet quark, gluon, non-singlet quark and photon distribution functions in the virtual photon (we put the symbol Δ for polarized quantities), and

$$\Delta \vec{C}^{\gamma} = \begin{pmatrix} \Delta C_S^{\gamma} \\ \Delta C_G^{\gamma} \\ \Delta C_{NS}^{\gamma} \\ \Delta C_{\gamma}^{\gamma} \end{pmatrix} \tag{7}$$

are the corresponding coefficient functions. The same relation holds for unpolarized structure function F_2^{γ} in terms of unpolarized parton distributions $\vec{q}$ and unpolarized coefficient functions $\vec{C}^{\gamma}$. In the leading order in QED coupling $\alpha = \frac{e^2}{4\pi}$, the photon distribution function can be taken as $\Delta \Gamma^{\gamma}(x, Q^2, P^2) = \delta(1 - x)$. Therefore we have the following inhomogeneous DGLAP evolution equation for $\Delta \boldsymbol{q}^{\gamma} = (\Delta q_S^{\gamma}, \Delta G^{\gamma}, \Delta q_{NS}^{\gamma})$:

$$\frac{d\Delta \boldsymbol{q}^{\gamma}(x, Q^2, P^2)}{d \ln Q^2} = \Delta \boldsymbol{K}(x, Q^2) + \int_x^1 \frac{dy}{y} \Delta \boldsymbol{q}^{\gamma}(y, Q^2, P^2) \times \Delta P(\frac{x}{y}, Q^2) \tag{8}$$

where $\Delta \boldsymbol{K}(x, Q^2)$ is the splitting function of the photon into quark and gluon, whereas $\Delta P(x/y, Q^2)$ is the 3×3 splitting function matrix.

Factorization Scheme Dependence

The solution to the DGLAP evolution equation can be given by

$$\Delta \vec{q}^{\gamma}(t) = \Delta \vec{q}^{\gamma(0)}(t) + \Delta \vec{q}^{\gamma(1)}(t), \quad t \equiv \frac{2}{\beta_0} \ln \frac{\alpha_s(P^2)}{\alpha_s(Q^2)} \tag{9}$$

where the first (second) term corresponds to LO (NLO) approximation. The initial condition we impose is the following,

$$\Delta \vec{q}^{\gamma(0)}(0) = 0, \quad \Delta \vec{q}^{\gamma(1)}(0) = \frac{\alpha}{4\pi} \vec{A}_n \tag{10}$$

where $\vec{A}_n$ is the constant which depends on the factorization scheme to be used. Or equivalently in the language of OPE, this constant appears as a finite matrix element of the operators, $\vec{O}_n$ renormalized at $\mu^2 = P^2$ between the photon states:

$$\langle \gamma(p) \mid \vec{O}_n(\mu) \mid \gamma(p) \rangle|_{\mu^2=P^2} = \frac{\alpha}{4\pi} \vec{A}_n \tag{11}$$

This scheme dependence arises from the freedom of multiplying the arbitrary finite renormalization constant Z_a and its inverse Z_a^{-1} in the n-th moment of (5):

$$g_1^{\gamma}(n, Q^2, P^2) = \Delta \vec{q}^{\gamma} \cdot \Delta \vec{C}^{\gamma} = \Delta \vec{q}^{\gamma} Z_a \cdot Z_a^{-1} \Delta \vec{C}^{\gamma} = \Delta \vec{q}^{\gamma}|_a \cdot \Delta \vec{C}^{\gamma}|_a \tag{12}$$

where the resulting $\Delta \vec{q}^{\gamma}|_a$ and $\Delta \vec{C}^{\gamma}|_a$ are the distribution function and the coefficient function in the a-scheme. The explicit expressions for the n-th moment of the parton distributions can be found in ref. [10].

Transformation from $\overline{\text{MS}}$ to a-scheme

Under the transformation from one factorization scheme to another, the coefficient functions as well as anomalous dimensions will change. Of course when they are combined together, we get the factorization-scheme independent structure function g_1^γ. Since $\overline{\text{MS}}$ is the only scheme in which both 1-loop coefficient functions and 2-loop anomalous dimensions are actually computed, we study the transformation rule from the $\overline{\text{MS}}$ to a new factorization scheme-a. We have considered the several different factorization schemes; 1) chirally invariant (CI) scheme, 2) Adler-Bardeen (AB) scheme, 3) off-shell (OS) scheme, 4) Altarelli-Ross (AR) scheme and 5) DIS$_\gamma$ scheme. (For the detailed description of each factorization scheme see [10].) The transformation rule for the singlet-quark coefficient function, for example, is given by

$$\Delta C_{S,\ a}^{\gamma,\ n} = \Delta C_{S,\ \overline{\text{MS}}}^{\gamma,\ n} - \langle e^2 \rangle \frac{\alpha_s}{2\pi} \Delta w(n, a) \tag{13}$$

where $\langle e^2 \rangle = \sum_i e_i^2 / N_f$ and $\Delta w(n, a)$ is the transformation functions, the explicit expressions for which as well as other coefficient functions together with the similar transformation rules for 2-loop anomalous dimensions are given in ref. [10].

The prescriptions to treat the axial anomaly are different from scheme to scheme. For example, the axial anomaly resides in the quark distribution in $\overline{\text{MS}}$ schme, whereas it exists in the gluon and photon coefficient functions in the CI scheme. These factorization schemes are also characterized by the behavior of the parton distribution functions near $x = 1$. We can study their analytic behaviors by the large n limit of their moments. Here we also note that the gluon distribution function is factorization-scheme independent in the class of factorization schemes considered here. By performing the inverse Mellin transform of the moments, the parton distributions as functions of x are reproduced numerically. We present our results for singlet-quark for various schemes and gluon in Figure 2. Note that the real photon's g_1^γ was studied in [11,12], which are consistent with present analysis.

The similar scheme dependence of the parton distributions inside the unpolarized virtual photon was studied for the $\overline{\text{MS}}$, OS and DIS$_\gamma$ schemes [13].

CONCLUDING REMARKS

We have studied the virtual photon's spin structure functions, $g_1^\gamma(x, Q^2, P^2)$ and the polarized parton distributions for the kinematical region $\Lambda^2 \ll P^2 \ll Q^2$, which are perturbatively calculable up to the NLO in QCD. The first moment of g_1^γ is non-vanishing in contrast to the real photon case, where we have vanishing sum rule. NLO QCD corrections are significant at large x as well as at low x. We also studied factorization-scheme dependence of parton distribution functions.

Future subjects to be studied are as follows. First of all we should understand how the transition occurs from vanishing 1st moment for real photon ($P^2 = 0$) to

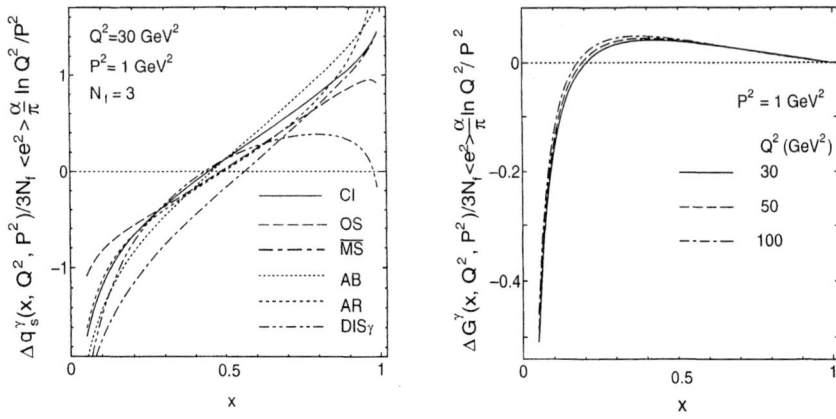

FIGURE 2. Polarized singlet-quark distribution function $\Delta q_S^\gamma(x, Q^2, P^2)$ in the several factorization schemes and the polarized gluon distribution $\Delta G^\gamma(x, Q^2, P^2)$ for $Q^2 = 30$ GeV^2 and $P^2 = 1$ GeV^2 with $N_f = 3$

non-vanishing one for virtual photon ($P^2 \gg \Lambda^2$). Secondly, another structure function $g_2^\gamma(x, Q^2, P^2)$ is yet to be computed where we also have twist-3 contribution. Furthermore, the power corrections due to target mass effects and higher-twist effects should be investigated. More reliable treatment for small-x behaviors of polarized p.d.f. should be studied in the framework of BFKL like approach.

REFERENCES

1. Krawczyk, M., these proceedings.
2. Söldner-Rembold, S., these proceedings.
3. Mertig, R., and van Neerven, W. L., *Z. Phys.* **C70**, 637 (1996).
4. Vogelsang, W., *Phys. Rev.* **D54**, 2023 (1996).
5. Uematsu, T., and Walsh, T. F., *Nucl. Phys.* **B199**, 93 (1982).
6. Bass, S. D., Brodsky, S. J., and Schmidt, I., Phys. Lett. **B437**, 417 (1998).
7. Sasaki, K., and Uematsu, T., *Phys. Rev.* **D59**, 114011 (1999).
8. Narison, S., Shore, G. M., and Veneziano, G., *Nucl. Phys.* **B391**, 69 (1993).
9. Sasaki, K., and Uematsu, T., *Phys. Lett.* **B473**, 309 (2000).
10. Sasaki, K., and Uematsu, T., hep-ph/0007055 (2000).
11. Sasaki, K., *Phys. Rev.* **D22**, 2143 (1980); *Prog. Theor. Phys. Suppl.* **77**, 197 (1983).
12. Stratmann, M., and Vogelsang, W., *Phys. Lett.* **B386**, 370 (1996).
13. Sasaki, K., and Uematsu, T., *Nucl. Phys. B (Proc. Suppl.)* **89**, 162 (2000).

Partonic structure of γ_L^* in hard collisions

Jiří Chýla*

* Research Center for Particle Physics, Institute of Physics of the Academy of Sciences
18221 Na Slovance 2, Prague 8, Czech Republic

Abstract. Manifestation of QCD improved partonic structure of longitudinally polarized virtual photons in hard collisions is discussed. As an example, dijet production in ep collisions at HERA is investigated in detail.

INTRODUCTION

In this talk I discuss phenomenological consequences of QCD improved partonic structure of longitudinally polarized virtual photons (γ_L^*), concentrating on LO QCD calculations of dijet production in ep collisions at HERA. Some of the results presented here are discussed in detail in [1–3]. The role of resolved γ_L^* in NLO QCD calculations will be addressed elsewhere [4]. I start by recalling the virtue of extending the concept of partonic "structure" to virtual photons [5,6]:

- In principle, the concept of partonic structure of virtual photons can be dispensed with as higher order QCD corrections to cross sections of processes involving virtual photons in the initial state are well–defined and finite even for massless partons.

- In practice, however, the concept of *resolved virtual photon* is extraordinarily useful as it allows us to include the resummation of higher order QCD effects that come from physically well–understood region of (almost) parallel emission of partons off the quarks and antiquarks coming from the primary $\gamma^* \to q\bar{q}$ splitting and subsequently participating in hard processes.

For the virtual photon, as opposed to the real one, its parton distribution functions (PDF) can therefore be regarded as "merely" describing higher order perturbative effects and not their "genuine" structure. Although this distinction between the content of PDF of real and virtual photons exists, it does not affect the *phenomenological* usefulness of PDF of the virtual photon. As shown in [5] the nontrivial part of the contributions of resolved γ_T^* to NLO calculations of dijet production at HERA is large and affects significantly the conclusions of phenomenological analyses of existing experimental data. Taking into account resolved γ_L^* turns out to be phenomenologically important as well.

CP571, *PHOTON 2000*, edited by A. J. Finch

PARTON DISTRIBUTION FUNCTIONS OF γ_L^* IN QCD

Most of the present knowledge of the structure of the photon comes from experiments at ep and e^+e^- colliders, where the incoming leptons act as sources of transverse (γ_T^*) and longitudinal (γ_L^*) virtual photons of virtuality P^2 and momentum fraction y. To order α their respective unintegrated fluxes are given as

$$f^{\gamma_T^*}(y, P^2) = \frac{\alpha}{2\pi} \left(\frac{1 + (1-y)^2}{y} \frac{1}{P^2} - \frac{2m_e^2 y}{P^4} \right), \tag{1}$$

$$f^{\gamma_L^*}(y, P^2) = \frac{\alpha}{2\pi} \frac{2(1-y)}{y} \frac{1}{P^2}. \tag{2}$$

Phenomenological analyses of interactions of virtual photons and their PDF have so far concentrated on γ_T^*. Neglecting longitudinal photons is a good approximation for $y \to 1$, where $f^{\gamma_L^*}(y, P^2) \to 0$, as well as for small virtualities P^2, where PDF of γ_L^* vanish by gauge invariance. But how small is "small" in fact? For instance, should we take into account the contribution of γ_L^* to jet cross sections in the region $E_T \gtrsim 5$ GeV, $P^2 \gtrsim 1$ GeV^2, where most of the data on virtual photons obtained in ep collisions at HERA come from? The present paper is devoted primarily to addressing this question.

In pure QED and to order α the probability of finding inside γ_L^* of virtuality P^2 a quark with mass m_q, charge e_q, momentum fraction x and virtuality $\tau \leq M^2$, is given, in units of $3e_q^2 \alpha/2\pi$, as [5]

$$q_L^{\mathrm{QED}}(x, m_q^2, P^2, M^2) = \frac{4x^2(1-x)P^2}{\tau^{\mathrm{min}}} \left(1 - \frac{\tau^{\mathrm{min}}}{M^2} \right), \tag{3}$$

where $\tau^{\mathrm{min}} = xP^2 + m_q^2/(1-x)$. The quantity defined in (3) has a clear physical interpretation: it describes the flux of quarks that are almost collinear with the incoming photon and "live" longer than $1/M$. For $\tau^{\mathrm{min}} \ll M^2$ (3) simplifies to

$$q_L^{\mathrm{QED}}(x, m_q^2, P^2, M^2) = \frac{4x^2(1-x)P^2}{xP^2 + m_q^2/(1-x)},$$

which for $x(1-x)P^2 \gg m_q^2$ further reduces to

$$q_L^{\mathrm{QED}}(x, 0, P^2, M^2) = 4x(1-x), \tag{4}$$

whereas for $x(1-x)P^2 \ll m_q^2$

$$q_L^{\mathrm{QED}}(x, m_q^2, P^2, M^2) \to \frac{P^2}{m_q^2} 4x^2(1-x)^2.$$

QCD corrections to QED expressions for PDF of γ_L^* have been derived in leading-logarithmic approximation in the region $1 \lesssim P^2 \ll M^2$ in [2]. By "leading–log" I

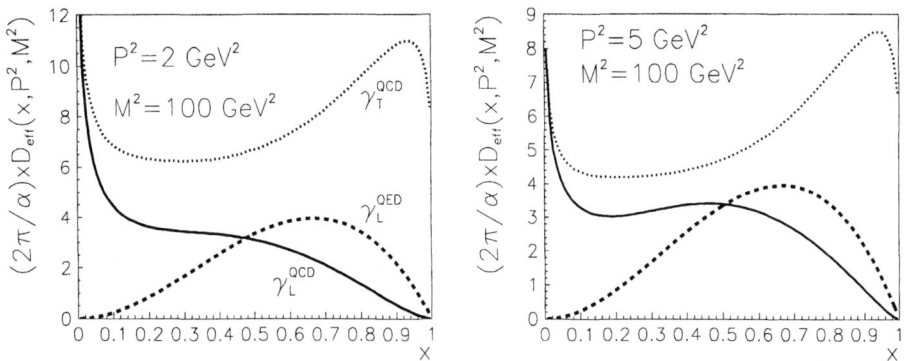

FIGURE 1. Comparison of the contributions of resolved γ_T^* and γ_L^* to D_{eff} defined in (5) for $M^2 = 100$ GeV^2 and $P^2 = 2, 5$ GeV^2. QED and QCD formulae discussed in the text were used for γ_L^* and SaS1D parameterization for γ_T^*.

mean resummation of the terms $(\alpha_s \ln M^2)^k$ at each order k of perturbative QCD. Note that for γ_T^* there is one power of $\ln M^2$ more at each order of α_s, the additional one coming from the primary QED $\gamma^* \to q\bar{q}$ splitting. In the case of γ_L^* the analogous splitting gives rise to "constant" term (4), hence the absence of this log. The resulting expressions exhibit typical hadronic form of scale dependence and contain Λ_{QCD} as the only free parameter. QCD effects thus suppress quark distribution functions $q_L^{\text{QED}}(x, P^2, M^2)$ at large x and enhance it on the other hand for $x \lesssim 0.4$. Moreover, they generate sizable gluon distribution function, absent in QED. The presence of the term proportional to $\ln M^2$ in the expressions for q_T in both QED and QCD implies the dominance of γ_T^* at large M^2, but one would have to go to very large M^2 for γ_L^* to become really negligible with respect to γ_T^*. For fixed M^2 the relative importance of γ_L^* with respect to γ_T^* grows with P^2, but to retain clear physical meaning of PDF we stay throughout this paper in the region where $P^2 \ll M^2$. The lower bound 1 GeV$^2 \lesssim P^2$ ensures that hadronic parts of QCD improved PDF of γ_L^*, which have not been taken into account in [2], can be safely neglected.

γ_L^* IN HARD COLLISIONS

The relevance of resolved γ_L^* in hard collisions of virtual photons depends on the theoretical framework we are working in. In this talk I will stay within the framework of LO QCD calculations of dijet production at HERA. The measurement of dijet cross sections in ep (and e$^+$e$^-$) collisions offers currently the best way of investigating interactions of virtual photons [7,8]. In general the corresponding cross sections are given as sums of contributions of all possible parton level subprocess. The simplest way of demonstrating the importance of the contributions of resolved γ_L^* employs the approximation [9] in which dijet cross sections are expressed in terms of single *effective parton distribution function* of the photon (γ_T^* or γ_L^*)

$$D_{\text{eff}}(x, P^2, M^2) \equiv \sum_{i=1}^{n_f} \left(q_i(x, P^2, M^2) + \bar{q}_i(x, P^2, M^2) \right) + \frac{9}{4} G(x, P^2, M^2), \qquad (5)$$

where the factorization scale M is conventionally identified with (a multiple of) jet E_T: $M = \kappa E_T$. In Fig. 1 the contributions to D_{eff} of γ_L^*, evaluated with both QED and QCD formulae for its PDF, are compared to those of γ_T^* using SaS1D parameterization [10]. The comparison is performed for two pairs of P^2 and M^2 typical for HERA experiments. In addition to softening effects at large x, QCD improved PDF of γ_L^* give sizable contribution to D_{eff} at small x that comes from the gluon content of γ_L^*. Fig. 1 moreover suggests that in the region accessible at HERA the contributions of resolved γ_L^* are numerically important, particularly after incorporating QCD effects in its PDF.

After this simple but approximate estimate of the contributions of resolved γ_L^*, I now turn to the evaluation of dijet cross sections at HERA using complete LO QCD formalism as implemented in HERWIG 5.9 event generator. To include the effects of resolved γ_L^* I have added the option of generating the flux of γ_L^* combined with the call to QED or QCD improved PDF of γ_L^*. For γ_T^* the SaS1D PDF [10] were used. The dijet cross sections were evaluated for $0.05 \leq y \leq 0.95$ in three windows of P^2

$$1.4 \leq P^2 \leq 2.4 \text{ GeV}^2, \ 2.4 \leq P^2 \leq 4.4 \text{ GeV}^2, \ 4.4 \leq P^2 \leq 10 \text{ GeV}^2$$

and imposing the following cuts on jet E_T (all quantities are in $\gamma^* p$ cms)

$$E_T^{(1)}, E_T^{(2)} \geq E_T^c, \ E_T^c = 5, 10 \text{ GeV}.$$

The effects of H1 and ZEUS detector acceptances were taken into account by performing all calculations without any restrictions on η as well for $-3 \leq \eta \leq 0$.

The results presented in Figs. 2 and 3 correspond to parton level calculations in the first window of P^2, without and with the mentioned cuts on η. The characteristic dependence of the contributions of resolved γ_L^* on y is illustrated by plotting for each of the distributions in η, E_T and x_γ also its ratio to that of γ_T^* for the whole interval $0.05 \leq y \leq 0.95$, as well as for three indicated subintervals. Except for x_γ close to 1, QCD improved PDF of γ_L^* enhance its contributions to dijet cross sections compared to those based on the purely QED. For $y \lesssim 0.5$ they amount to about 50% of those of γ_T^*. For $x_\gamma \lesssim 0.2$ this number increases further up to about 70%. Reducing the range of η to $-3 \leq \eta \leq 0$ affects mainly the distribution $d\sigma/dx_\gamma$ by suppressing it at both edges of the phase space. The ratia of the contributions of γ_L^* and γ_T^* are, however, affected only little by this cut.

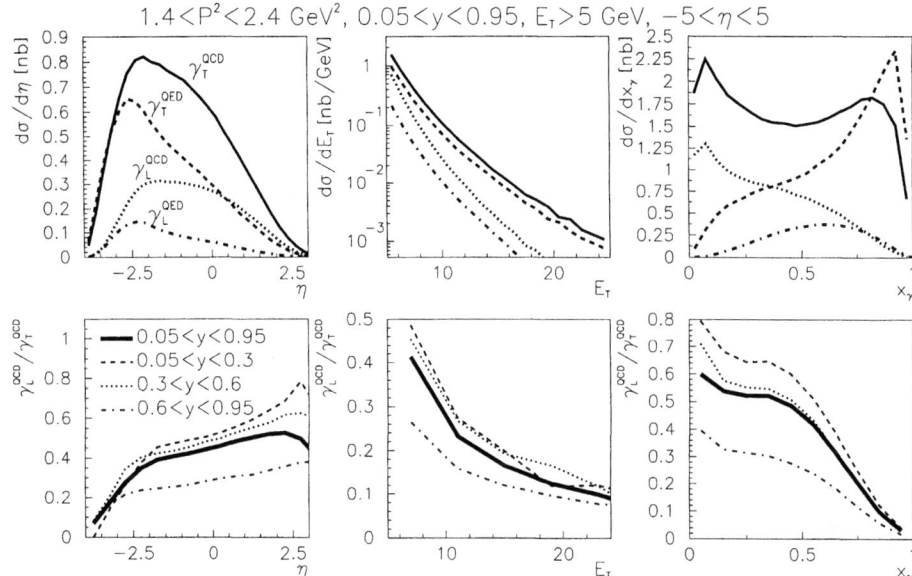

FIGURE 2. Upper three plots: dijet cross sections, corresponding to resolved γ_T^* and γ_L^* plotted as functions of η, E_T and x_γ for $1.4 \leq P^2 \leq 2.4$ GeV^2, $0.05 \leq y \leq 0.95$, $E_T \geq 5$ GeV, without any restriction on η. Lower three plots: the corresponding ratia of the contributions of γ_L^* and γ_T^*, integrated over the whole interval $0.05 \leq y \leq 0.95$, as well as in three indicated subintervals.

Increasing the photon virtuality P^2 enhances, approximately uniformly in the whole phase space, the relative importance of γ_L^* with respect to γ_T^*. On the contrary, rising the threshold E_T^c from 5 GeV to 10 GeV reduces it by a factor of about 2, since large E_T require large x_γ, where quarks from γ_T^* dominate.

The effects of hadronization on parton level results discussed above have been studied in detail in [8]. They are reasonably small ($\lesssim 10 - 20\%$) and model independent in the region $-2.5 \lesssim \eta$ but turn large and model dependent below that value. For the comparison with theoretical calculations the lower limit on accessible range of η enforced by H1 and ZEUS acceptances presents therefore no real restriction. On the other hand, it would be very useful to push the upper limit on η above $\eta \simeq 0$ since the relevance of γ_L^* grows with η.

Summarizing the message of this Section, we conclude that for $\Lambda^2 \ll P^2 \ll E_T^2$:

- The contributions of γ_L^* are substantial, particularly for small y, large P^2, low E_T and small x_γ.

- The cuts enforced by H1 and ZEUS acceptances reduce the sensitivity to γ_L^*, but its contributions still make up typically 50% of those of γ_T^* and can be identified by their characteristic y and P^2 dependencies.

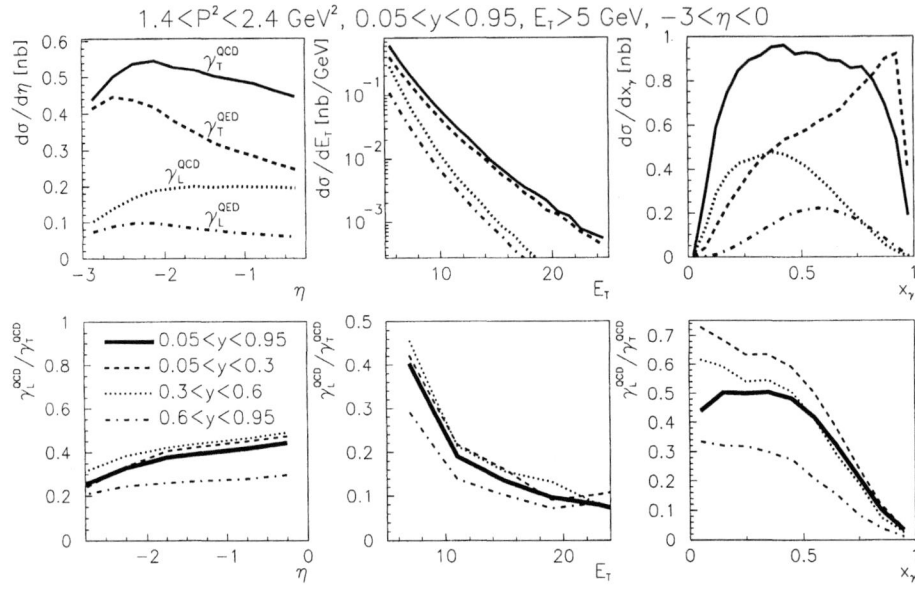

FIGURE 3. The same as in Fig. 2, but the restricted region $-3 \leq \eta \leq 0$.

CONCLUSIONS

The contributions of resolved γ_L^* to dijet production in ep collisions at HERA were evaluated using QCD improved PDF of γ_L^* constructed recently. In the region accessible at HERA they turn out to sizable, amounting typically to 50% of those from γ_T^*, and depend sensitively of y, E_T and x_γ.

Work performed under the project LN00A006 of the Ministry of Education of the Czech Republic

REFERENCES

1. Chýla, J., and Taševský, M., *Eur. Phys. J.* **C16**, 471 (2000).
2. Chýla, J., *Phys. Lett.* **B488**, 289 (2000).
3. Friberg, C., and Sjöstrand, T., LU TP 00-31, hep-ph/0009003.
4. Chýla, J., and Taševský, M., in preparation.
5. Chýla, J., and Taševský, M., *Phys. Rev.* **D**, (2000) in print, hep-ph/9912245.
6. Friberg, C., and Sjöstrand, T., *Eur. Phys. J.* **C13**, 151 (2000).
7. Adloff, C., et al. (H1 Collab.), *Eur. Phys. J.* **C13**, 397 (2000).
8. Taševský, M., PhD Thesis, Charles University, Prague, 1999.
9. Combridge, B.V., and Maxwell, C.J., *Nucl. Phys.* **B239**, 429 (1084).
10. Schuler, G., and Sjöstrand, T., *Z. Phys.* **C68**, 607 (1995).

QCD analysis of $F_2^\gamma(x, Q^2)$: an unconventional view

Jiří Chýla*

* Center for Particle Physics, Institute of Physics of the Academy of Sciences
18221 Na Slovance 2, Prague 8, Czech Republic

Abstract. Elements of the alternative approach to hard collisions of photons, proposed recently by the author, are reviewed, with particular attention to QCD analysis of F_2^γ. This approach is based on clear separation of genuine QCD effects from those of pure QED origin and does not rely on the assumption that parton distribution functions of the photon behave as α/α_s. It differs significantly from the conventional one, as illustrated on the example of charm contribution to F_2^γ, recently measured at LEP.

INTRODUCTION

In [1] I have proposed an alternative approach to QCD analysis of $F_2^\gamma(x, Q^2)$, and by implication to any hard process involving initial photon, which differs substantially from the conventional one. It builds in part on arguments advocated for a long time by the authors of [2] and agrees with the approach to calculations of direct photon production at HERA pursued in [3]. This alternative approach is based on two ingredients:

- Clear and systematic separation of genuine QCD effects from those of pure QED origin, which leads to unambiguous and universal definition of the concepts "leading" and "next–to–leading" order of QCD.

- Acknowledgement of the fact that parton distribution functions (PDF) of the photon are proportional to α and not, as assumed in the conventional approach, to α/α_s.

The general expression for $F_2^\gamma(x, Q^2)$ has the following structure

$$
\frac{1}{x} F_2^\gamma(x, Q^2) = q_{\mathrm{NS}}(M) \otimes C_q(Q/M) + \frac{\alpha}{2\pi} \delta_{\mathrm{NS}} C_\gamma + \langle e^2 \rangle \Sigma(M) \otimes C_q(Q/M) +
$$
$$
\frac{\alpha}{2\pi} \langle e^2 \rangle \delta_\Sigma C_\gamma + \langle e^2 \rangle G(M) \otimes C_G(Q/M) \tag{1}
$$

CP571, *PHOTON 2000*, edited by A. J. Finch

where quark nonsinglet and singlet and gluon distribution functions [1] satisfy the evolution equations

$$\frac{d\Sigma(x, M)}{d\ln M^2} = \delta_\Sigma k_q + P_{qq} \otimes \Sigma + P_{qG} \otimes G, \tag{2}$$

$$\frac{dG(x, M)}{d\ln M^2} = k_G + P_{Gq} \otimes \Sigma + P_{GG} \otimes G, \tag{3}$$

$$\frac{dq_{NS}(x, M)}{d\ln M^2} = \delta_{NS} k_q + P_{NS} \otimes q_{NS}, \tag{4}$$

where $\delta_{NS} = 6n_f \left(\langle e^4 \rangle - \langle e^2 \rangle^2\right), \delta_\Sigma = 6n_f \langle e^2 \rangle$. The splitting and coefficient functions k_i, C_j can be expanded in powers of α_s $(a(M) \equiv \alpha_s(M)/2\pi)$ as

$$k_q(x, M) = \frac{\alpha}{2\pi} \left[k_q^{(0)}(x) + a(M)k_q^{(1)}(x) + a^2(M)k_q^{(2)}(x) + \cdots \right], \tag{5}$$

$$k_G(x, M) = \frac{\alpha}{2\pi} \left[a(M)k_G^{(1)}(x) + a^2(M)k_G^{(2)}(x) + \cdots \right], \tag{6}$$

$$P_{ij}(x, M) = a(M)P_{ij}^{(0)}(x) + a^2(M)P_{ij}^{(1)}(x) + \cdots, \tag{7}$$

$$C_q(x, Q/M) = \delta(1 - x) + a(M)C_q^{(1)}(x, Q/M) + a^2(M)C_q^{(2)}(x, Q/M) + \cdots, \tag{8}$$

$$C_G(x, Q/M) = a(M)C_G^{(1)}(x, Q/M) + a^2(M)C_G^{(2)}(x, Q/M) + \cdots, \tag{9}$$

$$C_\gamma(x, Q/M) = C_\gamma^{(0)}(x, Q/M) + a(M)C_\gamma^{(1)}(x, Q/M) + a^2(M)C_\gamma^{(2)}(x, Q/M) \cdots, \tag{10}$$

where the lowest order coefficient functions $k_q^{(0)}$ and $C_\gamma^{(0)}$

$$k_q^{(0)}(x) = (x^2 + (1 - x)^2), \tag{11}$$

$$C_\gamma^{(0)}(x, Q/M) = \left(x^2 + (1 - x)^2\right) \left[\ln \frac{M^2}{Q^2} + \ln \frac{1 - x}{x} \right] + 8x(1 - x) - 1 \tag{12}$$

are unique [2] and due entirely to QED coupling of the initial photon to the primary $q\bar{q}$ pair. It is their presence what distinguishes hard collisions of photons from those of hadrons. In [1] I have discussed conceptual as well as numerical differences between the results obtained within the conventional and alternative approaches for the pointlike part of F_2^γ in the nonsinglet channel and under the simplifying assumption $\beta_1 = 0$. In this talk I will concentrate on proper definitions of the concepts "leading order" (LO) and "next–to–leading order" (NLO) of QCD and on the discussion of their implications for charm contribution $F_{2,c}^\gamma(x, Q^2)$ to photon structure function, recently measured at LEP [4,5].

---

[1] For their definitions as well as other details of the alternative approach, see [1].
[2] Throughout this paper we restrict ourselves to the case of real target photon, although the basic conclusions hold for the virtual photon as well. For the latter, however, the expression for $C_\gamma^{(0)}$ differs slightly from (12).

DEFINING LEADING AND NEXT-TO-LEADING ORDERS OF QCD

Although the definition of the concepts "LO" and "NLO" in hard collisions of photons is a matter of convention, it is preferable to define them in a way that retains the same basic content these concepts have in collisions of leptons and hadrons. Let me recall, for instance, their meaning in the case of the familiar ratio

$$R_{e^+e^-}(Q) \equiv \frac{\sigma(e^+e^- \to \text{hadrons})}{\sigma(e^+e^- \to \mu^+\mu^-)} = \left(3\sum_{i=1}^{n_f} e_i^2\right)(1 + r(Q)). \qquad (13)$$

The prefactor $R_{\text{QED}} \equiv 3\sum_{i=1}^{n_f} e_i^2$ multiplied by unity in the brackets of (13) comes purely from QED, whereas genuine QCD effects are contained in $r(Q)$, which is given as expansion in powers of α_s as

$$r(Q) = \frac{\alpha_s(M)}{\pi}\left[1 + \frac{\alpha_s(M)}{2\pi}r_1(Q/M) + \cdots\right]. \qquad (14)$$

For the quantity (13) it is a generally accepted practice to include R_{QED} in theoretical expressions but disregard it when defining the "LO" and "NLO" of QCD. For instance, the NLO approximation of (13) implies retaing *first two* terms in (14), i.e. the *first three* terms in (13). Let me emphasize that only if at least first two nontrivial powers of α_s in (13) are taken into account can this trunctated expansion be associated with a well-defined renormalization scheme. And it is this association what makes in my view the essential feature of the concept "NLO QCD approximation". The same convention should be adopted for all physical quantities getting contributions from pure QED. I think it is preferable to use the terminology that avoids potential confusion which might arise from mixing orders of α_s and α. Note that for both $R_{e^+e^-}$ and $F_{2,c}^{\gamma}(x, Q^2)$, discussed in the next Section, the purely QED contributions R_{QED} and $F_{2,c}^{\gamma,\text{QED}}$ are finite and unique and there is thus no reason why they should be treated differently as far as the definitions of the "LO" and "NLO" of QCD are concerned. The implications of the above considerations for $F_2^{\gamma}(x, Q^2)$ are the following:

- As shown in [1] the LO QCD expression for F_2^{γ} contains in addition to terms included in the conventional LO analysis of F_2^{γ} (i.e. those proportional to $k_q^{(0)}$ and $P_{ij}^{(0)}$), also terms involving $k_q^{(1)}, C_q^{(1)}, C_{\gamma}^{(0)}$ and $C_{\gamma}^{(1)}$. As all these functions are known, there is, however, no obstacle to performing such an analysis. As shown in [1] for the pointlike part of $F_2^{\gamma}(x, Q^2)$ in the nonsinglet channel, the results in these two approaches are numerically significantly different, the single most important contribution to this difference coming from $C_{\gamma}^{(1)}$. The coefficient function $C_q^{(1)}$ enters F_2^{γ} already at the LO [3] due to the fact that it does so in the convolution with purely QED part of quark distribution function of the photon, which has no analogue in hadronic collisions.

---

[3] Contrary to hadronic collisions, where it appears first at the NLO.

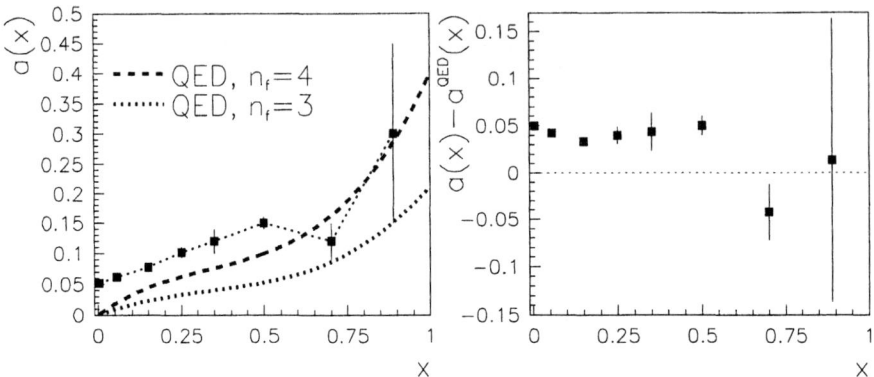

FIGURE 1. The derivative (15) taken from from the fit to LEP data (solid squares) compared to pure QED formula $a^{\mathrm{QED}}(x) = 3\sum_{i=1}^{n_f} e_i^4 x(1 + (1-x)^2)$ for $n_f = 3, 4$. Both data and curves in units of α.

- The NLO QCD analysis of $F_2^\gamma(x, Q^2)$ requires the knowledge of two quantities, $k_q^{(2)}$ and $C_\gamma^{(2)}$, that have not yet been calculated and thus is at the moment impossible to perform. Note, that in the conventional approach $C_\gamma^{(0)}$, which has nothing to do with QCD, appears only at the "NLO", and $C_\gamma^{(1)}$, which involves evaluation of Feynman diagrams with a single QCD vertex is not used even there!

For clear and unambiguous definition of the terms "LO" ("NLO") it is thus vital to agree on the basic criterion, namely that they refer to perturbative expansions of physical quantities retaining the first (first two) *nontrivial* powers of α_s. The purely QED contribution to $F_2^\gamma(x, Q^2)$ is irelevant from this point of view, but may be retained for comparison with experiment as it actually dominates scaling violations of $F_2^\gamma(x, Q^2)$ in most of accessible range of x. To identify genuine QCD effects one has to look for subtler effects than the dominant $\ln Q^2$ rise, like, for instance, the x-dependence of the slope

$$a(x) \equiv \frac{\mathrm{d}F_2^\gamma(x, Q^2)}{\mathrm{d}\ln Q^2}. \tag{15}$$

Compared to $F_2^p(x, Q^2)$, for which scaling violations are due entirely to QCD effects, the nonzero slope (15) is by itself no sign of QCD effects, as these are given by the difference $\Delta(x) \equiv a(x) - a^{\mathrm{QED}}(x)$. Fig. 1, based on numbers taken from [6], shows that for $x \gtrsim 0.5$ the precision of currently available data is insufficient to identify genuine QCD effects, although some indication of the turnover to negative $\Delta(x)$, expected theoretically, is visible there.

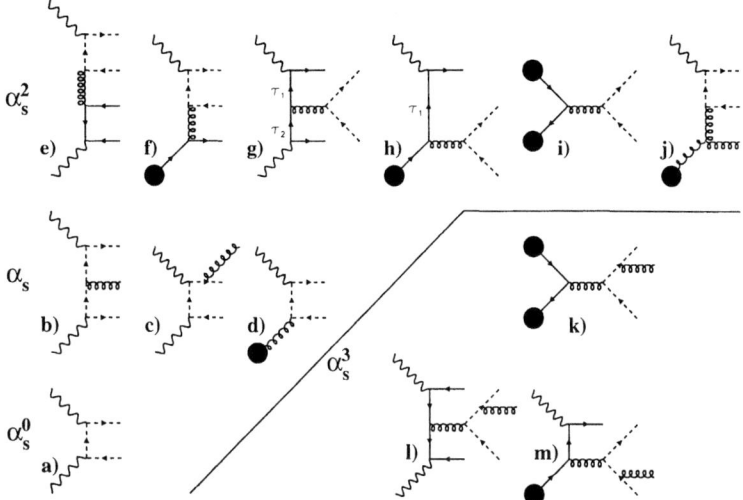

FIGURE 2. Examples of diagrams up to order α_s^3 contributing to heavy quark production in photon-photon collisions. The upper (lower) wavy lines corresponds to the probing (target) photon. Light quarks are denoted by solid, heavy quarks by dashed lines. For $F_{2,c}^\gamma$ the diagrams with blobs in the upper vertices are absent as only the target photon may be "resolved". The initial state singularities of the direct and single resolved photon contributions coming from the vertex $\gamma \to q\bar{q}$ are understood to be subtracted and put into PDF of the photon, described by solid blobs. As a results, the (subtracted) direct and resolved photon diagrams, as well as the associated PDF acquire dependence on the factorization scale M.

CHARM CONTRIBUTION TO $F_2^\gamma(X, Q^2)$

The QCD analysis of $F_{2,c}^\gamma(x, Q^2)$ in the region $x \gtrsim 0.1$ provides possibly the simplest illustration of the differences between the conventional and alternative approches to hard collisions of photons [7], since $F_{2,c}^\gamma$ is dominated in this region by the direct photon contribution, $F_{2,c}^{\text{dir}}$, which does not involve any PDF. It can be written as

$$F_{2,c}^{\text{dir}}(x, Q^2) = F_{2,c}^{\text{QED}}(x, Q^2) + \alpha_s(Q)F_{2,c}^{(1)}(x, Q^2) + \alpha_s^2(Q)F_{2,c}^{(2)}(x, Q^2) + \cdots, \quad (16)$$

where the coefficients $F_{2,c}^{(k)}, k \geq 1$ are calculable in perturbative QCD and the lowest order term $F_{2,c}^{(0)} \equiv F_{2,c}^{\text{QED}}$ comes from pure QED diagram in Fig. 2a. In the conventional approach the "NLO" approximation of the direct photon contribution is defined [8] by taking the first two terms in (16) including the purely QED contribution $F_{2,c}^{\text{QED}}$. However, as argued in the preceding Section, this definition does not have the basic attribute of genuine NLO QCD approximation. The inclusion of direct photon contributions of the order $\alpha^2\alpha_s^2$, coming from diagrams like those in Fig. 2e,g is vital not only for establishing the genuine NLO character of the direct photon contribution itself, but also for ensuring [7] factorization scale invariance of the full expression for $F_{2,c}^{\gamma}$. The latter involves adding the resolved photon contributions up to the order $\alpha^2 \alpha_s^2$, coming from diagrams like those in Fig. 2f,h,j.

SUMMARY AND CONCLUSIONS

The alternative approach to QCD analysis of F_2^{γ}, proposed recently by the author, differs substantially and in a number of aspects from the conventional one. It satisfies factorization scale invariance in a way that does not rely on physically untenable assumption that quark distribution functions of the photon behave as $\mathcal{O}(\alpha/\alpha_s)$. The simplest implications of this difference are illustrated on the case of charm contribution to $F_2^{\gamma}(x, Q^2)$, recently measured at LEP.

To be useful phenomenologically the proposed approach needs to be further elaborated by extending it to the singlet sector and merging it with the hadronic contributions. Work on this program is in progress. The NLO QCD analysis in the proposed approach requires evaluation of several so far unknown quantities and is thus currently impossible to perform. In view of the quality and number of experimental data on F_2^{γ}, this is at the moment no serious drawback and a complete LO QCD analysis seems sufficient for phenomenological purposes.

Work supported by the Ministry of Education of the Czech Republic under the project LN00A006.

REFERENCES

1. J. Chýla, JHEP**04**, 007 (2000).
2. J.H. Field, F. Kapusta, L. Poggioli, Phys. Lett. B **181**, 362 (1986).
   J.H. Field, F. Kapusta, L. Poggioli, Z. Phys. C **36**, 121 (1987).
   F. Kapusta, Z. Phys. C **42**, 225 (1989).
3. M. Krawczyk, Acta Physica Polonica **21**, 999 (1990)
   M. Krawczyk, A. Zembruski, Proc. *29th International Conference on High Enerhy Physics*, Vancouver, World Scientific 1999.
4. OPAL Collab., G. Abbiendi et al., CERN-EP/99-157
5. R. Nisius, This Conference, hep-ex/0010020.
6. R. Nisius, Phys. Rep. C **332**, 165 (2000).
7. J. Chýla, hep-ph/0010140.
8. S. Frixione, M. Krämer, E. Laenen, Nucl. Phys. B **571**, 169, (2000)

DIFFRACTION

Introduction to Diffractive Photoprocesses

Graham Shaw*

*Department of Physics and Astronomy, University of Manchester, Manchester M13 9PL, U.K.

Abstract. The objectives of my talk are to provide a very brief introduction to diffractive photoprocesses in general and the colour dipole model in particular; and to comment on possible gluon saturation effects at HERA and beyond.

INTRODUCTION

Diffraction exchange is the study of vacuum exchange at high energies. It is frequently divided into elastic, singly-dissociative and double dissociative processes as illustrated in Figure 1, where A and B may be photons or hadrons and X and Y may be single particles or an inclusive sum over $n \geq 1$ particle states. The wiggly

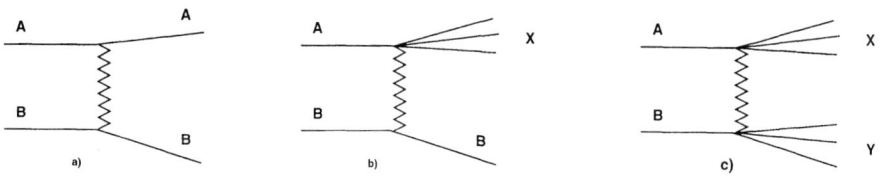

FIGURE 1. (a) elastic (b) singly dissociative and (c) double dissociative diffractive processes. (Figure from [1]).

line indicates an exchange of energy and momentum, but no non-zero colour or flavour quantum numbers may be exchanged. High energy means that the square of the centre of mass energy $s = W^2$ is much larger than any other energy scale:

$$s \gg t, m_X^2, \ldots \ .$$

For diffractive processes initiated by virtual photons, the latter include the virtuality Q^2, implying

$$x = Q^2/s \ll 1 \ .$$

CP571, *PHOTON 2000*, edited by A. J. Finch

We note that t, m_X^2, Q^2 can themselves become large, provided they remain much smaller than s.

Experimentally, diffractive processes are characterized by two distinctive features: **rising cross-sections** and **rapidity gaps**. The two groups of final state particles in Figure 1 emerge in roughly the forward and backward directions in the centre of mass frame; and are well-separated in rapidity or pseudo-rapidity

$$\eta = -\ln\tan(\frac{\theta}{2})$$

where θ is the polar angle with respect to the beam direction. Such rapidity gaps are characteristic of colour singlet exchange, in contrast to the hadronization strings associated with colour exchange. They occur not only in diffractive processes, but in, for example, colour singlet meson exchange processes. However meson exchange gives rise to cross-sections which fall rapidly with increasing energy, in contrast to diffractive processes which have constant or rising cross-sections. Nonetheless at finite energies one may need to take account of small contributions from the exchange of flavour singlet meson exchange contributions, which can in general interfere with the dominant diffractive process.

Diffractive processes, so defined, are copious and varied. For example the singly dissociative inclusive reaction

$$\gamma^* + p \rightarrow X + p \ , \tag{1}$$

where X is an inclusive sum over hadronic states, accounts for 10-20 % of the $\gamma^* p$ total cross-sections at low x. (Here and throughout, γ^* indicates either a real or virtual photon, while γ refers exclusively to real photons.) This reaction has stimulated an enormous literature already [2] and new data will be presented here [3]. Of particular interest is the behaviour for $M_X^2 \gg Q^2$, which explores aspects of diffraction which are not easily studied in other processes. Exclusive processes discussed at the conference include: elastic virtual Compton scattering

$$\gamma^* + p \rightarrow \gamma^* + p \ ,$$

which is not measured directly, but is related to the $\gamma^* p$ total cross-sections and hence the deep inelastic structure functions via the optical theorem; deeply virtual Compton scattering(DVCS)

$$\gamma^* + p \rightarrow \gamma + p \ , \tag{2}$$

for which the first data are presented at this conference [4]; and the vector meson production processes [5]

$$\gamma^* + p \rightarrow \rho + p \tag{3}$$

$$\gamma^* + p \rightarrow J/\Psi + p \tag{4}$$

where measuring the vector meson decay products the enables the spin structure of the interaction and the separate contributions from longtitudinal and transverse photons to be studied. In addition, the J/ψ mass introduces an at least moderately large scale into the problem even for real photons. Perturbative aspects of diffraction can also be enhanced by working at high t [6] and/or by the study of diffractive jet production [7]. Finally some of the first results on diffraction in $\gamma^*\gamma^*$ collisions are also reported [8].

THEORETICAL FRAMEWORK

Diffraction involves an interplay of perturbative and non-perturbative effects which presently defies a rigorous treatment in QCD. Rather there are innumerable models which throw light on different aspects of the problem with varying degrees of success. Here we try to provide a simple framework which can be used to classify and compare the various models and hopefully avoid confusion. To do this we emphasize two features.

Vacuum exchange.

The first thing to consider is the way the model implements vacuum exchange. There are three main approaches, which we will list for the moment and illustrate later.

- **Regge models**, in which the vacuum exchange is usually described by the exchange of one or more Regge poles with vacuum quantum numbers, called pomerons.

- **Gluon exchange models** in which the vacuum exchange is modelled by the exchange of two or more gluons in a colour singlet state.

- **Quasi-optical models** in which the projectile is regarded as a superposition of "scattering eigenstates" which are either absorbed or scatter unchanged at fixed impact parameter on traversing the "target."

Reference frames

Different reference frames are conveniently chosen to emphasize different aspects of the physics and caution is required in comparing dynamical models formulated in different frames. Popular choices for discussing γ^*p collisions include:

- The **infinite momentum frame** in which, for large Q^2 at least, the parton distribution functions(pdfs) have a simple interpretation and the photon is regarded as pointlike.

- The **laboratory frame** in which the incoming photon is typically absorbed a long distance, of order $1/(Mx)$ from the proton target and the intermediate states into which it converts are usually regarded as constituents of the photon.

HARD AND SOFT DIFFRACTION

The study of diffraction has been transformed by the discovery of hard diffraction in γ^*p collisions at HERA. Here we summarize this discovery and some of the questions it raises.

Diffraction in hadron physics

Before discussing diffractive photoprocesses, it is useful to comment on the "soft diffraction" observed in purely hadronic processes. At high energies $s \gg t$, hadronic scattering is well-described by Regge pole exchange, as illustrated in Figure 2 for the charge exchange reaction $\pi^-p \to \pi^0 n$. If a single pole i dominates, the differential

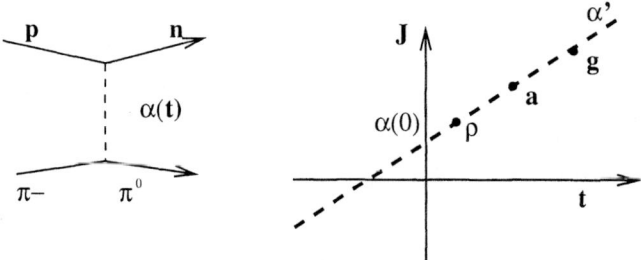

FIGURE 2. Regge pole exchange for the reaction $\pi^-p \to \pi^0 n$ and the associated meson trajectory (7)(Figure from [9]) .

cross-section for any $2 \to 2$ reaction satisfies

$$\frac{d\sigma}{dt} \propto \left(\frac{s}{s_0}\right)^{2\alpha_i(t)-2} ,$$
(5)

where s_0 is a convenient scale, usually taken to be 1 GeV^{-2}, and the *Regge trajectories*

$$\alpha_i(t) = \alpha_i(0) + \alpha_i' t$$
(6)

are found to be approximately linear. They relate the observed energy dependence in the scattering region $t \leq 0$ to the exchanged mesons at $\alpha(t = m_j^2) = j$, where j, m_j^2 are the spin and mass of the meson respectively. The picture applies to baryon as well as meson exchange, with an approximately universal *slope parameter* $\alpha' \approx 1$

GeV^{-2}. In contrast the *intercept* $\alpha_i(0)$ depends on the flavour exchange quantum numbers i, with

$$\alpha_M(t) \approx 0.5 + t \tag{7}$$

for the leading non-strange meson trajectories, leading to cross-sections (5) which fall roughly like $1/s$.

The above picture accounts remarkably well for reactions with non-zero flavour exchange, including other features - shrinkage, factorisation, dips - not mentioned here. It can be extended successfully to vacuum exchange processes by adding a single additional Regge pole to describe diffraction, called the pomeron. Specifically the available data on a wide range of different reactions is consistent with the same universal trajectory [10]

$$\alpha_P(t) \approx 1.08 + 0.25t \ . \tag{8}$$

where the high value of the intercept $\alpha_P(0)$ reflects the fact that diffractive cross-sections rise slowly with energy. In addition, the pomeron slope $\alpha'_P \approx 0.25$ GeV^{-2} differs markedly from the approximately universal slope $\alpha'_P \approx 1$ observed for all $q\bar{q}$ meson and qqq baryon Regge poles, suggesting the pomeron is not associated with $q\bar{q}$ meson exchange. It is rather assumed to be associated with the exchange of gluons, so that particles lying on the pomeron trajectory are presumably glueballs. The lightest glueball on the trajectory (8) is a 2^+ particle with a predicted mass of around 1.9 GeV. This is not unreasonable, although it must be said that little is known from experiment about the glueball spectrum and the situation may well be more complicated.

Finally, before leaving hadronic diffraction, we highlight two points about Regge theory whose importance cannot be overemphasized:

- the trajectory function (6) for any given Regge pole depends only on t and is independent of the energy range and the reaction considered; and

- the exchange of two or more Regge poles leads to more complicated terms - *Regge cuts* - which are neglected in most applications, but which must be present at some level of accuracy.

Diffraction in $\gamma^* p$ reactions

The above picture of soft pomeron exchange works quite well for some real photoprocesses like the total photoabsorption cross-section $\sigma_t(\gamma^* p)$ or ρ, ω or ϕ photoproduction. However a steeper rise with energy is observed if s is very large (or x is very small) and an at least moderately hard scale enters the process. More generally, if different data sets are parameterized by a single Regge pole exchange formula, the intercept is found to vary roughly in the range

$$1.08 \leq \alpha_{eff}(0) \leq 1.4 \ .$$

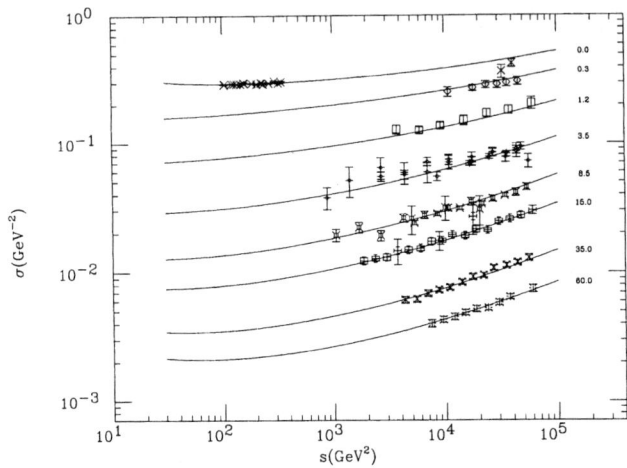

FIGURE 3. Representative sample of data points for the total cross-section $\sigma_{\gamma p}^{tot}$ together with curves calculated from a colour dipole model(see below) (from [13])

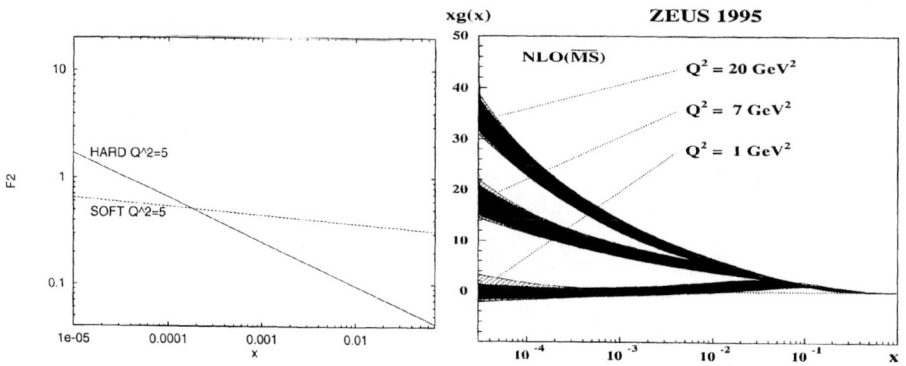

FIGURE 4. Above left: relative magnitudes of hard and soft pomeron contributions at $Q^2 = 5$. As Q^2 increases the range of hard pomeron dominance extends to larger x. (Figure from [11])
FIGURE 5. Above right: the gluon density $xg(x, Q^2)$ extracted from next leading order fits to the ZEUS $F_2(x, Q^2)$ data. (Figure from [15])

The particular value obtained depends on the reaction and on the ranges of Q^2 and x (or equivalently s) considered. This is illustrated in Figure 3, which shows the total cross-section $\sigma_{\gamma p}^{tot}$ as a function of s at various fixed Q^2, where the increase in $\alpha_{eff}(0)$ at high Q^2 and high s, corresponding to low x, is clearly seen.

It follows from the above that diffractive photoprocesses can not be described by a single Regge pole exchange, since this requires a universal energy dependence in all cases. The obvious interpretation is that there is a new phenomenon - "hard diffraction" - which becomes dominant for hard enough scales and large enough energies. If one assumes that this can also be approximated by a Regge pole one is led to the hypothesis of two pomerons: the soft pomeron (8) which dominates in hadronic diffraction and some "soft" photoprocesses; and a second "hard pomeron" which dominates for hard enough scales and large enough energies. This hypothesis has been explored by Donnachie and Landshoff [11] who obtain an excellent fit to data on the proton structure function, the charmed structure function and on J/ψ production (4) for a hard pomeron trajectory

$$\alpha_P(t) \approx 1.42 + 0.10t \ .$$

The varying energy dependence arises from the varying relative importance of the two contributions, which is illustrated in Figure 4 for the proton structure function. Alternatively, Gotsman will discuss [12] a two component model in which the "hard component" is described by a model based on perturbative QCD. In both cases, between the regions of soft and hard diffraction at $Q^2 = 0$ and high Q^2 respectively, there is an extensive transition region in which both can be important.

Hard diffraction and QCD

The discovery of hard diffraction at HERA opens up the subject to perturbative methods in QCD. These can be based on resummations of the perturbative expansion retaining leading terms in $\ln(Q^2/\Lambda^2)$ (DGLAP) or leading terms in $\ln(1/x)$ (BFKL), since both these quantities become large in the relevant kinematic region. The former is the most familiar and has it's best known application in the DGLAP evolution of the structure functions. In particular, it has long been known [14] that this can generate an increasingly steep low-x behaviour as Q^2 increases, and a good fit to the data can be obtained for $Q^2 \geq 1$ GeV^2 with the gluon distributions $xg(x, Q^2)$ shown in Figure 5 [15]. The gluon densities at $Q^2 \approx 1$ GeV^2 are unstable, implying that the DGLAP picture can not be trusted at such low Q^2 values, but the success at higher Q^2 values is impressive. This "DGLAP picture," based on the dominance of gluon ladder exchanges(see e.g. the talk of Gotsman [12]), can be extended to other diffractive processes at high Q^2, especially for those processes where, in lowest order at least, one can prove factorisation into terms describing the fluctuation of the initial photon into $q\bar{q}$ pairs; the formation of the final particle from the said pairs; and the interaction of the $q\bar{q}$ with the proton [16]. However the gluon distributions required are "skewed" parton distributions [17], which take into account the fact that the incoming and outgoing protons in inelastic processes like (1, 2, 3, 4) have different momenta, even in the forward direction. The empirical study of skewed parton corrections has only just begun [4] [18].

A potential problem with leading, next leading .. $\ln Q^2$ approximations is that as $x \to 0$, neglected terms might become important because although they are lower order in $\ln Q^2$, they are leading order in $\ln(1/x)$. Thus one might expect to see a breakdown of DGLAP at very small-x - but how small? This gives rise to the alternative BFKL approach of leading $\ln(1/x)$ resummation. In leading order this approach predicted the hard pomeron intercept with apparent success, but it runs into serious difficulties beyond leading order. This topic will be discussed by Ross [19] while a succinct comparison of the Regge, BFKL and DGLAP approaches and their relation to each other may be found in the recent review of Ball and Landshoff [20].

In the rest of this talk we will concentrate on two more phenomenological questions:

- Can we find a unified description of both hard and soft diffraction and of the wide variety of diffractive processes?

- When can we expect to see so-called "gluon saturation" effects at small x?

These are conveniently addressed in the context of the *colour dipole model*, to which we immediately turn.

THE COLOUR DIPOLE MODEL

Singly dissociative diffractive γ p processes are conveniently described in the rest frame of the hadron using a picture in which the incoming photon intially dissociates into a $q\bar{q}$ pair a long distance - typically of order of the "coherence length" $1/Mx$ - from the target proton. Assuming that the resulting partonic/hadronic state evolves slowly compared to the size of the proton or nuclear target, it can be regarded as frozen during the interaction. In this approximation, the process will factorize into a probability for the photon to have evolved into a given state $|\alpha>$, times the amplitude for that state to interact with the target. In the colour dipole model, [21,22] the dominant states $|\alpha>$ are assumed to be $q\bar{q}$ states of given transverse size. Specifically

$$|\gamma\rangle = \int dz d^2r \ \psi(z,r)|z,r\rangle + \dots , \qquad (9)$$

where r is the transverse size of the pair, z is the fraction of light cone energy carried by the quark and $\psi(z,r)$ is the *light cone wave function* of the photon. Assuming that these states are scattering eigenstates (i.e. that z, r remain unchanged in diffractive scattering) the elastic scattering amplitude for $\gamma^*p \to \gamma^*p$ is specified by Figure 6. This leads via the optical theorem to

$$\sigma_{T,L}^{\gamma^* p} = \int \mathrm{d}z \mathrm{d}^2 r \ |\psi_\gamma^{T,L}(z,r)|^2 \sigma(s,r,z) \ , \tag{10}$$

for the $\gamma^* p$ total cross-section in deep inelastic scattering, where $\sigma(s,r,z)$ is the total cross-section for scattering dipoles of specified (z,r) from a proton at fixed $s = W^2$. This "dipole cross-section" is a universal quantity for singly-dissociative diffractive processes on a proton target, playing a similarly fundamental role in, for example, open diffraction (1), exclusive vector meson production (3) and (4) and deeply virtual Compton scattering (2).

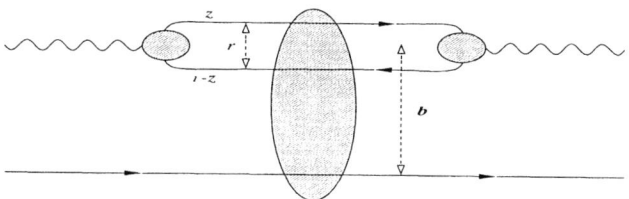

FIGURE 6. The colour dipole model for $\gamma^* p \to \gamma^* p$.

The dipole cross-section has been evaluated by several groups [23]. Although the assumptions made to do this vary, there are some features in common. The dipole cross-section at a given energy is assumed to be approximately "geometrical", i.e. to depend on the transverse size r of the dipole, but not to depend on z. In addition, approximate QCD behaviour(colour transparency) for small dipoles $r \to 0$ and "hadronic behaviour"

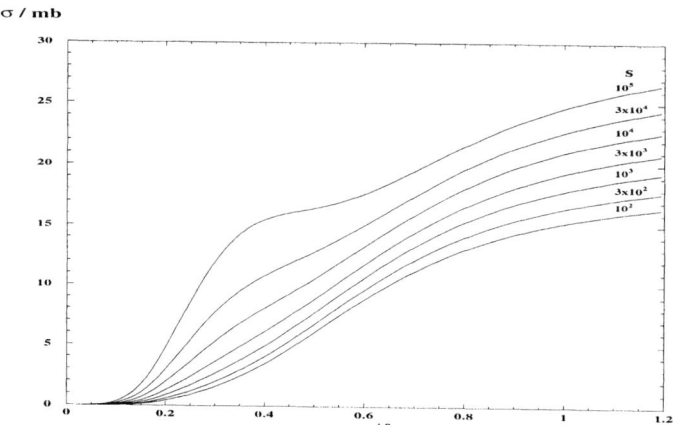

FIGURE 7. The dipole cross-section as a function of s in the HERA range.

for large dipoles $r \approx 1$fm are incorporated in varying degrees of detail[1]. A useful summary and comparison of the various approaches may be found in the recent review of McDermott [23]. From now on I shall present results from Forshaw, Kerley and Shaw [13], [24] - [26] who have extracted the dipole cross-section from DIS and real photoabsorption data assuming a form with two terms with a Regge type s dependence:

$$\sigma(s,r) = a_{soft}(r)s^{\lambda_S} + a_{hard}(r)s^{\lambda_H} \tag{11}$$

where the values $\lambda_S \approx 0.08$, $\lambda_H \approx 0.42$ resulting from the fit are characteristic of the soft and hard pomeron respectively. The functions $a_{soft}(r)$, $a_{hard}(r)$ are chosen so that for small dipoles the hard term dominates yielding a behaviour $\sigma \to r^2(r^2 s)^{\lambda_H}$ as $r \to 0$ in accordance with colour transparency ideas; while for large dipoles $r \approx 1$ fm the soft term dominates with a hadronlike behaviour $\sigma \approx \sigma_0(r^2 s)^{\lambda_S}$. Correspondingly the photon wavefunction is assumed to be perturbative for small dipoles, with a simple ansatz for confinement effects at large r. The resulting dipole cross-section, determined from DIS and real photoabsorption data, is shown in Figure 7 for various energies in the HERA region.

The above dipole cross-section, determined from DIS and real photoabsorption data, can be used to predict results for other diffractive processes. Successful predictions have been obtained for:

- the charmed structure function [13] by retaining only the charmed quark loop in Figure 6;

- open diffraction (1) from Figure 8, together with an additional contribution from intermediate $q\bar{q}g$ states which is important for large diffractive masses $m_X^2 \gg Q^2$, but small elsewhere [24].

- virtual Compton scattering (2), by replacing the final state photon in Figure 6 by a real photon [26].

The same dipole cross-section can also be used to predict vector meson production reactions like (3, 4), but in this case the vector meson wavefunctions are also required.

Saturation

The dipole model is particularly useful for discussing saturation effects, since it incorporates both soft and hard diffraction, associated with small and large dipoles respectively. There are actually two types of saturation effect, which are quite distinct and should not be confused.

---

[1] Very large dipoles $r \gg 1$fm make a negligible contribution, since the wavefunction factor in (10) decreases exponentially at large r.

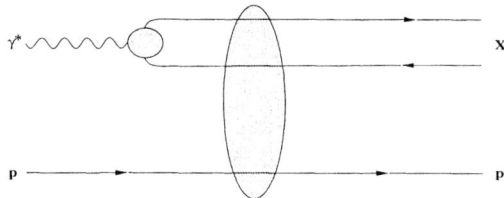

FIGURE 8. The dipole contribution to open diffraction (1).Figure from [25].

Low Q^2 saturation. As can be seen in Figure 7, the dipole cross-section increases rapidly as a function of the dipole size r at small r, but then "saturates" to a slowly varying cross-section of hadronic size at larger r values. This change - and the fact that it shifts to smaller r as s increases -is crucial to describe the form of the change from approximate scaling to the observed $Q^2 \to 0$ (and hence $x \to 0$) behaviour at fixed s. To see this we note that the Q^2 dependence in (10) arises entirely from the wavefunction. As Q^2 decreases, larger r values are explored and the slowly varying dipole cross-section results in a weakening Q^2 dependence for $\sigma_{\gamma*p}$. When $Q^2 \ll 4m_q^2$, where m_q is the constituent quark mass, the wavefunction and $\sigma_{\gamma*p}$ become independent of Q^2 so that $F_2 \propto Q^2$ as $Q^2 \to 0$ as required.

Gluon saturation. For high enough energies, the assumed s^λ ($\lambda > 0$) behaviours assumed above must be tamed by unitarity effects, especially for the hard term with $\lambda_H \approx 0.4$. At fixed Q^2, $x \to 0$ as $s \to \infty$ and the resulting softening of the corresponding $x^{-\lambda_H}$ behaviour is associated with gluon saturation in the quark-parton language. Gluon saturation can be incorporated into dipole and other closely related models by hand [27,28] or using the eikonal approximation [30,31] but are not included in (11). Hence the fact that an excellent fit is obtained to the DIS data using (11) means that the current HERA data are not at sufficiently high s to *require* the saturation effects that are built into some other dipole models [27,28]. We note that our model agrees with the standard Caldwell plot Figure 9, where the turn over as x decreases occurs because Q^2 is also decreasing and is understood as a low Q^2 saturation effect. No such effect is predicted in our model if x is decreased at fixed Q^2, as confirmed by the preliminary ZEUS97 data [32].

A strong indication of when saturation effects will be needed is given in Figure 7. As can be seen, the cross-section for small dipoles is initially small but increases rapidly and at the top of the accessible HERA range is becoming commensurate with the slowly increasing "hadronic" behaviour af the large dipoles. It is at this point that saturation effects are expected to become important; if they don't, the cross-section for small dipoles will exceed that for large dipoles at higher energies and the dipole cross-section will paradoxically decrease with increasing size r. Saturation effects are therefore expected to play an important role just beyond beyond the HERA range, in the planned THERA region with $s_{max} \approx 10^6$ GeV^2.

73

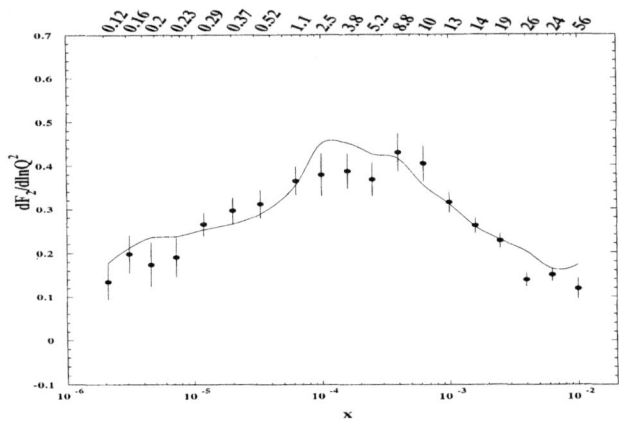

FIGURE 9. The Caldwell plot. Predictions are made at the Q^2 values of the data points and roughly interpolated. (Figure from [25])

REFERENCES

1. M. Arneodo, "Diffraction at HERA - an Introduction," Proceedings of Photon97.
2. see for example the review of A. Hebecker, Phys. Rep. **331** (2000) 1 and references therein.
3. See the talks by H. Mahlke-Krueger and by C. Johnson, these proceedings.
4. See the talk by R. Stamen, these proceedings.
5. See the talks by S.Kananov and by A.Savin, these proceedings.
6. See the talk by K. Klimek, these proceedings.
7. See the talks by A. Wyatt and by B. List, these proceedings.
8. See the talks by M. Wadhwa, M.Przybycien, N.Zimin and M.Kienzle, these proceedings.
9. N. Cartiglia, "Diffraction at HERA" lanl hep-ph/9703245.
10. A.Donnachie and P.V.Landshoff, Phys.Lett. **B296** (1992) 139.
11. A.Donnachie and P.V.Landshoff, Phys.Lett. **B437** (1998) 408.
12. See the talk by E. Gotsman, these proceedings.
13. J.R. Forshaw, G. Kerley and G. Shaw, Phys. Rev. **D60** (1999) 074012.
14. A. De Rujula et al, Phys. Rev.**D10** (1974) 1649.
15. ZEUS collaboration, Eur.Phys.J. **C7** (1999) 609
16. J.C. Collins, L. Frankfurt and M. Strikman, Phys. Rev. **D56** (1997) 2982; J.C. Collina and A. Freund, Phys.Rev. **D59** (1999) 074009
17. For a recent review, see Zhang Chen hep-ph/0008015.
18. L. Frankfurt, A. Freund and M. Strikman, Phys.Rev. **D58** (1998) 114001; erratum **D59** (1999) 119901.
19. D.A. Ross, these proceedings. See also G.Salam, lanl hep-ph/0005304
20. R.D. Ball and P.V. Landshoff J.Phys. **G26** (2000) 672-682 hep-ph/9912445.

21. N.N. Nikolaev and B.G. Zakharov, Z. Phys. **C49** (1991) 607; Z. Phys. **C53** (1992) 331

22. A. H. Mueller, Nucl. Phys. **B415** (1994) 373; A. H. Mueller and B. Patel, Nucl. Phys. **B425** (1994) 471

23. See the review of M.F.McDermott DESY report 00-126, hep-ph/0009086 and references therein.

24. J.R. Forshaw, G. Kerley and G. Shaw, Nucl.Phys. A675 (2000) 80; hep-ph/9910251.

25. J.R. Forshaw, G. Kerley and G. Shaw, Proc. of DIS2000, to be published. hep-ph/0009235.

26. J.R. Forshaw, R. Sandapen and G. Shaw, to be published.

27. K. Golec-Biernat and M. Wüsthoff, Phys. Rev. **D59** (1999) 014017, **D60** (1999) 114023.

28. M. McDermott, L. Frankfurt, V. Guzey and M. Strikman, hep/ph9912547v2.

29. E. Gotsman, E. Levin and U. Maor, Phys. Lett. **B425** (1998) 369; Eur. Phys. J **C5** (1998) 303.

30. E. Gotsman, E. Levin, U. Maor and E.Naftali Nucl.Phys. **B539** (1999) 535; Eur. Phys. J **C10** (1999) 689.

31. A. Capella, E.G. Ferreiro, C.A. Selgado and A.B. Kaidalov. hep-ph/0006233.

32. R. Yoshida; talk for the ZEUS collaboration at DIS99. Nucl. Phys. **B79** (1999) 83.

The Scattering of Photons on Proton and Photon Targets

E. GOTSMAN
In collaboration with E. Levin, U. Maor and E. Naftali

School of Physics and Astronomy,
Raymond and Beverly Sackler Faculty of Exact Sciences
Tel Aviv University, Tel Aviv, Israel.
E-mail: gotsman@post.tau.ac.il

Abstract. Using an approach based on Gribov's ideas for the photon-proton inter-action, we are able to describe data over a wide range of Q^2 and W. We introduce a separation parameter in the integrals over the virtual $q\bar{q}$ masses in the double dispersion relation, which allows us to deal with the long and short range forces individually. The long distance or "soft" component is calculated in the addative quark model, while the short distance or "hard" component is evaluated using pQCD techniques. We require different values of the separation parameter for longitudinal and transverse components. We also generalize Gribov's approach and apply it to photon-photon scattering.

FORMALISM FOR $\gamma^* P$ SCATTERING

The fact that at high energies a photon (real of virtual) can fluctuate into a hadronic system (i.e. a $q\bar{q}$ pair at lowest order) with fluctuation time $\tau_f \gg \tau_i$ (interaction time), has lead to a remarkable simplification of deep inelastic scattering (DIS) processes as has been suggested by Gribov [1]. This allows us to view DIS as a two step process:

- γ^* fluctuates into a $q\bar{q}$ pair long before it interacts with the target

- later the $q\bar{q}$ pair interacts with the target

This enabled Gribov to express the two stage process explicitly in terms of a double dispersion relation

$$\sigma(\gamma^* N) = \frac{\alpha_{em}}{3\pi} \int \frac{\Gamma(M^2)dM^2}{(Q^2 + M^2)} \sigma(M^2, M'^2, s) \frac{\Gamma(M'^2)dM'^2}{(Q^2 + M'^2)}, \tag{1}$$

where M and M' are the invariant masses of the incoming and outgoing quark-antiquark pairs, $\sigma(M^2, M'^2, s)$ is the cross section of a $q\bar{q}$ interaction with the

CP571, *PHOTON 2000*, edited by A. J. Finch

target, and the vertices $\Gamma^2(M^2)$ and $\Gamma^2(M'^2)$ are given by $\Gamma^2(M^2) = R(M^2)$, where $R(M^2)$ is the ratio:

$$R(M^2) = \frac{\sigma(e^+e^- \to \text{hadrons})}{\sigma(e^+e^- \to \mu^+\mu^-)}, \tag{2}$$

is taken from experimental data.

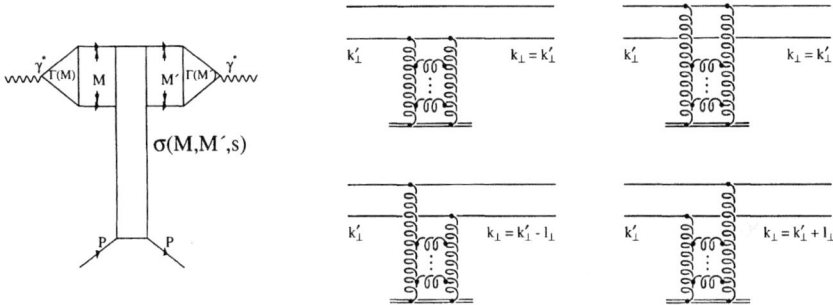

FIGURE 1. The generalized Gribov's formula.

FIGURE 2. Diagrams which contribute to the scattering of a $q\bar{q}$ pair off the target proton.

To evaluate the integrals in Eq.(1) we introduce a separation parameter $M_0 \approx$ 1 GeV in the integrals over M and M'. For $M, M' > M_0$ the quark - antiquark pair are produced at short distances ($r_\perp \propto \frac{1}{M} < \frac{1}{M_0}$) and we can evaluate this ("hard") contribution using pQCD. While for $M, M' < M_0$ the distance between quark and antiquark is large ($r_\perp \propto \frac{1}{M} > \frac{1}{M_0}$), and we have to use non-perturbative techniques i.e. the additive quark model (AQM) [2] to evaluate this "soft" contribution.

"Soft" component

For the "soft" interaction $M, M' < M_0$ we use the AQM in which

$$\sigma(M^2, M'^2, s) = \sigma_N^{soft}(M^2, s)\,\delta(M^2 - M'^2)$$
$$= (\,\sigma_{tot}(qN) + \sigma_{tot}(\bar{q}N)\,)\,\delta(M^2 - M'^2) \tag{3}$$

As in the AQM, the cross section is related to the imaginary part of the forward elastic quark scattering amplitude we have $M = M'$ and Eq.(1) simplifies to

$$\sigma(\gamma^*N) = \frac{\alpha_{em}}{3\pi} \int \frac{R(M^2)M^2 dM^2}{(Q^2 + M^2)^2} \sigma_N(M^2, s) \; ; \tag{4}$$

Using AQM to calculate $\sigma_N(M^2, s)$ and the Donnachie-Landshoff [3] parameterization we have

$$\sigma_T^{soft} = \frac{\alpha_{em}}{3\pi} \int \frac{R(M^2)M^2 dM^2}{(Q^2 + M^2)^2} \cdot [A(\frac{1}{x_M})^{\alpha_P(0)-1}) + B(\frac{1}{x_M})^{\alpha_R(0)-1})] \qquad (5)$$

with $\alpha_P = 1.079$, $\alpha_R = 0.55$, A = 13.1mb, B= 41.08mb and $x_M = \frac{M^2}{s}$.

"Hard" component

We use the two gluon exchange model to evaluate the "hard" contribution. We need to include the four diagrams shown in Fig.2.

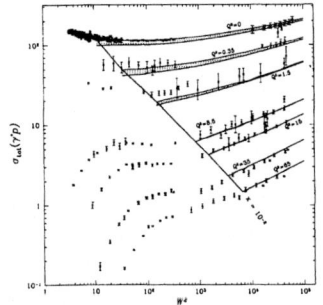

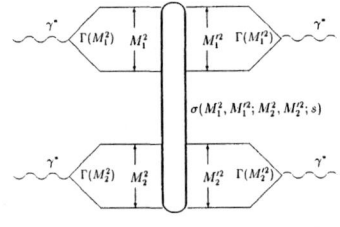

FIGURE 3. W dependence of the cross section $\sigma(\gamma^* p)$.

FIGURE 4. Gribovs approach for $\gamma^* - \gamma^*$.

The production amplitude can be written in a factorized form

$$\mathcal{M}_{\lambda\lambda'}(k_\perp, z) = \sqrt{N_c} \int d^2 k_\perp' \int_0^1 dz' \psi_{\lambda\lambda'}(k_\perp', z') \mathcal{T}_{\lambda\lambda'}(k_\perp', z'; k_\perp, z). \qquad (6)$$

where $\psi_{\lambda\lambda'}$ denotes the wave function of the $q\bar{q}$ system in the γ^*, $\mathcal{T}_{\lambda\lambda'}$ is the amplitude for scattering of the $q\bar{q}$ off the proton.

We need to include both the transverse and longitudinal components of the γ^*: For the transverse component of γ^* we have

$$\psi_{\lambda\lambda'}^{\pm} = -\delta_{\lambda,-\lambda'} Z_f e[z(1-2z)\lambda \mp 1]\frac{1}{\bar{Q}^2 + k_\perp^2}, \qquad (7)$$

while for the longitudinal γ^* we have

$$\psi_{\lambda\lambda'}^{L} = -2\delta_{\lambda,-\lambda'} Z_f e Q z(1-z)\frac{1}{\bar{Q}^2 + k_\perp^2}, \qquad (8)$$

We can also obtain the contribution of the heavy quarks by making the appropriate substitutions.

In our formalism the ratio of the contributions from AQM and pQCD depend on the value of the seperation parameter M_0. We find that we obtain good fits to the experimental data for σ^T for $0.7 \leq M_{0,T}^2 \leq 0.9\,GeV^2$ and for σ^L with $M_{0,L}^2 \leq 0.4\,GeV^2$.

Comparison with data for $\sigma(\gamma^* p)$

We can express

$$\sigma(\gamma^* p) = \sigma_T^{\text{soft}} + \sigma_T^{\text{hard}} + \sigma_{T,\bar{Q}Q}^{\text{hard}} + \sigma_L^{\text{soft}} + \sigma_L^{\text{hard}} + \sigma_{L,\bar{Q}Q}^{\text{hard}} . \tag{9}$$

In Fig.3 we display $\sigma_{tot}(\gamma^* p)$ as a function of W^2 for different values of Q^2 compared with the relevant experimental data.

Lack of space prevents us displaying other results such as $\frac{\sigma_L}{\sigma_T}$, $\frac{\sigma^{soft}}{\sigma^{hard}}$ and $\frac{\sigma^{heavy}}{\sigma^{light}}$, these can be seen in [4].

THE $\gamma^* \gamma^*$ CROSS SECTION

Gribov's treatment of DIS processes can easily be extended to $\gamma^* - \gamma^*$ interactions. Pictorially this is illustrated in Fig.4 (cf. Fig.2 for $\gamma^* p$ processes).

Thus instead of Eq.(1) we have the double dispersion relation for $\gamma^* \gamma^*$

$$\sigma(\gamma^* \gamma^*) = \left(\frac{\alpha_{\text{em}}}{3\pi}\right)^2 \int dM_1^2 \, dM_2^2 \, dM_1'^2 \, dM_2'^2 \frac{\Gamma(M_1^2)}{(Q_1^2 + M_1^2)} \frac{\Gamma(M_2^2)}{(Q_2^2 + M_2^2)} \times$$

$$\sigma(M_1^2, M_1'^2; M_2^2, M_2'^2; s) \frac{\Gamma(M_1'^2)}{(Q_1^2 + M_1'^2)} \frac{\Gamma(M_2'^2)}{(Q_2^2 + M_2'^2)} \tag{10}$$

Each of the two γ^*'s can be "soft" or "hard", for simplicity we will refer to the upper photon as harder than the lower one, hence we have three sectors to evaluate.

Hard-Hard Interaction

For the case where both photons are hard, we calculate the interaction between the two $q\bar{q}$ pairs using pQCD, and we need to include all 16 diagrams contributing. (For details see [5]).

Soft-Soft Interaction

We use the AQM in the same manner as was done in the soft-soft sector of γ p interactions previously, with the provisor of multiplying by the factor of $\frac{2}{3}$ which is necessary since the cross section for two photon interactions is related to $\frac{2}{3}\sigma(\pi p)$.

Hard-Soft Interaction

For the case of the interaction of a hard photon (upper) with a soft photon (lower). Applying factorization we can relate this process to the hard-soft one in the $\gamma^* p$ and adapt the results already calculated for that case.

Numerical calculations for $\gamma^* \gamma^*$

Our final expression for the total $\gamma^* \gamma^*$ cross section is the sum of the terms from the three sectors, and can be expressed

$$\sigma(\gamma^* \gamma^*) = \sigma(\text{Hard-Hard}) + \sigma(\text{Hard-Soft}) + \sigma(\text{Soft-Soft}). \qquad (11)$$

where each of the components in Eq.(11) includes terms from all possible polarizations, with two contributionns for each "non-diagonal" term.

Using the same values for the parameters as determined in the γp case, we display some of the results of our calculation compared to the relevant data [6–9].

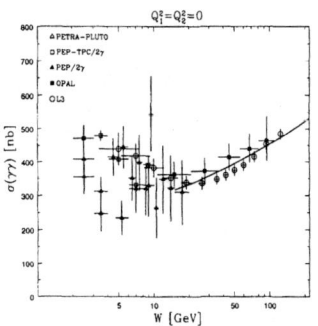

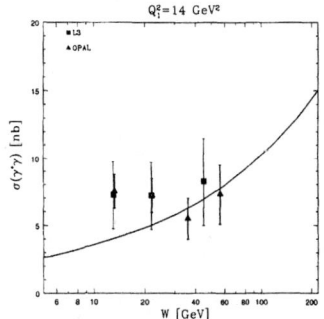

FIGURE 5. Our calculation for $\sigma(\gamma\gamma)$ and the experimental data

FIGURE 6. Our calculation for $\sigma(\gamma^* \gamma)$ and the experimental data for $Q_1^2 = 14. \, GeV^2$ and $Q_2^2 = 0.$

CONCLUSIONS

- Using the Gribov approach we obtain a successful description of experimental data for both $\gamma^* p$ and $\gamma^* \gamma^*$ interactions over a wide range of Q^2 and W.

- The key assumption in our approach is that pQCD (short range) and nonperturbative (long range) interactions in Gribov's formalism can be seperated by a parameter M_0.

- By fitting to $\gamma^* p$ data we find that the values $0.7 \leq M_{0,T}^2 \leq 0.9\,GeV^2$ and $M_{0,L}^2 \leq 0.4\,GeV^2$ produce excellent agreement with the data.

- The Additive Quark model provides a successful representation of the non perturbative component.

- Seperating the $\gamma^* \gamma^*$ interaction into three sectors soft-soft, soft-hard and hard-hard allows us to investigate the interplay between the long and short distance processes.

- More complete details are contained in [4,5].

REFERENCES

1. N. N. Gribov, Sov. Phys. JETP 30 (1970) 709.
2. E. M. Levin and L. L. Frankfurt, JEPT Letters 3 (1965) 652;
   H. J. Lipkin and F. Scheck, Phys. Rev. Lett. 16 (1966) 71:
   J. J. Kokkedee, "The Quark Model", W. A. Benjamin New York (1969).
3. A. Donnachie and P. V. Landshoff, Nucl. Phys. B 244 (1984) 322; Nucl. Phys. B 267 (1986) 690; Phys. Lett. B296 (1992) 227, Z. Phys. C 61 (1994) 139.
4. E. Gotsman, E. Levin, U. Maor and E. Naftali, Eur. Phys. J C 10 (1999) 689.
5. E. Gotsman, E. Levin, U. Maor and E. Naftali, Eur. Phys. J. C 14 (2000) 511.
6. PLUTO Collaboration, Phys. Lett. B 149 (1984) 421; Z. Phys. C 26 (1984) (353); TPC/2γ Collaboration, Phys. Rev. D 41 (1990) 2667; Phys. Rev. Lett. 54 (1985 763; OPAL Collaboration, hep-ex/9906039;
   L3 Collaboration, Phys. Lett B 408 (1997) 450.
7. OPAL Collaboration: F. Wäckerle, Proc. XXVIII int. symp. On Multiparticle Dynamics, Frascati, 1997.
8. L3 Collaboration: M. Acciarri etal, L3 Note 2400, 1999.
9. L3 Collaboration: M. Acciarri etal, Phys. Lett. B 453 (1999) 333.

The BFKL Equation Beyond Leading Order

Douglas Ross

Department of Physics & Astronomy
University of Southampton
Southampton SO15 1BJ.
U.K.

Abstract. Progress towards the understanding of the large corrections to the BFKL prediction for the $s-$ behaviour of diffractive processes is discussed.

INTRODUCTION

The BFKL equation gives us the amplitude for diffractive processes in perturbative QCD. For diffraction we require the $t-$channel exchange of an 'object' known as a "pomeron", which has the quantum numbers of the vacuum (in particular it is a colour singlet).

I THE BFKL EQUATION IN FIRST AND SECOND ORDER

This can be achieved by the exchange of two gluons and at large s such contributions dominate.

In this way, all diffractive processes are dominated by the amplitude, $f(s, \mathbf{k}, \mathbf{k}')$, for the scattering of two gluons with centre-of-mass energy $\sqrt{s}$ and transverse momenta $\mathbf{k}$ and $\mathbf{k}'$ respectively. For diffractive processes $s \gg \mathbf{k}^2, \mathbf{k}'^2$, but for perturbation theory to be valid we also assume $\mathbf{k}^2, \mathbf{k}'^2 \gg \Lambda_{QCD}^2$.

The BFKL equation determines the $s-$ dependence of this amplitude in terms of an integral equation involving a kernel $\mathcal{K}(\mathbf{k}, \mathbf{k}')$:

$$\frac{\partial}{\partial \ln s} f(s, \mathbf{k}, \mathbf{k}') = \int d^2 \mathbf{l} \, \mathcal{K}(\mathbf{k}, \mathbf{l}) f(s, \mathbf{l}, \mathbf{k}'), \tag{1}$$

which has a solution, which is written in terms of an integral over a parameter ν:

$$f(s, \mathbf{k}, \mathbf{k}') \sim \int d\nu \left(\frac{\mathbf{k}}{\mathbf{k}'}\right)^{i\nu} \left(\frac{s}{\mathbf{k}\,\mathbf{k}'}\right)^{\bar{\alpha}_s \chi(\nu)} \tag{2}$$

CP571, *PHOTON 2000*, edited by A. J. Finch
© 2001 American Institute of Physics 0-7354-0010-5/01/$18.00

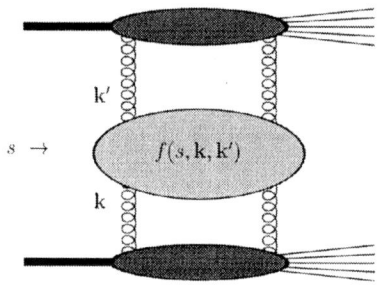

FIGURE 1. QCD description of a diffractive process

$$\bar{\alpha}_s \equiv \frac{3}{\pi}\alpha_s$$

The characteristic function $\chi(\nu)$ has a perturbative expansion

$$\chi(\nu) = \chi_1(\nu) + \bar{\alpha}_s\chi_2(\nu),$$

where $\chi_1(\nu)$ is given by the leading order BFKL equation [1], which sums the leading $\ln s$ terms to all orders and has a small ν expansion

$$\chi_1(\nu) = 2.77 - 16.8\,\nu^2 + \cdots,$$

whereas $\chi_2(\nu)$ comes from the next to leading order expansion [2] (which took twenty years to calculate) and has a small ν expansion

$$\chi_2(\nu) = -18.3 + 147\,\nu^2 + \cdots.$$

It can be seen from eq.(2) that the $s-$ dependence of the amplitude is dominated by the value of χ at $\nu = 0$, for which we have

$$\bar{\alpha}_s\frac{\chi_2(0)}{\chi_1(0)} = -6.6\,\bar{\alpha}_s. \tag{3}$$

We see that even for small values of α_s of order 0.2 this has a correction of greater than 100 %.

Rather than abandoning the entire programme at this stage, we should note that such a comparison is misleading [3]. The $s-$ dependence given in eq.(2) involves an integral over all ν and although the correction is indeed huge for $\nu = 0$ it rapidly becomes more modest at higher values of ν (see Figure 2), so the correction, although still large, does not turn out to be as bad as that suggested by eq.(3).

II SUMMING THE LARGE CORRECTIONS

In addition to this there have been several attempts to tame the expansion by identifying the origin of the substantial part of the higher order correction such that this part can be summed, leaving a modest remaining higher order corrections. Such re-summation programs include:

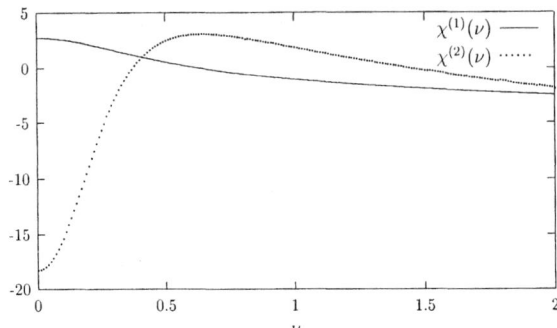

FIGURE 2. Plots of $\chi_1(\nu)$ and $\chi_2(\nu)$ against ν

- Accounting for the fact that the amplitude for the gluon-gluon scattering involves the emission of a ladder of gluons and since we are in the diffractive regime, it is reasonable to require a minimum rapidity gap between the emission of sequential gluons. I shall not comment further on this approach.

- Accounting for the fact that the expression for the amplitude must be compatible with the DGLAP equations when applied to deep-inelastic scattering at low-x.

- Accounting for the running of the coupling.

III   COMPATIBILITY WITH DGLAP

The diffractive amplitude $f(s, \mathbf{k}, \mathbf{k}')$, also controls the deep-inelastic structure functions at low-x which are dominated by the gluon distribution. The quantity s is (at low-x) replaced by Q^2/x.

In this case the transverse momentum $\mathbf{k}'^2 \sim Q^2$, whereas the lower transverse momentum $\mathbf{k}^2 \sim \mu^2$. μ is a hadronic scale determined by the primordial transverse momenta of partons inside the target hadron, and is usually in the infrared regime where perturbation theory is invalid. For the purposes of this discussion, we will assume that this is not the case and that μ is sufficiently large that perturbative calculations are still reliable [1].

The BFKL equation gives an expression for the structure functions at low-x, whose Q^2 dependence may be written

$$F_2(Q^2, x) \;\sim\; \int d\nu \left(\frac{Q^2}{\mu^2}\right)^{i\nu/2 + \bar{\alpha}_s \chi(\nu)} y(x, \nu) \tag{4}$$

---

[1] There are experimental scenarios, which are currently under investigation at HERA for which this is indeed the case.

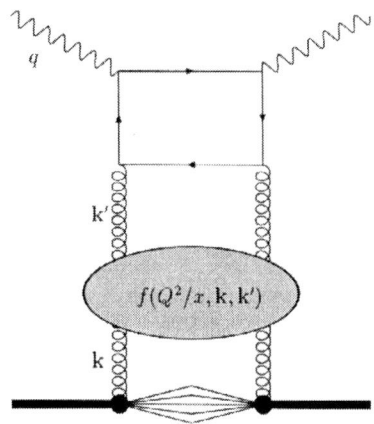

FIGURE 3. BFKL applied to deep-inelastic scattering at low=x

($y(x, \nu)$, being independent of Q^2). From the DGLAP evolutions we know that to order α_s^n, we can only have n powers of $\ln Q^2$. In particular, this means that the characteristic function $\chi(\nu)$ can only have simple poles in ν - higher order poles have residues which involve higher powers of $\ln Q^2$.

This allows a partial summation. For example, $\chi_1(\nu)$ has simple poles at $\nu = \pm i/2$

$$\lim_{\nu \to \pm i/2} \chi_1(\nu) = \frac{4}{(1 + 4\nu^2)} \tag{5}$$

whereas $\chi_2(\nu)$ has triple poles at $\nu = \pm i/2$

$$\lim_{\nu \to \pm i/2} \chi_2(\nu) = \frac{64}{(1 + 4\nu^2)^3} \tag{6}$$

The re-summed expression is

$$\lim_{\nu \to \pm i/2} \chi(\nu) = \frac{\sqrt{(1 + 4\nu^2)^2 + 16\bar{\alpha}_s} - (1 + 4\nu^2)}{2\bar{\alpha}_s}, \tag{7}$$

which reproduces eqs.(5) and (6) upon expansion in $\bar{\alpha}_s$.

As can be seen from Figure 4, for $\bar{\alpha}_s = 0.2$, such a re-summation leaves a small remnant correction.

This technique has been developed to take into account other known features of the DGLAP evolution. [6]

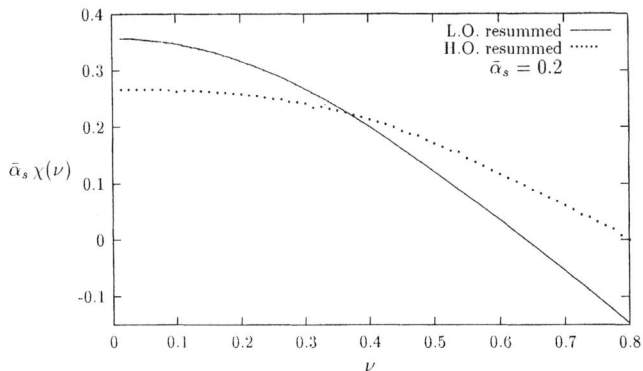

FIGURE 4. Plots of $\chi_1(\nu)$ and $\chi_2(\nu)$ against ν after re-summation

IV ACCOUNTING FOR THE RUNNING OF THE COUPLING

The BFKL kernel has an expansion in the strong coupling. As can be seen from eq.(1) the equation involves an integral over *all* transverse momenta and the coupling should therefore be allowed to run with the magnitude of the transverse momentum, rather than keeping it fixed as we have done so far.

A careful analysis of such a running [7] yields an $s-$ dependence for the amplitude, which is given by

$$f(s, \mathbf{k}, \mathbf{k}') \sim \int d\nu \left(\frac{\mathbf{k}}{\mathbf{k}'}\right)^{i\nu} \int d\omega s^\omega \exp\left(-\frac{1}{b\omega}\int^\nu d\nu' \chi(\nu')\right), \qquad (8)$$

where b is proportional to the the first term in the β-function, such that to leading order

$$\bar{\alpha}_s(\mathbf{k}^2) = \frac{1}{b\ln(\mathbf{k}^2/\Lambda_{QCD}^2)}.$$

This expression is even more complicated than the one given in eq.(2), where the coupling was assumed to be fixed. The important feature of this expression is that

$$\lim_{s\to\infty}\int d\omega s^\omega \exp\left(-\frac{1}{b\omega}\int^\nu d\nu' \chi(\nu')\right) \sim \exp\left(A\sqrt{\ln s}\right),$$

which grows slower than any power of s.

In other words, when the running of the coupling is taken into account in am amplitude that can be calculated purely perturbatively, *the power-behaviour prediction of BFKL completely disappears*.

On the other hand, if we were to simulate non-perturbative effects by arguing that the coupling freezes at some critical magnitude of the transverse momentum

k_0, such that for $k^2 < k_0{}^2$, the coupling was given by

$$\alpha_s(k^2) = \frac{1}{b \ln(k_0{}^2/\Lambda^2_{QCD})},$$

we would expect that the power-like behaviour should be reinstated for sufficiently low transverse momenta. Work is in progress [8] [9] to study in detail the extrapolation between the large transverse momentum regimes, in which there is no power-like $s-$behaviour and the small transverse momentum regime in which one does indeed expect to see a power-like behaviour.

V SUMMARY

- Higher order corrections to the BFKL solution are large, but it is misleading simply to compare that ratio of the values of the characteristic functions $\chi_1(\nu)$ and $\chi_2(\nu)$ at $\nu = 0$.

- Requiring that the $\ln Q^2$ terms in any perturbative order should be consistent with the DGLAP equations allows one to resum the major part of the higher order corrections, leaving a small remnant higher order correction.

- When the running of the coupling is correctly accounted for in the purely perturbative calculation, the power-like behaviour of the $s-$dependence completely disappears.

- Work is in progress to understand the extrapolation between the running coupling regime and the infrared regime of fixed coupling, where the power-like behaviour is expected to be reinstated.

REFERENCES

1. V.S. Fadin, E.A. Kurayev and L.N. Lipatov, *Sov. Phys. JETP* **44** 443 (1978)
   Y.Y. Balitski and L.N. Lipatov, *Sov. J. Nucl. Phys.* **28** 822 (1978)
2. V.S. Fadin and L.N. Lipatov, *Phys. Lett.* **B429** 127 (1998)
3. D.A. Ross, *Phys. Lett.* **B431** 165 (1998)
4. C.R. Schmidt, *Phys. Rev.* **D60** 074003 (1999)
5. C.P. Salam *JHEP* **9807:019** (1998)
6. M. Ciafaloni and D. Colferai, *Phys. Lett* **B453** 372 (1999)
7. R.S. Thorne, *Phys. Rev.* **D60** 054031 (1999); R.S. Thorne, *Phys. Lett.* **B474** 372 (2000);
8. J.R. Forshaw, D.A. Ross, and A. Sabio-Vera *in preparation*
9. M. Ciafaloni, D. Colferai and G.P. Salam, *JHEP* **9910:017** (1999); *JHEP* **0007:054** (2000);

Photoproduction at HERA with a Leading Proton

Hanna Mahlke-Krüger

H1 Collaboration, DESY

Abstract. The total cross-section for the semi-inclusive photoproduction process with a leading proton in the final state has been measured at centre-of-mass energies W between 91 and 231 GeV. The measured cross-sections refer to the kinematic range with transverse momenta of the scattered proton restricted to $p_T \leq 0.2$ GeV and $0.66 \leq z \leq 0.90$, where $z = E'_p/E_p$ is the scattered proton energy normalized to the beam energy. The cross-section $\mathrm{d}\sigma_{\gamma p \to p' X}(W)/\mathrm{d}z$ is observed to be independent of the photon-proton centre-of-mass energy and the scattered proton momentum within the measurement errors and amounts to $8.00 \pm 0.42(stat) \pm 1.07(syst)\,\mu$b on average. The measured cross-sections are compared with semi-inclusive deep inelastic scattering data with a leading proton in the final state subjected to the same kinematic selection criteria. The semi-inclusive photoproduction cross section is suppressed more strongly than the semi-inclusive DIS cross section with respect to the inclusive data.

I INTRODUCTION

The HERA electron proton collider has allowed semi-inclusive cross sections of the form $\gamma^{(*)}p \to Xp'$ to be studied in detail for both real and virtual photons. In this paper, a measurement of the semi-inclusive cross-section of the photoproduction process is presented, which is combined with the corresponding data from deep inelastic scattering (DIS) processes [1]. The process under study is sketched in Figure 1.

The transition in reactions $\gamma^{(*)}p \to X$ from finite Q^2 to $Q^2 = 0$ is conveniently described by introducing a total virtual photon proton cross-section

$$\sigma_{\gamma^{(*)}p}^{tot}(W, Q^2) \approx \frac{4\pi^2\alpha}{Q^2} F_2(x, Q^2) \tag{1}$$

where $x = Q^2/(2p_p \cdot q)$ is the Bjorken scaling variable and p_p and q are the four-momenta of the incident proton and photon, respectively. α is the fine-structure constant and F_2 the proton structure function. W denotes the total γ^*p centre-of-mass energy, which for the present application of small values of the scaling variable x can be approximated by $W^2 \approx Q^2/x$. The inelasticity of the process, y,

CP571, *PHOTON 2000*, edited by A. J. Finch
© 2001 American Institute of Physics 0-7354-0010-5/01/$18.00

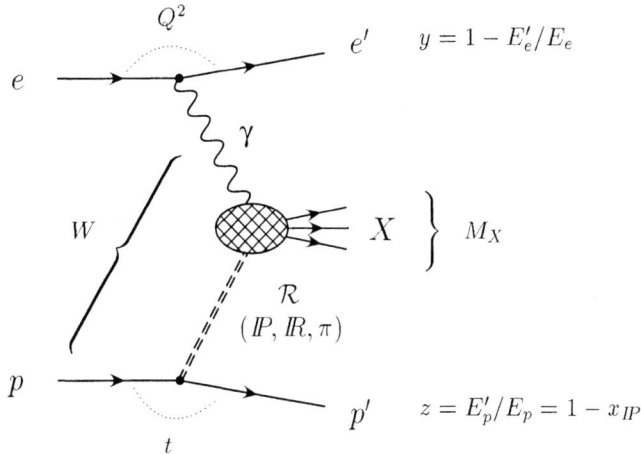

FIGURE 1. *The process $ep \to e'p'X$. At the proton vertex, $z = E'_p/E_p$ is the fractional momentum kept by the proton.*

is defined as $y = (p_p \cdot (p'_e - p_e))/(p_p \cdot p_e)$ with p_e and p'_e the four-momenta of the incoming and outgoing electron. In photoproduction, y can be measured using the relation $y = 1 - E'_e/E_e$, where E_e and E'_e indicate the energies of the incoming and outgoing electron, respectively. Here, for $\gamma^{(*)}p \to p'X$, the situation is viewed as follows: The photon dissociates to a system of mass M_X. The proton remains intact leaving the interaction with a fractional energy $z = E'_p/E_p$, where E'_p and E_p are the energies of the incident and scattered proton, respectively[1]. In the publication [1] DIS data in the kinematic range $0.71 \leq z \leq 0.90$ were shown to be well described by a Regge model in which the virtual photon interacts with colourless constituents of the proton. In Regge terminology, this corresponds to a hadronic exchange $\mathcal{R}$ colliding with the photon at fixed energy $(1 - z)E_p$. In the kinematic range of this study, the pomeron contribution to $\mathcal{R}$ is suppressed relative to the exchange of pion or the degenerate trajectories of the ρ, ω, f, and a_2 exchange in this paper denoted by $I\!R$. The variables describing this process are $\beta = x/(1 - z)$ instead of x and the invariant mass of the $\gamma\mathcal{R}$ system $M_X = W\sqrt{1 - z}$. Note that for each value of z, the choice of W fixes M_X.

A comparison of semi-inclusive data with a leading proton in the final state at fixed z and finite values of Q^2 with data at $Q^2 = 0$ can be viewed as investigating the limit $Q^2 \to 0$ for $\gamma I\!R$ or $\gamma\pi$ reactions at fixed $\gamma\mathcal{R}$ CM energies. Assuming the proton and photon vertices can be factorised, the flux factors describing the probability of exchanging $\mathcal{R}$ at fixed z can be considered as a common normalisation parameter for photoproduction and DIS.

---

[1] In some publications, $x_{I\!P} = 1 - z$ is used instead.

II APPARATUS AND DATA SELECTION

The H1 detector is described in detail elsewhere [2]. The direction of the incoming proton beam is labelled the forward direction and taken to be the $+z$ axis, the x axis points towards the centre of the storage rings, and the y axis points upwards relative to the plane of the HERA electron ring. The selection on the final state is performed by identifying the scattered proton, the scattered electron, and a charged track in the central region of the H1 detector. The analysis is based on a data set from the 1996 running period, where HERA collided 820 GeV protons with 27.5 GeV positrons, the integrated luminosity amounts to 3.3 pb^{-1}.

The Forward Proton Spectrometer[2] (FPS) [3] is used to measure the energy and scattering angle at the interaction point of the outgoing proton. The error on the energy measurement is less than 8 GeV. The detector size, the distance of the detector stations from the interaction point, and the proton beam size restrict the acceptance in the proton's transverse momentum to $p_T \leq 0.2$ GeV. For the present analysis the energy range was restricted to $0.66 \leq z = E'_p/E_p \leq 0.90$, divided into five intervals. The measurement integrates over the accessible p_T range.

An electron was required to be found in one of the electron calorimeters, belonging to the luminosity system, which is used also for determination of the luminosity and background rejection due to Bethe-Heitler events. The acceptance is experimentally restricted to values of $Q^2 < 0.01$ GeV^2. The data sample was binned according to the value of the inelasticity $y = 1 - E'_e/E_e$, where E'_e is equal to the energy measured in the tagger calorimeters. The values for the mean γp centre-of-mass energy W, calculated as $W = \sqrt{ys}$, for the three bins are: $W_1 = 91$ GeV, $W_2 = 187$ GeV, $W_3 = 231$ GeV.

Furthermore, at least one reconstructed track inside the main H1 detector with $p_T^{\mathrm{trk}} > 500$ MeV, $20° < \theta < 160°$ was required. The effect of this acceptance cut was corrected using the PHOJET Monte Carlo program [4].

Further background rejection cuts were imposed.

III CROSS SECTION MEASUREMENT

The differential cross section $d\sigma_{\gamma p \to Xp'}(W^2, E'_p)/dz$ can be determined using the relation

$$\frac{d^3\sigma_{ep \to eXp'}(W^2, Q^2, z)}{dy\,dQ^2\,dz} = \mathcal{F}_{\gamma/e}(y, Q^2)\frac{d\sigma_{\gamma p \to Xp'}(W^2, z)}{dz}, \tag{2}$$

where $\mathcal{F}_{\gamma/e}(y, Q^2)$ is the photon flux in the Equivalent Photon Approximation [5]. The data are corrected for acceptances and efficiencies. Further details can be found in [6].

---

[2] The Forward Proton Spectrometer was supported by the INTAS93-43 project and the NATO contract PST.CLG.975100.

A   Background

The following sources of background were studied:

- The contribution arising from events in which the proton signal in the final state is faked in the FPS is estimated to be on the percent level and neglected.

- Tracks in the central H1 detector produced by beam gas interactions: This background fraction has been determined to vary between 2.2 and 3.4%, depending on the y bin, and was subtracted.

- Bethe Heitler events $ep \rightarrow ep\gamma$: By experimental means it is verified that these events contribute only at the level of 0.25%. The background has been subtracted.

B   Systematic errors

Three types of systematic errors are distinguished. These are errors common to events in the same y interval, errors related to the proton energy (or z) interval and global errors.

The following sources contribute to errors common to all events in the same y bin: background from proton beam gas interactions in the central region of the H1 detector, errors on the acceptance of the electron taggers, errors on the selection efficiency for tracks in the H1 tracking chambers, and migration correction between adjacent y bins. These errors vary between 6.5% and 8.5%.

Depending on the z bin, uncertainties between 2.6% and 14.1% are obtained, related to correction factors for migration and acceptance correction. The factors were determined using the PHOJET [4] and POMPYT [7] Monte Carlo generators.

Further contributions to the systematic error given by the uncertainty on the luminosity measurement, a value for FPS systematic errors such as alignment errors, uncertainty on the hodoscope efficiencies, and other effects not covered by the simulation, and additionally systematic errors on the vertex cut efficiency and positron beam related background give rise to 5.3% of systematic error to all bins.

C   Results

The cross section $d\sigma_{\gamma p \rightarrow p'X}(W)/dz$ for $W = 91$, 187, and 231 GeV and five values of z for $0.66 \leq z \leq 0.90$ is shown in Figure 2. The errors shown are the square root of the quadratic sum of the systematic errors as described above and the statistical error. The total errors vary between 9.2% and 21.0%. No radiative corrections were applied to the data since they are expected to be negligibly small.

It is observed that in all z bins the measured cross-sections for different values of the γp centre-of-mass energy W agree with each other within the experimental errors. Furthermore the z dependence is only weak without a clear trend so that the present data may be represented by one average cross-section value of $d\sigma(W,z)/dz = 8.00 \pm 0.42(stat) \pm 1.07(syst)\,\mu b$ (the measured values ranging between 7.2 and 9.4 μb) which corresponds to $d\sigma(W, E'_p)/dE'_p \sim 9.76\,nb/\,GeV$ for the kinematic range covered by this experiment. The restriction in the transverse momentum of the final state proton to $p_T \leq 0.2\,GeV$ should be kept in mind, which is equivalent to the statement that the measured cross-section represents only about $(24 \pm 2)\%$ of the total semi-inclusive photoproduction cross-section if a slope of $b = (6.8 \pm 0.8)\,GeV^{-2}$ in $d\sigma/dp_T^2 \sim e^{-bp_T^2}$ is assumed as measured by the ZEUS collaboration [8] for the kinematic range under study. The semi-inclusive photoproduction process with a proton proton of $0.66 \leq z \leq 0.90$ and $p_T < 200\,MeV$ contributes 1% to the total photoproduction cross-section.

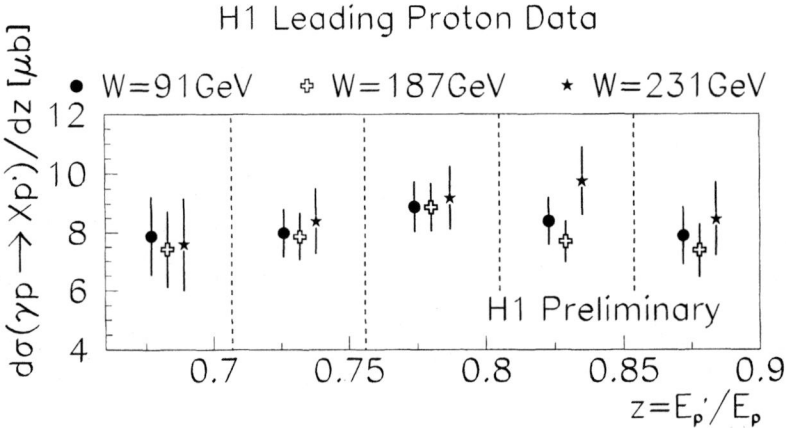

FIGURE 2. *The cross-section $d\sigma_{\gamma p \to p'X}/dz$ as a function of z for three values of W. The inner error bar is the statistical and the outer the total error (statistical and systematical error added in quadrature). The dashed lines show the limits of the z intervals over which the differential cross-sections are averaged.*

IV LEADING PROTON PRODUCTION IN DIS AND PHOTOPRODUCTION

Applying Equation (1), inclusive deep inelastic scattering and inclusive photoproduction data from the reaction $\gamma^{(*)}p \to X$ have been combined in order to study the transition between the two kinematic regimes in [9]. In the present paper, the same investigation is performed for the process $\gamma^{(*)}p \to p'X$.

For the semi-inclusive case, the deep inelastic structure function $F_2^{LP(3)}$ [1] is related to the leading proton photoproduction data as indicated below. The data points are displayed in Figure 3:

$$\frac{\mathrm{d}\sigma_{\gamma^{(*)}p \to p'X}(W^2, Q^2, z)}{\mathrm{d}z} = \frac{4\pi^2\alpha}{Q^2}F_2^{LP(3)}(W^2, Q^2, z). \tag{3}$$

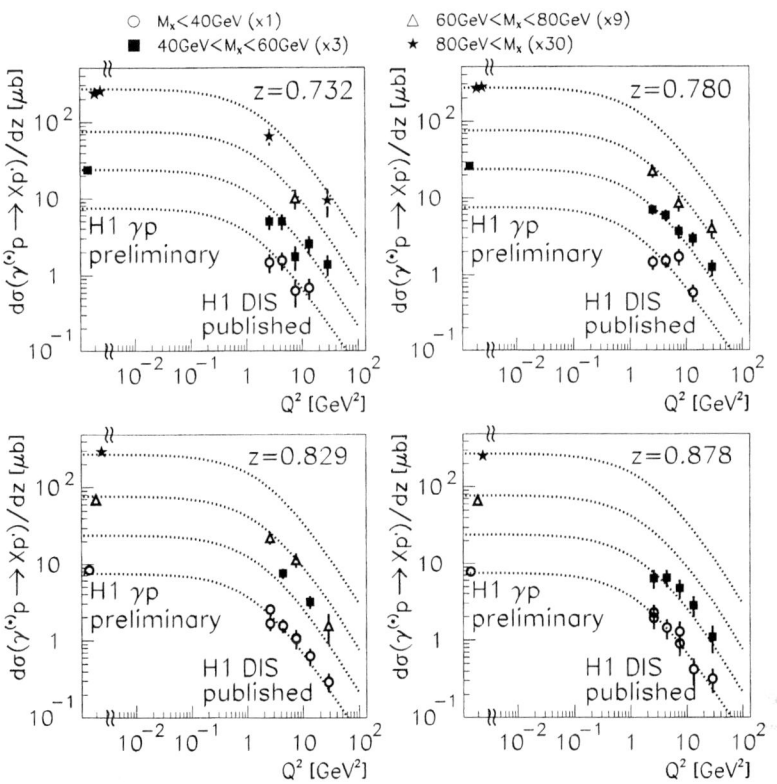

FIGURE 3. *The cross-section* $\mathrm{d}\sigma_{\gamma^{(*)}p \to p'X}/\mathrm{d}z$ *for four bins of* z *and four bins of* M_X *as a function of* Q^2. *The curves are the result of the fit described in the text.*

In [9] the inclusive data have been parametrized by a function inspired by the model described in [10], where the proton structure is assumed to consist of two components: F_2^{VDM} dominant at low values of the photon virtuality describing the contribution of vector mesons coupling to the virtual photon [11], and a partonic structure function F_2^{part}, governing the region $Q^2 > 1\,\mathrm{GeV}^2$ [10], which can be taken from a structure function measurement.

The model of [10] has not been developed to cover reactions with a leading proton, which according to the diagram of Figure 1 corresponds to a situation where the real or virtual photon interacts with a colourless constituent of the proton and not with the proton. However, an argument can be made that the exchanged object itself has a partonic structure, such that the VDM part describes a vector meson on reggeon or pion scattering process. A similar ansatz is used here. Therefore, making use of the observation that the deep inelastic semi-inclusive cross-section is also nearly independent of W and z, two additional parameters are introduced. A common scale factor F absorbs the flux factor for the probability to find $\mathcal{R}$ in the proton. This factor is assumed to be independent of Q^2 at fixed z and is a compensation for the reduction in phase space due to the limited transverse momentum of the scattered protons. A further normalization factor C_{VM}^{LP} for the vector meson contribution is assumed. The variable replacing x of the inclusive case is $\beta = x/(1 - z)$. The parameters introduced in [9] were used with the values determined in that paper.

For the structure function of $\mathcal{R}$, we use the VDM and QCD parameterisation for the proton, implicitly assuming that hadronic structure functions have the same behaviour at low x, up to a normalisation. This normalisation is absorbed into the common scale factor, for which a value $F = 0.124 \pm 0.011(stat) \pm 0.010(syst)$ is obtained. The result for the relative normalization factor is $C_{VM}^{LP} = 0.16 \pm 0.08$.

It is thus observed that the VDM term is suppressed by a factor of 0.16 with respect to the inclusive case. This is significantly smaller than the factor of 2/3 between the total cross-sections for pion proton and proton proton scattering, which one would expect to be applicable if cross-sections were related to the number of valence quarks involved in a particular reaction.

The fits are displayed in Figure 3 together with the DIS and photoproduction data points. As in the case of inclusive reactions, the fits suggest a transition near $Q^2 \approx 1 \, \mathrm{GeV}^2$, where the DIS cross-section turns over into a non-scaling regime and approaches the photoproduction value.

V SUMMARY

Photoproduction reactions with a final state proton of $p_T \leq 0.2 \, \mathrm{GeV}$ observed in the H1 Forward Proton Spectrometer have been analysed in the kinematic range $0.66 \leq z \leq 0.90$ at γp centre-of-mass energies $W = 91, 187$ and $231 \, \mathrm{GeV}$. The cross-section $d\sigma(W)/dz$ was determined to be on average $8.00 \pm 0.42(stat) \pm 1.07(syst) \, \mu\mathrm{b}$, independent of z and W within the experimental errors.

The combination of data from photoproduction and deep inelastic scattering with a final state proton observed in the FPS, $\gamma^{(*)}p \to p'X$, in the kinematic range $0.71 \leq z \leq 0.90$ shows a qualitatively similar behaviour as observed when going from virtual to real photons in the inclusive reaction $\gamma^{(*)}p \to X$. A parametrization where the contributions to the cross-section are divided into a VDM and a partonic component shows a fair description of the data, provided the overall normalization and the relative normalization between the two components are adjusted. The fit indicates that the semi-inclusive cross-section for photoproduction is more strongly suppressed than the semi-inclusive DIS cross-section with respect to the inclusive data.

REFERENCES

1. H1 Collaboration, C. Adloff et al., *Eur. Phys. J.* **C 6** (1999) 587
2. H1 Collaboration, I. Abt et al., *Nucl. Instrum. Methods* **A 386** (1997) 310
3. P. Van Esch et al., *Nucl. Instrum. Methods* **A 446** (2000) 409
4. R. Engel, "PHOJET manual (Program version 1.05c, June 96)", University of Siegen preprint 95-05 (1995) (revised Feb. 96)
5. S. Frixione et al., *Phys. Lett.* **B 319** (1993) 330
6. H. Mahlke-Krüger, Dissertation Universität Hamburg, 1999
   H1 Collaboration, *paper in preparation*
7. P. Bruni, A. Edin, G. Ingelman, "POMPYT version 2.6 – A Monte Carlo Program to Simulate Diffractive Hard Scattering Processes", unpublished program manual (1996), see `http://www3.tsl.uu.se/thep/pompyt`
8. ZEUS Collaboration, in "Proceedings of ICHEP98 Conference, Vancouver", p. 924, World Scientific A. Astbury et al. ed., World Scientific (1998)
9. H1 Collaboration, C. Adloff et al., *Nucl. Phys.* **B 497** (1997) 3
10. B. Badełek, J. Kwieciński, *Phys. Lett.* **B 295** (1992) 263
11. J.J. Sakurai, D. Schildknecht, *Phys. Lett.* **B 40 No. 1** (1972) 121

RAPIDITY GAPS BETWEEN JETS AT LARGE RAPIDITY SEPARATIONS IN PHOTOPRODUCTION AT HERA

Angela Wyatt

for the H1 Collaboration
Dept. of Physics and Astronomy, University of Manchester, Manchester M13 9PL, England

Abstract. Dijet events in photon-proton collisions in which there is a rapidity separation between the two highest E_t jets $\Delta\eta > 2.5$ are studied with the H1 detector at HERA. Jets are defined using the inclusive k_t algorithm. Rapidity gap events are defined as events in which the sum of the transverse energy of all jets between the two highest E_t jet axes is less than E_t^{cut}, for $0.5 < E_t^{cut} < 5.0$ GeV. The gap fraction, defined as the fraction of dijet events with a rapidity gap, is plotted for different E_t^{cut} values for the inclusive dijet sample, and differentially in the dijet separation, $\Delta\eta$, and in x_γ^{jets} and x_p^{jets} the fraction of the photon's and proton's momentum entering the hard scatter. A large excess of events with rapidity gaps at low values of E_t^{cut} is observed above the expectation from standard photoproduction processes.

I INTRODUCTION

Events with large rapidity gaps in the hadronic final state and a perturbatively hard scale at both ends of the colourless exchange have been observed at both the Tevatron [1,2] and HERA [3]. Such events are characterised by two hard jets separated by a large rapidity interval and an unusually small energy flow between the jets. We tag such events by measuring the energy flow and define a gap event in terms of this energy, introducing a further scale.

The rate of such events depends not only on the cross-section for colour-singlet exchange but also on the gap survival probability: the probability that a gap formed in the hard scatter may later be destroyed by further parton-parton interactions. This interplay can be treated as a perturbative exchange, such as the BFKL calculation, plus an overall gap survival probability or it may be possible to take a purely perturbative approach if the gap is defined in terms of a minimum energy which is greater than Λ_{QCD} [4]. Alternatively, the production process can be treated non-perturbatively, such as in soft color models [5].

CP571, *PHOTON 2000*, edited by A. J. Finch
© 2001 American Institute of Physics 0-7354-0010-5/01/$18.00

II EVENT SELECTION

The data for this analysis were collected with the H1 detector [6] during the 1996 running period, when HERA collided 27.6 GeV positrons with 820 GeV protons, with an integrated luminosity of 6.2 pb^{-1}. Photoproduction events were tagged by measuring the scattered electron, restricting the photon virtuality to $Q^2 < 0.01$ GeV^2. In order to select areas of high tagger acceptance the photon-proton centre of mass energy is restricted to the range $165 < W < 233$ GeV.

The inclusive k_t clustering algorithm [7] was applied to the final state, assigning every particle excluding the scattered electron to a jet. An inclusive event sample was selected by keeping events with 2 or more jets, where the 2 highest E_t jets in the event fulfilled:

$$E_t^{jet1} > 6.0 \text{ GeV} \tag{1}$$

$$E_t^{jet2} > 5.0 \text{ GeV} \tag{2}$$

$$\eta^{jet1}, \eta^{jet2} < 2.8 \tag{3}$$

$$\Delta\eta > 2.5 \tag{4}$$

Any other jets in the event are refered to as mini-jets.

In standard QCD processes the colour connection between the jets results in an energy flow in the entire space between them; this is suppressed in colour-singlet exchange. To distinguish between these two processes we measure the transverse energy between the jets, $E\!\!\!/_t^{jets}$. This is calculated by summing the transverse energy of all mini-jets whose jet-centre is between the axes of the two highest E_t jets in rapidity:

$$E\!\!\!/_t^{jets} = \sum E_t^{minijets} \quad , \quad \eta_{forward}^{jet} > \eta_{minijet} > \eta_{backward}^{jet} \tag{5}$$

A rapidity gap event is then defined as an event in which $E\!\!\!/_t^{jets} < E_t^{cut}$, where E_t^{cut} is varied between 0.5 GeV and 5.0 GeV. This definition is infra-red safe because it is only in terms of the sum of k_t jets. In addition, a gap is also relatively insensitive to hadronisation effects because any radiation at its edge which may destroy the gap is likely to be included in a jet which has its centre outside the gap. Further details can be found in [8].

This definition uses the maximum area of $\eta - \phi$ space which can be included in the rapidity gap search. We use the whole region in order to reduce the possibility that a colour-octet exchange can form a rapidity gap by a statistical fluctuation in the number of particles in this region. This reduction in the colour-octet contribution allows a significant excess of rapidity gap events to be seen without having to increase the cut on $\Delta\eta$. As the $\Delta\eta$ distribution falls rapidly this produces a large increase in the available statistics. Two scaling variables x_γ^{jets} and x_p^{jets} are also defined,

$$x_\gamma^{jets} = \frac{\sum_{jets1,2}\left(E^{jet} - p_z^{jet}\right)}{\sum_{objects}\left(E^{obj} - p_z^{obj}\right)} \tag{6}$$

$$x_p^{jets} = \frac{\sum_{jets1,2} \left(E^{jet} + p_z^{jet} \right)}{2E_p} \qquad (7)$$

where E_p is the energy of the incoming proton and the sum in the denominator of x_γ^{jets} is over all the objects in the final state, excluding the scattered electron. In leading order jet production x_γ^{jets} and x_p^{jets} are equal to the fraction of the photon's and proton's momentum entering the hard scatter.

III RESULTS AND MODEL COMPARISONS

The PYTHIA 5.7 [9] and HERWIG 5.9 [10] Monte Carlo event generators were used to correct the data for detector acceptance and bin migration effects. In order to describe the data both included a colour-singlet exchange model. The resolved photoproduction events in PYTHIA included a model of multiple interactions, but no multiple interactions were included in HERWIG . The difference between the two correction methods was assigned as a systematic error.

PYTHIA was used with $P_t^{min} = 2.2$ GeV and $P_t^{mi} = 1.5$ GeV, P_t^{mi} is the minimum P_t of the multiple scatters. The rate of multiple scatters is set by P_t^{mi} and this is chosen such that the jet pedestals are well described. When making comparisons to the data HERWIG was used with JIMMY [11] which simulates multiple interactions and with $P_t^{min} = 1.8$ GeV and $P_t^{mi} = 1.8$ GeV. These parameters were tuned to give the best description of the jet pedestals and other H1 data.

HERWIG incorporates the BFKL [12] leading logarithmic approximation calculation of Mueller and Tang [13], with the choice of fixing $\alpha_s = 0.17$. This is not yet implemented in PYTHIA , consequently to describe the data high-t photon exchange with the cross section multiplied by a factor of 1200 was added.

The gap fraction, defined as the fraction of dijet events with a rapidity gap, is plotted for different E_t^{cut} values for the inclusive dijet sample in figure 1. The predictions of both PYTHIA and HERWIG with JIMMY underestimate the observed gap fraction. The excess in the data over the models is for low values of E_t^{cut} where colour-singlet exchange events are expected to be. The predictions are very sensitive to the way that the multiple interactions are modeled. In JIMMY the shape of the curve was found to vary greatly with P_t^{mi}. The data at higher E_t^{cut} are shown here as they are useful in fitting a value of P_t^{mi}.

The addition of a colour-singlet exchange provides a better description of the data. The cross-section for the photon-exchange in PYTHIA is scaled to fit this inclusive gap fraction. The normalisation of the BFKL calculation added to HERWIG is set by the choice of fixing $\alpha_s = 0.17$ and provides a reasonable description of the data. This is a significant observation; such a choice was also found to fit the Tevatron gaps between jets data [14]. However, there is a large uncertainty in the prediction of the gap fraction in the standard photoproduction processes, both from the difference in the predictions from HERWIG and PYTHIA and in the choice of P_t^{mi}. This results in an uncertainty in the cross-section of the colour-singlet exchange.

The gap fraction differential in the dijet separation, $\Delta\eta$, is shown in figure 2, for $E_t^{cut} = 0.5, 1.0, 1.5$ and 2.0 GeV. For this plot alone the $\Delta\eta$ cut is relaxed to $\Delta\eta > 2.0$. A significant excess in the gap fraction over the predictions from the Monte-Carlo models without any colour-singlet exchange is seen, particularly at large $\Delta\eta$. Both colour-singlet models are able to describe this.

The gap fraction differential in x_γ^{jets} is shown in figure 3, for $E_t^{cut} = 0.5, 1.0, 1.5$ and 2.0 GeV. The colour singlet events are distributed across the whole x_γ^{jets} range, while the background photoproduction sample is strongly peaked at high x_γ^{jets}. This peak is because in direct photoproduction the t-channel exchange is a quark which has a far lower probability to radiate into the gap than an exchanged gluon which dominates the resolved contribution. Both colour-singlet models provide a good description of the data.

The gap fraction differential in x_p^{jets} is shown in figure 4, for $E_t^{cut} = 0.5, 1.0, 1.5$ and 2.0 GeV. The high-t photon gap fraction rises with increasing x_p^{jets}, whilst the BFKL sample is flat or falling. This can be understood since a t-channel photon can only couple to quarks in the colliding hadrons, whilst the BFKL pomeron couples more strongly to gluons than to quarks due to the enhanced colour factor. This distribution may then be sensitive to the underlying dynamics of the exchange, but more statistics are needed to draw firm conclusions from the data.

REFERENCES

1. S. Abachi et al (D0 collaboration), Phys. Rev. Lett. 72 (1994) 2332; Phys. Rev. Lett. 76 (1996) 734; B. Abbott et al (D0 collaboration), Phys. Lett. B440 (1998) 189.
2. F. Abe et al (CDF collaboration), Phys. Rev. Lett. 74 (1995) 855; Phys. Rev. Lett 80 (1998) 1156; Phys. Rev. Lett. 81 (1998) 5278.
3. M. Derrick et al (ZEUS collaboration), Phys. Lett. B369 (1996) 55.
4. G. Oderda and G. Sterman, Phys. Rev. Lett. 81, 3591 (1998)
5. O. J. Eboli, E. M. Gregores and F. Halzen, Phys. Rev. **D58** (1998) 114005; hep-ph/9908374.
6. H1 Collab.; I. Abt et al., Nucl. Instrum. Meth. **A386** (1997) 310; 348.
7. S. D. Ellis and D. E. Soper, Phys. Rev. **D48** (1993) 3160.
8. B. E. Cox, M. H. Seymour and A. C. Wyatt, in preparation.
9. T. Sjöstrand and M. van Zijl, Phys. Rev. **D36** (1987) 2019.
10. G.Marchesini et al., Comp.Phys.Comm. 67 (1992) 465.
11. J. M. Butterworth, J. R. Forshaw and M. H. Seymour, Z. Phys. **C72** (1996) 637-646.
12. E.A.Kuraev, L.N.Lipatov and V.S.Fadin, Sov.Phys.JETP 45 (1977) 199.
   Ya.Ya.Balitsky and L.N.Lipatov, Sov.J.Nucl.Phys 28 (1978) 822.
   L.N.Lipatov, Sov.Phys.JETP 63 (1986) 904.
13. A.H.Mueller and W.-K.Tang, Phys. Lett. B284, (1992) 123.
14. B.E.Cox, J.R. Forshaw, L.Lönnblad, JHEP10 (1999) 023.

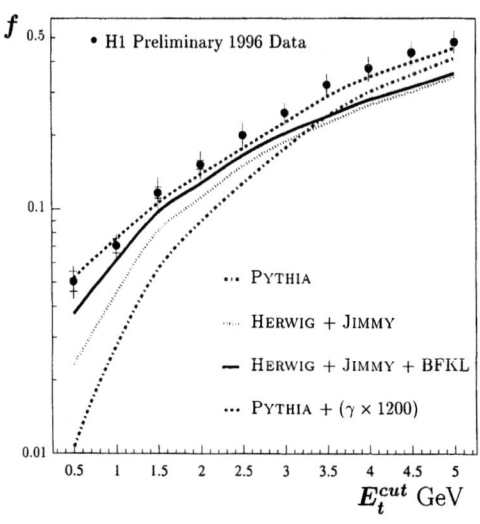

FIGURE 1. The gap fraction for the inclusive sample, compared to different Monte Carlo models (see text).

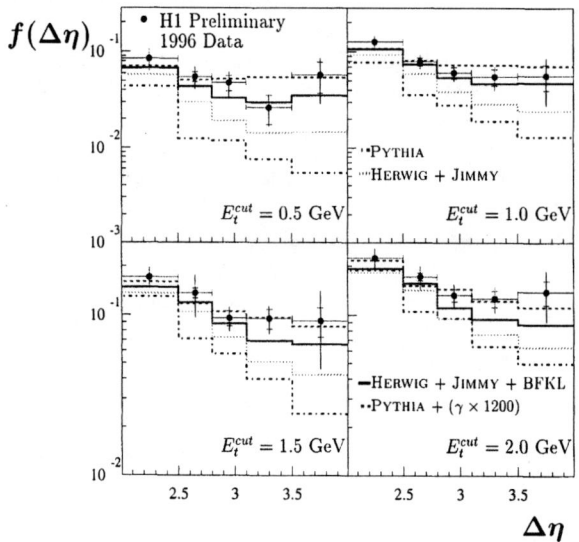

FIGURE 2. The gap fraction differential in $\Delta\eta$, for $E_t^{cut} = 0.5, 1.0, 1.5, 2.0$ GeV , compared to different Monte Carlo models (see text).

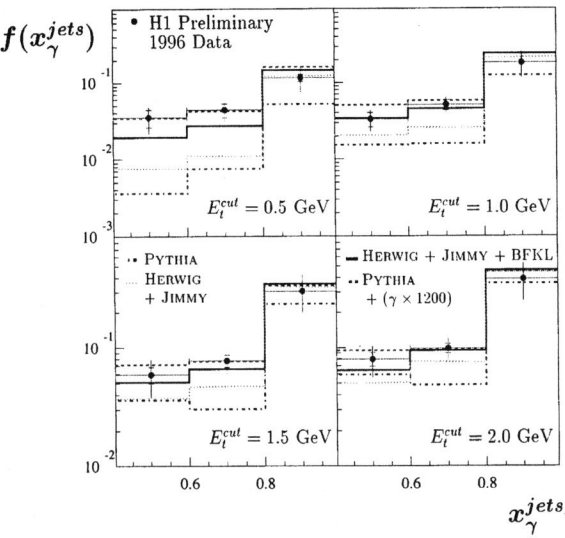

FIGURE 3. The gap fraction differential in x_γ^{jets} for $E_t^{cut} = 0.5, 1.0, 1.5, 2.0$ GeV , compared to different Monte Carlo models (see text).

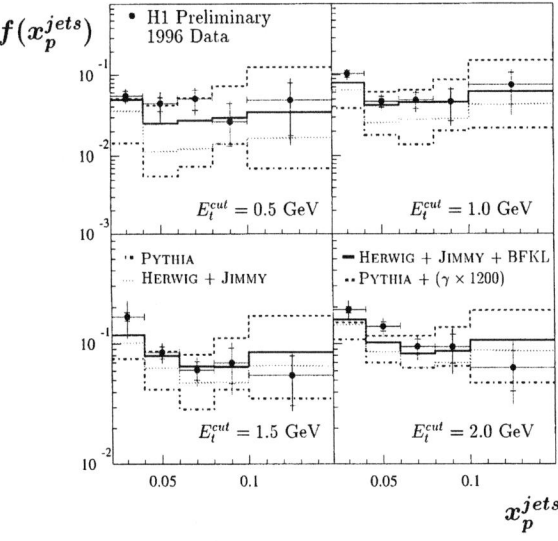

FIGURE 4. The gap fraction differential in x_p^{jets}, for $E_t^{cut} = 0.5, 1.0, 1.5, 2.0$ GeV , compared to different Monte Carlo models (see text).

Inclusive Diffraction at HERA

Carrie L. Johnson

Department of Physics and Astronomy, University of Birmingham,
Edgbaston, Birmingham, B15 2TT, U.K.

Abstract. Results from the HERA collaborations H1 and ZEUS on the diffractive structure function are presented. Intermediate Q^2 data are successfully described by a two-component Regge-motivated model, and QCD analysis reveals the pomeron to have a large gluonic content. Recent low Q^2 data confirm the rise of the effective pomeron intercept with Q^2, which is unexpected for a simple Regge pole. The qualitative features of the data are successfully described by the "saturation" model.

INTRODUCTION

The diffraction of virtual photons in e^+p collisions has been a subject of great interest. The H1 and ZEUS collaborations study the diffractive process $ep \rightarrow eXY$, as shown in figure 1, which is mediated by the exchange of a net colour singlet state with vacuum quantum numbers, the pomeron ($\mathbb{P}$). X and Y are hadronic systems produced at the $\gamma^*\mathbb{P}$ and $p\mathbb{P}$ vertices respectively separated by a large gap in rapidity, a characteristic signature of diffractive deep-inelastic scattering (DIS) interactions. Taking the 4-vectors of the incoming(outgoing) lepton, proton and virtual photon to be $e(e')$, $p(p')$ and q respectively, the DIS kinematic variables are,

$$s = (e + p)^2 \qquad W^2 = (q + p)^2 \qquad Q^2 = -q^2 = (e - e')^2, \qquad (1)$$

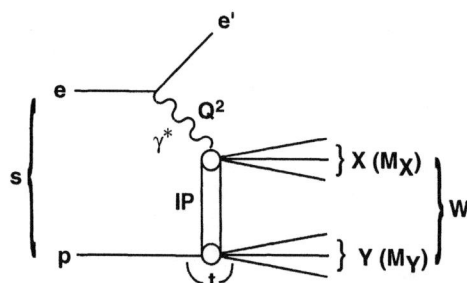

FIGURE 1. Kinematics of inclusive diffractive DIS at HERA.

CP571, *PHOTON 2000*, edited by A. J. Finch
© 2001 American Institute of Physics 0-7354-0010-5/01/$18.00

where s is the ep centre of mass energy squared, W^2 the $\gamma^* p$ centre of mass energy squared and Q^2 the photon virtuality. The scaling variables are $x = (-q^2)/(2p \cdot q)$ and $y = (p \cdot q)/(p \cdot e)$. The diffractive variables are,

$$t = (p - p')^2 \qquad x_{\mathbb{P}} = \frac{q \cdot (p - p')}{q \cdot p} \qquad \beta = \frac{-q^2}{2q \cdot (p - p')}, \tag{2}$$

where t is the 4-momentum transfer squared at the proton vertex and $x_{\mathbb{P}}$ and β may be interpreted as the fraction of the proton's 4-momentum transferred to the pomeron and the fraction of the 4-momentum of the pomeron carried by the struck quark respectively. Note that $x = x_{\mathbb{P}} \beta$.

The process has previously been well measured at medium [1] [2] and high [3] Q^2. New data have been released by ZEUS [4] in the transition region between photoproduction and DIS, which previously was neither well measured nor understood.

Soft hadron-hadron and photoproduction experiments showed diffraction to be universally well described in terms of a pomeron trajectory [5],

$$\alpha_{\mathbb{P}}(t) = \alpha_{\mathbb{P}}(0) + \alpha'_{\mathbb{P}} t, \tag{3}$$

where $\alpha_{\mathbb{P}}(0) = 1.08$ and $\alpha'_{\mathbb{P}} = 0.25$ GeV^{-2}. In comparison DIS is well described by the hard processes of QCD. There therefore exists a region of transition where the interplay of soft and hard processes can be studied.

INTERMEDIATE Q^2 MEASUREMENTS

The diffractive structure function $F_2^{D(3)}$ is defined as,

$$\frac{d^3\sigma(ep \to epX)}{dx d\beta dQ^2} = \frac{4\pi\alpha_{em}^2}{\beta^2 Q^4}\left(1 - y + \frac{y^2}{2}\right) F_2^{D(3)}(\beta, Q^2, x_{\mathbb{P}}), \tag{4}$$

where the longitudinal photon contribution F_L^D is neglected. A measurement of the inclusive diffractive proton structure function was extracted from H1 1994 data [1] via detection of large rapidity gap events. The kinematic region covered was $4.5 < Q^2 < 75$ GeV^2, $0.0002 < x_{\mathbb{P}} < 0.04$ and $0.04 < \beta < 0.9$. The measurement was also constrained to the region $M_Y < 1.6$ GeV and $|t| < 1$ GeV^2. The measured $x_{\mathbb{P}}$ dependence of $x_{\mathbb{P}} F_2^{D(3)}$ is shown in bins of β and Q^2 in figure 2.

The data show an overall trend for $x_{\mathbb{P}} F_2^{D(3)}$ to fall with increasing $x_{\mathbb{P}}$. However at low values of β, the data rise at the highest values of $x_{\mathbb{P}}$. These features of the data are successfully described by a two component Regge model comprising a leading pomeron and a subleading meson component. Assuming Regge factorisation, the diffractive structure function can be represented as,

$$F_2^{D(3)}(x_{\mathbb{P}}, \beta, Q^2) = f_{\mathbb{P}/p}(x_{\mathbb{P}}) F_2^{\mathbb{P}}(\beta, Q^2) + f_{\mathbb{R}/p}(x_{\mathbb{P}}) F_2^{\mathbb{R}}(\beta, Q^2), \tag{5}$$

where,

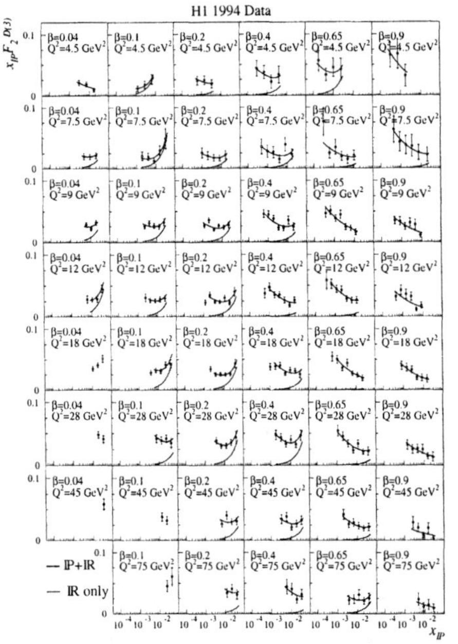

FIGURE 2. Measurement of the diffractive structure function, $x_{I\!P} F_2^{D(3)}$ as a function of $x_{I\!P}$.

$$f_{I\!P/p} = \int_{t_{cut}}^{t_{min}} \frac{e^{B_{I\!P} t}}{x_{I\!P}^{2\alpha_{I\!P}(t)-1}} dt \qquad f_{I\!R/p} = \int_{t_{cut}}^{t_{min}} \frac{e^{B_{I\!R} t}}{x_{I\!P}^{2\alpha_{I\!R}(t)-1}} dt, \qquad (6)$$

where $B_{I\!P}$ and $B_{I\!R}$ are slope parameters. Regge fits to the data yield values of,

$$\alpha_{I\!P}(0) = 1.203 \pm 0.020 \text{ (stat.)} \pm 0.013 \text{ (syst.)} \, ^{+0.030}_{-0.035} \text{ (model)}, \qquad (7)$$
$$\alpha_{I\!R}(0) = 0.50 \pm 0.11 \text{ (stat.)} \pm 0.11 \text{ (syst.)} \, ^{+0.09}_{-0.10} \text{ (model)}. \qquad (8)$$

Hence although the value of $\alpha_{I\!R}(0)$ agrees with that from total hadron cross section analyses (for f_2 exchange for example), the measured value of $\alpha_{I\!P}(0)$ is larger than that of the soft pomeron, indicating that the physics processes differ in the photoproduction and DIS regimes.

Scaling Violations

The Q^2 dependence of $F_2^{D(3)}$ as measured from 1994-1995 data [3] is shown in figure 3(a). The low value of $x_{I\!P}(= 0.005)$ at which the study is performed implies that subleading contributions are negligible, hence the dependence on Q^2 of the pomeron structure function is directly visible. Figure 3(a) shows $x_{I\!P} F_2^{D(3)}$ as a function of Q^2 and positive scaling violations are visible, persisting to high β values, implying a large gluonic component in the pomeron.

Assuming Regge factorisation, the structure functions $F_2^{\mathbb{P}}(\beta, Q^2)$ and $F_2^{\mathbb{R}}(\beta, Q^2)$ describe the structure of the pomeron and meson exchanges respectively. Assuming a parameterisation of the parton distributions of the pomeron and meson exchanges at a starting scale Q_0^2, fits have been performed to the data assuming DGLAP evolution.

Parton Distributions

Figure 3(b) shows the parton distributions which result from a QCD fit involving both light quarks and gluons with $Q_0^2 = 3.0$ GeV^2, where $f(z)$ is the parton distribution function and z is the fraction of the momentum of the pomeron carried by the parton [1]. The momentum fraction carried by gluons decreases with increasing Q^2, from approximately 90% at $Q^2 = 4.5$ GeV^2 to about 80% at $Q^2 = 75$ GeV^2. Hence the data favour a large gluonic component in the pomeron and boson-gluon fusion to be the dominant mechanism in diffractive DIS.

LOW Q^2 $F_2^{D(3)}$

Recently ZEUS measured $F_2^{D(3)}$ (1997 data) [4] using their Beampipe Calorimeter in the region $0.22 \lesssim Q^2 \lesssim 0.7$ GeV^2, a previously unexplored region of phase space. Diffractive events were selected by two different methods, the "ln M_X^2" method,

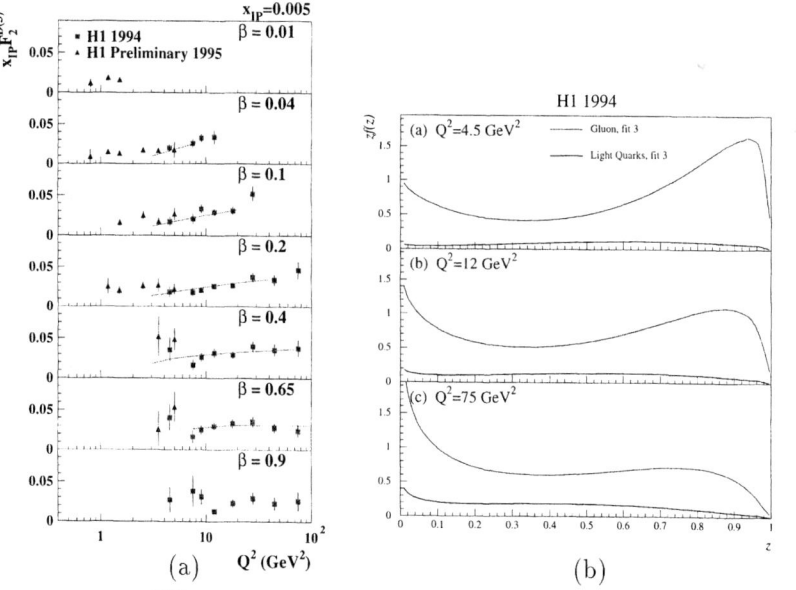

FIGURE 3. (a) $x_{\mathbb{P}} F_2^{D(3)}$ as a function of Q^2 showing the rising scaling violations and (b) Pomeron parton distributions in bins of Q^2.

where the uncorrected $\ln M_X^2$ distribution is separated into diffractive and non-diffractive parts, and the "LPS" method, in which diffractive events are identified via the diffractive peak near unity in the x_L distribution where $x_L = E_p/E_{p'}$ and E_p and $E_{p'}$ are the energies of the proton before and after scattering respectively.

For the former sample the diffractive differential cross section was measured as a function of W in bins of Q^2 and M_X. The cross section is parameterised at fixed M_X and Q^2 as,

$$\frac{d\sigma^{diff}}{dM_X} \propto (W^2)^{2(\bar{\alpha}_{I\!P}-1)}, \tag{9}$$

in Regge theory where,

$$\bar{\alpha}_{I\!P} = \alpha_{I\!P}(0) - \alpha'_{I\!P} \cdot |\bar{t}|. \tag{10}$$

A fit of the form of equation 9 was performed, yielding the value of $\bar{\alpha}_{I\!P}$ to be,

$$\bar{\alpha}_{I\!P} = 1.126 \pm 0.012 \text{ (stat.)} \,^{+0.027}_{-0.032} \text{ (syst.)}. \tag{11}$$

Assuming $\alpha'_{I\!P} = 0.25$ GeV^{-2} and using $|\bar{t}| = 0.13$ GeV^{-2}, $\alpha_{I\!P}(0)$ was extracted via equation 10. The result is shown in figure 4 where the effective $\alpha_{I\!P}(0)$ is shown as a function of Q^2, with other results from diffractive and total cross section measurements. There is evidence for a rise of $\alpha_{I\!P}(0)$ with Q^2 in the diffractive case, unexpected from simple Regge pole theory. The diffractive and inclusive results are observed to be compatible at low Q^2 but inconsistent at higher values.

The differential cross section $d\sigma/dM_X$ has been measured as a function of Q^2 at fixed M_X and W. A turnover is observed near $Q^2 = 1$ GeV^2. The relation between the differential cross section and the diffractive structure function is given by,

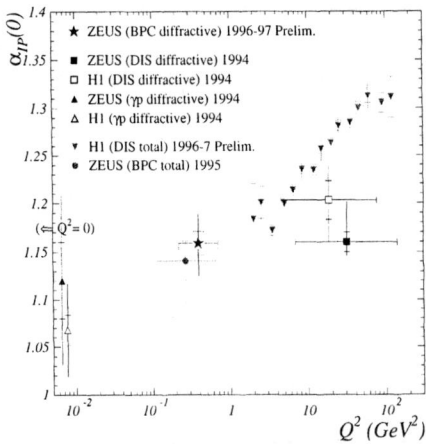

FIGURE 4. Q^2 dependence of the effective $\alpha_{I\!P}(0)$ for inclusive and diffractive ep processes.

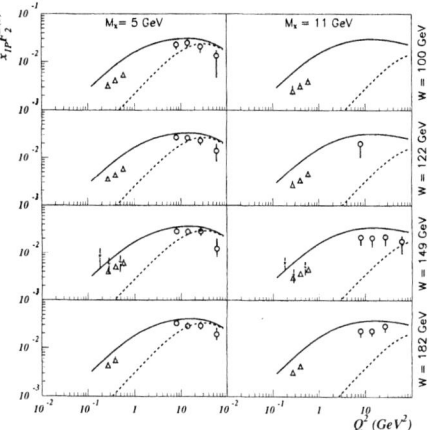

FIGURE 5. Q^2 dependence of $F_2^{D(3)}$ showing the transition at low Q^2. The triangles are from the $\ln M_X^2$ analysis, the stars are LPS data and the circles are from a higher Q^2 analysis [2]. The dashed and solid curves show the saturation model predictions with $q\bar{q}$ and $q\bar{q}g$ terms respectively.

$$\frac{d\sigma^{diff}}{dM_X} \sim 4\pi^2\alpha\frac{2M_X}{Q^2+M_X^2}\frac{x_{I\!P}F_2^{D(3)}}{Q^2},\qquad(12)$$

so at low values of Q^2 and fixed M_X, $x_{I\!P}F_2^{D(3)}$ is proportional to Q^2. Figure 5 shows $x_{I\!P}F_2^{D(3)}$ plotted against Q^2 in bins of M_X and W. The predictions in figure 5 are from the "saturation" model [6] [7] which assumes the virtual photon fluctuates into $q\bar{q}$ and $q\bar{q}g$ states before an interaction with the proton via a two gluon exchange. The dipole cross section for this interaction increases as Q^2 decreases and "saturates" at low Q^2. Figure 5 shows that the qualitative features (although not the normalisation) of the low Q^2 data are well reproduced. The "saturation" model is however successful at describing data at higher values of Q^2.

REFERENCES

1. H1 Collaboration, *Z. Phys.* **C76**, 613-629 (1997)
2. ZEUS Collaboration, *Eur. Phys. J.* **C6**, 43-66 (1999)
3. H1 Collaboration, "Measurement of the diffractive structure function $F_2^{D(3)}$ at low and high Q^2 at HERA", Abstract Number 571 Submitted to ICHEP98
4. ZEUS Collaboration, "Measurement of the difractive cross section at $Q^2 < 1$ GeV^2 at HERA", Abstract Number 875 Submitted to ICHEP2000
5. Donnachie, A. and Landshoff, P., *Nucl. Phys.* **B231**, 189-204 (1984)
6. Golec-Biernat, K. and Wusthoff, M., *Phys. Rev.* **D59**, 014017 (1999)
7. Golec-Biernat, K. and Wusthoff, M., *Phys. Rev.* **D60**, 114023 (1999)

Jet Production in Diffractive *ep* scattering

Benno List*

representing the H1 and ZEUS Collaborations

*Institute for Particle Physics IPP[1]
ETH Zürich, CH–8093 Zürich, Switzerland

Abstract. Jet production in deep inelastic scattering events with a large rapidity gap is a powerful way to test our understanding of the nature of diffraction. The ZEUS and H1 collaborations have measured cross sections for two and three jet production in diffractive deep inelastic scattering and compared them to predictions of various theoretical models. No model is able to describe all aspects of the data. The resolved Pomeron model still provides the best description of the experimantal results, but two gluon exchange models start to become competitive.

INTRODUCTION

The electron–proton collider HERA provides a unique opportunity to study deep inelastic *ep* collisions at center–of–mass energies up to $\sqrt{s} = 320\,\text{GeV}$ and to observe the hadronic final states in these interactions. Diffractive deep–inelastic scattering (DIS) events were first observed in 1993 [1]. These events are distinguished from "ordinary" DIS events by the absence of hadronic activity around the direction of the outgoing proton remnant (called "forward direction" at HERA), so that a large interval in rapidity between the unobserved proton remnant and the current jet is void of hadrons. This interval is termed the "rapidity gap". Subsequent studies have shown that in these large rapidity gap events the proton remnant consists mostly of a single proton [2], or a low–mass excitation of the nucleon.

MODELS AND MONTE CARLO GENERATORS

The closer investigation of this class of events started with the measurement of the semi–inclusive structure function $F_2^{D(3)}$, which parametrizes the *ep*–cross section for events with a large rapidity gap [3,4]. Today a QCD factorization theorem exists [5] which proves that this is indeed a valid approach.

---

[1] Supported by the Swiss National Science Foundation.

The Resolved Pomeron Model: Rapidity gap events can be interpreted as the scattering of the electron off a colorless object that is emitted by the proton and behaves rather like a hadron. In the framework of Regge theory [6], this object is identified with the pomeron, as was predicted already in 1985 by Ingelman and Schlein [7]. Their so-called resolved Pomeron model, which describes diffractive events as the emission of a Pomeron[2], whose interaction with the electron in DIS events the eIP cross section is parametrized by a structure function, is today still the model that best describes the properties of large rapidity gap events. It has been implemented in the Monte Carlo generator RAPGAP [8], which additionally incorporates also the models of Bartels et al., the saturation model, and the semiclassical model of Buchmüller et al. that are described below.

Two–Gluon and Color–Dipole Models: Attempts to describe diffraction in terms of QCD started with a proposal from Low and Nussinov [9] that Pomeron exchange may be modeled as the exchange of two gluons.

Different calculations exist for the process $ep \rightarrow e'p'X$, where X is a hadronic system separated in rapidity from the final state proton p', in the framework of two–gluon models. Bartels, Jung, Lotter, and Wüsthoff [10] calculate the cross section for processes where the photon splits into a $q\bar{q}$ pair or a $q\bar{q}g$ state before the interaction with the proton, using the unintegrated gluon distribution in the proton[3]. To avoid divergencies, a lower cutoff must be applied on the gluon p_t in $q\bar{q}g$ states, which has to be tuned to the data.

Ryskin [11] also considers contributions from $q\bar{q}$ and $q\bar{q}g$ states, but only in the leading log approximation, taking only k_t ordered contributions into account. This model is implemented in the Monte Carlo generator RIDI [12].

The saturation model from Golec–Biernat and Wüsthoff [13] models the gluon k_t distribution down to 0, and can therefore be applied to diffractive and non–diffractive DIS and photoproduction with few free parameters, which are all fixed by non–diffractive DIS.

Semiclassical Model: The semiclassical model of Buchmüller, Gehrmann, and Hebecker [14] treats the interaction of the $q\bar{q}(g)$ state with the proton as an interaction with a color field, similar to the saturation model. This leads to a prediction for the gluon distribution the proton and hence to a prediction of the rate of diffractive events.

Soft Color Interactions: Somewhat similar to the semiclassical model, the model of soft color interactions by Edin, Ingelman, and Rathsman [15] links diffractive and non–diffractive topologies in ep scattering. It is assumed that $q\bar{q}$ states originating from photon–gluon fusion can be transformed from a color octett to a singlet state by the exchange of soft gluons between the $q\bar{q}$ system and the proton remnant. This model, which was recently further developed [16], is implemented in the Monte Carlo generator LEPTO [17].

---

[2] In certain kinematic regions additional Regge trajectories, especially four degenerate trajectories termed collectively "Reggeon," must also be included in this picture.

[3] Compared to the "normal" gluon distribution, which depends on Bjorken x and Q^2, the unintegrated gluon distribution additionally depends on the k_t of the gluon.

H1 MEASUREMENTS OF DIFFRACTIVE DI– AND THREE-JET PRODUCTION

The H1 Collaboration has used data taken in 1996–97, when HERA collided 820 GeV protons with 27.5 GeV positrons, to measure cross sections for diffractive production of two or three jets [18], extending a previous measurement [19]. The data correspond to an integrated luminosity of $18\,\mathrm{pb}^{-1}$. They use a cone algorithm in the γp center–of–mass system to identify jets with a transverse momentum p_T greater than 4 GeV. The kinematic range covered by the analysis is $4 < Q^2 < 80\,\mathrm{GeV}^2$ and $x_P < 0.05$. Diffractive events are identified by requiring a rapidity gap visible in the detector and no activity in the forward direction. This restricts the mass of the proton remnant to values of $M_Y < 1.6\,\mathrm{GeV}$.

Comparison to the Resolved Pomeron Model

The left side of Fig. 1 shows a comparison of the measured spectrum of z_P with predictions from the resolved Pomeron model. The quantity z_P is defined as

$$z_P = \frac{Q^2 + M_{12}^2}{Q^2 + M_X^2},$$

where M_{12} and M_X are the mass of the dijet system and the diffractive hadronic system (*i. e.* the γIP system), respectively. It can be interpreted as the fractional

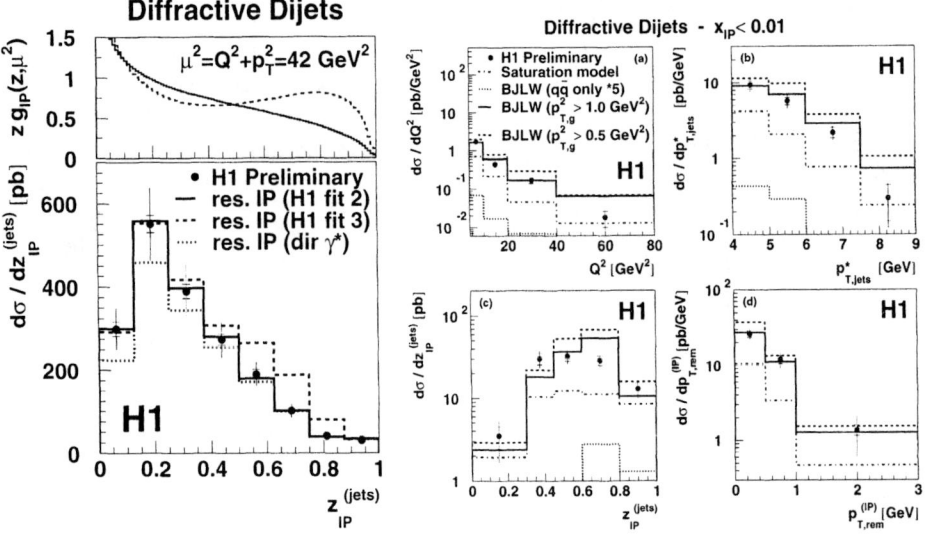

FIGURE 1. H1 dijets events. Left: Comparison of the z_P distribution to the resolved Pomeron model for various gluon densities in the Pomeron. Right: Comparison of the H1 dijet data to the two–gluon exchange model of Bartels et al. and the saturation model.

momentum of the parton, particularly the gluon, in the Pomeron that enters the hard interaction with the photon. The measurement is compared to predictions for two gluon density parametrisations of the Pomeron, derived in Ref. [4]: "H1 fit 2" is a gluon density which is flat at the starting scale Q_0^2, whereas "H1 fit 3" is strongly peaked at $z_P = 1$ for Q_0^2 (at Q_0^2, the H1 fit 3 gluon looks like a single gluon, which is accompanied or "dressed" by a very soft gluon). Both these gluon distributions fit the inclusive diffractive data from H1 equally well, whereas the dijet data are directly sensitive to the gluon distribution and prefer the "fit 2" gluon density, as seen in Fig. 1.

Comparison to Two–Gluon Models

In order to make a comparison of the H1 dijet data with the two–gluon model of Bartels et al. (BJLW) and the saturation model, an additional cut of $x_P < 0.01$ was applied to the data, to restrict the comparison to a kinematic range which is expected to be dominated by gluon–exchange. The right side of Fig. 1 shows a comparison of the data to these two models. For the BJLW model, curves are shown for two different values (0.5 GeV^2 and 1.0 GeV^2 of the cutoff for the gluon p_T^2, plus a curve where only $q\bar{q}$ states are taken into account in the calculation. Shown are the distributions of Q^2, the transverse momentum $p_{T,\text{jets}}^*$ of the jets, of z_P, and the quantity $p_{T,\text{rem}}^{(P)}$, which is the total transverse momentum of all particles in the Pomeron hemisphere that do not belong to one of the jets.

As can be seen, the BJLW model describes the data reasonably well in shape and normalization for a cutoff value of the gluon $p_t^2 > 1.0\,\text{GeV}^2$, although the $z_P^{(\text{jets})}$ distribution shows some deviations from the data. Clearly the consideration of $q\bar{q}$ states alone leads to jet rates more than an order of magnitude too low. The saturation model describes the shapes of the distributions comparably well; however, with the present set of parameters the normalization is low by a factor of about 2.

The ability of these two models to describe the shape and partially also the normalization of the data can be regarded as a major success of these calculations.

Comparison to the Soft Color Interaction and Semiclassical Models

Fig. 2 shows a comparison of the Soft Color Interaction (SCI) model (in its original form and with the generalized area law) and the Semiclassical model to the data. Plotted are the distributions of $p_{T,\text{jets}}^*$, M_X, $\log_{10} x_P$, and z_{pom}. The Semiclassical and the original SCI model give very similar predictions, which are both too low by about a factor of 2 in normalization; the generalized area law version of the SCI model predicts approximately the correct rate of two–jet events, but fails to correctly describe any of the kinematic distributions.

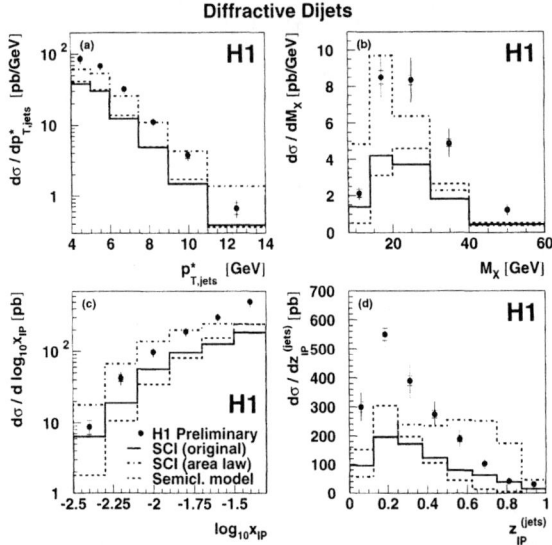

FIGURE 2. H1 dijet data compared to the soft color interaction and the semiclassical Model.

Three–Jet Events

Apart from about 2500 dijet events, also 130 three–jet events were found in the H1 data sample. The invariant mass M_{123} of the three jets and $z_{\mathbf{P}}$ distributions for these events are shown in Fig. 3, in comparison with the resolved Pomeron and the BJLW models. The resolved Pomeron model predicts about a factor of 2 too few events, regardless of whether parton showers (MEPS) or color dipole radiation (CDM) are used to approximate higher order contributions to jet production. The BJLW model is about a factor of 10 too low.

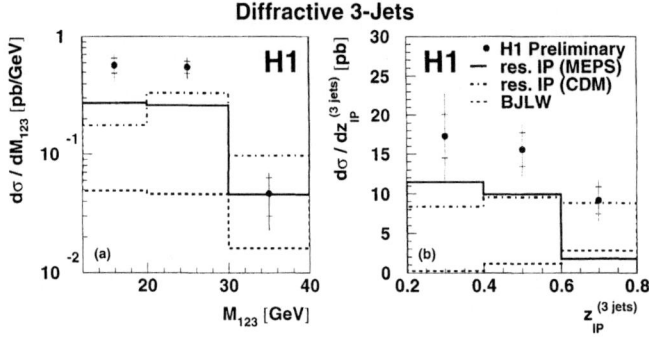

FIGURE 3. H1 three–jet event data, compared to the predictions from the resolved Pomeron model and the two–gluon model by Bartels et al.

ZEUS MEASUREMENT OF DIFFRACTIVE THREE–JET PRODUCTION

The ZEUS Collaboration has analysed [20] data taken in 1998–2000, when HERA collided 920 GeV protons with 27.5 GeV electrons or positrons, extending an earlier measurement [21]. The data correspond to an integrated luminosity of about 39 pb^{-1}. Diffractive events were identified by requiring a rapidity gap in the calorimeter and no activity in the forward direction, especially the newly installed forward plug calorimeter [22]. The kinematic range covered is given by $5 < Q^2 < 100$ GeV, $200 < W < 250$ GeV (W is the γp invariant mass), $23 < M_X < 40$ GeV, and $x_P < 0.025$. Jets are identified in the center–of–mass system of the observed hadronic final state using the Durham (k_t) algorithm [23]. Studying the rates of two versus three jet events gives an insight into the relative importance of $q\bar{q}$ and $q\bar{q}g$ parton configurations; additionally, the jet topologies are sensitve to the p_t of the partons in the pomeron and therefore allow to test various models in a new domain.

Fig. 4 shows the dependency of the reconstructed number of jets on the jet resolution parameter y_{cut}, for data and three Monte Carlo models: the resolved Pomeron model (RAPGAP), the saturation model (SATRAP), and the two–gluon model by Ryskin (RIDI). A resolution parameter of 0.05 gives a good jet–parton correlation, with about 20 % three–jet events, and was therefore chosen for the analysis. A total number of 678 three-jet events were such found.

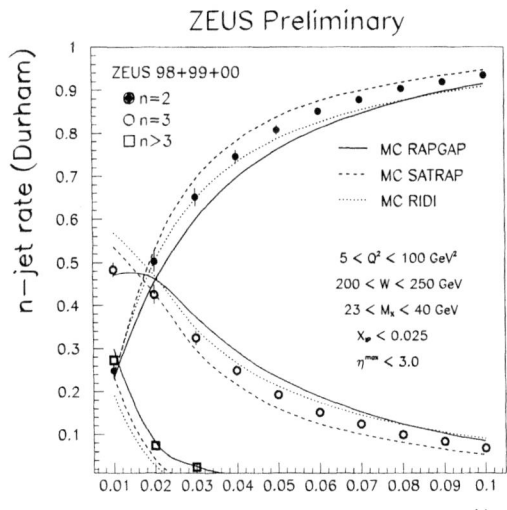

FIGURE 4. ZEUS measurement of jet rates as function of the jet resolution parameter y_{cut}, compared to the predictions from the resolved Pomeron model (RAPGAP), the saturation model (SATRAP), and the two–gluon model by Ryskin (RIDI).

Three–Jet Final–State Topology

The topology of a three–jet configuration can be described by introducing the variables $x_i = 2\,E_i/M_X$, where E_i ($i = 1, 2, 3$) are the jet energies, ordered such that $E_1 > E_2 > E_3$. Two of these three variables are sufficient to describe the jet topology. Following studies of three–jet production in e^+e^- collisions at PETRA [24], the quantities x_1 versus $(x_2 - x_3)/x_1$ were plotted in Fig. 5. As can be seen, all possible three-jet topologies are present in the data. The right side of Fig. 5 shows the energy flow in the three–jet plane in two regions of the $(x_1, (x_2 - x_3)/x_1)$ plane, namely $(x_2 - x_3)/x_1 > 0.4$ (upper right) and $x_1 < 0.8, (x_2 - x_3)/x_1 < 0.4$ (lower right). The resolved Pomeron model (RAPGAP) and the saturation model (SATRAP) give a good description of the observed energy flow, while the RIDI model, although somewhat too peaked, is still reasonable.

Jet Rapidity Distribution

The ZEUS Collaboration has also measured the differential jet cross section $\mathrm{d}\sigma/\mathrm{d}\eta^{\mathrm{jet}}$ as function of the pseudorapidity η^{jet} of the jets in the hadronic center–of–mass system (the direction of η^{jet} is defined such that $\eta^{\mathrm{jet}} < 0$ corresponds to the

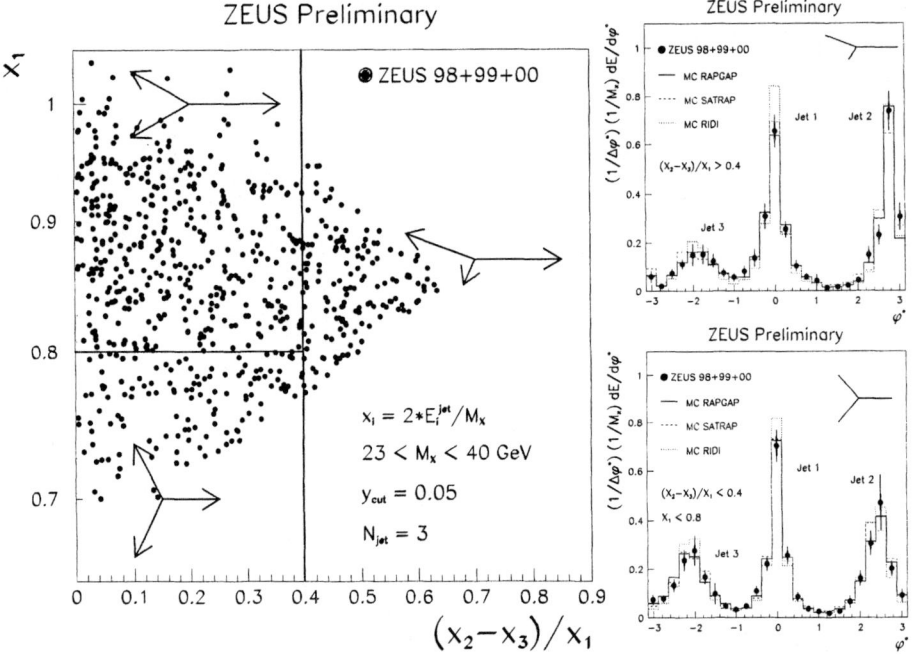

FIGURE 5. Three jet topologies as observed by the ZEUS collaboration and energy flow in the three–jet plane for two different in the $(x_1, (x_2 - x_3)/x_1)$ plane.

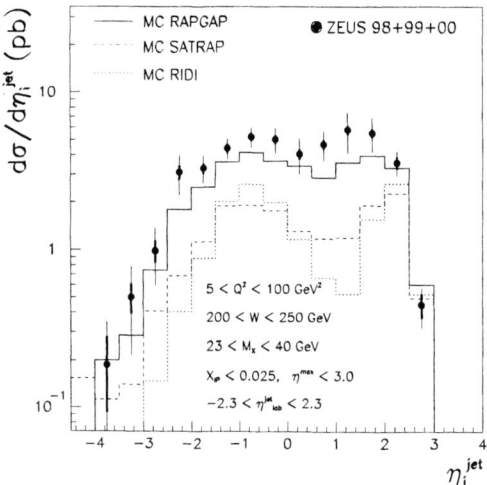

FIGURE 6. The differential jet cross section $d\sigma/d\eta^{\text{jet}}$ as a function of the pseudorapidity η^{jet} of the jets in the hadronic center–of–mass system, measured by ZEUS, in comparison to the Monte Carlo models.

direction of the virtual photon). The result is shown in Fig. 6. Only the resolved Pomeron model (RAPGAP) is able to describe the observed data in shape and normalization; it is still low by a factor of about 1.5, which may be attributable to the remaining proton dissociation events in the data, which are not modeled in the Monte Carlo sample.

By comparison, both the saturation model (SATRAP) and the Ryskin model (RIDI) lie about a factor of 3 below the data. Moreover, the RIDI model shows a marked dip around $\eta^{\text{jet}} = 1$. It seems that in the RIDI model the Pomeron remnant is too much aligned with the Pomeron direction, which may be an indication that the gluons that enter the hard interaction carry too little p_t in this model.

SUMMARY AND CONCLUSIONS

Both the H1 and ZEUS Collaborations have measured jet production cross sections in diffractive deep inelastic scattering. The data demonstrate clearly that the Pomeron is a gluon–dominated object, and are sensitive to the gluon distribution in the Pomeron. No single model is able to perfectly describe all aspects of the data. The resolved Pomeron model is still "the model to beat," as it provides overall the most satisfactory description of the data. However, models based on perturbative QCD that rest on firmer ground than the resolved pomeron model are becoming more successful. This may open the door to a solid theoretical understanding of diffraction in terms of QCD.

REFERENCES

1. ZEUS Collab., M. Derrick et al., *Phys. Lett.* **B315**, 481 (1993).
   H1 Collab., T. Ahmed et al., *Nucl. Phys.* **B429**, 477 (1994).
2. ZEUS Collab., J. Breitweg et al., *Eur. Phys. J.* **C1**, 81 (1998).
3. ZEUS Collab., M. Derrick et al., *Z. Phys.* **C68**, 569 (1995), *ibid.* **C70**, 391 (1996),
   Eur. Phys. J. **C6**, 43 (1999).
   H1 Collab., T. Ahmed et al., *Phys. Lett.* **B348**, 681 (1995).
4. H1 Collab., C. Adloff et al., *Z. Phys.* **C76**, 613 (1997).
5. J. C. Collins, *Phys. Rev.* **D57**, 3051 (1998).
6. T. Regge, *Nuov. Cim.* **14**, 951 (1959), *ibid.* **18**, 947 (1960).
   P. D. B. Collins, *Introduction to Regge theory and high energy physics*, Cambridge:
   Cambridge Univ. Press 1977.
7. G. Ingelman, and P. E. Schlein, *Phys. Lett.* **152B**, 256 (1985).
8. H. Jung, *Comp. Phys. Commun.*. **86**, 147 (1995).
9. F. E. Low, *Phys. Rev.* **D12**, 163 (1975).
   S. Nussinov, *Phys. Rev. Lett.* **34**, 1286 (1975).
10. J. Bartels, H. Lotter, and M. Wüsthoff, *Phys. Lett.* **B379**, 239 (1996), erratum *ibid.*
    B382, 449 (1996).
    J. Bartels, H. Jung, and M. Wüsthoff, *Eur. Phys. J.* **C11**, 111 (1999).
11. M. G. Ryskin, *Sov. J. Nucl. Phys.* **52**, 529 (1990).
12. M. G. Ryskin, and A. Solano, in: A. T. Doyle et al. (eds.), *Monte Carlo generators
    for HERA physics*, Hamburg: DESY (DESY–PROC–1999-02), 1999, p. 386.
13. K. Golec–Biernat, and M. Wüsthoff, *Phys. Rev.* **D59**, 014017 (1998),
    ibid. **D60**, 114023 (1999).
14. W. Buchmüller, T. Gehrmann, and A. Hebecker, *Nucl. Phys.* **B537**, 477 (1999).
15. A. Edin, G. Ingelman, and J. Rathsman, *Phys. Lett.* **B366**, 371 (1996),
    Z. Phys. **C75**, 57 (1997).
16. J. Rathsman, *Phys. Lett.* **B452**, 364 (1999),
17. A. Edin, G. Ingelman, and J. Rathsman, *Comp. Phys. Commun.* **101**, 108 (1997).
18. H1 Collaboration, *Diffractive jet production in deep–inelastic* e+p *collisions at
    HERA*, contributed paper 960 to the 30th International Conference on High–Energy
    Physics ICHEP2000, Osaka, Japan, July 27 – August 2, 2000.
19. H1 Collab., C. Adloff et al., *Eur. Phys. J.* **C6**, 421 (1999).
20. ZEUS Collaboration, *Three-jet production in diffractive deep inelastic scattering at
    HERA*, contributed paper 872 to the 30th International Conference on High–Energy
    Physics ICHEP2000, Osaka, Japan, July 27 – August 2, 2000.
21. ZEUS Collab., M. Derrik et al., *Phys. Lett.* **B332**, 228 (1994).
22. ZEUS FPC Group, A. Bamberger et al., *DESY 99-194 (1999)*.
23. S. Catani et al., *Phys. Lett.* **B269**, 432 (1991).
24. S. L. Wu, *Phys. Rep.* **107**, 59 (1984),
    B. Naroska, *Phys. Rep.* **148**, 67 (1987).

Deeply Virtual Compton Scattering at HERA

Rainer Stamen*

on behalf of the H1 and ZEUS Collaborations

*Universität Dortmund[1]
Germany
IIHE-Université Libre de Bruxelles
Belgium

Abstract. Measurements of Deeply Virtual Compton Scattering at the ep collider HERA by the H1 and ZEUS Collaborations are presented and the cross section is shown as a function of Q^2 and W. The results are compared to LO QCD calculations based on the two gluon exchange model.

INTRODUCTION

Deeply Virtual Compton Scattering (DVCS) [1,2] is the hard diffractive scattering of a virtual photon off a proton. The QCD interpretation of this process, based on the two gluon exchange model, is shown in figure 1. The virtual photon emitted by the incoming electron interacts via a quark loop with two gluons from the proton resulting in a proton, a photon and an electron in the final state.

DVCS offers a new and comparatively clean way to study diffraction at HERA. In comparison to vector-meson production it is free from the large uncertainties on the vector-meson wave-function. The largest interest arises from the access it may provide experimentally to the skewed parton distributions of the proton [3].

DVCS contributes to the reaction $e^+p \longrightarrow e^+\gamma p$, whose total cross section is dominated by the purely electromagnetic Bethe–Heitler process [2] where the final state photon originates from initial or final state radiation off the in– or outgoing electron. The diagrams of this process are shown in figure 2.

In the presence of a large scale the DVCS process can be calculated in perturbative QCD. In the present case the scale is given by the photon virtuality Q^2 which is above a few GeV^2. It has been shown that for the DVCS process factorisation

---

[1] This work was supported by the Graduiertenkolleg Elementarteilchenphysik of the Deutsche Forschungsgemeinschaft at the Universität Dortmund, the Deutscher Akademischer Austauschdienst, Bonn and the Martin Schmeißer Stiftung at the Universität Dortmund
[2] also called QED Compton process

CP571, *PHOTON 2000*, edited by A. J. Finch

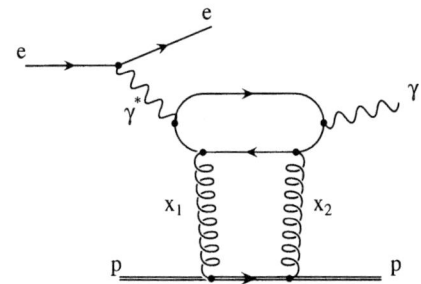

FIGURE 1. DVCS diagram in the QCD picture.

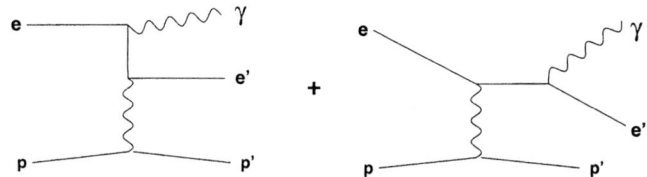

FIGURE 2. LO Diagrams contributing to the Bethe–Heitler process.

holds which means that the scattering amplitude factorises in a hard scattering part which can be calculated in perturbative QCD and a parton density which absorbs non–perturbative effects [4]. In a naive model the gluon content of the proton can be modeled by using the square of the ordinary gluon distribution $|xg(x,Q^2)|^2$ in the calculations. Since for the DVCS process the virtual photon is scattered onto mass shell in the final state, it is necessary to transfer longitudinal momentum from the photon to the proton, i.e. forcing a non–forward kinematic situation. In the picture of the two gluon exchange the fractional momenta of the gluons x_1 and x_2 are therefore not equal which leads to the concept of skewed parton distributions [3].

LO QCD calculations, based on the two gluon exchange model exist [2]. These calculations, including the interference between the DVCS and the Bethe–Heitler process, are embedded in two independent MC generators used by the two experiments at the HERA ep collider. The DVCS process provides in principle the possibility to extract these distributions from experimental data [5].

ANALYSIS STRATEGY

Around the interaction region both experiments, H1 [6] and ZEUS [7], are equipped with tracking devices which are surrounded by calorimeters. Since the proton escapes the main detector through the beam pipe only the scattered electron and the photon are detected. Therefore the event selection is based on the requirement of two electromagnetic clusters, one in the backward and one in the central or forward part of the detector where the backward direction is defined as the direction of the incoming electron. No other cluster above a certain noise threshold must be found in the calorimeter.

If a track can be reconstructed in the event it has to be associated to one of the clusters and determines the electron candidate. To enhance the DVCS contribution w.r.t. the Bethe–Heitler process the phase space is restricted by requiring the photon candidate in the forward or central part of the detector. In case of the H1 analysis the elastic component is specifically selected by additional use of the forward detectors placed close to the beam pipe in order to identify particles originating from proton dissociation processes.

ZEUS RESULTS

The ZEUS analysis [8] is based on a luminosity of $37.5 \, \text{pb}^{-1}$ taken in the 1996/97 data taking period. A photon virtuality $Q^2 > 6 \, \text{GeV}^2$ is demanded. In figure 3a the polar angular distribution of the electron and photon candidates measured in the central part of the calorimeter is shown. A clear signal above the expectations for the Bethe–Heitler process is observed. If one takes the full calculations including the DVCS process a good description of the experimental data is achieved. Figure 3b shows the polar angular distribution for the photon candidates only. Here the difference between the data and the expectation for the Bethe–Heitler process is larger since the photon is more likely to be found in the central part for the DVCS process than for the Bethe-Heitler process. This can also be seen in figure 3c which shows the angular distribution of the ratio of the number of photon candidates over the sum of photon and electron candidates. If the photon energy cut is increased from 2 to $5 \, \text{GeV}$, the clear signal persists (not shown). In all cases the data are quite well described by the Monte Carlo model.

As a cross check a shower shape analysis has been performed showing that the signal originates from single photons and not from π^0 background which could arise from low multiplicity hadron production. Such processes are included in commonly used DIS MC generators. Since these processes have, however, never been explicitly measured at HERA energies the predictions of the available MC programs, tuned for high multiplicity events, are unreliable. Hence π^0 background has been studied using real data.

A cluster shape analysis is performed using the variables Z_{width} and f_{MAX}, where Z_{width} is the energy weighted sum over all distances z between each cluster cell and the reconstructed center of the electromagnetic cluster; f_{MAX} represents the fraction of energy in the most energetic cell w.r.t. the total energy of the cluster. Since the π^0 decays into two photons with a finite opening angle, a broader distribution of the variable Z_{width} is expected compared to single photons. For the same reason the variable f_{MAX} should be well below one for π^0 induced clusters whereas for single

119

ZEUS 1996/97 Preliminary

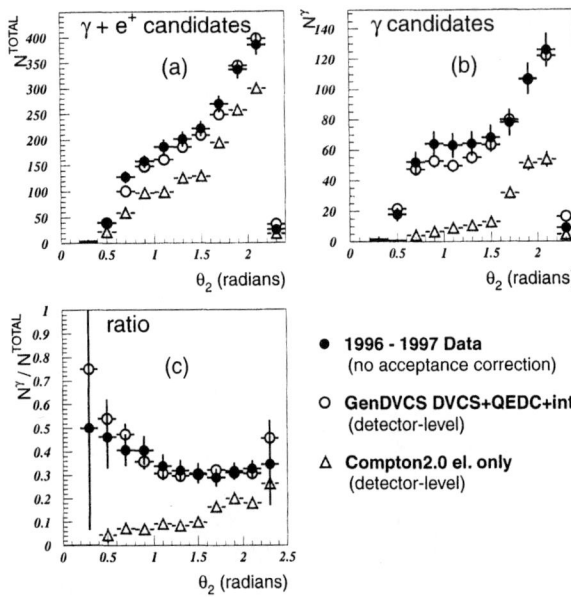

FIGURE 3. (a) Distribution of the polar angle of the photon and electron candidates in the central part of the calorimeter with an energy $E > 2\,\text{GeV}$; (b) the polar angle of the photon candidates with no track associated with the electromagnetic cluster; (c) the ratio of the distributions in (b) and (a). The data are shown as solid points; the (DVCS + QEDC + Int.) MC prediction is shown as open circles and the one from the QED Compton MC as open triangles.

photons a value close to one is expected.

The results of these studies are summarised in figure 4. The data are compared to the MC prediction of the full QCD calculation and to the predictions from the RAPGAP [10] and the DJANGOH [11] programs. In figure 4 a) and b) the angular distributions for the photon candidates are shown. For the shower shape analysis only candidates with $\theta_2 < 1.5$ are used where the DVCS contribution to the cross section is expected to dominate. The resulting distributions for the shower shape variables are shown in figures 4 c)–f) and compared to the different MC predictions. A good description of the distributions by the DVCS MC is achieved while the π^0 prediction differs significantly from the data indicating that the data cannot be explained as hadronic background from low–multiplicity DIS events.

H1 RESULTS

The H1 analysis [9] is based on a luminosity of $8\,\text{pb}^{-1}$ taken in the 1997 data taking period. The event sample was divided into two parts:

- A **control sample** where the electron is required to be in the central part of the detector. The electron is identified by a track measured with the central tracking system associated to an electromagnetic cluster in the fine grained calorimeter.

120

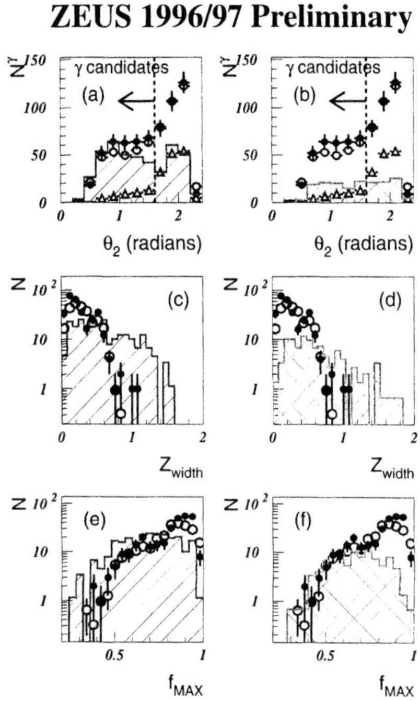

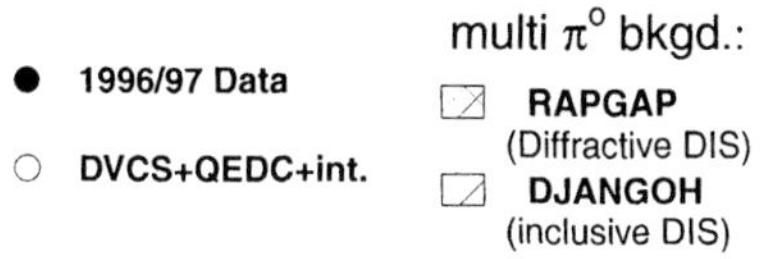

ZEUS 1996/97 Preliminary

multi π^o bkgd.:

● 1996/97 Data

○ DVCS+QEDC+int.

☐ RAPGAP
(Diffractive DIS)

☐ DJANGOH
(inclusive DIS)

FIGURE 4. (a,b) Distribution of the polar angle of the photon candidates with an energy $E > 2\,\text{GeV}$; (c,d) distribution of the variable Z_{width}, expressed in units of the BCAL EMC cell width and (e,f) distribution of variable f_{MAX}. The ZEUS data are shown as solid points; the (DVCS + QEDC + INT) MC simulated events are shown as open circles; the QEDC Monte Carlo simulated events are shown as open triangles. In (a,c,e) the data are compared to a low-multiplicity Monte Carlo simulated sample from DJANGOH, while (b,d,f) show the comparison with RAPGAP-generated events. The MC predictions are normalised to the same luminosity as the data.

- A **signal sample** where the photon is required to be in the central part of the detector. Here the photon is identified by the absence of a track in front of the electromagnetic cluster measured in the calorimeter.

The control sample is dominated by the Bethe–Heitler process, while the DVCS contribution is negligible. In addition two further processes were found to contribute to this event signature: the diffractive production of ρ-mesons and the elastic dilepton production. Both processes contribute if one of the outgoing particles leaves the main detector through the beam pipe. The processes have been studied using the DIFFVM [13] and GRAPE [14] MC programs respectively. In figure 5, the uncorrected event distributions are seen to be well described by the sum of the three MC predictions, assuring that the detector response is well understood.

The acceptance, initial state radiation of real photons and detector effects have been estimated by MC to extract the elastic $e^+p \longrightarrow e^+\gamma p$ cross section which has been measured in the kinematic region $2 < Q^2 < 20\,\text{GeV}^2$, $|t| < 1\,\text{GeV}^2$ and $30 < W < 120\,\text{GeV}$ using the signal sample. The event kinematics is calculated using the measured angles of the final state electron and photon. The proton dissociation background is estimated to 10% and has been subtracted. The main systematic error arises from the measurement of the polar angles of the electron and photon used to determine the kinematics. The full error is however dominated by the limited statistics.

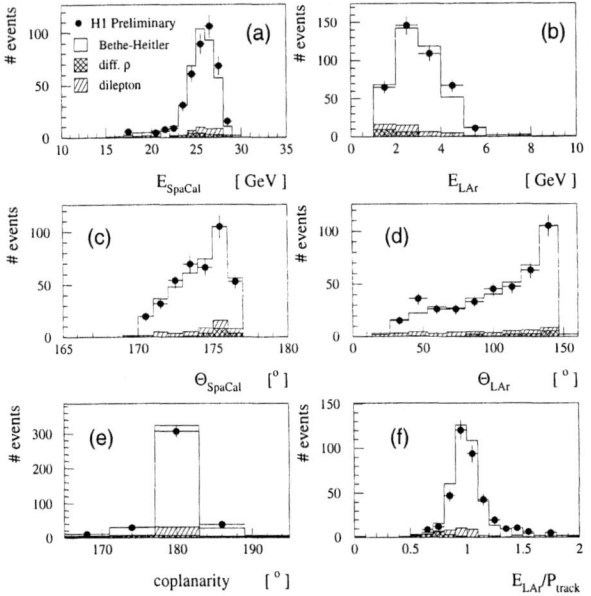

FIGURE 5. Event distributions (uncorrected) of the control sample (see text). The data are compared to the sum of the prediction for the Bethe–Heitler process, elastic dilepton production and diffractive ρ–production. (a) energy of the cluster in the backward calorimeter (SpaCal), (b) energy of the cluster in the central calorimeter (LAr), (c) polar angle of the cluster in the SpaCAL, (d) polar angle of the cluster in the LAr calorimeter, (e) coplanarity, i.e. the difference in the azimuthal angle of the two clusters, (f) ratio of the energy of the cluster and the momentum measured by the tracking detector for the electron candidate.

Figures 6 and 7 show the $e^+p \longrightarrow e^+\gamma p$ differential cross section as a function of Q^2 and of W. The data are compared with the Bethe–Heitler prediction alone and with the full calculation including Bethe–Heitler and DVCS. One finds a good description of the data by the full calculations.

Finally it must be noted that the t-slope cannot be calculated at low t leading to a free parameter in the theory which can only be extracted from experimental data. This has not yet been done. In the QCD comparisons the t dependence is assumed to be $e^{-b|t|}$. The free parameter b leads mainly to a normalisation uncertainty of the prediction. In the figures the t-slope b is assumed to be between 7 and 10 GeV^{-2}. Hence the data should be compared to the prediction only w.r.t. the shape and the absolute normalisation should be taken with care.

CONCLUSION

The DVCS process has been observed by the H1 and ZEUS Collaborations and the first $e^+p \longrightarrow e^+\gamma p$ cross section measurement has been presented. The experimental results are well described by the LO QCD calculations of Frankfurt et al. [2] although a dedicated t-slope measurement is needed to determine the normalisation for the calculations.

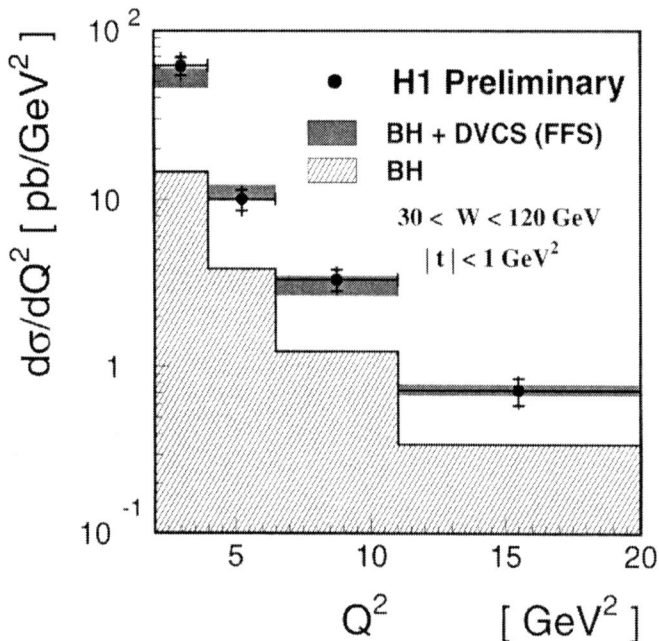

FIGURE 6. The measured cross sections of the reaction $e^+p \longrightarrow e^+\gamma p$ as a function of Q^2 is shown and compared to theoretical prediction (FFS)[2] (shown as band). The cross section is measured in the kinematic range $30 < W < 120\,\mathrm{GeV}, |t| < 1\,\mathrm{GeV}^2$. The uncertainty in the theoretical prediction for the DVCS process is dominated by the unknown slope of the t-dependence of the DVCS part of the cross section, assuming $7 < b < 10\,\mathrm{GeV}^{-2}$. In addition the prediction for the Bethe–Heitler process is shown separately (hatched).

REFERENCES

1. X. Ji, Phys. Rev. **D55**(1997) 7114.
2. L. L. Frankfurt, A. Freund, M. Strikman, Phys. Rev. **D58** (1998) 114001 and erratum Phys. Rev. **D59** (1999) 119901E.
3. A. V. Radyushkin Phys. Rev. **D56** (1997) 5524.
4. J. C. Collins, A. Freund, Phys. Rev. **D59**(1999) 074009.
5. A. Freund, Phys. Lett. **B 472** (2000) 412.
6. H1 Collaboration, I. Abt et al., *Nucl. Instrum. Methods* A **386** (1997) 310 and 348.
7. ZEUS Collab., M. Derrick *et al.* Status Report 1993, DESY (1993).
8. P. R. B. Saull, hep-ex/0003030 and ZEUS Collab., paper cont. EPS 99 http://www-zeus.desy.de/~schlenst/eps99_paper.html.
9. H1 Collab.,paper cont. ICHEP 2000, 966.
10. RAPGAP 2.08/00: H. Jung, Comp. Phys. Commun. **86** (1995) 147.
11. A. Kwiatkowski, H. Spiesberger and H. J. Möhring,

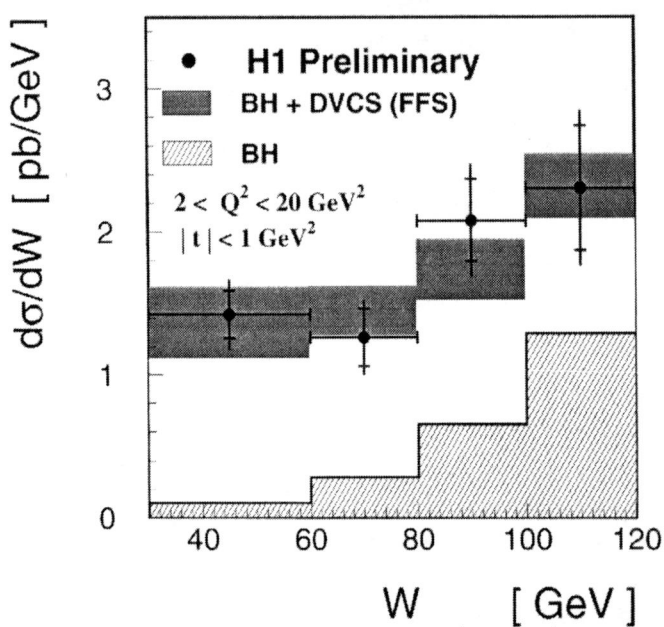

FIGURE 7. The measured cross section of the reaction $e^+p \longrightarrow e^+\gamma p$ as a function of W is shown and compared to the theoretical prediction (FFS)[2] (shown as band). The cross section is measured in the kinematic range $2 < Q^2 < 20\,\text{GeV}^2$, $|t| < 1\,\text{GeV}^2$. The uncertainty in the theoretical prediction for the DVCS process is dominated by the unknown slope of the t-dependence of the DVCS part of the cross section, assuming $7 < b < 10\,\text{GeV}^{-2}$. In addition the prediction for the Bethe–Heitler process (BH) is shown separately (hatched).

Comp. Phys. Commun. **69** (1992) 155
K. Charchula,G. A. Schuler and H. Spiesberger,
Comp. Phys. Commun. **81** (1994) 381
DJANGOH version 1.1: H. Spiesberger, available from
http://www.desy.de/~hspiesb/djangoh.html.
12. A. Courau and P. Kessler, Phys. Rev. **D46** (1992) 117.
13. B. List, Diploma Thesis, Techn. Univ. Berlin, (1993).
14. T. Abe et al., in Proc. *Monte Carlo Generators for HERA physics*, Eds. A. T. Doyle, G. Grindhammer, G. Ingelman, H. Jung, DESY-PROC-1999-02, p. 566.

Proton-dissociative Diffractive Photoproduction of Vector Mesons at Large $|t|$ at HERA

Katarzyna Klimek representing the ZEUS Collaboration

Institute of Nuclear Physics, High Energy Departments
ul. Kawiory 26 A, 30-055 Kraków, Poland
E-mail: Katarzyna.Klimek@ifj.edu.pl

Abstract. Photoproduction of vector mesons, $\gamma p \to V N$, where N is the system into which the proton dissociates, has been measured in $e^+ p$ interactions with the ZEUS detector at HERA using an integrated luminosity of 24 pb^{-1}. The data were taken at a photon-proton centre-of-mass energy $W \simeq 100$ GeV and extend up to $-t = 12$ GeV^2, where t is the square of the four-momentum transferred to the vector meson. The differential cross section $d\sigma/dt$, and the results of a decay-angle analysis as a function of t for ρ^0, ϕ and J/ψ production are compared to the expectations of phenomenological models.

INTRODUCTION

This paper describes measurements of proton-dissociative vector meson production, $ep \to eVN$, where $V = \rho^0, \phi, J/\psi$ and N is the proton dissociative system. Our aim is to examine whether the four-momentum transfer squared, t, between the photon and the final-state vector meson may also serve, like the photon virtuality Q^2 or the vector meson mass, as a hard QCD scale at large values, as predicted in [1]. In particular, the differential cross section $\frac{d\sigma}{dt}$ is compared to the predictions of the pQCD-based models; so are the decay-angle distributions of the final-state mesons.

The scattered positron was detected in an electromagnetic calorimeter 44 m from the interaction point in the direction of the outgoing positron, ensuring that the virtuality of the exchanged photon is $Q^2 < 0.02$ GeV^2. This allows a precise determination of t from the measurement of the meson transverse momentum with respect to the beam direction, p_T (here $t = -p_T^2$ is a very good approximation).

The data used in the present analysis were collected in 1996 and 1997 with the ZEUS detector at HERA and correspond to an integrated luminosity of 24 pb^{-1}.

CP571, *PHOTON 2000*, edited by A. J. Finch
© 2001 American Institute of Physics 0-7354-0010-5/01/$18.00

ANALYSIS

The ZEUS detector has been described in detail elsewhere [2]. The main components for this analysis are the high-precision uranium-scintillator calorimeter (CAL) [3] and the central tracking detector (CTD) [4].

The study was restricted to events with 1) two well-measured tracks in the CTD having an invariant mass consistent with the mass of the vector meson under study, 2) no energy in the CAL other than that associated with either the two tacks or the dissociated-proton system N and 3) the scattered positron detected in the 44m tagger. The selected events exhibit a diffractive topology with a gap of at least two units of rapidity between the dissociated nucleonic system and the tracks. The γp centre-of-mass energy was restricted to the range $80 < W < 120$ GeV (where the tagging efficiency is well known). From the trigger conditions $-t > 1.2$ GeV^2.

For the simulation of the reaction $ep \rightarrow eVN$, the EPSOFT Monte Carlo generator [5] was used, reweighted to describe the data.

RESULTS

Differential cross sections $\frac{d\sigma}{dt}$ for the ρ^0, ϕ and J/ψ mesons

Figure 1 shows the differential cross sections for the ρ^0 and ϕ mesons together

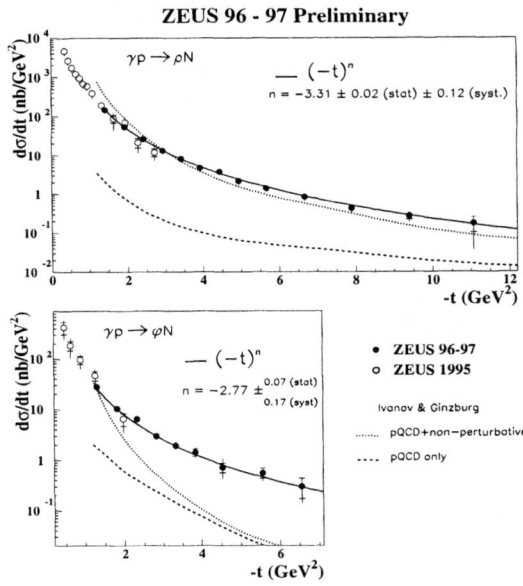

FIGURE 1. The measured differential cross sections $d\sigma/dt$ for proton-dissociative photoproduction of ρ^0 and ϕ mesons, in the range $80 < W < 120$ GeV. The inner error bars indicate the statistical error, the outer bars the statistical and systematic uncertainties added in quadrature.

126

with results obtained in our earlier study of the lower $-t$ region [6]. For $-t > 1.2$ GeV^2, the cross sections are well described by a power dependence on t, as expected for a hard production mechanism. A fit to the data with the function $A \cdot (-t)^{-n}$ for ρ^0 production gives $n = 3.31 \pm 0.02$ (stat.)± 0.12 (syst.) and $n = 2.77 \pm 0.07$ (stat.)± 0.17 (syst.) for ϕ production. The data are compared to a QCD calculation [7] which estimates both perturbative and non-perturbative contributions. The calculation underestimates the cross sections for the ϕ meson over most of the t range covered. The magnitude of the perturbative contribution is smaller than the measured cross section by more than one order of magnitude for both mesons.

The differential cross section for J/ψ photoproduction is shown in Fig. 2, together with the ZEUS measurement at smaller $-t$ [6]. A fit with the function $A \cdot (-t)^{-n}$ gives $n = 1.6 \pm 0.2$ (stat.)± 0.2 (syst.). A calculation based on the BFKL-formalism [8] compares satisfactorily with the data for values of $-t$ as low as 1 GeV^2. It should be noted, however, that the uncertainties due to the choice of the model parameters are significant.

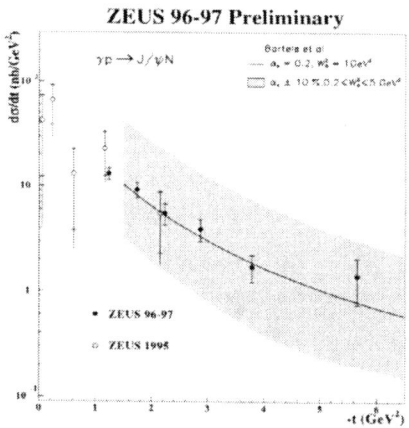

ZEUS 96-97 Preliminary

FIGURE 2. The differential cross section $d\sigma/dt$ for proton-dissociative J/ψ photoproduction. The inner error bars indicate the statistical error, the outer bars the statistical and systematic uncertainties added in quadrature.

Ratios of cross sections for vector meson photoproduction

The ratios of the differential cross section $\frac{d\sigma}{dt}(\gamma p \to \phi N)/\frac{d\sigma}{dt}(\gamma p \to \rho^0 N)$ and $\frac{d\sigma}{dt}(\gamma p \to J/\psi N)/\frac{d\sigma}{dt}(\gamma p \to \rho^0 N)$ are shown as a function of $-t$ in Fig. 3, together with previous ZEUS results [6]. A rise of the ϕ/ρ^0 ratio up to $-t \approx 3.5$ GeV^2 is observed, approaching the SU(4) value [1]. A clear increase of the $(J/\psi)/\rho^0$ ratio with increasing $-t$ is also observed, although the ratio is significantly smaller than

---

[1] The SU(4) prediction, which ignores the meson mass differences, is that the coupling strengths of the photon to the ρ^0, ω, ϕ and J/ψ mesons should be in the ratios $9 : 1 : 2 : 8$.

the SU(4) value of 8:9 up to $-t$ values of 4 GeV^2. It is interesting to compare

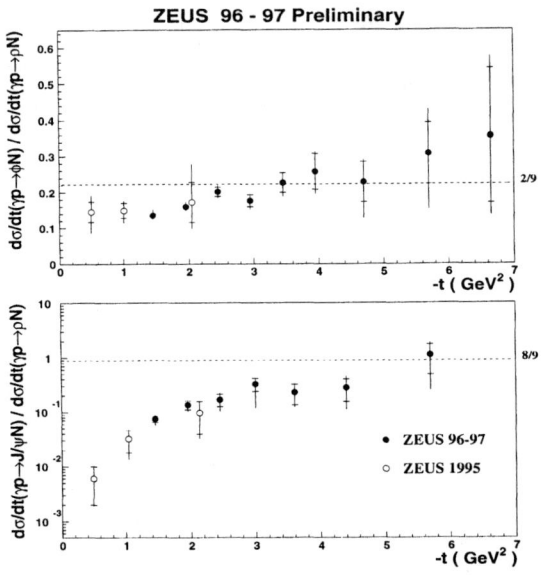

FIGURE 3. The ratios of the cross sections $d\sigma/dt$ for ϕ to ρ^0 and J/ψ to ρ^0 for proton-dissociative photoproduction. The inner error bars indicate the statistical error, the outer bars the statistical and systematic uncertainties added in quadrature. The dashed lines correspond to the SU(4) predictions.

the t and Q^2 dependences of the cross section ratios for vector meson production. At large Q^2 the ϕ/ρ^0 ratio [9] approaches the SU(4) value for $Q^2 \geq 6$ GeV^2, while in photoproduction it does already for $-t > 3.5$ GeV^2. For $-t \approx Q^2 \approx 3.5$ GeV^2 the ψ/ρ^0 ratios are significantly smaller than the SU(4) prediction, but the photoproduction ratio is a factor of five larger than that for DIS [10]. It is clear that cross section ratios grow faster with t than with Q^2, implying that t and Q^2 cannot be treated as equivalent scales.

Decay-angle analyses for the ρ^0 and ϕ mesons

The angular distributions of the meson decay products were used to determine some of the ρ^0 and ϕ spin-density matrix elements. From the decay angular distribution $W(\cos\Theta, \varphi, \Phi)$ [11], where Θ, φ and Φ are the decay angles defined in the s-channel helicity frame [2], two matrix elements were evaluated: r_{00}^{04}, corresponding to the probability that the meson is produced in the helicity 0 state and r_{1-1}^{04}, which

---

[2] The s-channel helicity frame is defined as the rest frame of the meson in which the meson direction in the photon-proton centre-of-mass frame is taken as the quantization axis.

is related to the probability of helicity double-flip. If s-channel helicity conservation (SCHC) holds, the matrix elements r_{00}^{04} and r_{1-1}^{04} should be zero. In Fig. 4

ZEUS 96-97 Preliminary

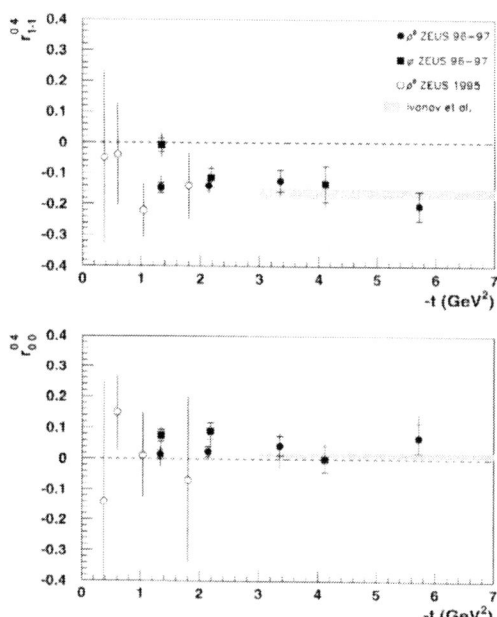

FIGURE 4. The fitted values of r_{00}^{04} and r_{1-1}^{04} for proton-dissociative ρ^0 (circles) and ϕ (squares) photoproduction as a function of $-t$. The inner error bars indicate the statistical error, the outer bars the statistical and systematic uncertainties added in quadrature. The gray bands indicate the predictions of the pQCD calculations.

the values of r_{00}^{04} and r_{1-1}^{04} for the ρ^0 and ϕ mesons are displayed as function of t, together with earlier ZEUS measurements at lower $-t$ [6]. The measurements of r_{00}^{04} show that the helicity single-flip contribution is small over the entire range in t. A clear indication of a non-zero helicity double-flip contribution is shown by the measurements of r_{1-1}^{04} at large $-t$ for both mesons. The results are in very good agreement with recent pQCD calculations [12].

SUMMARY

Diffractive photoproduction of ρ^0, ϕ and J/ψ mesons with proton dissociation has been measured at $W \approx 100$ GeV for $-t$ values up to 12 GeV^2. The results are consistent with the ZEUS 1995 measurements [6].

A parameterization of the differential cross sections as a power function, $d\sigma/dt \propto (-t)^{-n}$, yields a value for n which decreases with increasing mass of the meson.

The resulting powers n are: 3.31 ± 0.02 (stat.)± 0.12 (syst.) for ρ^0 production, 2.77 ± 0.07 (stat.)± 0.17 (syst.) for ϕ production, and 1.6 ± 0.2 (stat.)± 0.2 (syst.) for J/ψ production. A comparison of the measured differential cross sections with QCD models shows that the perturbative part of the calculations for ρ^0 and ϕ production [7] at the t values covered in this analysis is well below the data. However, the perturbative QCD prediction [8] is in satisfactory agreement with the J/ψ data.

The ratio of the ϕ to ρ^0 production cross sections increases with $-t$ and approaches the SU(4) value at $-t \approx 3.5$ GeV^2. The ratio of the J/ψ to ρ^0 production cross sections, on the other hand, is a factor of two below the SU(4) value up to $-t$ values of 4 GeV^2. From the comparison of cross section ratios dependence on t and Q^2, we see that these ratios increase faster with t than with Q^2, indicating that t and Q^2 are not equivalent scales.

The results of the decay-angle analysis for the ρ^0 and ϕ mesons demonstrate a clear deviation from SCHC. The measured values of r_{00}^{04} and r_{1-1}^{04} are in agreement with pQCD predictions [12].

In conclusion, a complete picture of vector meson production at large t is still lacking. Based on the measurements of the J/ψ differential cross section, the vector meson cross section ratios, and the decay-angle distributions of the ρ^0 and ϕ, there are indications that large $-t$ can provide a suitable hard scale for pQCD. However, the available pQCD calculations do not describe the results for light vector mesons cross sections quantitatively, so $-t$ cannot be consistently treated as a hard scale.

REFERENCES

1. L. Frankfurt and M. Strikman, Phys. Rev. Lett. 63 (1989) 1914;
   A.H. Mueller and W-K. Tang, Phys. Lett. B 284 (1992) 123.
2. ZEUS Collaboration, M. Derrick et al., The ZEUS Detector Status Report 1993, DESY.
3. M. Derrick et al., Nucl. Instr. and Meth. A 309 (1991) 77;
   A. Anderson et al., Nucl. Instr. and Meth. A 309 (1991) 101;
   A. Caldwell et al., Nucl. Instr. and Meth. A 321 (1992) 356;
   A. Bernstein et al., Nucl. Instr. and Meth. A 336 (1993) 23.
4. N. Harnew et al., Nucl. Instr. and Meth. A 279 (1989) 290;
   B. Foster et al., Nucl. Phys. B 32 (Proc. Suppl.) (1993) 181;
   B. Foster et al., Nucl. Instr. and Meth. A 338 (1994) 254.
5. M. Kasprzak, PhD thesis, Warsaw University, DESY F35D-96-16 (1996).
6. ZEUS Collaboration, J. Breitweg et al., Eur. Phys. J. C 14 (2000) 213.
7. D.Yu. Ivanov, Phys. Rev. D 53 (1996) 3564.
8. J. Bartels et al., Phys. Lett. B 375 (1996) 301.
9. H1 Collaboration, C. Adloff et al., DESY Report 00-070 (2000);
   ZEUS Collaboration, J. Breitweg et al., DESY Report 00-084 (2000).
10. ZEUS Collaboration, J. Breitweg et al., Eur. Phys. J. C 6 (1999) 603.
11. K. Schilling and G. Wolf, Nucl. Phys. B 61 (1973) 381.
12. D.Yu. Ivanov et al., Phys Lett. B 478 (2000) 101.

Total Photonic and Hadronic Cross-sections

R.M. Godbole*, A. Grau†, G. Pancheri**

*Centre for Theoretical Studies, Indian Institute of Science, Bangalore 560 012, India
† Departamento de Física Teórica y del Cosmos, Universidad de Granada, Spain
** INFN - Laboratori Nazionali di Frascati, Via E. Fermi 40, 100044 Frascati, Italy

Abstract. We discuss total cross-sections within the context of the QCD calculable mini-jet model, highlighting its successes and failures. In particular we show its description of $\gamma\gamma \rightarrow hadrons$ and compare it with OPAL and L3 data. We extrapolate this result to $\gamma\,p$ total cross-sections and propose a phenomenological ansätz for virtual photon cross-sections. We point out that the good agreement with data obtained with the Eikonal Minijet Model should not hide the many uncertainties buried in the impact parameter distribution. A model obtained from Soft Gluon Summation is briefly discussed and its application to hadronic cross-sections is shown.

I INTRODUCTION

This talk will be a short review of the status of the calculation of total cross-sections using a QCD driven mini-jet model [1], with particular emphasis on recent measurements and results of theoretical calculations of $\gamma\gamma$ cross-sections.

One of the aims of QCD is to calculate cross-sections for hadronic processes. We find that the level of available experimental information on total hadronic cross-sections has now reached a stage so as to allow, for the first time, definite progress in the calculation of the one quantity which has so far escaped a complete quantitative understanding in a QCD framework, namely the total hadronic cross-section. We have a complete set of processes, pp, $p\bar{p}$, γp and $\gamma\gamma$ measured in a common energy range, $\sqrt{s} = 1 \div 100$ GeV, with the purely hadronic processes measured up to $\sqrt{s} \approx 3 \times 10^5$ GeV [2]. The latter allows for very good parametrizations in a large energy range, the other two cross-sections to test the QCD content of hadrons versus the one in the photons. In Figure 1 we show a compilation of all presently available photon and proton total cross-sections, scaling them so as to be able to compare one to each other. For $\gamma\,p$ processes, the scale factor is the product of a Quark Parton Model factor $3/2$ multiplied by a Vector Meson Dominance type factor $1/240$ [1], for $\gamma\gamma$ we just square this factor.

The plan of this paper is as follows :

CP571, *PHOTON 2000*, edited by A. J. Finch
© 2001 American Institute of Physics 0-7354-0010-5/01/$18.00

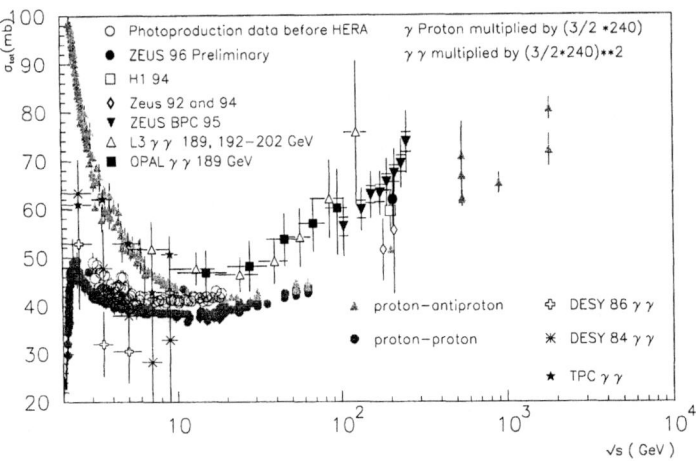

FIGURE 1. Energy dependence of σ_{ab}^{tot} for various choices of a, b as indicated in the figure. The cross-sections for the photon-induced processes have been scaled as indicated on the figure.

1. In Sect. II we describe our recent results for $\gamma\gamma$ and extrapolate them to $\gamma\, p$

2. In sect. III we discuss a possible ansätz on an extension of the minijet model to $\gamma^*\gamma^*$

3. In Sect. IV we present recent results for pp and $p\bar{p}$, using an impact parameter distribution for partons in the hadrons obtained from the Bloch-Nordsieck summation technique, and discuss its possible extensions to real and virtual photons.

II $\gamma\gamma$ **AND** $\gamma\ proton$

While for quite some time the photon-photon data at LEP exhibited a discrepancy between different collaborations (although within the experimental errors), we now have two sets of data points which are in excellent agreement with each other. A theorist can then start his/her work. We show in Figure 2 the description of L3 [3] and OPAL data [4] using the Eikonal Minijet Model (EMM). The theoretical context in which this curve was obtained is discussed in [1,5] : the EMM uses the eikonal approximation [6] to calculate the total cross-section, i.e.

$$\sigma_{tot} = 2P_{had}^{ab} \int d^2\vec{b}[1 - e^{i\chi(b,s)}] \tag{1}$$

and approximates the eikonal function neglecting $\Re e\chi(b, s)$ and putting

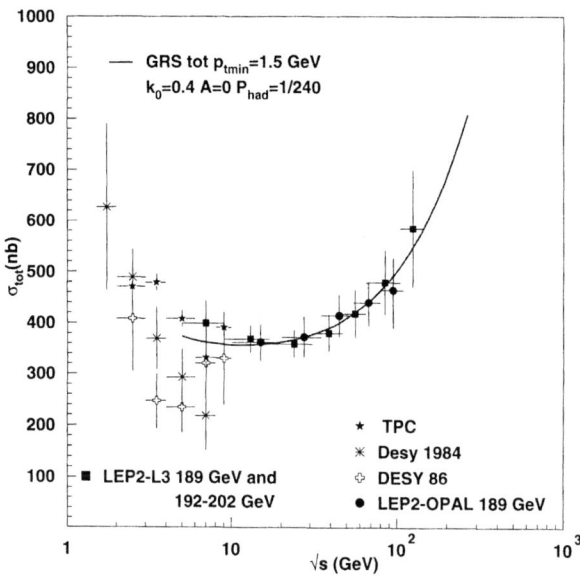

FIGURE 2. The total photon-photon cross-section as described by the EMM (see text)

$$2\Im m\chi(b,s) = n(b,s) = A(b)[\sigma_{soft}(s) + \frac{\sigma_{jet}(s,p_{tmin})}{P_{had}^{ab}}] \qquad (2)$$

where the QCD calculable jet cross-section is the quantity which drives the rise [7] in all total cross-sections. This function is defined as

$$\sigma_{jet} = \int d^2\vec{p}_t \frac{d\sigma^{QCD}}{d^2\vec{p}_t} = \sum \int dx_1 dx_2 f^{i/a}(x_1) f^{j/b}(x_2) \int d^2\vec{p}_t \frac{d\hat{\sigma}^{ij,kl}}{d^2\vec{p}_t} \qquad (3)$$

where $\hat{\sigma}$ is the parton-parton cross-section for the subprocess $ij \to kl$, the sum runs over i,j,k,l,=parton type and the integration covers a region from a minimum p_t to the entire phase space. The quantity σ_{jet} is higly dependent upon the regulator p_{tmin} and one of the aims of a complete QCD calculation is to eliminate this dependence. Presently this is not yet possible, but one can nonetheless expect p_{tmin} to be the smallest momentum exchanged between partons such that perturbative QCD can be applied, namely not less than 1 GeV and probably not more than 2 GeV. The parameter $P_{had}^{ab} \equiv P_{had}^a P_{had}^b$ is in principle energy dependent and P_{had}^a can be interpreted as the probability for particle a to behave like a hadron, i.e. $P_{had}^{hadron} \equiv$ 1 and, typically, $P_{had}^\gamma \approx \mathcal{O}(\alpha)$ [8,9]. Phenomenologically, in order to obtain a description of the measured total cross-sections, there are other input quantities which need to be fixed, namely σ_{soft} and the b-distribution function $A(b)$. One way to proceed has been, previously, to determine all the parameters from the process γp. However, the EMM model is not working very well for the proton case, as discussed in [10], probably because of uncertainties in the hadronic transverse momentum distributions, and we have opted, in this note, for a different approach,

133

namely we obtain the parameters from γp, extrapolate them to $\gamma\gamma$ varying them within at most 10% to fit the $\gamma\gamma$ cross-section, and then revert back and see what the best description of $\gamma\gamma$ will produce, when applied to photoproduction. The result of this *modus operandi* is presented here. For Figure 2, we have chosen to describe the partonic matter distribution inside the hadrons through a function inspired by the pion electromagnetic form factor, namely

$$A(b) = \frac{1}{(2\pi)^2} \int d^2\vec{q}\, e^{i\vec{q}\cdot\vec{b}} \left(\frac{k_0^2}{q^2 + k_0^2}\right)^2 \tag{4}$$

The scale k_0 has been let to vary, according to an intrinsic transverse momentum ansätz [1]. Thus, Figure 2 corresponds to $k_0 = 0.4\ GeV$, $P_{had}^\gamma = 1/240$, $\sigma_{soft} = (21 + \frac{42}{s})\ mb$, $p_{tmin} = 1.5\ GeV$ and GRS [11] type densities for the photon. Next we extrapolate this curve to γp processes, putting $\sigma_{soft}^{\gamma p} = 3/2\sigma_{soft}$, the proton form factor with dipole type expression instead of one of the photon type monopole expressions, GRV type densities for the proton [12] and GRS for the photon. The result is the upper curve shown in Figure 3 and compared with old and recent data [13–16]. For completeness and as a reference to our previous work, we also show a band, with the lowest curve corresponding to GRV densities for both proton and photon, $p_{tmin} = 2\ GeV$, $k_0 = 0.66\ GeV$ for the photon, same P_{had}^γ as the upper curve, and $\sigma_{soft}^{\gamma p} = (31 + \frac{10}{\sqrt{s}} + \frac{38}{s})\ mb$.

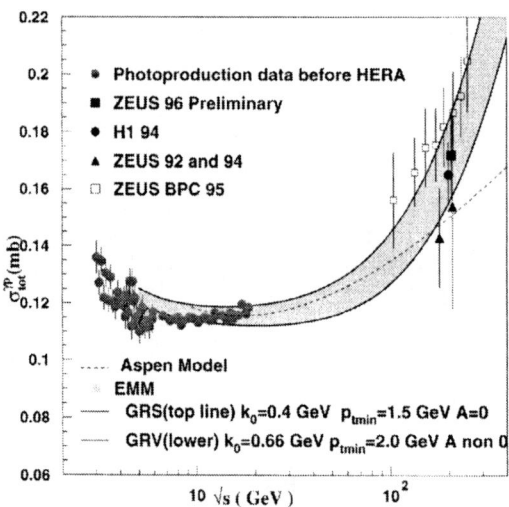

FIGURE 3. Comparison between the eikonal minijet model predictions and data for total $\gamma\, p$ cross-section as well as BPC data extrapolated from DIS [15]. Predictions from [17] are also shown.

III THE CASE OF THE VIRTUAL PHOTON

Hadronic interactions of the virtual photon in e^+e^- and $\gamma\gamma$ processes have been a subject of interest for some time now [18]. The appearance of experimental data on jet production with virtual photons at HERA [19] and total $\gamma^*\gamma^*$ cross-sections at LEP [20], has given added impetus to develop a model which will describe the total cross-sections for virtual photons, especially the 'resolved' part [21].

We want to develop a model to understand these cross-sections in the context of EMM. A possibility is that the virtual photon description be the same as for the case of real photons except that the intrinsic transverse momentum ansätz is complemented with a factor inspired by Extended Vector Meson Dominance [22]. For $\gamma\gamma$ processes, the function $A(b)$ would be modified to become

$$A(b, Q_1^2, Q_2^2) = \frac{m_\rho^2}{m_\rho^2 + Q_1^2} \frac{m_\rho^2}{m_\rho^2 + Q_2^2} A(b) \tag{5}$$

Such a factor is invoked only for the part of the cross-section which has a hadronic content. In Figure 4 we show our predictions for the hadronic content of the $\gamma^*\gamma^*$ cross-section, excluding for the time being the direct and single resolved contribution. Similar results can also be obtained using factorization in the context of the Aspen [17] model.

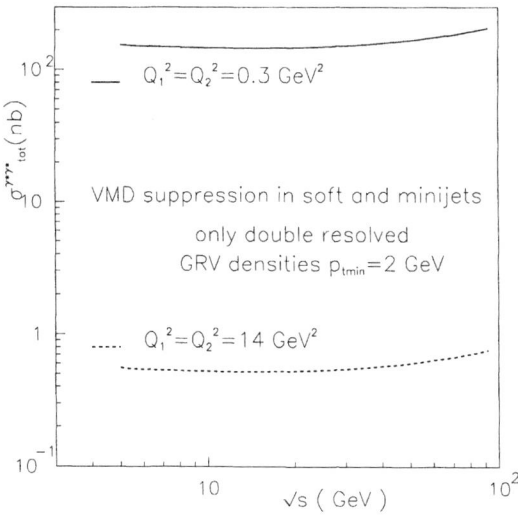

FIGURE 4. The EMM predictions for energy dependence of the double resolved part of virtual photons cross-section. Q_i^2 corresponds to the virtual photon mass

IV THE BLOCH NORDSIECK MODEL FOR THE IMPACT PARAMETER DISTRIBUTION

It is evident from the previous discussion, and we have pointed this out in many papers, that it is not possible to have a QCD description of the total cross-section without understanding the transverse momentum distribution of partons in the colliding hadrons, or, in the eikonal language, without understanding the impact parameter distribution. We have attempted such description, and will show in the following the main highlights. Some very recent results in the case of proton proton and proton antiproton are shown in Figure 5, where the two curves have

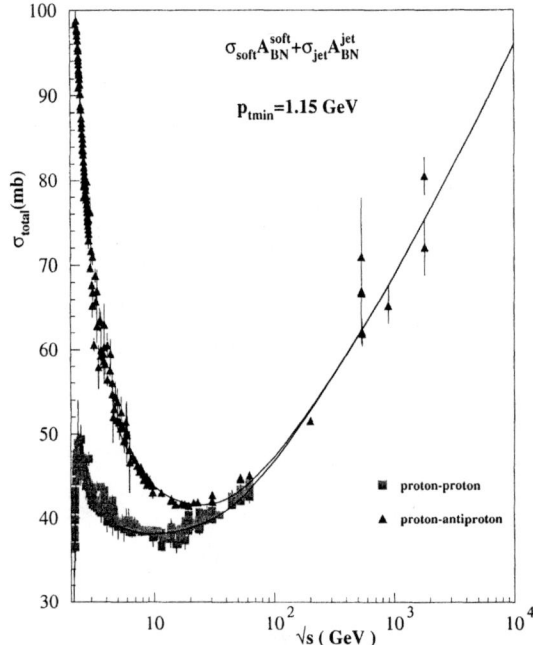

FIGURE 5. Total proton-proton and proton-antiproton cross-sections, compared with an eikonal minijet model which incorporates soft gluon summation

been obtained using the eikonal formula of Eq.(1), the jet cross-sections with GRV densities and $p_{tmin} = 1.15\ GeV$, $\sigma_{soft}^{pp} = 48\ mb$, $\sigma_{soft}^{p\bar{p}} = 48(1 + \frac{2}{\sqrt{s}})mb$. Unlike the curves obtainable in a straightforward application of the EMM with b-distribution determined by the proton form factor, this figure uses an s-dependent b-distribution given by

$$A(b, s) = \frac{\int d^2\vec{q}\,e^{i\vec{b}\cdot\vec{q}}\Pi(q, s)}{(2\pi)^2\Pi(0, s)} = \frac{e^{-h(b,s)}}{\int d^2\vec{b}\,e^{-h(b,s)}} \qquad (6)$$

where $\Pi(q, s)$ is the transverse momentum distribution of colliding partons generated by soft gluon radiation from the initial state and, to leading order,

$$h(b, s) = \frac{8}{3\pi} \int_0^{q_{max}} \alpha_s(k_t) \frac{dk_t}{k_t} [1 - J_0(bk_t)] \log \frac{q_{max} + \sqrt{q_{max}^2 - k_t^2}}{q_{max} - \sqrt{q_{max}^2 - k_t^2}} \qquad (7)$$

The functions $h(b, s)$ and $A(b, s)$ are energy dependent inasmuch as the kinematics of a given subprocess determine how much energy is available to single soft gluon emission. The energy dependence appears through the quantity q_{max} which is a function of the c.m. energy $\sqrt{s}$ and p_{tmin}. In addition $h(b, s)$ depends upon the single soft gluon distribution function $\alpha_s(k_t) dk_t / k_t$. Since $\Pi(q, s)$ is summed over all possible gluon distributions, the infrared divergence is of course cancelled, there remains, however, the problem of evaluating α_s in the infrared limit. We have used a phenomenological ansätz, namely a singular but integrable α_s and the result shown in Figure 5 is discussed in [10] for the rising part of the total proton cross-sections. The main characteristics of this treatment is that, as the minijet cross-section rises with energy, soft gluon emission produces an acollinearity of the partons and reduces the probability of collisions. This affects the cross-sections in two ways : at low energy it produces a very soft decrease in σ^{pp} and contributes to the faster decrease in $\sigma^{p\bar{p}}$, at high energy it tames the rise due to σ^{jet}. It is then possible to have a very small p_{tmin} to see the very beginning of the rise around $10 \div 20 \, GeV$, without having too large a cross-section when the energy climbs into the TeV range and beyond.

The above discussion illustrates also the way to study virtual and real photon interactions : the aim is to obtain an energy and momentum dependence in the impact parameter distribution through the kinematics characterizing real or virtual photon process, as it has been done for the proton. At high Q_γ^2, one can expect q_{max} in Eq.(7) to increase with Q_γ^2, thus producing the same suppression effect that high c.m. energy values have in the softening of the rise due to the jets. The overall effect may be similar to the EVMD factor discussed in Sect. III. Work is in progress to apply the Bloch-Nordsieck formalism to these collisions.

V   CONCLUSION

We have discussed total cross-sections for protons and photons, real and virtual, using QCD calculable mini-jets cross-sections as the physical process which drives the rise of all total cross-sections. We have used the eikonal formalism to unitarize the cross-section and discussed possible ways to reduce the arbitrariness introduced in this formalism by the impact parameter distribution. A QCD model using soft gluon summation has been presented and compared with data for purely proton processes.

VI ACKNOWLEDGEMENT

Two of us, A.G. and G.P., acknowledge the support of EEC TMR-CT98-0169.

REFERENCES

1. A. Corsetti, R.M. Godbole and G. Pancheri, *Phys.Lett.* **B435**, 441 (1998).
2. M.M. Block, F. Halzen and T. Stanev, *Phys.Rev.* **D62**, 077501 (2000). M.M. Block, F. Halzen, G. Pancheri and T. Stanev, hep-ph/0003226, *25th Pamir-Chacaltaya Collaboration Workshop*, Lodz, Poland, November 1999.
3. L3 Collaboration, Paper 519 submitted to *ICHEP'98*, Vancouver, July 1998. M. Acciari et al., *Phys. Lett.* **B 408**, 450 (1997); L3 Collaboration, A. Csilling, *Nucl. Phys. Proc. Suppl.* **B82**, 239 (2000). L3 Note 2548, Submitted to the *OSAKA Conference*.
4. OPAL Collaboration. G. Abbiendi et al., *Eur. Phys. J.* **C14**, 199 (2000). F. Waeckerle, *Nucl. Phys. Proc. Suppl.* **B71**, 381 (1999), *Multiparticle Dynamics 1997* Eds. G. Capon, V. Khoze, G. Pancheri and A. Sansoni; Stefan Söldner-Rembold, hep-ex/9810011, To appear in the proceedings of the *ICHEP'98*, Vancouver, July 1998.
5. R.M. Godbole and G. Pancheri, e-print Archive hep-ph/0010104.
6. L. Durand and H. Pi, *Phys. Rev. Lett.* **58**, 58 (1987). A. Capella, J. Kwiecinsky, J. Tran Thanh, *Phys. Rev. Lett.* **58**, 2015 (1987). M.M. Block, F. Halzen, B. Margolis, *Phys. Rev.* **D 45**, 839 (1992). A. Capella and J. Tran Thanh Van, *Z. Phys.* **C 23**, 168 (1984). P. l'Heureux, B. Margolis and P. Valin, *Phys. Rev.* **D 32**, 1681 (1985).
7. D. Cline, F. Halzen and J. Luthe, *Phys. Rev. Lett.* **31**, 491 (1973). G. Pancheri and C. Rubbia, *Nucl. Phys.* **A 418**, 117c (1984). T.Gaisser and F.Halzen, *Phys. Rev. Lett.* **54**, 1754 (1985). G.Pancheri and Y.N.Srivastava, *Phys. Lett.* **B 158**, 402 (1986).
8. J.C. Collins and G.A. Ladinsky, *Phys. Rev.* **D 43**, 2847 (1991).
9. R.S. Fletcher , T.K. Gaisser and F. Halzen, *Phys. Rev.* **D 45**, 377 (1992); erratum *Phys. Rev.* **D 45**, 3279 (1992).
10. A. Grau, G. Pancheri and Y.N. Srivastava, *Phys. Rev.* **D60**, 114020 (1999).
11. M. Glück, E. Reya and I. Schienbein, *Phys. Rev.* **D60**, 054019 (1999), Erratum-ibid. **D 62**, 019902 (2000).
12. M. Glück, E. Reya and A. Vogt, *Zeit. Physik* **C 67**, 433 (1994).
13. ZEUS Collaboration, *Phys. Lett.* **B 293**, 465 (1992); *Zeit. Phys.* **C 63**, 391 (1994).
14. H1 Collaboration, *Zeit. Phys.* **C 69**, 27 (1995).
15. ZEUS Collaboration, J. Breitweg et al., DESY-00-071, e-print Archive: hep-ex/0005018.
16. ZEUS Collaboration (C. Ginsburg et al.), Proc. 8th International Workshop on Deep Inelastic Scattering, April 2000, Liverpool, to be published in World Scientific.
17. M.M. Block, E.M. Gregores, F. Halzen and G. Pancheri, *Phys. Rev.* **D58**, 17503 (1998); M.M. Block, E.M. Gregores, F. Halzen and G. Pancheri, *Phys. Rev.* **D60**, 54024 (1999).
18. See for example, M. Drees and R.M. Godbole, *Phys. Rev.*, **D 50**, 3124 (1994), e-print Archive: hep-ph/9403229; *Proceedings of PHOTON-95, incarporating the*

Xth International workshop on Gamma-Gamma collisions and related processes , Sheffield, April 6-10, 1995, pp. 123-130, e-print Archive: hep-ph/9506241.

19. C. Adloff *et al.*, H1 Collaboration, *Phys. Lett.*, **B 415**, 418 (1997), e-print Archive: hep-ex/9709017; *Eur. Phys. J.*, **C 13**, 397 (2000), e-print Archive: hep-ex/9812024.

20. L3 Collaboration, M. Acciari *et al.*, *Phys. Lett.*,**B453**, 333 (1999).

21. Ch. Friberg and T. Sjöstrand, *Eur. Phys. J.*, **C 13**, 151 (2000), e-print Archive: hep-ph/9907245; *JHEP*,**09**, 10 (2000) e-print Archive: hep-ph/0007314.

22. B. Surrow, DESY-THESIS-1998-004; A. Bornheim, In the Proceedings of the *LISHEP International School on High Energy Physics*, Brazil, 1998, hep-ex/9806021.

Double Tag Events in Two-Photon Collisions at LEP

Maneesh Wadhwa*

*University of Basel, Klingelbergstrasse 82,
CH-4056 Basel, Switzerland
E-mail: Maneesh.Wadhwa@cern.ch

Abstract. Double tag events in two photon collisions are studied using the L3 detector at the LEP center of mass energies $\sqrt{s} \simeq 189 - 202$ GeV. The cross-section of $\gamma^*\gamma^*$ collisions is measured at an average photon virtuality $\langle Q^2 \rangle = 15$ GeV^2. The results are in agreement with Monte Carlo predictions based on perturbative QCD, while the Quark Parton Model alone is insufficient to describe the data. The measurements are compared to the LO and the NLO BFKL calculations.

INTRODUCTION

In this paper we present new results on double-tag two-photon events $e^+e^- \rightarrow e^+e^-$hadrons. The data, collected at centre-of-mass energies $\sqrt{s} \simeq 189 - 202$ GeV, correspond to an integrated luminosity of 401 pb^{-1}. Both scattered electrons [1] are detected in the small angle electromagnetic calorimeters. The virtuality of the two photons, Q_1^2 and Q_2^2, is in the range of 4 GeV$^2 < Q_{1,2}^2 < 40$ GeV^2.

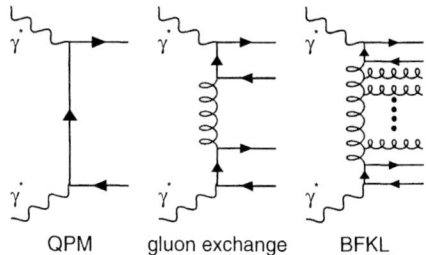

FIGURE 1. Diagrams for the a) QPM, b) one-gluon exchange and c) BFKL Pomeron processes in a $\gamma^*\gamma^*$ interaction.

The centre-of-mass energy of the two virtual photons, $\sqrt{\hat{s}} = W_{\gamma\gamma}$, ranges from 5 GeV to 90 GeV. The cross-section measurement of the two virtual photons is

---

[1] Electron stands for electron or positron throughout this paper.

considered as "golden" process to test the BFKL dynamics [2]. For this scheme the $\gamma^*\gamma^*$ interaction can be seen as the interaction of two $q\bar{q}$ pairs scattering off each other via multiple gluon exchange. (Fig. 1c). In the leading order approximation (LO), the cross-section in the saddle point approximation for the collision of two virtual photons is [3,4]:

$$\sigma_{\gamma^*\gamma^*} = \frac{\sigma_0}{Q_1 Q_2 Y} \left(\frac{s}{s_0}\right)^{\alpha_P - 1} \tag{1}$$

Here

$$\sigma_0 = \text{const}$$
$$s_0 = \frac{K Q_1 Q_2}{y_1 y_2} \;,\;\; Y = \ln(s/s_0) \tag{2}$$
$$y_i = 1 - (E_i/E_b)\cos^2(\theta_i/2)$$

where E_b is the beam energy, E_i and θ_i are the energy and polar angle of the scattered electrons and α_P is the "hard Pomeron" intercept; K is a scale factor which accounts for uncertainity in the BFKL energy scale s_0. The centre-of-mass energy of the two-photon system is related to the e^+e^- centre-of-mass energy s by $\hat{s} = W_{\gamma\gamma}^2 \approx s y_1 y_2$. In leading order $(\alpha_P - 1) = (4\ln 2)N_c\alpha_s/\pi$, where N_c is the number of colours. Using $N_c = 3$ and $\alpha_s = 0.2$, $(\alpha_P - 1) \simeq 0.53$. The born cross-section of one gluon exchange (see Fig. 1b) is independent of $W_{\gamma\gamma}$. Recently, effort has been devoted to improve the exact leading order calculation [2] by studying the effect of charm mass and the contribution of longitudinal photon polarization states [5]. Still these effects are not sufficient to describe our previous measurement [6]. One needs next to leading order corrections(NLO). It turns out that the NLO corrections [7] to the intercept "$\alpha_P - 1$" are negative for $\alpha_s > 0.16$. Different techniques [8–14] have been proposed to improve the NLO calculations in a suitable renormalization scheme thus giving values of $(\alpha_P - 1)$ in the range $0.17-0.33$.

EVENT SELECTION

The double-tag two-photon hadronic events are mainly triggered by two independent triggers: the central track [18,19] and the single and double tag energy [20] triggers. The central track trigger requires at least two charged particles, each with $p_t > 150$ MeV, back-to-back in the transverse plane. The single–tag energy trigger requires an high energy cluster in one of the small angle electromagnetic calorimeters, in coincidence with at least one track in the central tracking chamber. For the double–tag trigger the back-to-back small angle electromagnetic calorimeters must have a high energy cluster and no coincidence with the central detector is required. The total trigger inefficiency of the selected events is less than 1%.

Two-photon hadronic event candidates, $e^+e^- \rightarrow e^+e^-$hadrons, are selected using the following cuts:

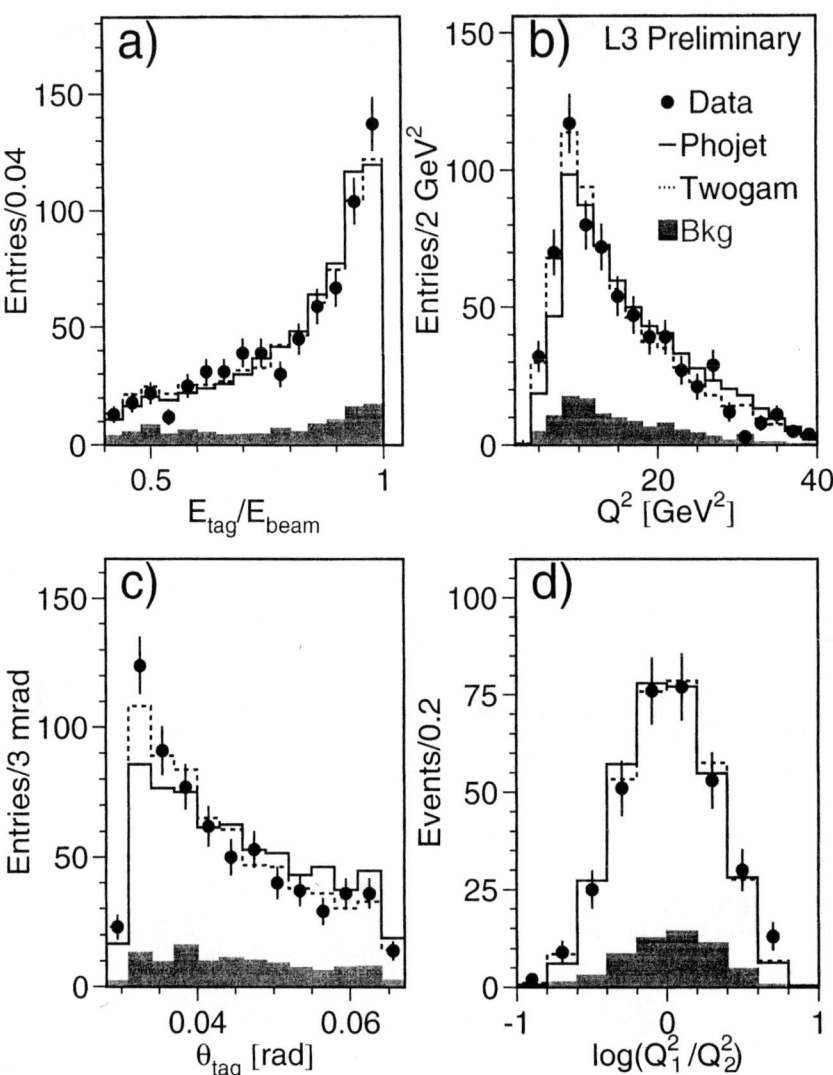

FIGURE 2. Distributions of a) E_{tag}/E_{Beam}, the energy of a tagging electron to the beam energy, b) Q^2 , c) θ_{tag} and d) the ratio Q_1^2/Q_2^2 of the scattered electrons. The data are compared to the Monte Carlo predictions, scaled to the number of data events. The backgrounds are mainly due to $e^+e^- \rightarrow e^+e^-\tau^+\tau^-$ and single tag two-photon hadronic events.

- There must be two identified electrons, forward and backward (double-tag), in the small angle electromagnetic calorimeters. Each electron is identified as the highest energy cluster in one of the calorimeters, with energy greater than 40 GeV. The polar angle of the two tagged electrons has to be in the range 30 mrad $< \theta_1 <$ 66 mrad and 30 mrad $< \pi - \theta_2 <$ 66 mrad.

- The number of particles, obtained from tracks and isolated calorimeter clusters in the polar angle region $20° < \theta < 160°$, must be greater than 5. The tracks are selected by requiring a transverse momentum, p_t, greater than 100 MeV and a distance of closest approach in the transverse plane to the interaction vertex smaller than 10 mm. Isolated energy clusters are required to have transverse energy greater than 100 MeV and no nearby track inside 35 mrad. These cuts are illustrated in

- The value of $Y = \ln (s/s_0)$ is required to be in the range $2 \le Y \le 7$.

- The opening angle between the scattered electrons must be smaller than 179.5°, to reject Bhabha events.

After these cuts, 336 events are selected with an estimated background of 31 events from misidentified single-tag and 21 events from $e^+e^- \to e^+e^-\tau^+\tau^-$. The contamination from annihilation processes (4 events) is negligible. More details regarding the selection can be found in ref [21].

Figure 2 shows the kinematic distributions of E_{tag}/E_{beam}, Q^2, θ_{tag} and of the ratio $\log(Q_1^2/Q_2^2)$ of the scattered electrons. PHOJET [15] and TWOGAM [16] Monte Carlos describe the data reasonably well. The Monte Carlo distribution shown are normalized to the number of data events.

DOUBLE-TAG CROSS-SECTION

The cross-sections are measured in the kinematic region limited by:

- $E_{1,2} > 30$ GeV, 30 mrad $< \theta_{\text{tag}} <$ 66 mrad and $2 \le Y \le 7$

The data is then corrected for efficiency and acceptance with two Monte-Carlo models; PHOJET [15] and Vermaseren(QPM) [17] respectively. The differential cross-sections $d\sigma(e^+e^- \to e^+e^- + \text{hadrons})/dY$ are measured in four ΔY intervals. As one can be seen in Table 1 and in Fig. 3, none of the models are sufficient to describe the data. The value of the cross-section at $5< Y < 7$ exceeds the Monte Carlo prediction by about 3.5 standard deviations.

From the measurement of the $e^+e^- \to e^+e^- + \text{hadrons}$ cross-section, σ_{ee}, we extract the two-photon cross-section, $\sigma_{\gamma^*\gamma^*}$, by using only the transverse photon luminosity function, $\sigma_{ee} = L_{TT} \cdot \sigma_{\gamma^*\gamma^*}$. In Fig. 4 we show $\sigma_{\gamma^*\gamma^*}$, after subtraction of the QPM contribution as a function of Y. Using an average value of Q^2, $\langle Q^2 \rangle = 15$ GeV^2 at $\sqrt{s} \simeq 189 - 202$ GeV, we calculate the one-gluon exchange contribution with the asymptotic formula. The expectations are below the data. The leading

TABLE 1. The differential cross-section $d\sigma(e^+e^- \to e^+e^- + \text{hadrons})/dY$ in picobarn measured in the kinematic region defined in the text, at $\sqrt{s} \simeq 189 - 202$ GeV. The predictions of the PHOJET and the QPM Monte Carlo models are also listed. The first error is statistical and the second is systematic.

ΔY	DATA $d\sigma/dY$(pb)	PHOJET $d\sigma/dY$ (pb)	QPM $d\sigma/dY$ (pb)
$2.0 - 2.5$	$0.50 \pm 0.07 \pm 0.03$	0.40	0.32
$2.5 - 3.5$	$0.29 \pm 0.03 \pm 0.02$	0.29	0.17
$3.5 - 5.0$	$0.15 \pm 0.02 \pm 0.01$	0.14	0.05
$5.0 - 7.0$	$0.08 \pm 0.01 \pm 0.01$	0.03	0.006

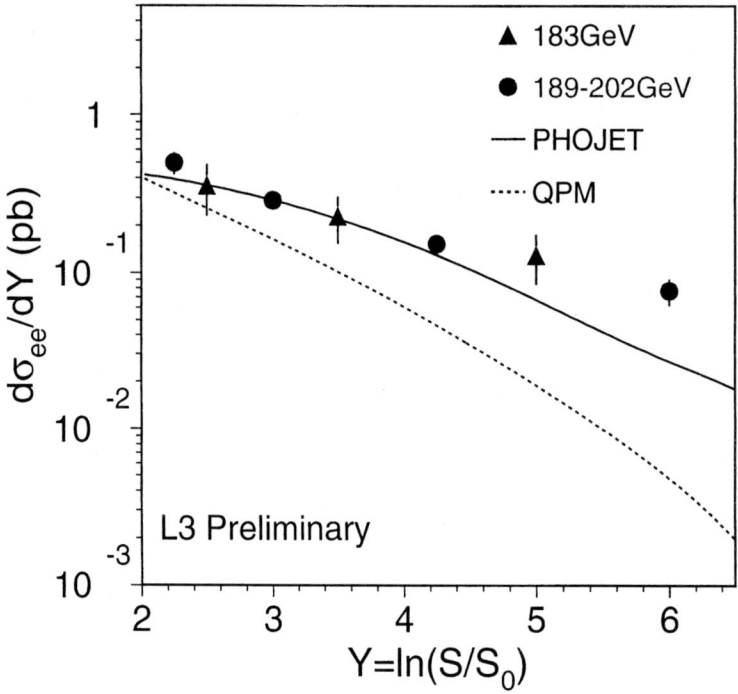

FIGURE 3. The cross-section of $e^+e^- \to e^+e^- hadrons$ as a function of Y in the kinematical region defined in the text at $\sqrt{s} \simeq 189 - 202$ GeV compared to our previous results $\sqrt{s} \simeq 183$ GeV. In the figure the predictions of PHOJET (continuos line) and of the QPM (dashed line) are indicated.

order expectations of the BFKL model,, shown as a dotted line in Fig. 4, are too high. By leaving α_P as a free parameter and $K = 1$, a fit to the data, taking into account the statistical, yields:

$$\alpha_P - 1 = 0.36 \pm 0.02, \quad \chi^2/d.o.f = 0.98/3$$

with $\chi^2/d.o.f = 0.98/3$ and if the energy scale factor K is a free parameter and $(\alpha_P - 1)=0.53$, a fit to data yields:

$$K = 6.4 \pm 1.0, \quad \chi^2/d.o.f = 1.34/3$$

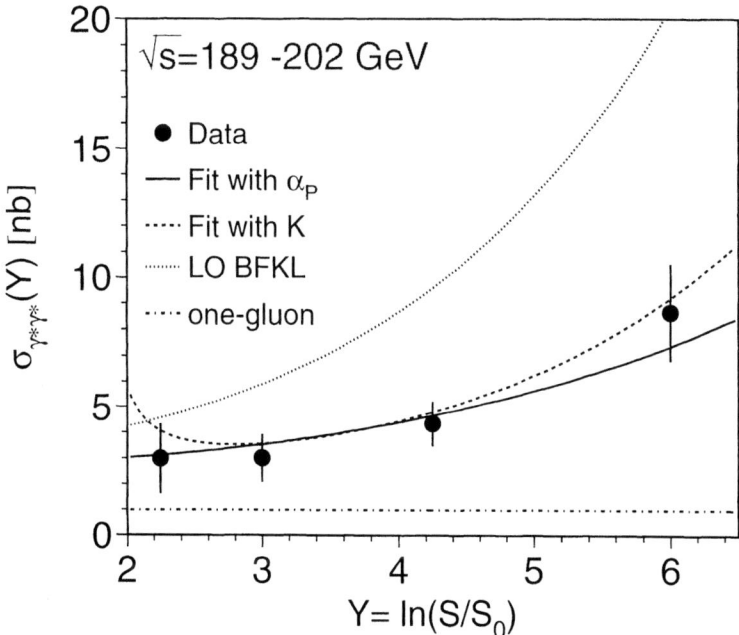

FIGURE 4. Two-photon cross-sections, $\sigma_{\gamma^*\gamma^*}$, after the subtraction of the QPM contribution at $\sqrt{s} \simeq 189 - 202$ GeV ($\langle Q^2 \rangle = 15$ GeV^2). The data are compared to the predictions of the LO BFKL calculation at saddle point approximation(eq.1)(dotted line) with K=1 and $(\alpha_P - 1) = 0.53$ and the solid line is the fit to the data of the LO BFKL (eq.1) with K=1 and the coefficient $(\alpha_P - 1)$ as a free parameter. The dashed line is the fit with $(\alpha_P - 1) = 0.53$ and the scale factor K as a free parameter.

These results are shown in Fig. 4 as a soild and dashed lines respectively. The value of $(\alpha_P - 1)$, smaller than expected from the LO BFKL calculation at the saddle point approximation, and the scale factor K much larger than unity indicate that higher order corrections are important. NLO calculations are in progress [14,9,13] which agree better with the experimental results.

ACKNOWLEDGEMENTS

I would like to thank C. H. Lin of his collaboration. This work is supported by the Swiss National Science Foundation.

REFERENCES

1. PLUTO Coll., C. Berger *et al.*, Phys. Lett. **B 142** (1984) 119; TPC/2γ Coll., D. Bintinger *et al.*, Phys. Rev. Lett. **54** (1985) 763; MD-1 Coll., S.E. Baru *et al.*, Z. Phys. **C 53** (1992) 219; TOPAZ Coll., R. Enomoto *et al.*, Phys. Lett. **B 368** (1996) 299

2. E.A. Kuraev, L.N. Lipatov and V.S. Fadin, Sov. Phys. JETP **45** (1977) 199; Ya.Ya. Balitski and L.N. Lipatov, Sov. J. Nucl. Phys. **28** (1978) 822

3. S.J. Brodsky, F. Hautmann and D.E. Soper, Phys. Rev. **D 56** (1997) 6957

4. J. Bartels, A. De Roeck and H. Lotter, Phys. Lett. **B 389** (1996) 742; J. Bartels, A. De Roeck, C. Ewerz and H. Lotter, hep-ph/9710500

5. J. Bartels, C. Ewerz, R. Staritzbichler, hep-ph/0004029

6. L3 Collab., M. Acciarri *et al.*, Phys. Lett. **B 453** (1999) 333

7. V. S. Fadin and L. N. Lipatov, Phys. Lett. **B 429** (1998) 127; G. Camici and M. Ciafaloni, Phys. Lett. **B 430** (1998) 349

8. V. S. Fadin and L. N. Lipatov, Proc. Theory Institute on Deep Inelastic Diffraction, ANL, Argonne, September 14 - 16, 1998; C. R. Schmidt, Phys. Rev. **D 60** (1999) 074003; J. R. Forshaw, D. A. Ross and A. Sabio Vera, Phys. Lett. **B 455** (1999) 273; S.J. Brodsky *et al.*, JETP Lett. **70** (1999) 15, hep-ph/99101229

9. G. Salam, JHEP **9807** (1998) 019

10. M. Ciafaloni *et al.*, Phys. Rev. **D 60** (1999) 114036; M. Ciafaloni and D. Colferai, Phys. Lett. **B 452** (1999) 372

11. R. S. Thorne, Phys. Rev. **D 60** (1999) 054031

12. G. Altarelli, R. D. Ball and S. Forte, hep-ph/0001157

13. V. T. Kim, L. N. Lipatov and G. B. Pivovarov, hep-ph/9911228 and hep-ph/9911242; V. Kim, private communication

14. N.N. Nikolaev,J. Speth and V.R. Zoller, hep-ph/0001120

15. PHOJET version 1.05c is used, R. Engel, Z. Phys. **C 66** (1995) 203; R. Engel and J. Ranft, Phys. Rev. **D 54** (1996) 4244

16. TWOGAM version 1.71 is used.
   L. Lönnblad *et al.*, "$\gamma\gamma$ event generators", in Physics at LEP2, ed. G. Altarelli, T. Sjöstrand and F. Zwirner, CERN 96-01 (1996), Volume 2, 224.
   S. Nova *et al.*, DELPHI Note 90-35 (1990).
   We thank our colleagues from DELPHI to make their program available to us

17. J.A.M. Vermaseren, Nucl. Phys. **B 229** (1983) 347

18. P. Béné *et al.*, Nucl. Inst. Meth. **A 306** (1991) 150

19. D. Haas *et al.*, Nucl. Inst. Meth. **A 420** (1999) 101

20. R. Bizzarri *et al.*, Nucl. Inst. Meth. **A 283** (1989) 799

21. Chih-Hsun Lin and M.Wadhwa, L3 Preprint 2568, June 2000

Measurement of the Cross-Section for the Process ee → eeγ*γ* → eeX at $\sqrt{s_{ee}} = 189 - 202$ GeV

M. Przybycień

CERN, CH-1211 Geneve 23, Switzerland
OPAL Collaboration

Abstract. The hadronic structure of interactions of virtual photons (γ^*) is investigated using the reaction $ee \to ee\gamma^*\gamma^* \to eeX$ based on data taken by the OPAL experiment at $\sqrt{s_{ee}}$ =189–202 GeV, where X represents the hadronic final state. The measured cross-sections are compared to predictions of the Quark Parton Model (QPM), to the leading order Monte Carlo model PHOJET, and to BFKL calculations. PHOJET describes the data reasonably well, the QPM predicted cross-section is too low and the cross-section prediction based on a LO BFKL calculation is too large.

I INTRODUCTION

The classical way to investigate the structure of the photon at e^+e^- colliders is the measurement of the process

$$e(p_1)e(p_2) \to e(p_1')e(p_2') \, \text{hadrons}, \tag{1}$$

proceeding via the exchange of two photons, which can be either quasi-real, γ, or virtual γ^*. The terms in brackets represent the four-vectors of the particles as shown in Fig. 1. Depending on the virtualities of the exchanged photons the scattered electrons[1] can be observed in the detector. In the analysis presented here both electrons are observed (double-tagged), which means they must be scattered at sufficiently large polar angles θ_i. This corresponds to the situation where both radiated photons, which take part in the hard scattering process, are highly virtual. Throughout the paper, $i = 1, 2$ denotes quantities which are connected with the upper and lower vertex in Fig. 1, respectively.

The results are based on data recorded by the OPAL experiment in the years 1998 and 1999 at e^+e^- centre-of-mass energies of $\sqrt{s_{ee}} = 189 - 202$ GeV, using events where both scattered electrons are observed in the small-angle silicon-tungsten (SW) luminometer. The measured differential cross-sections are compared

---

[1] Electrons and positrons are generically referred to as electrons.

CP571, *PHOTON 2000*, edited by A. J. Finch

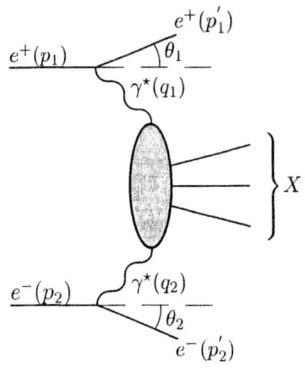

FIGURE 1. The diagram corresponding to the process $e^+e^- \rightarrow e^+e^-$ hadrons.

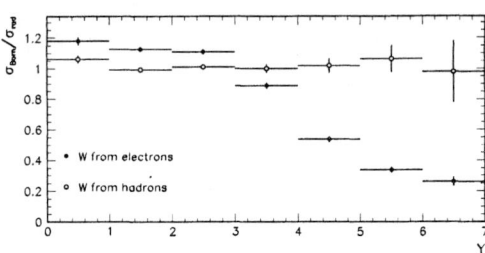

FIGURE 2. Radiative corrections for the process $e^+e^- \rightarrow e^+e^-$ hadrons as a function of Y for two different methods to calculate W: a combined electron/hadronic final state method (electrons), and a pure hadronic final state method (hadrons), as explained in the text.

to the prediction of the Quark Parton Model (QPM), to the PHOJET Monte Carlo model [1] and to BFKL calculations [2]. A similar analysis has been presented by the L3 Collaboration [3].

The e^+e^- centre-of-mass energy squared is given by $s_{ee} = (p_1 + p_2)^2$ and the hadronic invariant mass squared by $W^2 = (q_1 + q_2)^2$.

The kinematical variables Q_i^2, y_i and x_i are obtained from the four-vectors of the tagged electrons and the hadronic final state via:

$$Q_i^2 = 2E_b E_i (1 - \cos\theta_i), \qquad y_i = 1 - \frac{E_i}{E_b}\cos^2(\theta_i/2), \qquad x_i = \frac{Q_i^2}{Q_1^2 + Q_2^2 + W^2}, \quad (2)$$

where E_b refers to the beam energy, and the electron mass, m_e, has been neglected. In this analysis, the hadronic invariant mass, W, is obtained from the energies and momenta of final state hadrons.

For the comparison of the data to BFKL calculations the following kinematic quantity, which is a measure of the length of the gluon ladder, is defined [4]: $Y = \ln(s_{ee}/s_0)$, with $s_{ee}/s_0 = s_{ee}y_1y_2/\sqrt{Q_1^2 Q_2^2} \simeq W^2/\sqrt{Q_1^2 Q_2^2}$, where the last equality requires $W^2 \gg Q_i^2$.

The differential cross-section for the process of Eq. 1 in the limit $Q_i^2 = -q_i^2 \gg m_e^2$ and for small values of y_i is given by [5]:

$$d^6\sigma = \frac{d^3p_1' d^3p_2'}{E_1\, E_2} \mathcal{L}_{TT} \left(\sigma_{TT} + \sigma_{LT} + \sigma_{TL} + \sigma_{LL} + \frac{1}{2}\tau_{TT}\cos 2\bar{\phi} - 4\tau_{TL}\cos\bar{\phi} \right)$$

$$= \frac{d^3p_1' d^3p_2'}{E_1\, E_2} \mathcal{L}_{TT}\, \sigma_{\gamma^*\gamma^*}, \qquad \mathcal{L}_{TT} = \int \frac{d^3p_1' d^3p_2'}{E_1\, E_2} \mathcal{L}_{TT}, \quad (3)$$

where $\bar{\phi}$ is the angle between the two scattering planes of the electrons in the photon-photon centre-of-mass system. The cross-sections σ_{TT}, σ_{TL}, σ_{LT} and σ_{LL}

and the interference terms τ_{TT} and τ_{TL} correspond to specific helicity states of the interacting photons (T=transverse and L=longitudinal). The factor L_{TT} describes the flux of transversely polarized photons. For more details see [6]. The cleanest experimental quantity which can be extracted without making further assumptions, e.g. on the interference terms, is the cross-section for the reaction $e^+e^- \rightarrow e^+e^-$ hadrons as given in Eq. 3. By dividing with L_{TT} also the cross-section $\sigma_{\gamma^\star\gamma^\star}$ for the reaction $\gamma^\star\gamma^\star \rightarrow$ hadrons can be extracted with $\sigma_{\gamma^\star\gamma^\star}$ denoting the sum of all terms contained in the bracket of Eq. 3.

Recently, much attention has been given to the BFKL pomeron [2]. It has been argued [4,7–9] that e^+e^- colliders offer an excellent opportunity to test the BFKL prediction, through a measurement of $\sigma_{\gamma^\star\gamma^\star}$. For sufficiently large photon virtualities Q_1^2 and Q_2^2 (a few GeV^2), the BFKL calculation can be carried out without phenomenological input. The original LO BFKL calculations led to expect an increase of $\sigma_{\gamma^\star\gamma^\star}$ by a factor 20 or more compared to LO DGLAP calculations [10]. Meanwhile the LO calculations have been improved by including charm quark mass effects, running of the strong coupling constant α_s and the contribution of the longitudinal photon polarisation states. Recently, it has become clear that the Next-to-Leading Order (NLO) corrections to the BFKL equation are large. A phenomenological determination of the Higher Order (HO) effects was presented in Ref. [9] and the resulting BFKL scattering cross-sections were shown to increase by a factor 2-3 only, relative to the calculations without BFKL effects.

The size of the radiative corrections depends on the variables used to calculate the kinematics, and also to some extend on the Born cross-section, and was estimated using the BDK Monte Carlo [11]. Fig. 2 compares two methods to calculate the variable Y. The first method uses the hadronic final state variable W. The second method uses W calculated from the electrons, if at least one y_i is larger than 0.25, otherwise Y is calculated as for the first method. The QPM Born cross-section was reweighted to the leading order PHOJET cross-section, which agrees well with the measured cross-section. Fig. 2 shows the ratio of the Born to the full radiative (measured) cross-section. For the fully hadronic method the radiative corrections are small. However, for the electron method the corrections can be larger than 50% at large Y values. Obviously, measurements based on the electron kinematics cannot be compared readily with models or BFKL calculations in the region $Y > 4$, unless radiative corrections have been applied. Since the actual size of the radiative corrections also depends on the Born cross-section itself, an iterative procedure would be required to extract the Born cross-section. Since the present statistics does not permit such a procedure, only the hadronic variables have been used to calculate Y, for which the corrections are much smaller.

II EVENT SELECTION

The data sample used in this analysis corresponds to an integrated luminosity of 377.7 pb^{-1}. Double-tagged two-photon events were selected with the following cuts:

1. Two electron candidates, one in each SW detector, with energies $E_{1,2} > 0.4E_b$ and polar angles in the range $34 < \theta_{1,2} < 55$ mrad, should be observed.

2. There should be no single cluster in other subdetectors with an energy above $0.25E_b$.

3. At least 3 tracks (N_{ch}) have to be found in the tracking system.

4. The visible invariant mass, W_{vis}, is required to be larger than $5\,\text{GeV}$.

5. The z position of the primary vertex $|\langle z_0 \rangle|$ is required to be less than 4 cm, and the distance of the vertex from the beam axis should be less than 0.5 cm.

6. The z component of the total momentum vector of the event, is required to be less than 35 GeV and the total energy measured in the event should be less $2.2E_b$.

7. Remaining Bhabha events with random overlap of hadronic activity are tagged using the back-to-back topology of the scattered electrons, both having an energy larger than $0.7E_b$. Events are rejected if the difference in radius, ΔR and difference in azimuthal angle, $\Delta\phi$ of the position of the two clusters are $\Delta R < 0.5$ cm and $(\pi - 0.1) < |\Delta\phi| < \pi$ rad.

With these cuts 129 events are selected in the data. Among these events, we expect 12.8 background events coming from e^+e^- background processes. The dominant background stems from the processes $e^+e^- \to e^+e^-\tau\bar{\tau}$ and $e^+e^- \to e^+e^-e^+e^-$ and was estimated using the Vermaseren Monte Carlo program [12].

Upstream beam-gas interactions result in off-momentum electrons observed in the SW detectors faking final state electrons from the process $e^+e^- \to e^+e^-$ hadrons. This background was estimated using a sample of Bhabha events. Additional clusters in the SW detectors, which do not belong to the Bhabha event, are counted as off-momentum electrons. The off-momentum background estimate was used to calculate the contribution of fake double-tagged events, resulting from the overlap of one background cluster with a single-tag two-photon event and the overlap of two background clusters with an untagged event. In total 18.2 overlap events are predicted for the data. After subtraction of all backgrounds 98 events remain.

A PHOJET Monte Carlo sample is used to correct the data for acceptance and resolution effects. It has been checked that the sum of the signal as predicted by PHOJET and the estimated background from overlaps with off-momentum electrons and other physics processes gives a good description of all important kinematical variables. A possible exception is the low energy part of the electron spectrum, where some excess in data is seen. Note that PHOJET does not contain any effects from BFKL, which would show up exactly in that region.

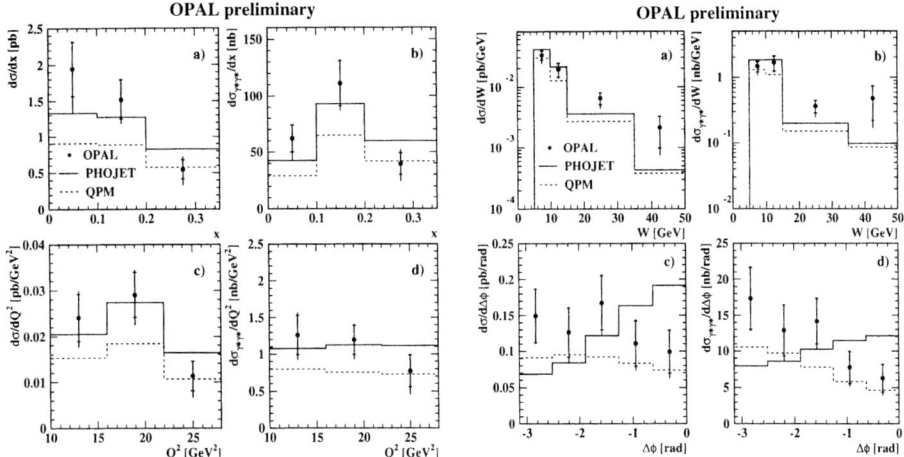

FIGURE 3. Cross-sections for the process $e^+e^- \rightarrow e^+e^-$ hadrons in the region $E_{1,2} > 0.4E_b$, $34 < \theta_{1,2} < 55$ mrad and $W > 5$ GeV, and for the process $\gamma^*\gamma^* \rightarrow$ hadrons for $\langle Q^2 \rangle = 17$ GeV^2, as functions of x (a,b) and Q^2 (c,d). Data are shown as full dots in the center of the bins. The inner error bars represent the statistical errors and the outer error bars represent statistical and systematic errors added in quadrature. Predictions of PHOJET are shown as solid lines, and those of QPM as dashed lines.

FIGURE 4. Cross-sections for the process $e^+e^- \rightarrow e^+e^-$ hadrons in the region $E_{1,2} > 0.4E_b$, $34 < \theta_{1,2} < 55$ mrad and $W > 5$ GeV, and for the process $\gamma^*\gamma^* \rightarrow$ hadrons for $\langle Q^2 \rangle = 17$ GeV^2, as functions of W (a,b) and $\Delta\phi$ (c,d). Symbols as in Fig. 3.

III   RESULTS

The cross-section for the process $e^+e^- \rightarrow e^+e^-$ hadrons has been measured in the kinematic region defined by the scattered electron energies $E_{1,2} > 0.4E_b$, the polar angles in the range $34 < \theta_{1,2} < 55$ mrad with respect to the beam direction, and $W > 5$ GeV. From the measurement of the cross-section of $e^+e^- \rightarrow e^+e^-$ hadrons we extract the cross-section $\gamma^*\gamma^* \rightarrow$ hadrons using L_{TT} (Eq. 3), calculated separately for each bin using Monte Carlo. The results for $\gamma^*\gamma^* \rightarrow$ hadrons are at an average $\langle Q^2 \rangle$ of 17 GeV^2. The cross-sections are presented as a function of x, Q^2 (Fig. 3), and W and the azimuthal correlation between the two electrons $\Delta\phi$ (Fig. 4),. Here Q^2 refers to the maximum of Q_1^2 and Q_2^2, and x is the corresponding value of x_i. For the comparison with a BFKL calculation we also present the cross-section as a function of Y (Fig. 5).

The main contribution to the systematic errors comes from varying the lower cuts on $\theta_{1,2}$ and W. The normalisation uncertainty due to the luminosity measurement is less than 1% and has been neglected.

The total measured cross-section for the process $e^+e^- \rightarrow e^+e^-$ hadrons in the previously defined phase space, is 0.40 ± 0.05 (stat) ± 0.05 (sys) pb. The expected

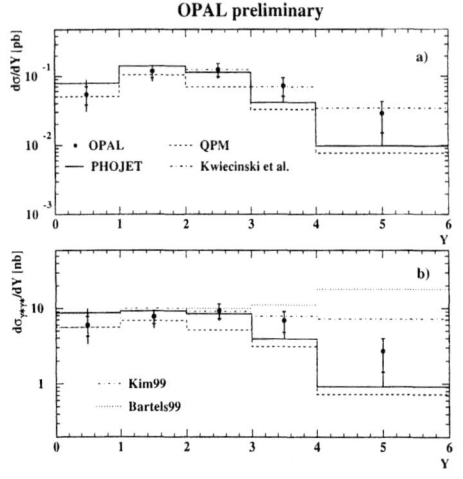

FIGURE 5. Cross-sections for the process $e^+e^- \to e^+e^-$ hadrons in the region $E_{1,2} > 0.4E_b$, $34 < \theta_{1,2} < 55$ mrad and $W > 5$ GeV, and the process $\gamma^*\gamma^* \to$ hadrons for $\langle Q^2 \rangle = 17$ GeV^2, as a function of Y. Data are shown as full dots in the center of the bins. The inner error bars represent the statistical errors and the outer error bars represent statistical and systematic errors added in quadrature. Predictions of PHOJET are shown as the solid lines, and those of QPM as dashed lines. Three BFKL calculations are shown: a LO one from Bartels et al. (Bartels99), NLO from Kim et al. (Kim99) and the calculation from Kwiecinski et al., using the consistency constraint.

cross-section from PHOJET is 0.39 ± 0.02 (stat) pb, while the prediction for QPM is 0.27 ± 0.02 (stat) pb.

PHOJET generally describes the data while the QPM prediction is too low. The model predictions are generally below the data for W values larger than 20 GeV. The data show that the distribution of $\Delta\phi$ is flat. PHOJET does not describe the $\Delta\phi$ distribution whereas QPM reproduces the shape of the distribution well. The BFKL predictions are shown for the LO calculation [14], a NLO calculation [15] and a (partial) HO calculation [16]. All BFKL predictions are shown for $Y > 2$, except for the NLO calculation which has been calculated for $Y > 1$. The QPM prediction is systematically below the data for $Y > 2$. The PHOJET model, which does not include BFKL-like effects, gives a good description of the data, with the possible exception of the highest Y values. In the region $Y > 2$ the $\gamma^*\gamma^* \to$ hadrons measured cross-section is 0.221 ± 0.039 pb with 0.165 pb (PHOJET) and 0.112 pb (QPM) expected.

The BFKL predictions can be calculated only for large values of Y. For all BFKL predictions shown the cross-section is significantly larger than the PHOJET prediction for $Y > 3$, and the differences increase with increasing Y. The LO BFKL calculation predicts a cross-section which is too large compared to the data. This LO BFKL calculation (Bartels99) [14] already incorporates improvements compared to the original results [4] by including effects of the charm quark mass, the running of the strong coupling constant α_s and contribution of longitudinal photon polarization states. Hence BFKL effects as large as predicted by the LO calculations are not confirmed by the data. BFKL cross-sections have been calculated to NLO (Kim99) [15], using the so-called BLM [17] optimal scale setting. At the highest Y value the NLO-BFKL cross-section is a factor seven larger than the PHOJET prediction. The data lies in between these two predictions. Finally, the calculation (Kwiecinski) [16] contains the dominant contribution of the higher order corrections via the so called consistency constraint, to all orders. Its prediction in the highest reachable Y range is about a factor two lower than for the NLO calculation, and agrees with the data in the full range.

In short, the LO BFKL cross-sections are too large, however the higher order calculations, which predict smaller BFKL effects, are close to the measurements. The measurements lie in between the predictions with and without BFKL effects, but limited statistics precludes firmer conclusions.

Acknowledgements. I am very greatful to Albert de Roeck and Richard Nisius for the fruitfull collaboration on the results presented in this paper.

REFERENCES

1. R. Engel, Z. Phys. **C66** (1995) 203;
   R. Engel and J. Ranft, Phys. Rev. **D54** (1996) 4246.
2. E.A. Kuraev, L.N. Lipatov and V.S. Fadin, Sov. Phys. JETP **45** (1977) 199;
   Ia. Balitski and L.N. Lipatov, Sov. J. Nucl. Phys. **28** (1978) 822.
3. L3 Collaboration, M.Acciarri et al., Phys. Lett. **B453** (1999) 333;
   Maneesh Wadhwa, *Double tagged events in two photon collisions*, these proceedings.
4. J. Bartels, A. De Roeck and H. Lotter, Phys. Lett. **B389** (1996) 742;
   J. Bartels, A. De Roeck, H. Lotter and C. Ewerz, DESY preprint 97-123E, *The $\gamma^*\gamma^*$ Total Cross-Section and the BFKL pomeron at the 500 GeV e^+e^- Linear Collider*, hep-ph/9710500;
   S.J. Brodsky, F. Hautmann and D.E. Soper, Phys. Rev. **D56** (1997) 6957.
5. V.M. Budnev, I.F. Ginzburg, G.V. Meledin and V.G. Serbo, Phys. Rep. **15**(1975)181.
6. R. Nisius, Phys. Rep. **332** (2000) 165.
7. S.J. Brodsky, F. Hautmann and D.E. Soper, Phys. Rev. Lett. **78** (1997) 803.
8. A. Białas, W. Czyż and W. Florkowski, Eur. Phys. J. **C2** (1998) 683.
9. M. Boonekamp, A. De Roeck, C. Royon and S. Wallon, Nucl. Phys. **B555**(1999)540.
10. G. Altarelli and G. Parisi, Nucl. Phys. **B126** (1977) 298;
    V.N. Gribov and L.N. Lipatov, Sov. J. Nucl. Phys. **15** (1972) 438;
    L.N. Lipatov, Sov. J. Nucl. Phys. **20** (1975) 96;
    Y.L. Dokshitzer, Sov. Phys. JETP. **46** (1977) 641.
11. F.A. Berends, P.H. Daverveldt and R. Kleiss, Nucl. Phys. **B253** (1985) 421; Comp. Phys. Comm. **40** (1986) 271; Comp. Phys. Comm. **40** (1986) 285; Nucl. Phys. **B264** (1986) 243.
12. J.A.M. Vermaseren, Nucl. Phys. **B229** (1983) 347.
13. J. Bartels, V. Del Duca and M. Wüsthoff, Z. Phys. **C76** (1997) 75.
14. J. Bartels, C. Ewerz and R. Staritzbichler, *Effects of the Charm Quark Mass on the BFKL $\gamma^*\gamma^*$ Total Cross Section at LEP*, hep-ph/0004029.
15. S.J. Brodsky et al., JETP Lett. **70** (1999) 155;
    V. Kim, private communication.
16. J. Kwiecinski and L. Motyka, Phys. Lett. **B462** (1999) 203.
17. S.J. Brodsky, G.P. Lepage and P.B. Mackenzie, Phys. Rev. **D28** (1983) 228.

Total Cross Sections in $\gamma\gamma$ Collisions

Maria Novella Kienzle-Focacci*

*University of Geneva, Switzerland.
E-mail : Maria.Kienzle@cern.ch

Abstract. The reaction $e^+e^- \to e^+e^-\gamma^*\gamma^* \to e^+e^-$ *hadrons* is analysed for quasi-real photons using data collected by the L3 detector during the LEP high energy runs from $\sqrt{s}$ = 189 GeV to 202 GeV. The total cross sections $\sigma(e^+e^- \to e^+e^-$ *hadrons*) and $\sigma(\gamma\gamma \to$ *hadrons*) are measured in the interval 5 GeV $\leq W_{\gamma\gamma} \leq$ 145 GeV. The centre-of-mass energy dependence of the two-photon cross section is described by the Regge parametrisation. We observe a steeper rise with energy as compared to hadron-hadron cross sections. The observed energy dependence can be reproduced by introducing a hard Pomeron component or by QCD models which include an important contribution of the hard scattering of the partons inside the photon.

INTRODUCTION

At high energies the two-photon process $e^+e^- \to e^+e^-\gamma^*\gamma^* \to e^+e^-$ *hadrons* is a copious source of hadron production. In this reaction most of the initial energy is taken by the scattered electrons and positrons. As their scattering angle is close to the beam they often go undetected. A large fraction of the hadrons escape detection, due to the large diffractive cross section and to the Lorentz boost of the $\gamma\gamma$ system. For these events, the measured visible mass W_{vis} is smaller than the two photon effective mass $W_{\gamma\gamma}$.

Data have been collected with the L3 detector at $\sqrt{s}$= 189 GeV with a total integrated luminosity of 171.8 pb^{-1} during 1998 and at $192 \leq \sqrt{s} \leq 202$ GeV with a total integrated luminosity of 220.8 pb^{-1} during 1999. New results on total cross section $\sigma(e^+e^- \to e^+e^-$ *hadrons*) are presented for anti-tagged events. The two-photon cross section $\sigma(\gamma\gamma \to$ *hadrons*) is then derived in the interval $5 \leq W_{\gamma\gamma} \leq 145$ GeV. The analysis of the data taken at $\sqrt{s}$= 130 − 161 GeV [1] is superseded by the present analysis which covers a wider mass interval and has twenty times more statistics. The $\gamma\gamma \to$ *hadrons* processes are simulated with the Monte Carlo event generators PHOJET [1] [2] and PYTHIA [2] [3].

---

[1] PHOJET version1.05c
[2] PYTHIA version 5.718 and JETSET version 7.408

CP571, *PHOTON 2000*, edited by A. J. Finch

EVENT SELECTION

Hadronic two-photon events are selected by the following criteria :

- At least six particles must be detected (Fig.1a). A particle can be a track, a photon or an hadronic cluster in the hadron calorimeter or in the luminosity monitor.

- The total energy in the electromagnetic calorimeter is required to be greater than 500 MeV in order to suppress beam-gas and beam-wall backgrounds and smaller than 50 GeV, to exclude events of the type $e^+e^- \rightarrow \gamma(Z \rightarrow q\bar{q})$. The total energy deposited in the calorimeters must be smaller than 40 % of the centre of mass energy, to exclude annihilation events (Fig.1b).

- An anti-tag condition is imposed which excludes events with clusters having energy greater than 30 GeV in the luminosity monitor, in a fiducial region of $33-64$ mrad (Fig.1c).

After selection, the background from beam-gas and beam-wall interactions is found to be negligible. The visible effective mass of the event, calculated from the four-momentum vectors of the measured particles, is required to be $W_{\text{vis}} \geq 5$ GeV. Almost 2 million events are selected. In Fig.1d the W_{vis} spectrum is shown for the full data sample. The observed cross-section is ~ 4.6 nb.

Comparison of the data with Monte Carlo predictions are shown in Fig.1. The background is below 1% at low masses, dominated by the two-photon τ production. It increases at high masses, due mainly to annihilation processes and reaches a maximum of $\simeq 15\%$ at 130 GeV. Although the data are better reproduced by PHOJET, the analysis of specific hadronic channels, the diffractive ρ [4] and the inclusive charm [5] production, show some inadequacy of the PHOJET simulation. For this reason we use in this analysis both Monte Carlos, with equal statistics, to unfold and correct the data

MEASUREMENT OF CROSS SECTIONS

From the observed distribution of W_{vis} the true hadron mass $W_{\gamma\gamma}$ distribution is extracted, using the same method described in [1]. The result depends on the Monte Carlo used to unfold the data. Data unfolded with PYTHIA are up to 30% higher than if unfolded with PHOJET.

From the number of events, corrected with the Monte Carlo in each $W_{\gamma\gamma}$ bin, and the integrated e^+e^- luminosity, the cross section $d\sigma(e^+e^- \rightarrow e^+e^- hadrons)$ is measured. The differential cross section $d\sigma/dW_{\gamma\gamma}$ is shown in Fig.2a together with our measurements at lower LEP collision energy. The fast decrease of the cross section as a function of $W_{\gamma\gamma}$ is due to the two photon luminosity function, $\mathcal{L}_{\gamma\gamma}$, which depends on $\tau = W_{\gamma\gamma}^2/s$.

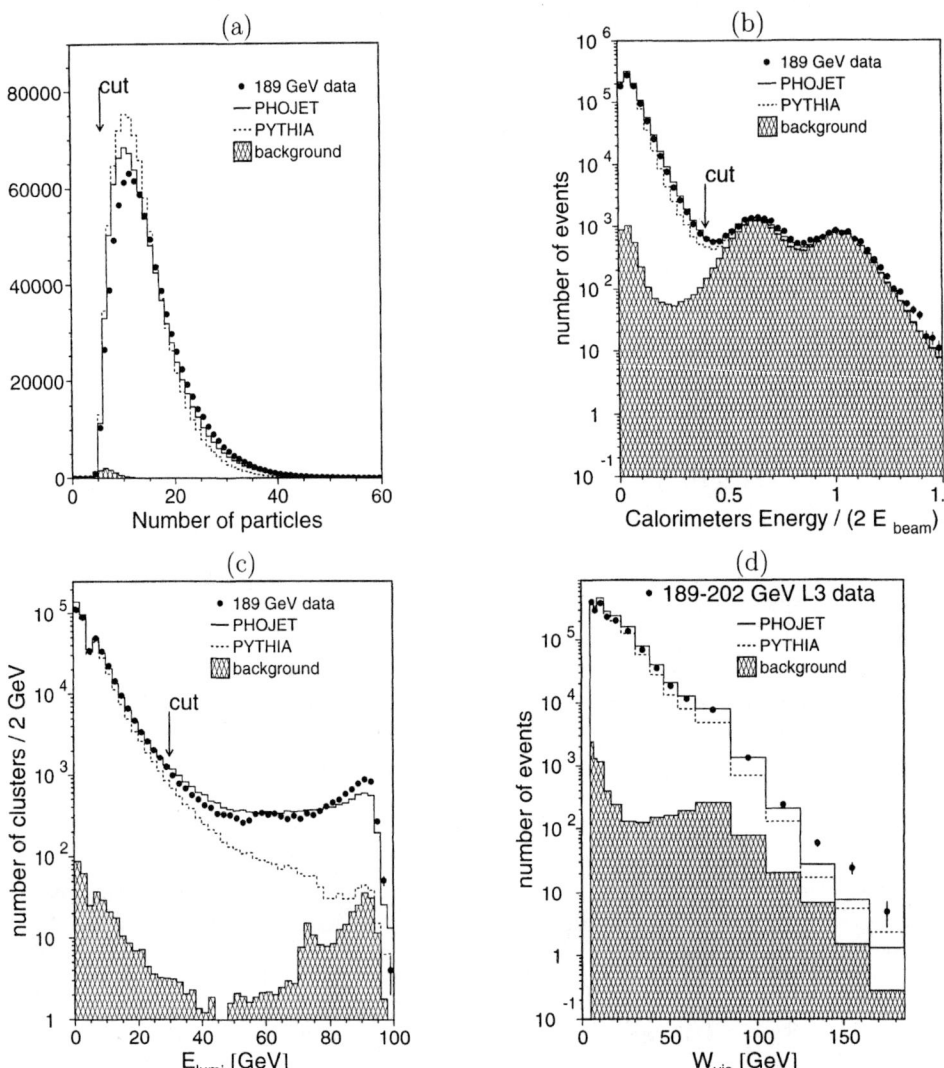

FIGURE 1. a) Distribution of the number of particles (pions and photons) in two-photon events. b) Energy in the electromagnetic and hadronic calorimeters normalised to the centre-of-mass energy. c) Energy in the luminosity monitor calorimeter, the cut at 30 GeV separates the untagged from the tagged sample. d) The measured hadronic mass W_{vis}. The data are compared to Monte Carlo predictions. The backgrounds due to e^+e^- annihilation and $e^+e^- \to e^+e^-\tau^+\tau^-$ are indicated as a shaded area.

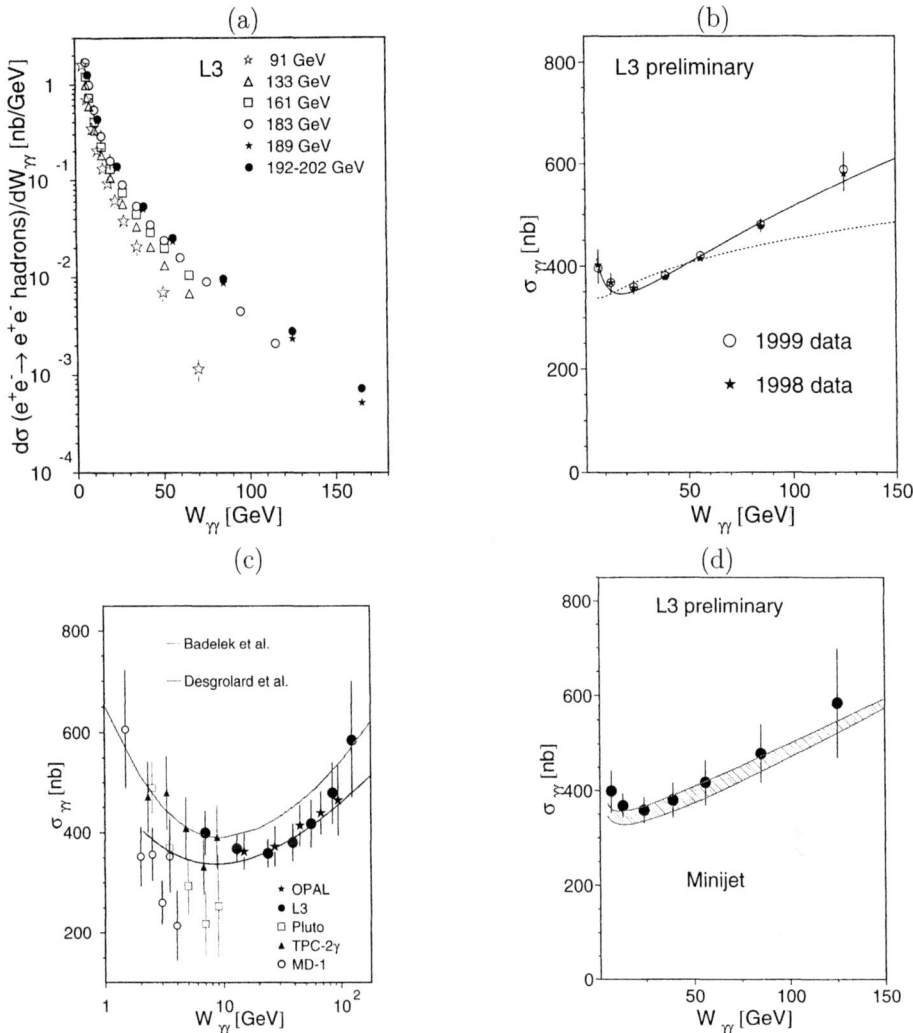

FIGURE 2. a) The cross section $d\sigma(\,e^+e^- \to e^+e^-hadrons)/dW_{\gamma\gamma}$ measured at various beam energies. b) The cross section $\sigma(\gamma\gamma \to hadrons)$ for real photons. The data of 1998 at $\sqrt{s}=189$ GeV are compared to the data of 1999 at $\sqrt{s}=192-202$ GeV. The Regge fits described in the text are superimposed to the data. The continous lines is the fit with the coefficient ϵ left as a free parameter, the dashed line is the fit with ϵ fixed to 0.095. c) The predictions of Ref. [9] and [10] are compared to all two-photon total cross section data. d) Predictions of the minijet model [15] compared to present data. In a) and b) only the statistic and experimental systematic errors are added in quadrature. In c) and d) also the model systematics are added.

The total cross section of two real photons is obtained by deconvoluting the photon flux $\mathcal{L}_{\gamma\gamma}$ [6] and by extrapolating the measurement to $Q^2 = 0$. This is done with an analytical calculation [7] as in [1]. An overall uncertainty of $\pm 5\%$ must be attributed to the extracted $\sigma_{\gamma\gamma}$ due to different hypothesis for the form factors.

The systematic errors are evaluated for each $W_{\gamma\gamma}$ bin. They are $\leq 8\%$, reaching the maximum value at the lowest and at the highest $W_{\gamma\gamma}$ bin. The main contributions are due to the cut on the particle multiplicity and to the antitag cut. The uncertainty related to the Monte Carlo model, evaluated by repeating the analysis with PYTHIA and PHOJET separately, is the largest individual contribution almost in all bins and is considered only when comparing the data to theoretical models.

The $\gamma\gamma \rightarrow hadrons$ cross sections at $\sqrt{s} = 189$ GeV and $\sqrt{s} = 196\text{-}202$ GeV are compared in Fig.2b. The two sets of data are compatible within the experimental systematic errors.

MODELS FOR $\gamma\gamma$ TOTAL CROSS SECTIONS

A Regge parametrisation of the form $\sigma_{\text{tot}} = A s^{\epsilon} + B s^{-\eta}$ can account for the energy behaviour of all hadronic and photoproduction total cross sections. A fit of all hadron total cross sections [8] gives a result compatible with a universal value of $\epsilon = 0.095 \pm 0.002$ and $\eta = 0.34 \pm 0.02$. The coefficients A and B are process and Q^2 dependent. If photons behave predominantly like hadrons, this expression may also be valid for the two-photon total hadronic cross section. A fit to our data, with ϵ and η fixed to the universal value, dashed line in Fig.2b, does not represent well the $\sigma_{\gamma\gamma}$ energy dependence. We therefore try a fit with A, B and ϵ as free parameters (full line in Fig.2b), we obtain: $\epsilon = 0.250 \pm 0.016$ with a confidence level of 30%. This value is more than a factor two higher than the universal value, independent on the Monte Carlo model used to correct the data.

Many models where recently compared to the L3 and OPAL [11] measurements. Often they constrain the two-photon cross section with the measurements of proton-proton and photoproduction total cross sections, by assuming factorization for the Pomeron residue: $\sigma_{\gamma\gamma} \approx \sigma_{\gamma\text{p}}^2 / \sigma_{\text{pp}}$ [13]. In general these models give an energy dependence of the cross section similar to the universal fit, two examples [9], [10] are shown in Fig.2c) together with the results of previous experiments [12]. Our measurement is in very good agreement with the OPAL measurement [11], this is true also if the two experiments unfold their data with only PYTHIA or PHOJET models.

Recently models with two Pomerons have been proposed [14] to explain the fast energy increase of deep inelastic scattering J/ψ photoproduction. In these schemes, corresponding to a colour-dipole picture for the high energy scattering amplitude, there are two trajectories carrying the vacuum quantum numbers with different fixed intercepts and with residues dependent on Q^2. The "soft Pomeron" has intercept $\alpha_{soft}(0) \sim 1.08$ and the "hard Pomeron" $\alpha_{hard}(0) \sim 1.4$. Because of the

$q\bar{q}$ component in the photon wave-function the hard Pomeron contributes to the two-photon cross-section also at $Q^2 = 0$. This model predicts a more rapid energy dependence for $\sigma_{\gamma\gamma}$ than for a normal hadron, as observed in our data.

The energy dependence of the total cross sections is also calculated by eikonalized mini-jet models. The calculations of R.M.Godbole and G.Pancheri [15], shown in Fig.2d as a band of values consistent with photoproduction data, agree well with our measurement.

ACKNOWLEDGEMENTS

I will like to express my gratitude to all the organizers and especially to Alex Finch for the lively atmosphere of this Conference.

REFERENCES

1. L3 Coll., M. Acciari et al.,*Phys. Lett.*B **408** (1997) 450.
2. R. Engel, Z. Phys. **C 66** (1995) 203 ;
   R. Engel and J. Ranft, *Phys. Rev.*D **54** (1996) 4246.
3. T. Sjöstrand, Comput. Phys. Commun. **82** (1994) 74.
4. Á. Csilling, Diffractive Vector Meson Production in Photon-Photon Interactions, Ph.D. thesis, Eötvös Univerity, Budapest, 1999, L3 internal note 2487.
5. L3 Collab., M. Acciarri et al., *Phys. Lett.*B 453 (1999) 83; and *Phys. Lett.*B 467 (1999) 137.
6. V.M. Budnev et al., Phys.Rep. **C 15** (1974) 181.
7. G.A. Schuler, Improving the equivalent-photon approximation in electron-positron collisions, hep-ph/9610406, CERN-TH/96-297.
8. Review of Particle Physics, Eur.Phys.J. **C3** (1998) 1.
9. B. Badelek et al., Acta Phys. Polon. 30(1999) 1807 and hep-ph/0001161 (2000).
10. P. Desgrolard et al., Eur. Phys. J **C 9** (1999) 623.
11. OPAL Coll.,K. Ackerstaff et al., Eur. Phys. J **C 14** (2000) 199.
12. PLUTO Coll., Ch. Berger et al., *Phys. Lett.*B 149 (1984) 421;
    TPC/2γ Coll., H. Aihara et al., *Phys. Rev.*D 21 (1990) 2667;
    MD1 Coll., S.E. Baru et al., Z.Phys. C 53 (1992) 219.
13. V.N.Gribov and I.Ya.Pomeranchuk, *Phys. Rev. Lett.*8 (1962) 343.
    S.J.Brodsky, J.Phys.Suppl. (Paris) **35** (1974) C2-69.
    K.V.L.Sarma and V. Singh , *Phys. Lett.*bf B 101 (1981) 201.
14. A. Donnachie, H.G. Dosch and M. Rueter *Phys. Rev.*D **59** (1999) 074011.
15. R.M.Godbole and G.Pancheri, *Nucl. Phys.* Proc. Suppl. **82** (2000) 246 and these proceedings.

INCLUSIVES

Introduction to high$-p_T$ inclusives

Matthew Wing[†]

[†] *McGill University, Physics Department,*
3600 University Street, Montreal,
Canada, H3A 2T8
E-mail: wing@mail.desy.de

Abstract. A selection of theoretical and experimental results are presented in the broader context of understanding QCD and its relation to photon physics. A phenomenological analysis of HERA data to constrain the gluon content of the proton and photon is discussed. Measurements from the Tevatron and fixed-target experiments are compared to theoretical predictions. Finally, the future of higher order pQCD calculations is addressed.

INTRODUCTION

In these proceedings, advances in understanding QCD are discussed in experiments which directly complement those where measurements of the photon structure are being made. For a generalised accelerator, where the incoming particles, I_1 and I_2 resolve into partons, the cross section at leading order, $d\sigma_{I_1 I_2 \to cd}$, can be written as,

$$d\sigma_{I_1 I_2 \to cd} = \sum_{ab} \int_{x_{I_2}} \int_{x_{I_1}} f_{I_2 \to b}(x_{I_2}, \mu_{I_2}^2) f_{I_1 \to a}(x_{I_1}, \mu_{I_1}^2) \mathcal{M}_{ab \to cd}^2, \qquad (1)$$

where $f_{I \to b}(x_I, \mu_I^2)$, is the parton density function for a give momentum fraction, x_I and factorisation scale, μ_I and $\mathcal{M}_{ab \to cd}$ is the $2 \to 2$ matrix element. This entails two or three unknowns; the perturbatively calculable matrix element and one or two structure functions. From equation (1) it can be seen that other experiments, such as the Tevatron, can provide complementary information on $\mathcal{M}_{ab \to cd}$ and therefore indirectly help measurements of the photon structure function.

HERA PHOTOPRODUCTION DATA

Improvements in the understanding of pQCD for jet photoproduction have led to the agreement bewteen independent calculations to within $5 - 10\%$ [1,2]. Considering the cross section as a function of pseudorapidity of one jet whilst restricting

CP571, *PHOTON 2000*, edited by A. J. Finch

the other jet in the dijet system to a smaller region in pseudorapidity provides a good test of dijet production and the structure of the photon [1].

In a recent paper [3], Aurenche et al. have considered using the HERA photoproduction data to constrain the gluon content of the photon and proton. The distribution in x_γ^{obs}, the fraction of the photon's energy participating in the production of the two highest energy jets;

$$x_\gamma^{obs} = \frac{\sum_{jet1,2} E_T^{jet} e^{-\eta^{jet}}}{2y E_e}, \tag{2}$$

where yE_e is the initial photon energy, was considered in different regions of pseudorapodity of the jet. Requiring two jets, $E_T^{jet1,2} > 12, 10$ GeV and the pseudorapidity, $0 < \eta^{jet} < 1$, the cross section shows a small dependence on the gluon distribution in the photon. However, when the jets are constrained to be more forward in pseudorapidity, $1 < \eta^{jet} < 2$, a significant sensitivity is seen at low$-x_\gamma^{obs}$. A change of 30% in the gluon density of the photon results in a 25% change in the cross section at $x_\gamma^{obs} = 0.2$.

The NLO predictions are then compared to published data from the ZEUS collaboration [1], in which two jets, $E_T^{jet1,2} > 14, 11$ GeV, are required to be within $-1 < \eta^{jet} < 2$. The comparison is shown in Figure 1, where one jet is restricted to be in the forward region, $1 < \eta^{jet} < 2$ (Figure 1a) and the rear direction $-1 < \eta^{jet} < 0$ (Figure 1b).

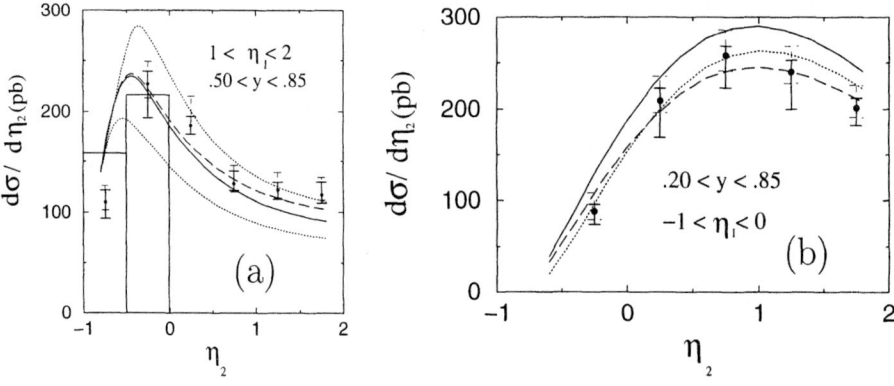

FIGURE 1. The NLO cross section $d\sigma/d\eta_2$ compared to ZEUS data. In (a) the central prediction is the solid line and increasing the gluon content of the photon is the dashed line. In (b) the central prediction is the solid line and decreasing the gluon content of the proton is the dashed line. The dotted lines in (a) and (b) are changing the y distribution (from [3]).

In Figure 1a, increasing the gluon density in the photon by a value of 20% improves the description of the data at large values of η_2, increasing the cross section by $\sim 10\%$. At negative values of η_2, the prediction remains the same and is larger than the data, however this region is subject to large hadronisation corrections. When the jet is in the rear direction as in Figure 1b, the cross section is less sensitive to the photon structure function and more so to the proton structure function. Decreasing the gluon density of the proton by 20% in this region improves the description of the data. However, it should again be noted that the hadronisation corrections increase with decreasing η_2.

Aurenche et al. conclude that the HERA data indicates a roughly 20% increase in the gluon content of the photon and a 20% decrease in the gluon content of the proton. The errors on the measurements are, however, as large as the change in cross section of these variations. Improved measurements are needed with larger statistics and are being investigated by both the H1 and ZEUS collaborations [4,5].

TESTS OF QCD

Tests of QCD from other experiments, such as those at the Tevatron, complement those at HERA and LEP, with the Tevatron having the advantage of a better constrained incoming particle, the proton. Roughly, where HERA finishes; $E_T^{\text{jet}} \sim 50$ GeV and $M_{\text{jj}} \sim 200$ GeV, the Tevatron starts, providing overlap between the experiments at the different colliders. A selection of results of relevance to high$-E_T$ measurements at HERA and LEP will be discussed. In particular, high$-E_T$ jet production, jet substructure and prompt photon production will all be discussed later in the proceedings from at least one of the HERA or LEP experiments [6–8].

High$-E_T$ jet production

Inclusive jet measurements at the Tevatron extend to values of $E_T^{\text{jet}} \sim 450$ GeV [9], falling by ~ 7 orders of magnitude as in Figure 2. The NLO prediction describes the data from the D0 collaboration very well, with the CDF data lying above the prediction at high transverse energies, which could be a signal for new physics such as quark substructure. However, the proton structure function is less well constrained at these large scales and the question of whether the data can be accomodated in a change of the proton PDF arises.

In Figure 3, the data is compared to the same calculation with three different proton structure functions. The MRST PDF gives a generally poor description of both the CDF and D0 measurements at low and high transverse energies. The CTEQ4M proton PDF gives a good description of the D0 data but inadequately describes the CDF data at high transverse energies. The CTEQ4HJ PDF includes the CDF data from Run IA in its fit, so would be expected to describe the Run IB data shown better than the other structure functions. This modified PDF describes well the D0 and the newer CDF data.

The comparison of data and theory clearly displays how understanding the parton density functions is essential in interpreting measurements in terms of new physics.

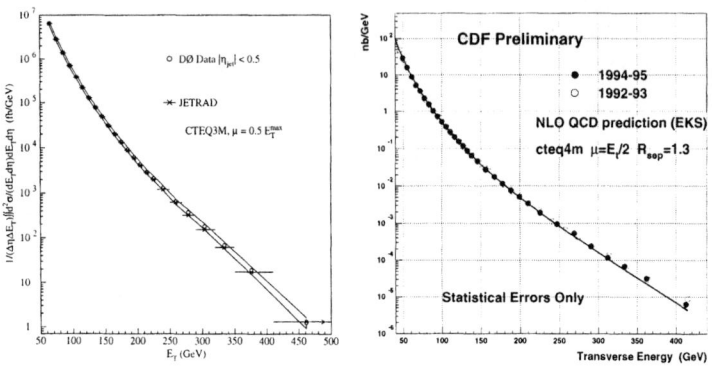

FIGURE 2. Inclusive jet cross sections from the (a) D0 (from [10]) and (b) CDF (from [11]) collaborations at $\sqrt{s}$ = 1800 GeV. The measurements are compared to NLO predictions.

Data to be and already analysed at HERA will allow the proton PDF to be further constrained at higher scales and hence more accurate predictions for the Tevatron measurements in Run II.

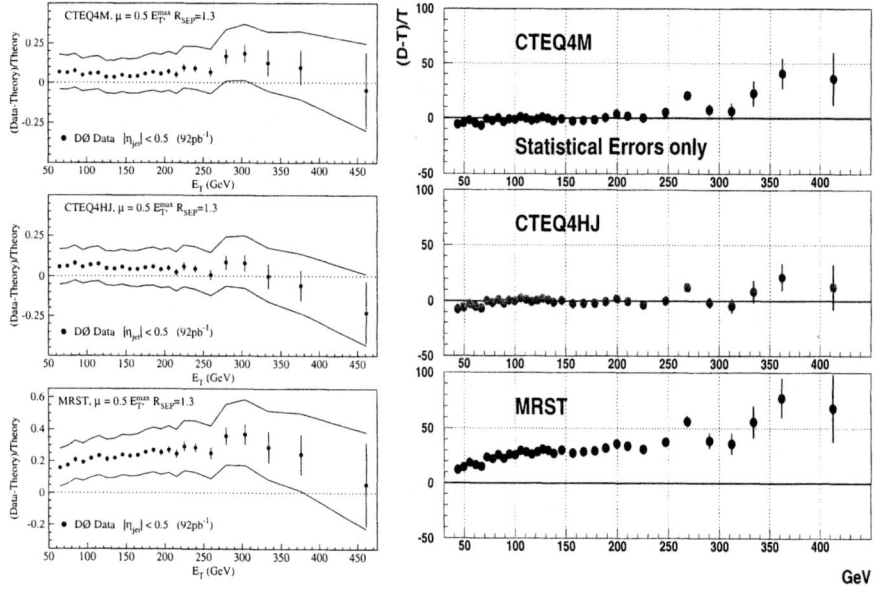

FIGURE 3. Relative difference of data and theory for inclusive jet cross sections from D0 [10] and CDF (from [11]) at $\sqrt{s}$ = 1800 GeV. The data is compared to three proton PDF's.

The Tevatron also produced data at centre-of-mass energies, $\sqrt{s}$ = 630 GeV, lower than the nominal $\sqrt{s}$ = 1800 GeV. Comparing the cross sections at the two different energies provides a test of QCD whilst reducing systematic uncertainties on the measurement and the sensitivity to the choice of proton PDF. Figure 4a shows the ratio of the inclusive jet cross sections as a funtion of $x_T \equiv \frac{2E_T}{\sqrt{s}}$ for both CDF and D0 data compared to NLO predictions. The reduction in the uncertainty in the choice of proton PDF can be seen in the difference of the prediction given by the three lines. At moderate and large x_T, the data agree well between the two collaborations, however at low x_T, the data sets diverge. At these lower values of x_T, the data suffer from possible problems of understanding the soft underlying event, which, both experiments correct for. The data also lie consistently below the prediction at the scale, μ, chosen. Reasons for the discrepancy are unclear, with possible interpretations being the use of different renormalistion scales or k_T effects, although neither of these are attractive explanations.

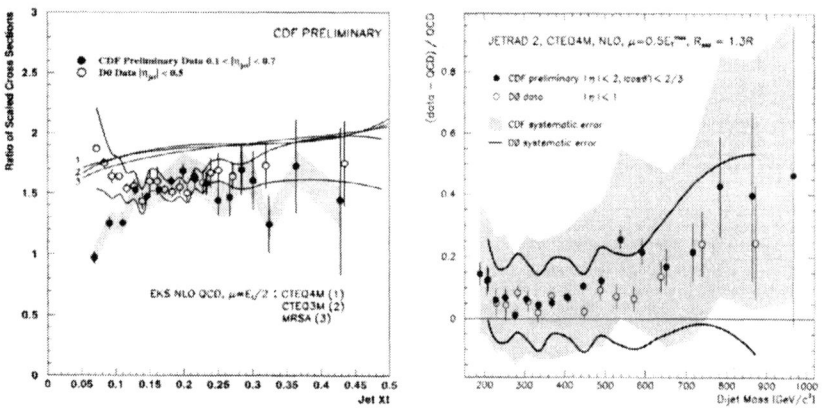

FIGURE 4. (a) Ratio of scaled cross sections ($\sqrt{s} = 630$ GeV/$\sqrt{s} = 1800$ GeV) compared to an NLO calculation with three proton PDF's. (b) Relative difference of dijet invariant mass, measured by CDF and D0, compared with an NLO prediction (from [11]).

Consideration of the invariant mass of the dijet system, like the inclusive jet cross section, provides both a test of QCD and the opportunity to search for new physics. Figure 4b shows the comparison of both experiments, CDF and D0, with an NLO prediction, where the two measurements are defined with jets in different angular regions. The measured data agree very well between experiments and with the theory. There is a tendency for the data to deviate from the prediction at high masses, but the systematic errors are too large to make any firm conclusions.

Jet substructure

Jet substructure has been studied at the Tevatron by considering both the jet shape and subjet multiplicity. Rerunning the k_T algorithm on those particles assigned to jets and stopping the clustering when all values of d_{ij} satisfy $d_{ij} > y_{\text{cut}} E_T^2$ gives numbers of subjets as a function of y_{cut}. Figure 5 show two different measurements of subjets using the k_T algorithm [12,13], where Figure 5a shows a comparison of the number of subjets in data and different MC's and Figure 5b shows how the subjet multiplicity can be used to differentiate between quark and gluon jets. The jets measured in Figure 5a are above 250 GeV, which means that the scale being studied at the lowest y_{cut} is about 1 GeV. When one considers this range in scale, the description of the data by HERWIG is extremely good. Figure 5b demonstrates a method for separating quark and gluon jets based on a statistical subtraction for events at $\sqrt{s} = 630$ GeV and $\sqrt{s} = 1800$ GeV. As expected, gluon jets show more activity than quark jets. A method for distinguishing quark and gluon jets has also been developed by the ZEUS collaboration [8].

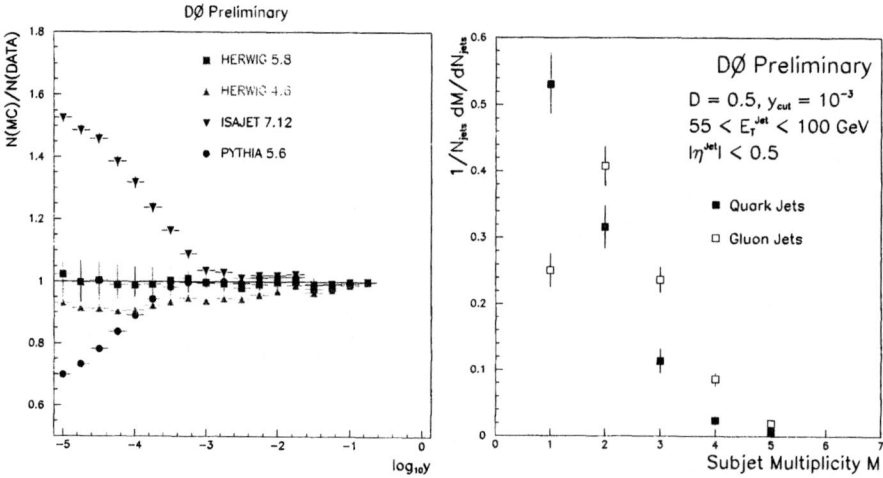

FIGURE 5. (a) Ratio of the multiplicity of subjets for data and MC for $E_T^{\text{jet}} > 250$ GeV (from [12]). (b) Corrected subjet multiplicity in q and g jets extracted from D0 data (from [13]).

Prompt photon production

Extensive measurements have been made and much theoretical work performed on prompt photon production at the Tevatron and fixed target experiments. Standard NLO perturbative calculations are unable to adequately describe the data at the Tevatron as shown in Figure 6a. Here, it can be seen that the data rise dramatically at low$-p_T$, the shape of which cannot be reproduced with a simple variation of the scale. The data can be described, however, by including an "intrinsic k_T" for the partons in the proton. Figure 6b shows the calculation rising at low$-p_T$ when a value of $< k_T > = 3.5$ GeV is used [14]. The use of intrinsic k_T is somewhat unsatisfactory, differing between data sets and centre-of-mass energies.

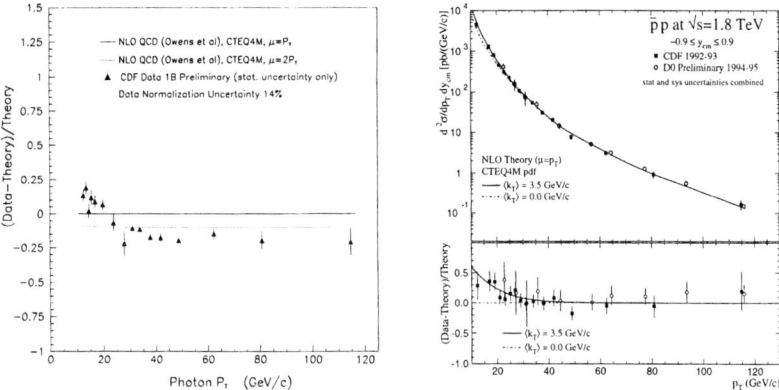

FIGURE 6. (a) CDF prompt photon production data compared to NLO predictions (from [11]). (b) CDF and D0 data compared to NLO predictions with intrinsic k_T and a value of $< k_T > = 3.5$ GeV (from [14]).

Prompt photon production has been extensively measured in fixed target experiments which have been compared with NLO predictions [15]. The NLO calculation from Aurenche et al [15] is able to describe all the fixed target data above a cut-off, $E_T \geq 4 - 5$ GeV except that from E706. The ratio of data to theory is shown in Figure 7a, where the E706 data is shown to dramatically rise at low $x_T = 2p_T/\sqrt{s}$.

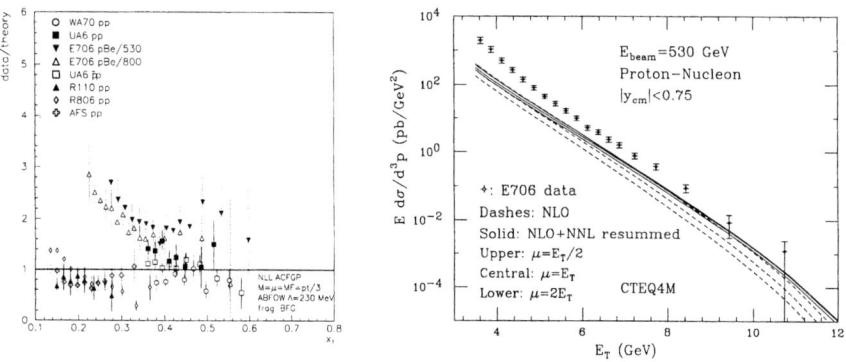

FIGURE 7. (a) Fixed target prompt photon data compared to NLO (from [15]). (b) Data from E706 compared to NLO predictions which have resummed large logarithms in x_T (from [16]).

Calculations resumming large logarithms in x_T [16] are also unable to describe the E706 data as shown in Figure 7b, although one would expect the resummation to improve the high$-p_T$ and not low$-p_T$ region. Figure 7b also demonstrates how

169

the resummation reduces the scale uncertainty in the calculation. Calculations in which Q_T, the nett transverse momentum of the final state γq pair, is resummed [17] improve the description of the E706 data at low p_T, although the NLO prediction is still too low. HERA data can provide useful informationon on intrinsic k_T, by filling the energy "gap" between the fixed target and Tevatron experiments [7].

HIGHER ORDER CALCULATIONS

The last few years have seen large advances in producing higher order calculations for the production of two or three jets [18,19]. Calculations of NLO 3−jet hadroproduction are becoming available and NNLO 2−jet hadroproduction will be produced sometime in the future. The needs for a 3−jet NLO calculation are many [18]. Measuring the 3−jet to 2−jet production ratio and hence α_s will be possible with a 3−jet NLO calculation. As well as testing QCD, the calculation will provide a better understanding of backgrounds to new physics processes. It is also an important step towards calculating NNLO 2−jet production.

Figure 8 shows the predicted cross section for the highest-transverse-energy jet in 3−jet hadroproduction for both LO and NLO and their respective estimtaions of the scale uncertainty. The central NLO and LO predictions are of similar value, however, NLO shows a large reduction in the scale uncertainty over LO.

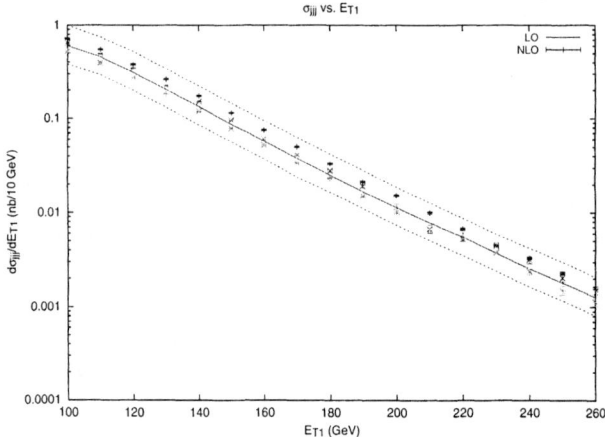

FIGURE 8. Predicted cross section for the highest-transverse-energy jet for 3−jet hadroproduction ($\sqrt{s} = 1800$ GeV) where, $E_T^{jet1} > 100$ GeV, $E_T^{jet2,3} > 50$ GeV and $|\eta^{jet}| < 4$. The LO prediction with an estimtaion of the scale uncertainty is given by the lines and the NLO prediction with the corresponding estimtaion of the scale uncertainty is given by the points (from [18]).

Progress on calculations of NNLO 2−jet hadroproduction has been good over the last two years and the interested reader is referred to recent talks on the subject [19].

A NNLO calculation for the production of two jets is anticipated in a few years where a reliable estimate of the error on the cross section will be acheived.

CONCLUSIONS

From the results presented, it can be seen that the theory of pQCD broadly describes the theory of the strong interaction. However, more detail both experimentally and theoretically is required to test QCD to great precision. This can also be acheived by considering the information from all experiments and what their results mean for each other.

REFERENCES

1. ZEUS Collab., J.Breitweg et al., *Euro. Phys. J.* **C11**, 427 (1999).
2. Frixione, S., *Nucl. Phys.* B **507**, 295 (1997); Frixione, S. and Ridolfi G., *Nucl. Phys.* B **507**, 315 (1997); Harris, B., Klasen M. and Vossebeld, J. hep-ph/9905348; Pötter B., hep-ph/9911221.
3. Aurenche, P., et al., hep-ph/0006011.
4. H1 Collab., H1prelim-00-052, Submitted to ICHEP2000 , Osaka, Japan. www-h1.desy.de/h1/www/publications/htmlsplit/H1prelim-00-052.long.html
5. ZEUS Collab., ICHEP−418, Submitted to ICHEP2000 , Osaka, Japan. http://www-zeus.desy.de/~schlenst/conf/osaka_paper/QCD/dijetpho.ps.gz; ZEUS Collab., EPS − 540, Submitted to the EPS High Energy Physics 99 conference, Tampere, Finland. http://www-zeus.desy.de/eps99/eps99_540.ps.gz
6. Surow, B., (these proceedings)
7. Terron, J., (these proceedings)
8. Glasman, C. (these proceedings)
9. CDF Collab., *Phys. Rev. Lett.* **77** 438.
10. D0 Collab., *Phys. Rev. Lett.* **82** (1999) 2451.
11. http://www-cdf.fnal.gov/physics/new/qcd/qcd99_blessed_plots.html
12. R. V. Astur, Proc. 10th Topical Workshop on Proton-Antiproton Collider Physics, 1995, eds. R. Raja and J. Yoh, p. 598.
13. D0 Collab., hep-ex/9907059, Submitted to the International Europhysics Conference on High Energy Physics, Tampere, Finland.
14. Apanasevich, L., et al., *Phys. Rev.* **D59** (1999) 074007.
15. Aurenche, P., et al., *Euro Phys. J.* **C9** (1999) 107.
16. Catani, S., et al., *JHEP* **9903** (1999) 25.
17. Laenen, E., Sterman, G. and Vogelsang, W., hep-ph/0006352 To appear in DIS2000 conference proceedings.
18. Kilgore, W. and Giele, W., "Hadronic three jet production at NLO", Talk presented at ICHEP2000, Osaka, Japan.
19. Glover, N., "Jet cross sections in NLO QCD and beyond", Talk presented at DIS2000, Liverpool, UK; Bern, Z., Dixon, L. and Kosower, D. "Recent progress in NNLO QCD calculation", Talk presented at ICHEP2000, Osaka, Japan.

Inclusive π^0 and K_S^0 in $\gamma\gamma$ reactions at L3

Pablo Achard *

University of Geneva, DPNC, 24 quai E. Ansermet, 1205 Geneva, Switzerland.
* on behalf of the L3 collaboration

Abstract. The $e^+e^- \rightarrow e^+e^-\pi^0 + X$ and $e^+e^- \rightarrow e^+e^-K_S^0 + X$ reactions are studied at LEP using data collected with the L3 detector at $\sqrt{s} = 189 - 202$ GeV. Preliminary results for the differential cross section $d\sigma/dp_T$ in the transverse momentum range 0.2 GeV $< p_T < 7.5$ GeV at central values of pseudo-rapidity, $|\eta| < 0.5$ or $|\eta| < 1.5$, and the differential cross section $d\sigma/d|\eta|$ for different values of p_T are presented. For $p_T \leq 1$ GeV, the p_T-dependence of the data are well described by an exponential fit. For $p_T > 1$ GeV the cross sections are compared to NLO QCD calculations and to LO Monte Carlo predictions.

INTRODUCTION

Two-photon collisions are the main source of hadrons at LEP via the process $e^+e^- \rightarrow e^+e^-\gamma^*\gamma^* \rightarrow e^+e^-$ *hadrons* . In the Vector Dominance Model (VDM), each photon can fluctuate into a vector meson with the same quantum numbers as the photon, thus initiating a strong interaction process with characteristics similar to hadron−hadron interactions. This process dominates in the "soft" region, where hadrons are produced with a low transverse momentum p_T . Hadrons with high p_T are produced by the QED direct process, $\gamma^*\gamma^* \rightarrow q\bar{q}$, or by QCD processes originating from the hadronic content of the photon. Depending on wether one or two photons fluctuate into a hadronic system, the QCD processes are called single−resolved or double−resolved respectively. In two-photon interactions, QCD calculations are available for single particle inclusive production at Next-to-leading order (NLO) [1].

In this paper the inclusive π^0 production and the inclusive K_S^0 production from quasi−real photons are measured. The center-of-mass energy of the two interacting photons, $W_{\gamma\gamma}$, is greater than 5 GeV for π^0 and greater than 10 GeV for K_S^0 . The π^0 is measured in the transverse momentum range $0.2 \leq p_T \leq 7.5$ GeV and in the pseudo−rapidity interval $-.5 \leq \eta \leq .5$, with $\eta = -\ln(\tan\theta/2)$. The angle θ is the polar angle of the π^0 , relative to the beam axis. The K_S^0 is measured in the range $0.2 \leq p_T \leq 4$ GeV and $-1.5 \leq \eta \leq 1.5$

The differential cross sections are compared to the Monte Carlo models PHOJET [2] and PYTHIA [3] and to analytical NLO QCD calculations [1].

CP571, *PHOTON 2000*, edited by A. J. Finch
© 2001 American Institute of Physics 0-7354-0010-5/01/$18.00

EVENT SELECTION

The data used for this analysis were collected by the L3 detector [4] in 1998 and 1999 at beam energies from 189 GeV to 202 GeV, with an average energy value of 194 GeV. The integrated luminosity is 414 pb^{-1}.

The $e^+e^- \rightarrow e^+e^-$ hadrons processes are simulated with the PHOJET[1] and PYTHIA[2] event generators. For the background the annihilation processes $e^+e^- \rightarrow$ hadrons(γ), $ZZ(\gamma)$, $Zee(\gamma)$, $We\nu(\gamma)$ are simulated with PYTHIA [3]; KORALZ [5] is used for $e^+e^- \rightarrow \tau^+\tau^-(\gamma)$ and KORALW [6] for $e^+e^- \rightarrow W^+W^-$. For the $e^+e^- \rightarrow e^+e^-\tau^+\tau^-$ channel the generator DIAG36 [7] is used. The events are simulated in the L3 detector using the GEANT [8] and GEISHA [9] programs and passed through the same reconstruction program as the data.

The selection of $e^+e^- \rightarrow e^+e^-$ hadrons events is based on information from the central tracking detectors and from the electromagnetic (BGO) and hadronic calorimeters. The following cuts are applied:

1. To reject annihilation events, the total energy in the calorimeters is required to be less than 40 % of the center-of-mass energy.

2. To exclude radiative events of the type $e^+e^- \rightarrow \gamma Z \rightarrow \gamma q\bar{q}$, the total energy in the electromagnetic calorimeter must be lower than 50 GeV.

3. Hadronic events are selected with at least 6 particles in the detector. These particles can be charged tracks or isolated clusters in the calorimeters.

4. Quasi−real two-photon interactions are selected by excluding events with a cluster of energy greater than 70 GeV and polar angle greater than 33 mrad.

5. Beam-gas and beam-wall events are suppressed by the requirement that at least 500 MeV must be deposited in the electromagnetic calorimeter.

6. The analysis is limited to events with a visible hadronic mass W_{vis}> 5 GeV , to allow comparisons with the Monte Carlo generators which generate events with $W_{\gamma\gamma}$> 3 GeV. W_{vis}, is the effective mass of the hadronic system, calculated from the four-momentum vector of all the measured particles[3].

After these cuts, the background (~ 1 %) is due to $e^+e^- \rightarrow$ hadrons and $e^+e^- \rightarrow e^+e^-\tau^+\tau^-$ events.

Inclusive π^0 production is studied via the π^0 decay into two resolved photons. A photon is defined as an electromagnetic cluster formed by the energy deposited in at least 2 BGO crystals, with energy greater than 100 MeV and separated by more than 200 mrad from the closest track. The number of photons detected is required to be at least 2 but not more than 20.

---

[1] PHOJET version1.05c
[2] PYTHIA version 5.718 and JETSET version 7.408
[3] All particles are considered to be pions, except for isolated clusters in the electromagnetic calorimeter which are considered to be photons.

Almost $2 \cdot 10^6$ events are selected, leading to more than $8 \cdot 10^7$ two photon combinations. The distribution of the mass of the reconstructed $\gamma\gamma$ system shows a narrow π^0 peak. A fit of this peak with a gaussian function over a Chebyshev polynomial background gives a resolution of 7.29 ± 0.03 MeV.

Inclusive K_S^0 production is studied via the K_S^0 decay into $\pi^+\pi^-$. The distance, in the transverse plane, between the secondary vertex and the e^+e^- interaction point is required to be greater than 3 mm. The angle between the flight direction of the K_S^0 candidate (taken as that of the line between the interaction point and the secondary vertex in the transverse plane) and the total transverse momentum vector of the two outgoing tracks must be less than 0.075 rad. After this cut, around $5 \cdot 10^5$ events are selected. A fit of the K_S^0 peak gives a resolution of 10 ± 1 MeV.

DATA ANALYSIS

To evaluate the number of π^0s or K_s^0s, on each bin of p_T and $|\eta|$, a fit is performed using a gaussian for the signal and a Chebyshev polynomial parametrisation of the background. In the π^0 case, to estimate the uncertainty on the background subtraction a side-band background subtraction is also performed. A three sigma region, "the signal region", is defined around the π^0 peak. Two "background regions" are defined on each side of the signal region with the same width. The average number of $\gamma\gamma$ combinations in the background regions is subtracted from the number of combinations in the signal region. Both methods give consistent results.

The reconstruction efficiency is evaluated using the two Monte-Carlo generators: PHOJET and PYTHIA. The kinematic limits of the cross section measurement are defined by imposing, at the generator level, $W_{\gamma\gamma} \geq 5$ GeV for π^0 and $W_{\gamma\gamma} \geq 10$ GeV for K_S^0. The virtuality of the photon is limited to $Q^2 \leq 1$ GeV^2 in PYTHIA, in PHOJET, a limit $Q^2 \leq 3$ GeV^2 for π^0 and $Q^2 \leq 8$ GeV^2 for K_S^0 is imposed. We have verified with PHOJET that the results do not depend on the Q^2 cutoff. The number of π^0 / K_S^0 reconstructed is obtained with the same procedure used for the data. As the two generators reproduce equally well the experimental distributions [10], a weighted average of the two is used to correct the data.

The level-1 trigger efficiency is evaluated by comparing the response of two independent triggers: the energy trigger [11] and the track trigger [12]. The efficiency is $90-95$ %, varying with the data taking conditions during the year. The efficiencies of higher level triggers, calculated using prescaled events, vary from 85% at low p_T to 100% at high p_T.

The main uncertainty on the measured cross sections comes from the choice of Monte Carlo generator used to calculate the reconstruction efficiency and from the background estimation. Half of the difference between the two generators and between the two estimations of the background are taken as systematic errors. The uncertainties due to cut variations are negligible.

Statistical and systematic errors are added in quadrature in the following. All the results are preliminary.

RESULTS

The differential cross sections $d\sigma/dp_T$ are shown in fig.1. Exponential fits $e^{-p_T/<p_T>}$ for 0.2 GeV $< p_T <$ 1.0 GeV reproduce well the data with $<p_T> =$ 233 $\pm$ 1 MeV for π^0 and $<p_T> = $ 357 $\pm$ 9 MeV for K_S^0. This behaviour is characteristic of hadrons produced by soft interactions and the fitted values of $<p_T>$ are similar to the one obtained in hadron$-$hadron or photon$-$hadron collisions [13].

Due to the existence of the direct process and of hard QCD interactions, $\gamma^*\gamma^*$ collisions start to exhibit a higher cross section at p_T values greater than $\sim$ 1.5 GeV. In fig.2 and 3 the $d\sigma/dp_T$ and $d\sigma/d|\eta|$ differential cross sections are compared to analytical NLO QCD predictions [1]. For this calculation, the flux of quasi$-$real photons is obtained using the Equivalent Photon Approximation, taking into account both transverse and longitudinal virtual photons. The interacting particles can be photons or partons from the $\gamma \to q\bar{q}$ quantum fluctuation, which evolves, via the Altarelli-Parisi equation, into quarks and gluons. The Gordon and Storrow [14] parton density functions are used. All elementary 2 $\to$ 2 and 2 $\to$ 3 processes are considered. The renormalization scale, the factorisation scale and the fragmentation scale are taken to be equal: $\mu = M = M_F = \xi p_T$. The uncertainty in the NLO calculation is estimated by varying the value of ξ from 0.5 to 2.0.

The measured cross sections are one order of magnitude above the direct process predictions. In the π^0 case, for $p_T >$ 2 GeV, the data are also higher than the predictions of NLO QCD calculations.

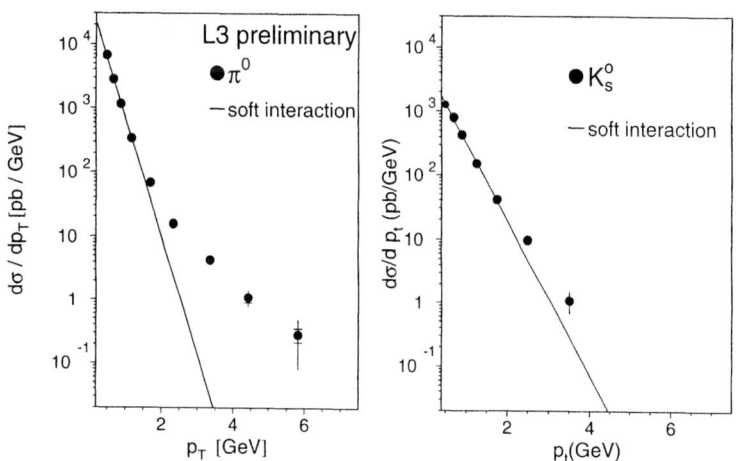

FIGURE 1. The differential cross sections $d\sigma/dp_T$ as a function of p_T for π^0 and K_S^0. The low p_T spectrum is well reproduced by an exponential fit, characteristic of soft interactions.

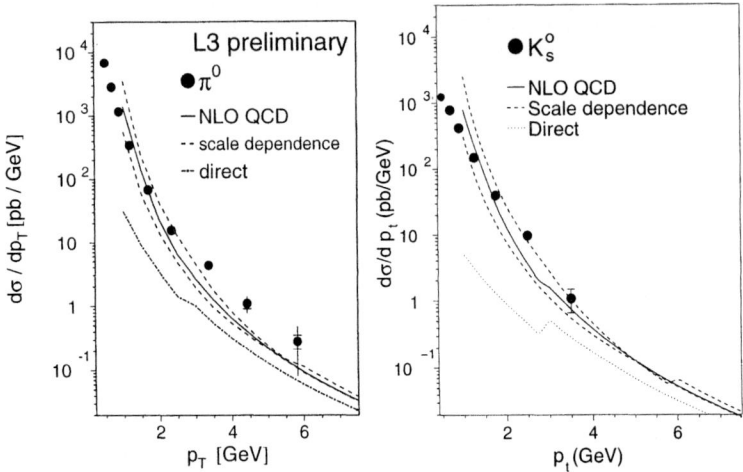

FIGURE 2. Comparison of the $d\sigma/dp_T$ distributions with NLO-QCD calculations. The central values (full lines) are the calculations with $\xi = 1$ (see text), the dashed lines refer to $\xi = 0.5$ (upper lines) and to $\xi = 2$ (lower lines). The contribution of the direct process alone are indicated with dash-dotted lines.

The cross sections are also compared to Monte Carlo predictions in fig.4. The high p_T region is better reproduced by PYTHIA than by PHOJET in the π^0 case. For the K_S^0 both Monte-Carlos reproduce quite well the data.

The differential cross sections, calculated for $W_{\gamma\gamma} > 10$ GeV, are compared to OPAL results [15] in fig.5. OPAL measured the inclusive production of charged hadrons (mainly $\pi^{\pm}$) and of K_S^0 in the range $|\eta| < 1.5$. The experiments agree within the experimental errors.

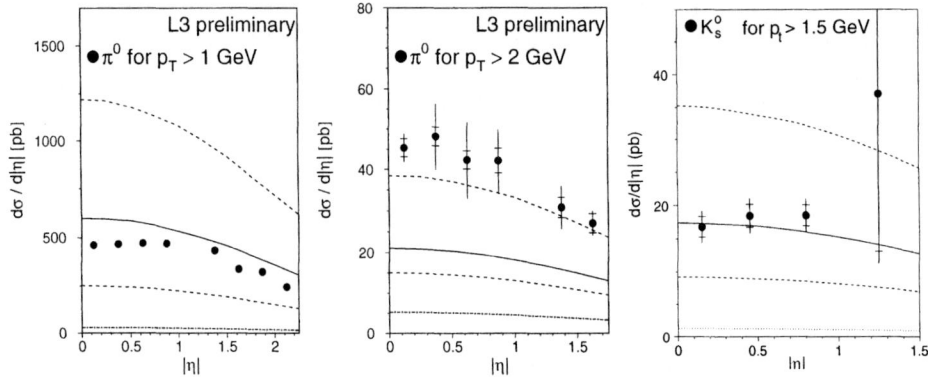

FIGURE 3. Comparison of the $d\sigma/d|\eta|$ distributions with NLO-QCD calculations. The lines are as in fig.2.

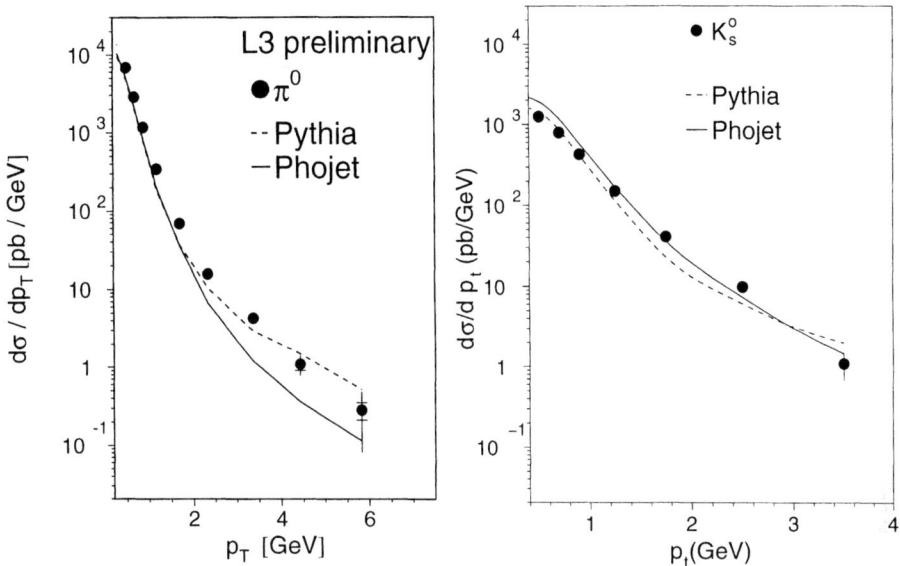

FIGURE 4. Comparison of the $d\sigma/dp_T$ distributions with Monte Carlo predictions.

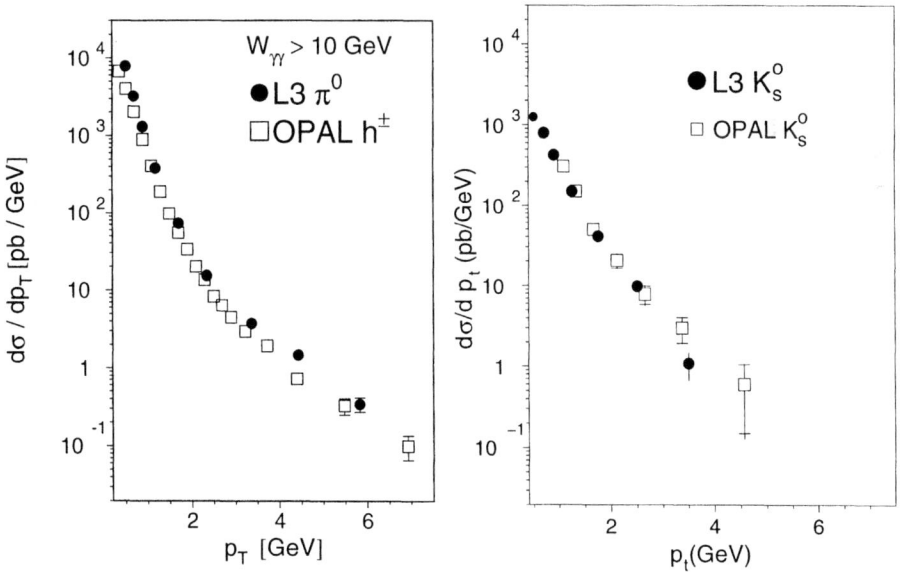

FIGURE 5. The L3 and OPAL $d\sigma/dp_T$ distributions compared. The OPAL data are scaled to the same center-of-mass energy and, in the π^0 vs. charged hadron case, are scaled also to the same pseudorapidity range and fragmentation function.

CONCLUSIONS

This analysis is the first study of inclusive π^0 production in two photon collisions at LEP. At low p_T, the differential cross sections $d\sigma/dp_T$ of π^0 and K_S^0 production follow an exponential law similar to that previous observed in hadron-hadron and hadron-photon collisions. At higher p_T, the contributions of $\gamma\gamma \to q\bar{q}$ and hard QCD processes are clearly visible. The agreement with previous measurement is very good, but there is only a qualitative agreement with the predictions of LO Monte Carlo and NLO QCD calculations.

ACKNOWLEDGMENTS

I would like to thank B.A. Kniehl for providing us the predictions of NLO QCD calculations and L. Gordon for very useful discussions.

Thanks a lot also to A. Finch and all the Photon 2000 team for the kindness and efficiency of the organization.

Many thanks, of course, to S. Braccini for providing me the K_S^0 results.

REFERENCES

1. Binnewies, J., Kniehl, B.A., and Kramer, G. *Phys. Rev.* **D 53**, 6110 (1996).
2. Engel, R. , *Z. Phys.* **C 66**, 203 (1995);
   Engel, R., and Ranft, J., *Phys. Rev.* **D 54**, 4246 (1996);
   And Engel, R., private communication.
3. Sjöstrand, T., *Comput. Phys. Commun.* **82**, 74 (1994).
4. L3 Coll., Adeva, B., et al., *N.I.M.* **A 289**, 35 (1990);
   L3 Coll., Acciarri, M., et al., *N.I.M.* **A 351**, 300 (1994).
5. Jadach, S., Ward, B. F. L., and Was, Z., *Comput. Phys. Commun.* **79**, 503 (1994).
6. Skrzypek, M., Jadach, S., Placzek, W. and Was, Z., *Comput. Phys. Commun.* **94**, 216 (1996).
7. Berends, F.A., Daverfeldt, P.H., and Kleiss, R., *Nucl. Phys.* **B 253**, 441 (1985).
8. Brun, R., et al., GEANT 3.15 *preprint CERN* DD/EE/84-1 (Revised 1987).
9. Fesefeldt, H., *RWTH Aachen report* PITHA 85/2 (1985).
10. Kienzle, M.N., *these proceedings*.
11. Bizzarri, R., et al. , *N.I.M.* **A 283**, 799 (1989).
12. Béné, P., et al., *N.I.M.* **A 306**, 150 (1991);
    Haas, D., et al., *N.I.M.* **A 420**, 101 (1999).
13. Perl, M.L., *High Energy Hadron Physics*, edited by J. Wiley, 1974.
14. Gordon, L.E., and Storrow, J.K., *Nucl. Phys.* **B 489**, 405 (1997).
15. OPAL Coll., Abbiendi, G., et al., *Eur. Phys. J.* **C 6**, 253 (1999).

Di-Jet Production in Photon-Photon Collisions at $\sqrt{s}_{ee}$ from 189 to 202 GeV

Bernd Surrow* and Thorsten Wengler[†]

* Max-Planck Institut für Physik (Werner-Heisenberg-Institut)
Föhringer Ring 6, D-80805 München
e-mail: bernd.surrow@cern.ch

† CERN, EP-Division
CH-1211 Geneva 23
e-mail: thorsten.wengler@cern.ch

Abstract. Di-jet production is studied in collisions of quasi-real photons radiated by the LEP beams at e^+e^- centre-of-mass energies $\sqrt{s}_{ee}$ from 189 to 202 GeV. The inclusive di-jet cross-section is measured as a function of the mean transverse energy $\bar{E}_T^{jet}$ of the two jets and as a function of x_γ for different regions of $\bar{E}_T^{jet}$. The results are compared to next-to-leading order perturbative QCD calculations and to the predictions of the leading order Monte Carlo generators PYTHIA and PHOJET.

INTRODUCTION

We have measured the production of di-jets in the collision of two quasi-real photons at $\sqrt{s}_{ee}$ from 189 to 202 GeV with a total integrated luminosity of 384 pb^{-1} collected with the OPAL detector at LEP. The jets are reconstructed using an inclusive $k_\perp$ clustering algorithm [1]. The two jets are used to estimate the fraction of the photon momentum participating in the interaction, which is a sensitive probe of the structure of the photon. Furthermore the transverse energy of the jets provides a hard scale which allows such processes to be calculated in perturbative QCD. Fixed order calculations at next-to-leading order (NLO) in the strong coupling constant α_s for di-jet production are available to be compared to the data, providing tests of the theory.

Inclusive jet cross-sections in photon-photon collisions have previously been measured at an e^+e^- centre-of-mass energy of $\sqrt{s}_{ee} = 58$ GeV at TRISTAN [2,3] and at $\sqrt{s}_{ee}$ from 130 to 172 GeV at LEP [4,5]. This paper extends the latter analyses to higher e^+e^- centre-of-mass energies, and provides an approximately twenty-fold increase in integrated luminosity.

CP571, *PHOTON 2000*, edited by A. J. Finch
© 2001 American Institute of Physics 0-7354-0010-5/01/$18.00

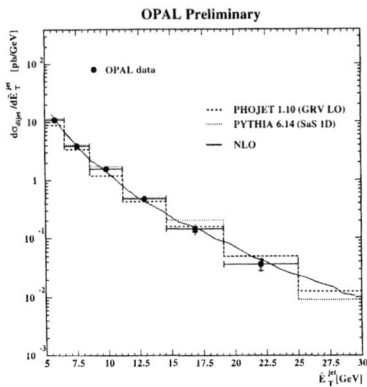

FIGURE 1. The di-jet cross-section as a function of the mean transverse energy $\bar{E}_T^{\text{jet}}$ of the di-jet system. The total of statistical and systematic uncertainties added in quadrature are shown. Predictions of the LO programs PHOJET and PYTHIA and an NLO calculation by Klasen et al. are compared with the data.

KINEMATICS, EVENT SELECTION AND JET FINDING

The underlying kinematics of the process ee → eeX at the e^+e^- collider LEP consists in the initial state of the incoming electron and positron, each of which - to lowest order in α_{em} - emits a virtual photon. The square of the four momentum transfer or the virtuality of the mediated virtual photons is denoted by $Q_1^2 = -q_1^2 = (k_1 - k_1')^2$ and $Q_2^2 = -q_2^2 = (k_2 - k_2')^2$ where k_i and k_i' ($i = 1, 2$) refer to the four-vectors of the electrons in the initial and final state, respectively. The invariant mass of the final state X is denoted by $W^2 = (q_1 + q_2)^2$.

If both final state electrons are not tagged, both virtualities Q_1^2 and Q_2^2 - depending on the detector acceptance - are small, and the photons can be considered as quasi-real, i.e. $Q_1^2 \simeq 0$ and $Q_2^2 \simeq 0$. This is the case of anti-tagged events. Since no tagged electrons are available in this analysis, properties of the hadronic final state are used to select photon-photon collision events.

In all parts of this analysis, a sum over the particles in the event or in a jet means a sum over two kinds of objects: tracks satisfying the quality cuts, and all calorimeter clusters, including the OPAL Forward calorimeter (FD). An algorithm is applied to avoid double-counting of particle momenta in the central tracking system and their energy deposits in the calorimeters [4].

Jets are reconstructed using the inclusive $k_\perp$ clustering algorithm as proposed in [1].

Jets entering the analysis have to have a minimum transverse energy $E_T^{\text{jet}} > 3$ GeV and a pseudorapidity $|\eta^{\text{jet}}| < 2$. In events with more than two jets, only

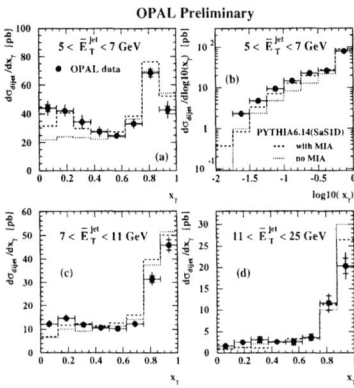

FIGURE 2. The di-jet cross-section as a function of x_γ and $\log 10(x_\gamma)$ for the regions of the mean transverse energy $\bar{E}_T^{\text{jet}}$ of the di-jet system indicated in the figures. The total of statistical and systematic uncertainties added in quadrature are shown. Predictions of the LO program PYTHIA are shown with and without multiple interactions (MIA) turned on.

the two jets with the highest E_T^{jet} values are taken. Throughout the paper, this procedure is used to define a di-jet event. We use data corresponding to an integrated luminosity of 384 pb^{-1} at $\sqrt{s_{ee}}$ from 189 GeV to 202 GeV. After applying all cuts and requiring at least two jets with $E_T^{\text{jet}} > 3$ GeV and $|\eta^{\text{jet}}| < 2$, 45509 events remain. The efficiency to trigger di-jet events in this region of phase space has been shown to be close to 100 % [5].

In LO QCD, neglecting multiple parton interactions, two hard parton jets are produced in $\gamma\gamma$ interactions. In single- or double-resolved interactions, the two hard parton jets are expected to be accompanied by one or two remnant jets. A pair of variables, x_γ^+ and x_γ^-, can be defined [6] which specify the fraction of the photon's momentum participating in the hard scattering:

$$x_\gamma^+ \equiv \frac{\sum\limits_{\text{jets}=1,2} (E + p_z)}{\sum\limits_{\text{hadrons}} (E + p_z)} \quad \text{and} \quad x_\gamma^- \equiv \frac{\sum\limits_{\text{jets}=1,2} (E - p_z)}{\sum\limits_{\text{hadrons}} (E - p_z)}, \tag{1}$$

where p_z is the momentum component along the z axis of the detector and E is the energy of the jets or hadrons. Ideally, for LO direct events without remnant jets, the total energy of the event is contained in the two jets with highest E_T^{jet}, i.e. $x_\gamma^+ = 1$ and $x_\gamma^- = 1$, whereas for single-resolved events either x_γ^+ or x_γ^- and for double-resolved events both values, x_γ^+ and x_γ^-, are expected to be smaller than 1. The asymmetry condition, $|E_{T,1}^{\text{jet}} - E_{T,2}^{\text{jet}}|/(E_{T,1}^{\text{jet}} + E_{T,2}^{\text{jet}}) < 1/4$, ensures asymmetric E_T^{jet} thresholds for the two jets of the di-jet system in connection with the minimum $\bar{E}_T^{\text{jet}}$ requirement, which is important in comparison to NLO QCD calculations [7].

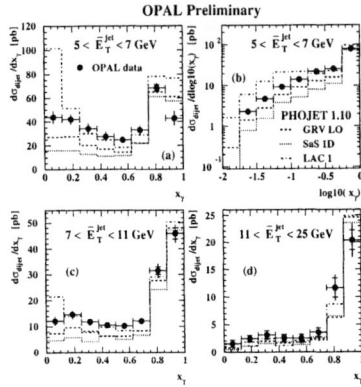

FIGURE 3. The di-jet cross-section as a function of x_γ and $\log 10(x_\gamma)$ for the regions of the mean transverse energy $\bar{E}_T^{\rm jet}$ of the di-jet system indicated in the figures. The total of statistical and systematic uncertainties added in quadrature are shown. Predictions of the LO program PHOJET are shown for three different parton density functions.

It has previously been used in [8]. In these definitions x_γ indicates that each event enters the distribution twice, at the value of x_γ^+ and the value of x_γ^-.

RESULTS

The di-jet cross-sections measured are compared below to NLO QCD calculations which predict jet cross-sections for partons, whereas the experimental jet cross-sections are measured for hadrons. The uncertainties due to the modelling of the hadronisation process have not been taken into account. Because the partons in the Monte Carlo models and the partons in the NLO calculations are defined in different ways it is not yet well defined to use the Monte Carlo to correct the data so that it can be compared with the NLO parton level predictions. The NLO QCD calculations also do not take into account the possibility of an underlying event which leads to an increased jet cross-section. The underlying event is simulated in the Monte Carlo models PYTHIA and PHOJET by multiple interactions.

The differential di-jet cross-section as a function of the mean transverse energy $\bar{E}_T^{\rm jet}$ of the di-jet system is shown in Figure 1. A perturbative QCD calculation at NLO [9] is compared to the measured distribution. This calculation uses the NLO GRV parameterisation of the parton distribution functions of the photon [10], and was repeated for the kinematic conditions of the present analysis. The calculation is in good agreement with the data. The predictions of the LO Monte Carlo models PHOJET 1.10 and PYTHIA 6.14 are also compared to the data using the LO GRV [10] and SaS1D [11] parton distribution functions of the photon, respectively. Both models are in good agreement with the $\bar{E}_T^{\rm jet}$-distribution measured.

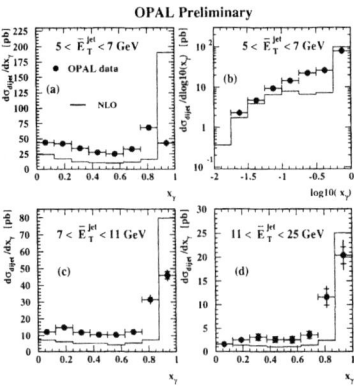

FIGURE 4. The di-jet cross-section as a function of x_γ and $\log10(x_\gamma)$ for the regions of the mean transverse energy $\bar{E}_T^{jet}$ of the di-jet system indicated in the figures. The total of statistical and systematic uncertainties added in quadrature are shown. Predictions of an NLO calculation by Klasen et al. are compared with the data.

In Figure 2, PYTHIA with multiple interactions switched on and off is compared to the measured differential di-jet cross-section as a function of x_γ. For the case of multiple interactions switched on, a lower cutoff parameter p_t^{mi} is used, which describes the transverse momentum of the parton and is set to 1.4 GeV. The SaS1D parton distributions are used. The effect is largest for the region of low $\bar{E}_T^{jet}$. Here the multiple interactions lead to an increase of the cross-section of up to 50 % at low x_γ. For higher $\bar{E}_T^{jet}$ on the other hand no significant influence of multiple interactions is visible.

Figures 3(a)-(d) demonstrate the sensitivity of the PHOJET model prediction to different parton density functions [11,10,12]. The already disfavoured LAC1 set is used to show the effect of the large gluon density assumed in this parton density function. The variation of the cross-section predicted is clearly larger than the data uncertainties. They are also larger than the uncertainties from the available model of multiple interactions as shown in Figure 2. This preliminary study demonstrates that, if uncertainties due to multiple interactions can be kept under control, then these distributions will give direct constraints on the gluon content of the photon.

Figures 4(a)-(d) show a comparison of the NLO calculation [9] to the measurement. In contrast to its good agreement with the $\bar{E}_T^{jet}$-distribution the NLO prediction is far too large for values of x_γ close to 1, while being too low by about a factor of two for $x_\gamma < 0.75$. At least part of the discrepancy at x_γ close to 1 may be attributed to the missing hadronisation in the NLO calculation. In the NLO calculation direct events often have x_γ precisely equal to 1. Hadronisation effects dilute this sharp peak when some of the resulting hadrons are not clustered into the two leading jets in the event. It is therefore a useful exercise to compare the

TABLE 1. The prediction of the NLO calculation compared to the data for x_γ larger and smaller than 0.75 for the three ranges of $\bar{E}_T^{\text{jet}}$. The errors on the data are statistical only.

		data [pb]	NLO [pb]
5 GeV < $\bar{E}_T^{\text{jet}}$ < 7 GeV	$x_\gamma > 0.75$	111.0±3.8	206.5
	$x_\gamma < 0.75$	205.5±4.8	84.4
7 GeV < $\bar{E}_T^{\text{jet}}$ < 11 GeV	$x_\gamma > 0.75$	77.4±2.6	87.0
	$x_\gamma < 0.75$	71.5±2.2	32.6
11 GeV < $\bar{E}_T^{\text{jet}}$ < 25 GeV	$x_\gamma > 0.75$	32.0±2.5	27.4
	$x_\gamma < 0.75$	15.8±1.7	7.5

NLO calculation to the data integrated for values of x_γ larger or smaller than 0.75. The results are shown in Table 1. The NLO calculation predicts much too large a cross section for $x_\gamma > 0.75$ and small $\bar{E}_T^{\text{jet}}$, while being too low by about a factor of two for $x_\gamma < 0.75$. With increasing $\bar{E}_T^{\text{jet}}$ the discrepancy for $x_\gamma > 0.75$ largely disappears, while for $x_\gamma < 0.75$ the NLO prediction remains too low by a factor of two for the highest $\bar{E}_T^{\text{jet}}$ considered. This suggests that the parton density functions used in the NLO calculation underestimate the gluon density in the photon.

REFERENCES

1. S. Catani, Yu.L. Dokshitzer, M.H. Seymour and B.R. Webber,
   Nucl. Phys. B406 (1993) 187;
   S.D. Ellis, D.E. Soper, Phys. Rev. D48 (1993) 3160.
2. AMY Collaboration, B.J. Kim et al., Phys. Lett. B325 (1994) 248.
3. TOPAZ Collaboration, H. Hayashii et al., Proc. 10th Workshop on
   Gamma-Gamma Collisions and Related Processes, Sheffield, UK (April 1995) 133;
   TOPAZ Collaboration, H. Hayashii et al., Phys. Lett. B314 (1993) 149.
4. OPAL Collaboration, K. Ackerstaff et al., Z. Phys. C73 (1997) 433.
5. OPAL Collaboration, G. Abbiendi et al., Eur. Phys. J. C10 (1999) 547.
6. L. Lönnblad and M. Seymour (convenors),
   $\gamma\gamma$ Event Generators, in 'Physics at LEP2',
   CERN 96-01, eds. G. Altarelli, T. Sjöstrand and F. Zwirner, Vol. 2 (1996) 187.
7. M. Klasen and G. Kramer, Phys. Lett. B366 (1996) 385.
8. H1 Collaboration, C. Adloff et al., Eur. Phys. J. C1 (1998) 97.
9. M. Klasen, T. Kleinwort and G. Kramer, Eur. Phys. J. Direct C1 (1998) 1.
10. M. Glück, E. Reya and A. Vogt, Phys. Rev. D45 (1992) 3986;
    M. Glück, E. Reya and A. Vogt, Phys. Rev. D46 (1992) 1973.
11. G.A. Schuler and T. Sjöstrand, Z. Phys. C68 (1995) 607.
12. H. Abramowicz, K. Charchula and A. Levy, Phys. Lett. B269 (1991) 458.

Measurements of di-jet cross sections in $\gamma\gamma \to$ *hadrons* with ALEPH

P. Hodgson*, M. Lehto*
representing the ALEPH Collaboration
and B. Pötter[†]

Department of Physics and Astronomy, University of Sheffield
Sheffield S3 7RH United Kingdom
[†]*Max-Planck-Institut für Physik, Theoretical Physics Division*
D-80805 München Germany

Abstract. Measurements of di-jet cross sections in anti-tagged and single -tagged $\gamma\gamma$ events have been performed. For the anti-tagged case the KTCLUS jet algorithm was used and the data corrected back to hadron level using the PHOJET and PYTHIA Monte Carlo models. In the single-tagged case a cone based jet algorithm was used together with the HERWIG Monte Carlo. In both studies good agreement between NLO calculations and the data was found.

INTRODUCTION

The study of di-jet production in photon photon collisions can be used to examine the structure of the photon and test QCD predictions. In the analyses presented here two distinct tagging modes are used: single-tagged, where one of the scattered electrons is observed, corresponding to the probing of a nearly real photon by a virtual photon, and untagged, where events with a scattered electron are explicitly excluded and thus both photons are real. Both analyses were based on data taken with the ALEPH [1] detector at $\sqrt{s} = 183$ GeV with an integrated luminosity of $\sim$55 pb^{-1}

EVENT SELECTION AND JET FINDING

The following common event selections were made for both analyses.

- The event must contain at least 3 charged tracks

- The invariant mass of the hadronic system must be greater than 3.0 GeV/c^2 (and ≤ 40.0 GeV/c^2 for untagged)

CP571, *PHOTON 2000*, edited by A. J. Finch

- The hadronic energy must be less than 50 GeV

- Vertex < 4 cm in z and < 5 cm radially from beam line

For the single-tagged case a cluster of more than 70 GeV energy was required in one of the forward luminosity calorimeters (LCAL or SiCAL) within the angular range 34 mrad $< \theta_{tag} <$ 155 mrad corresponding to a Q^2 range of $10 < Q^2 < 200$ GeV/c^2. In the untagged case the event must contain no clusters with energy greater than 40.0 GeV in LCAL or SiCAL. These selection cuts give samples of 1912 single-tagged events and 145952 untagged events. The main background was seen to be the $\gamma\gamma \to \tau^+\tau^-$ process. This was modelled using Vermaseren [2] and subtracted from the data.

The two analyses used different jet algorithms. In the single-tagged case a cone jet finder was used with the following requirements on the jets entering the analysis: $E_T^{jet} > 3.0$ GeV, $N_{particles}$ in jet ≥ 2 and $|\eta^{jet}| < 1.6$. The untagged analysis used KTCLUS [9] in the KTRECO mode with YCUT =0.018, the input to the jet finder being all tracks with $|\cos\theta| < 0.94$ ($|\eta| < 1.74$). A di-jet event was then defined as one with at least two jets with pseudorapidity between ± 1.4, in which at least one jet has p_T greater than 4 GeV/c and at least one other jet has p_T greater than 3 GeV/c.

THEORETICAL PREDICTIONS

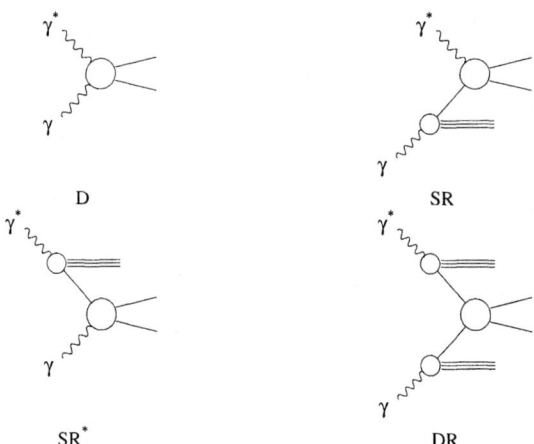

FIGURE 1. The four different interaction modes in $\gamma^*\gamma$-scattering.

The e^+e^--scattering cross sections considered contain the $\gamma^*\gamma$-scattering subprocess

$$\gamma_a^*(Q^2) + \gamma_b(P^2 = 0) \to \text{jet}_1 + \text{jet}_2 + X \tag{1}$$

The interaction of a virtual with a real photon can happen in four different ways, as shown in Fig. 1 depending on whether the photon interaction is direct or resolved. The variables y_a, y_b in the following describe the momentum fraction of the photon a, b relative to the electron or positron and x_a, x_b describes the momentum fraction of the parton in the photon a, b.

Taking into account both the transverse and longitudinal polarizations of the virtual photon, the cross section $d\sigma_{e^+e^-}$ for e^+e^--scattering is conveniently written as the convolution

$$
\frac{d\sigma_{e^+e^-}}{dQ^2 dy_a dy_b} = \sum_{a,b} \int dx_a dx_b F_{\gamma/e^-}(y_b) f_{b/\gamma}(x_b)
$$

$$
\times \frac{\alpha}{2\pi Q^2} \left[\frac{1 + (1 - y_a)^2}{y_a} f_{a/\gamma^*}^U(x_a) d\sigma_{ab} \right.
$$

$$
\left. + \frac{2(1 - y_a)}{y_a} f_{a/\gamma^*}^L(x_a) d\sigma_{ab} \right] \tag{2}
$$

The PDF's of the real and the virtual photon are $f_{b/\gamma}(x_b)$ and $f_{a/\gamma^*}^{U,L}(x_a)$, respectively, where U and L denote the unpolarized and longitudinally polarized photon contributions, respectively. The direct photon interactions are included in formula (2) through delta functions. For the direct virtual photon one has the relation $f_{a/\gamma^*}^{U,L} d\sigma_{ab} = \delta(1 - x_a) d\sigma_{\gamma^* b}^{U,L}$, whereas for the direct real photon the relation is $f_{a/\gamma} d\sigma_{ab} = \delta(1 - x_b) d\sigma_{\gamma b}$, where $d\sigma_{ab}$ refers to the partonic cross section. The function $F_{\gamma/e^-}(y_b)$ describes the spectrum of the real photons emitted from the electron according to the Weizsäcker-Williams approximation [3]. In the case of single-tagged reactions, the resolved virtual photon component will be more and more suppressed until eventually the modes with a direct virtual photon dominate.

The partonic cross sections in LO consist of two final state particles. The NLO corrections consist of the virtual and real corrections, which both exhibit characteristic divergencies. The NLO corrections for the different subprocesses D, SR, SR* and DR have been calculated with the phase-space slicing method [4]. The sum of real and virtual corrections is finite after factorization of singularities from the initial state. Of special importance here is the $\gamma^* \to q\bar{q}$ splitting term. In the limit of the $q\bar{q}$-pair being collinear, the logarithm

$$
M = \ln\left(1 + \frac{y_s s}{z Q^2}\right) P_{q \leftarrow \gamma}(z) \tag{3}
$$

(where y_s is the phase-space-slicing parameter and s is the partonic cms energy) can be factorized from the cross section, which is singular for $Q^2 = 0$. This singularity is absorbed into the PDF of the virtual photon with virtuality Q^2 in such a way that the $\overline{\text{MS}}$ factorization result of the real photon is obtained in the limit $Q^2 \to 0$. The different NLO subprocesses D, SR, SR* and DR have been implemented into the fixed order computer program JetViP [5].

In the case of single-tagged reactions, the resolved virtual photon component will be more and more suppressed for larger tagging angles until eventually the two modes with a direct virtual photon dominate. This corresponds to the kinematics of deep-inelastic scattering. For untagged reactions, both photons will be on-shell and the subtraction term described above, M, will yield the result for real photoproduction. In this way, the complete range in virtuality, from very low to very high Q^2, is covered with JetViP.

For the untagged case, an additional independant NLO calculation was performed using code provided by M Klasen [6]. This code implements the NLO perturbative QCD calculation of jet production in direct and resolved photon-photon collisions by Klasen, Kleinwort and Kramer [7,8]. Two and three parton final states are generated, using an invariant mass cutoff to isolate soft and collinear regions of phase space. A jet algorithm equivalent to KTCLUS was applied to the final state partons together with cuts corresponding as closely as possible to the experimental event selection described above. This independant NLO calculation proved to be in excellent agreement with the results from JetViP.

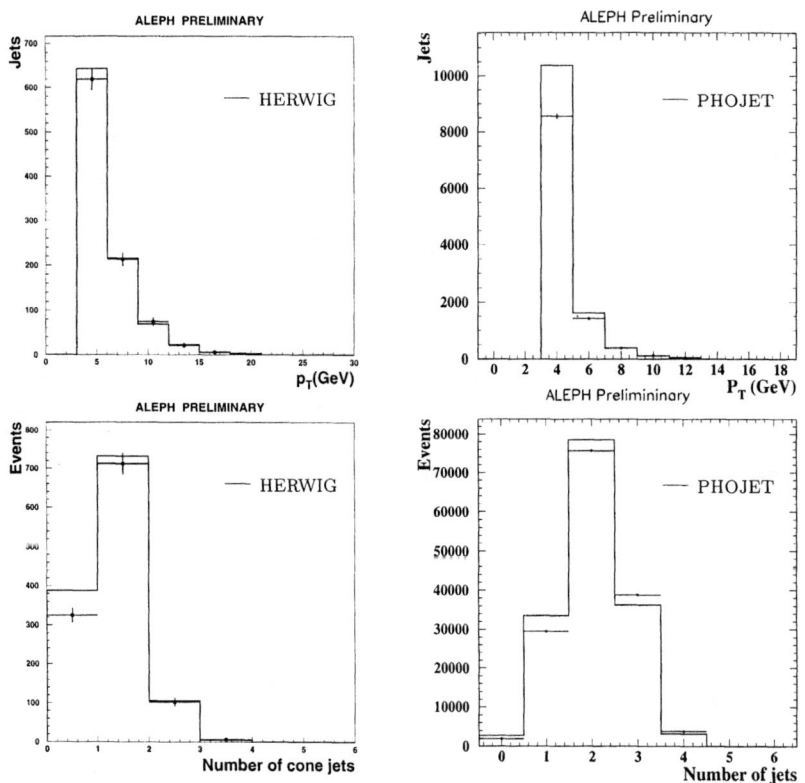

FIGURE 2. Data compared to fully simulated Monte Carlo.

188

EXTRACTING THE DI-JET CROSS SECTION

The theoretical calculations described above produce a parton level cross section. In this work we invoke local parton hadron duality and assume the hadron level is directly comparable to the parton level. We then need to account for detector effects correcting back to hadron level from the data. The correction requires a model that provides a good description of the data. For the single-tagged analysis the HERWIG [10] Monte Carlo was used for the correction, as it gives a good description of the data. Pythia 6.122 [11] was used to provide a conservative estimate of the model dependance. For the untagged analysis the PHOJET [12] Monte Carlo was used for the main correction and PYTHIA to estimate the systematic error from model dependance.

The experimentally selected multi-jet samples need several forms of correction before they can be compared with the theoretical predictions. The first must account for the fact that there are multi-jet events measured in the detector (after the application of the p_T requirements on the jets) that do not correspond to multi-jet events at hadron level after the same p_T requirements are imposed on the 'true' hadron-level jet p_T. These events may be regarded as a form of background impurity; the level of impurity is worst at low reconstructed p_T, but does not exceed 20%. The next correction must then account for the fact that not all multi-jet events at hadron level will also pass the multi-jet selection in the reconstructed detector events; this represents an inefficiency in the selection. It was seen that the reconstructed p_T is highly correlated with the hadron-level p_T, 86% for single-tagged and 96% for the untagged. The binning in p_T was chosen to be larger than the typical resolution, reducing the effect of bin to bin migration. A bin by bin correction factor was then calculated by taking the ratio in HERWIG or PHOJET of the p_T distribution for all multi-jet events passing the p_T restrictions at hadron level to the same distribution for the reconstructed and selected multi-jet events at detector level. The observed data entries in a given reconstructed p_T bin are then corrected according to the relation

$$n(p_T)_{corrected} = n(p_{T,recon.})_{data} \cdot [n(p_{T,hadron})/n(p_{T,recon.})]_{HERWIG,PHOJET}$$

Figure 3 shows the differential p_T distributions for di-jet production obtained as described above. The full error bars are the total error; the inner error bars show the statistical errors only. The results are compared to the relevant NLO QCD calculations.

CONCLUSION

We have extracted di-jet cross sections for the single-tagged and untagged cases of $\gamma\gamma \rightarrow hadrons$. We see that the Monte Carlos are in good agreement with the data. The HERWIG Monte Carlo was used to correct the data back to hadron level and we see that JetViP is in good agreement with the data. For the untagged

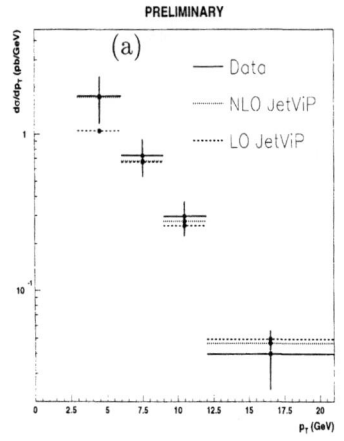
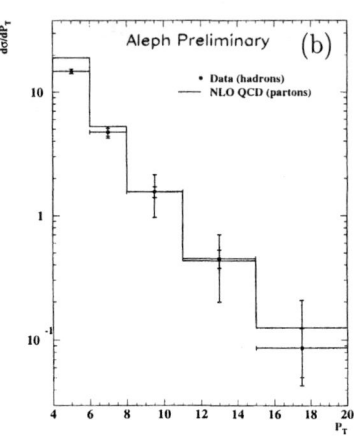

FIGURE 3. (a) Data vs. JetViP LO and NLO predictions for the single-tagged case (b)Data vs. JetViP and M. Klasen NLO predictions for the untagged case

case the PHOJET Monte Carlo was used to correct to hadron level and we see that both JetViP and M. Klasens NLO prediction are in good agreement with the data.

REFERENCES

1. D. Decamp et al., ALEPH collab. Nucl. Instr. Meth. **A294** (1990) 121.
2. J.A.M. Vermaseren, Proc. of the IV International Workshop on Gamma Gamma interactions, *eds* G. Cochard and P. Kessler (1980).
3. C.F. v. Weizsäcker, Z. Phys. **88** (1934) 612; E.J. Williams, Kgl. Danske Vidensk. Selskab. Mat-Fiz. Medd. **13** (1935) N4.
4. B. Pötter, Nucl. Phys. B559 (1999) 323; Nucl. Phys. B540 (1999) 382; Eur. Phys. J. Direct C5 (1999) 1
5. B. Pötter, Comp. Phys. Comm. 119 (1999) 45
6. M. Klasen private communication.
7. M. Klasen, T. Kleinwort and G. Kramer,Eur. Phys. J. direct **C1** (1998) 1 [hep-ph/9712256].
8. M. Klasen and G. Kramer, Z. Phys. **C76** (1997) 67 [hep-ph/9611450].
9. M. Seymour, Nucl.Phys.**B406** (1993) 187.
10. G. Marchesini et al. ,Comp Phys. Comm. **67** (1992) 465.
11. T. Sjöstrand, Computer Phys. Commun. **82** (1994) 74.
12. R. Engel, J. Ranft, Phys. Rev. **D54** 4246-4262 (1996)

Dijet Cross Sections in Photoproduction and Photon Structure

Stephen Maxfield[†]

H1 Collaboration

[†]*Dept. of Physics, University of Liverpool*

Abstract. The production of hard dijet events in photoproduction at HERA is dominated by resolved photon processes in which a parton in the photon with momentum fraction x_γ is scattered from a parton in the proton. These processes are sensitive to the quark and gluon content of the photon. The differential dijet cross-section $d\sigma/d\log(x_\gamma)$ is presented here, measured in tagged photoproduction at HERA using data taken with the H1 detector, corresponding to an integrated luminosity of 7.2 pb^{-1}. Using a restricted data sample at high transverse jet energy, $E_T^{jet} > 6$ GeV, the effective parton density $f_\gamma^{eff}(x_\gamma) = [q(x_\gamma) + \bar{q}(x_\gamma) + 9/4\, g(x_\gamma)]$ in the photon in leading order QCD is measured down to x_γ=0.05 from which the gluon density in the photon is derived.

INTRODUCTION

In leading order, the photoproduction of dijet events at HERA is expected to proceed through direct and resolved interactions of the type shown in Figure 1. In resolved processes, the photon structure is probed at a scale set by the transverse momentum, E_t^2, of the jets. The cross section is sensitive to the parton densities in the photon. In the kinematic region relevant here, the proton densities have been accurately measured so dijet cross section can be used to determine the photon parton densities. In this talk, I will describe an analysis [1] made by H1 which looks in particular at the low x_γ structure of the photon where the gluon density is expected to dominate. The particle probing the photon structure is coloured and so can couple directly to the gluons. The relatively high γp centre of mass energies (up to ~ 300 GeV) makes the low x_γ region accessible but it also helps to go to low jet E_t as $x_\gamma \sim \hat{p}_t^2/(yx_p s_{ep})$ where $\hat{p}_t$ is the transverse momentum of the outgoing partons giving rise to the high E_t jets. At such low E_t, the influence from additional interactions between the photon and proton remnants is significant. This enables us to study their effect but also forces us to find means of correcting the events before the parton densities can be disentangled. The analysis uses data collected in 1996 corresponding to an integrated luminosity of 7.2pb^{-1}.

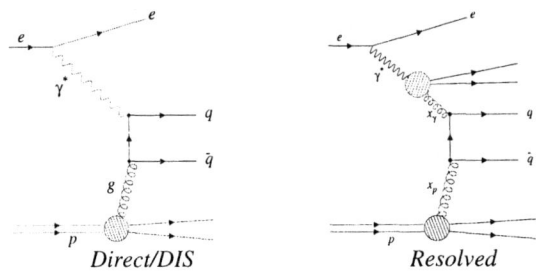

Direct/DIS   Resolved

FIGURE 1. The production of dijets in γp collisions by direct (left) and resolved (right) photoproduction processes

EVENT SELECTION, MODELS AND SIMULATIONS

Events were triggered by signals from the small angle electron tagger located 33 m downstream of the interaction vertex in coincidence with track activity in the central tracker [2]. Measurement of the scattered positron in the tagger ensures that the photon is quasi-real ($Q^2 < 0.01$ GeV^2). The inelasticity, y is restricted to the range $0.5 < y < 0.7$. Jets were found using a cone algorithm [3] with a small cone radius (0.7) which helps reduce the influence of underlying event energy. The events were accepted if they contained ≥ 2 jets with $E_{T,jet} > 4$ GeV with pseudorapidity in the region $-0.5 < \eta_{jet} < 2.5$ and jet-jet mass > 12 GeV. It is assumed that the two highest jets are associated with the hard scattering process. Background from events where this is not so is reduced by asking that the difference in η between the jets is < 1. The jets are used to estimate x_γ from:

$$x_{\gamma,jets} = \frac{E_{T,jet1}e^{-\eta_{jet1}} + E_{T,jet2}e^{-\eta_{jet2}}}{2yE_{ebeam}} \qquad (1)$$

The cuts on y and the jets limit the range to $x_{\gamma,jets} > 0.03$.

Leading order simulations, PHOJET [4] and PYTHIA [5] are used both to correct the data and to compare data and perturbative QCD predictions. The simulations are based on leading order matrix elements with parton showers and string fragmentation. Factorisation and renormalisation scales were set to $\hat{p}_t$ and GRV92-LO parton distribution functions [6] were used for the proton and photon. The simulations differ in the way they regulate the $2 \rightarrow 2$ cross section and the modeling of remnant-remnant interactions. PHOJET uses a fixed cut-off which, in this analysis, was set to $\hat{p}_t = 2.5$ GeV. PYTHIA was run with a damping factor $\hat{p}_t^2/(\hat{p}_t^2 + \hat{p}_{0t}^2)$ with $\hat{p}_{0t} = 1.55$ GeV. PHOJET implements the dual parton model which connects soft and hard interactions in a unitarisation scheme. PYTHIA simulates multiple parton interactions with LO matrix elements for the scattering of partons from the remnants. A cut-off of $\hat{p}_t = 1.2$ GeV is imposed on the final state partons.

THE DIJET CROSS SECTION FOR $E_T > 4$ GEV

Because the analysis extends to low jet E_t it is essential to understand the energy flow around the jets. Both simulations give a good description of the energy flow within the jet but differ in their predictions for the jet pedestals which are controlled partly by parton showering and hadronisation effects and partly by multiple interactions. To quantify this, the transverse energy flow per unit area in $\eta - \phi$ was measured in the strip $-1 < \eta - \eta_{jet} < 1$ excluding a cone of radius 1 about the jets. Figure 2(*left*) shows this as a function of η_{jet}. Note that the contribution to the E_t of the jet is considerable. This is especially so in the forward region where x_γ is typically small leaving a large energy fraction in the photon remnant for subsequent remnant-remnant interactions. Nevertheless the two simulations give a reasonable description of the flow and so can be confidently used to correct the data.

The correction for detector effects is done using a Bayesian unfolding technique [7] with correlations established by PYTHIA and PHOJET simulations. The mean result is taken and the difference ($10 - 15\%$, mostly from the differing pedestal predictions) is used as an estimate of systematic error. Scale uncertainties were estimated by varying the scales by a factor of two. The largest systematic effect ($\lesssim 24\%$) comes from a 4% uncertainty in the hadronic energy scale of the calorimeter. Figure 2(*right*) shows the resulting differential cross section $d\sigma/d\log(x_{\gamma,jets})$. The outer error bars represent systematic and statistical errors added in quadrature and the inner the statistical alone. The absolute predictions from PHOJET and PYTHIA are also shown. There is a large difference between these even though they use the same parton density functions and renormalisation scales. They should

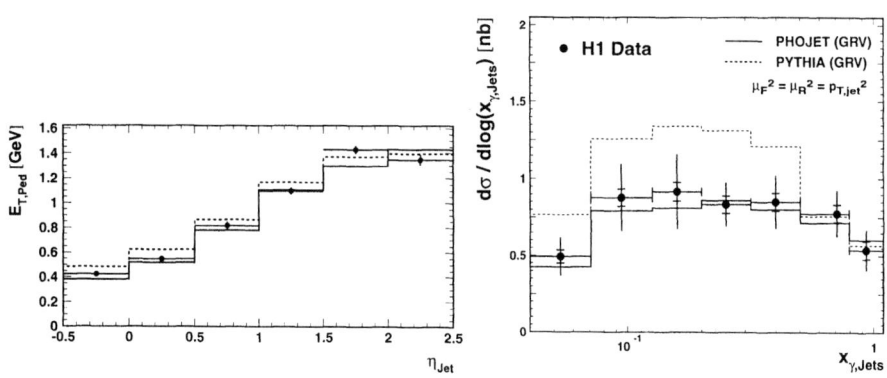

FIGURE 2. *Left*: The mean transverse energy density per unit area in $\eta - \phi$ outside the jets. The measured distribution (points) is compared with predictions from PYTHIA(dashed) and PHOJET(full). *Right*: The dijet cross section as a function of $x_{\gamma,jets}$ in the range $E_{T,jet} > 4$ GeV, $M_{1,2} > 12$ GeV, $-0.5 < \eta_{jet} < 2.5$, $|\eta_{jet1} - \eta_{jet2}| < 1$ and $0.5 < y < 0.7$.

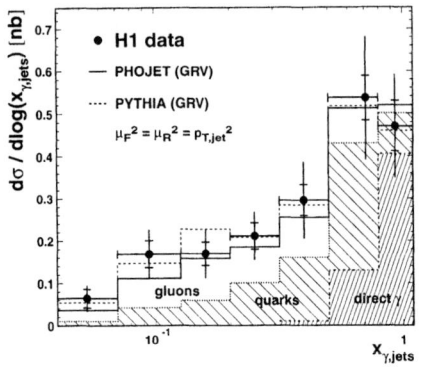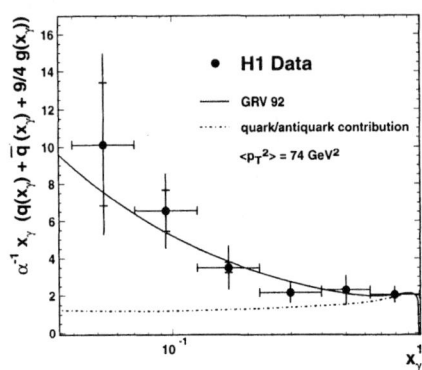

FIGURE 3. *Left*: The dijet cross section as a function of $x_{\gamma,jets}$ in the range $E_{T,jet} > 6$ GeV, $-0.5 < \eta_{jet} < 2.5$, $|\eta_{jet1} - \eta_{jet2}| < 1$, $\eta_{jets} > -0.9 - \ln(x_{\gamma,jets}$ and $0.5 < y < 0.7$. The cut on jet E_T was performed after subtraction of the average pedestal energy. *Right*: The effective parton density of the photon multiplied by $\alpha^{-1}x_\gamma$ as a function of x_γ and at a mean scale of $\hat{p}_t^2 = 74$ GeV^2.

be the same if hard scattering effects dominate. The difference comes from the parton p_t spectra in the region $2 < p_t < 4$ GeV where it is effected by the different regularisation procedures used. Such partons can give rise to measured jets because of the large underlying event energy.

THE EFFECTIVE PARTON DENSITY AND $\mathbf{g(x_\gamma)}$

Although the cross section measurement is sound and can be compared with theoretical predictions it cannot be used in a perturbative QCD analysis which requires data dominated by hard processes. To achieve this, the average expected pedestal energy as a function of η_{jet} was subtracted from the E_t of the jets. Monte Carlo studies show that this not only leads to good agreement between jet and parton energies on average but also improves the correlation in individual events. Jets are accepted if their E_t *after* pedestal subtraction exceeds 6 GeV.

The data is then unfolded as before. The result is shown in Figure 3(*left*). The LO QCD predictions are now similar and reasonably describe the data. For PHOJET, the quark and direct photon contributions are shown separately. It can be seen that, although the above procedure has depopulated the low x_γ region, there is still sensitivity to the gluon density down to $x_\gamma \sim 0.05$.

The cross section for dijet production in LO can be written

$$\frac{\mathrm{d}^4\sigma}{\mathrm{d}y\,\mathrm{d}x_\gamma\,\mathrm{d}x_\mathrm{p}\,\mathrm{d}\cos\theta^*} = \frac{1}{32\pi s_{\mathrm{ep}}}\,\frac{f_{\gamma/e}(y)}{y}\sum_{ij}\frac{f_{i/\gamma}(x_\gamma)}{x_\gamma}\,\frac{f_{j/\mathrm{p}}(x_\mathrm{p})}{x_\mathrm{p}}\,|M_{ij}(\cos\theta^*)|^2 \qquad (2)$$

where the sum is over all parton species. However, we cannot distinguish the individual $2 \to 2$ processes and instead rely on the fact that the dominant ones have almost identical angular dependence and differ only by colour factors. In this single effective sub-process approximation [8] the cross section in Equation 2 reduces to:

$$\frac{f_{\gamma/e}(y,Q^2)}{y}\,\frac{f_{eff,\gamma}(x_\gamma,P_t^2,Q^2)}{x_\gamma}\,\frac{f_{eff,\mathrm{p}}(x_\mathrm{p},P_t^2)}{x_\mathrm{p}}\,|M_{SES}(\cos\theta^*)|^2 \qquad (3)$$

With *effective parton densities* for the proton and photon:

$$f_{eff,A} \equiv (f_{q/A} + f_{\bar{q}/A}) + \frac{9}{4}f_{g/A}, \quad A = p,\gamma. \qquad (4)$$

A second unfolding is then used to correct for hadronisation and residual affects from the underlying event energy and to extract the effective parton density of the photon. The result is shown in Figure 3 (*right*). The dominant systematic error is from the pedestal subtraction procedure (15 to 20%). Note that the direct contribution has been subtracted. The average scale is $<\hat{p}_t^2>= 74$ GeV^2. A rise to low x_γ is clearly seen. The data are compared to the GRV92 LO parameterisation of the photon pdf. The quark contribution to this, including charm, is shown separately. It describes the data in the highest x_γ bin but fails at lower values. The difference can be attributed to the gluon contribution which is thus shown to be responsible for the rise at low x_γ. By subtracting the quark density, we obtain the gluon density shown in Figure 4. The total error includes a 30% contribution from the uncertainty in the quark densities (including that of the calculated charm contribution) increasing to 60% in the lowest bin.

The figure also shows (open boxes) the result of an earlier measurement from H1 [9] using tracks. The excellent agreement is significant given the different systematic effects. In particular the track measurement is much less sensitive to the underlying event. A variety of gluon densities [10] is shown. The GRV92-LO, used throughout this analysis, gives the best description.

CONCLUSIONS

H1 has measured two differential dijet cross sections $d\sigma/dx_{\gamma,jets}$. The kinematic region covered is strongly effected by the underlying event and uncertainties in the transition between hard and soft physics. This is reduced in the cross section with $E_t > 6$ GeV where the cut is applied after a pedestal subtraction, and the good correlation with the parton dynamics enables the subtraction of an effective parton density and gluon density for the photon. The latter is shown to rise strongly as x_γ decreases.

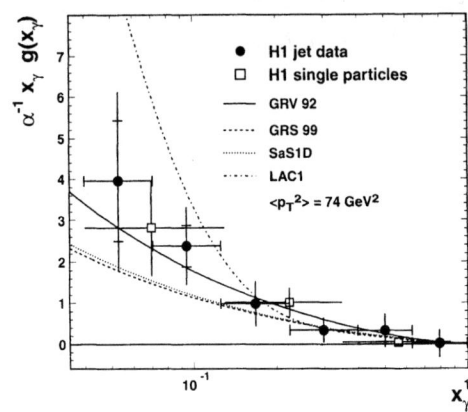

FIGURE 4. The gluon distribution of the photon multiplied by $\alpha^{-1}x_\gamma$ as a function of x_γ and at a mean scale of $\hat{p}_t^2 = 74$ GeV^2. The open squares show a previous measurement using single particles at a mean scale of $\hat{p}_t^2 = 38$ GeV^2.

REFERENCES

1. H1 Collab., C. Adloff *et al.*, *Phys. Lett.* **B 483**(2000) 36.
2. H1 Collab., I. Abt *et al.*, *Nucl. Instrum. Methods* **A 386**(1997) 310 and 348.
3. CDF Collab., F. Abe *et al.*, *Phys. Rev.* **D 45** (1992) 1448.
4. PHOJET Monte Carlo, R. Engel, *Z. Phys* **C 66** (1995) 203.
5. T. Sjöstrand, CERN-TH-6488 (1992), *Comput. Phys. Commun.* **82** (1994) 74.
6. M. Glück, E. reya, A. Vogt,*Phys. Rev.* **D 46** (1992) 1973;
   M. Glück, E. reya, A. Vogt,*Z. Phys.* **C 53** (1992) 127.
7. G. D'Agostini, *Nucl. Instrum. Methods* **A 362**(1995) 487.
8. B. L. Combridge and C. J. Maxwell, *Nucl. Phys.* **B 239** (1984) 429.
9. H1 Collab., C. Adloff *et al.*, *Eur. Phys. J.* **C 10**(1993) 363.
10. H. Abramowicz, K. Charchula, A. levy, *Phys. Lett.* **B 269** (1991) 458.
    M. Glück, E. Reya, I. Schienbein, *Phys. rev.* **D 60** (1999) 54019.
    G. A. Schüler, T. J. Sjöstrand, *Phys. lett.* **B 376** (1996) 193.

Total Cross Sections and Event Properties from Real to Virtual Photons

Christer Friberg

Theoretical Physics, Lund University, Sölvegatan 14A, SE-223 62 LUND Sweden

Abstract. A model for total cross sections with virtual photons is presented. In particular γ^*p and $\gamma^*\gamma^*$ cross sections are considered. Our approach extends on a model for photoproduction, where the total cross section is subdivided into three distinct event classes: direct, VMD and anomalous. With increasing photon virtuality, the latter two decrease in importance. Instead Deep Inelastic Scattering dominates, with the direct class being the $\mathcal{O}(\alpha_s)$ correction thereof. Hence, the model provides a smooth transition between the two regions. By the breakdown into different event classes, one may aim for a complete picture of all event properties.

INTRODUCTION

In this section we summarize the model presented in [1,2]. It starts from the model for real photons in [3], but further develops this model and extends it also to encompass the physics of virtual photons. The physics has been implemented in the PYTHIA generator [4], so that complete events can be studied under realistic conditions.

Photon interactions are complicated since the photon wave function contains so many components, each with its own interactions. To first approximation, it may be subdivided into a direct and a resolved part. (In higher orders, the two parts can mix, so one has to provide sensible physical separations between the two.) In the former the photon acts as a pointlike particle, while in the latter it fluctuates into hadronic states. These fluctuations are of $\mathcal{O}(\alpha_{em})$, and so correspond to a small fraction of the photon wave function, but this is compensated by the bigger cross sections allowed in strong-interaction processes. For real photons therefore the resolved processes dominate the total cross section, while the pointlike ones take over for virtual photons.

A MODEL FOR PHOTON INTERACTIONS

The fluctuations $\gamma \to q\bar{q} (\to \gamma)$ can be characterized by the transverse momentum $k_\perp$ of the quarks, or alternatively by some mass scale $m \simeq 2k_\perp$, with a

CP571, *PHOTON 2000*, edited by A. J. Finch
© 2001 American Institute of Physics 0-7354-0010-5/01/$18.00

spectrum of fluctuations $\propto dk_\perp^2/k_\perp^2$. The low-$k_\perp$ part cannot be calculated perturbatively, but is instead parameterized by experimentally determined couplings to the lowest-lying vector mesons, $V = \rho^0$, ω^0, ϕ^0 and J/ψ, an ansatz called VMD for Vector Meson Dominance. Parton distributions are defined with a unit momentum sum rule within a fluctuation [5], giving rise to total hadronic cross sections, jet activity, multiple interactions and beam remnants as in hadronic interactions. In interactions with a hadron or another resolved photon, jet production occurs by typical parton-scattering processes such as qq$'$ $\to$ qq$'$ or gg $\to$ gg.

States at larger $k_\perp$ are called GVMD or Generalized VMD, and their contributions to the parton distribution of the photon are called anomalous. Given a dividing line $k_0 \simeq 0.5$ GeV to VMD states, the anomalous parton distributions are perturbatively calculable. The total cross section of a state is not, however, since this involves aspects of soft physics and eikonalization of jet rates. Therefore an ansatz is chosen where the total cross section of a state scales like $k_V^2/k_\perp^2$, where the adjustable parameter $k_V \approx m_\rho/2$ for light quarks. The spectrum of GVMD states is taken to extend over a range $k_0 < k_\perp < k_1$, where k_1 is identified with the $p_{\perp\text{min}}(s)$ cut-off of the perturbative jet spectrum in hadronic interactions, $p_{\perp\text{min}}(s) \approx 1.5$ GeV at typical energies [4]. Above that range, the states are assumed to be sufficiently weakly interacting that no eikonalization procedure is required, so that cross sections can be calculated perturbatively without any recourse to Pomeron phenomenology. There is some arbitrariness in that choice, and some simplifications are required in order to obtain a manageable description.

A real direct photon in a γp collision can interact with the parton content of the proton: γq $\to$ qg (QCD Compton) and γg $\to$ q$\bar{q}$ (Boson Gluon Fusion). The $p_\perp$ in this collision is taken to exceed k_1, in order to avoid double-counting with the interactions of the GVMD states. In $\gamma\gamma$, the equivalent situation is called single-resolved, where a direct photon interacts with the partonic component of the other, resolved photon. The $\gamma\gamma$ direct process $\gamma\gamma \to$ q$\bar{q}$ has no correspondence in γp.

As an illustration of this scenario, the phase space of γp events is shown in Fig. 1a-b. (A corresponding plot can be made for $\gamma\gamma$, but then requires three dimensions.) Two transverse momentum scales are introduced, namely the photon resolution scale $k_\perp$ and the hard interaction scale $p_\perp$. Here $k_\perp$ is a measure of the virtuality of a fluctuation of the photon and $p_\perp$ corresponds to the most virtual rung of the ladder, possibly apart from $k_\perp$. As we have discussed above, the low-$k_\perp$ region corresponds to VMD and GVMD states that encompasses both perturbative high-$p_\perp$ and non-perturbative low-$p_\perp$ interactions. Above k_1, the region is split along the line $k_\perp = p_\perp$. When $p_\perp > k_\perp$ the photon is resolved by the hard interaction, as described by the anomalous part of the photon distribution function. This is as in the GVMD sector, except that we should (probably) not worry about multiple parton–parton interactions. In the complementary region $k_\perp > p_\perp$, the $p_\perp$ scale is just part of the traditional evolution of the proton PDF's up to the scale of $k_\perp$, and thus there is no need to introduce an internal structure of the photon. One could imagine the direct class of events as extending below k_1 and there being the low-$p_\perp$ part of the GVMD class, only appearing when a hard interaction at a larger

198

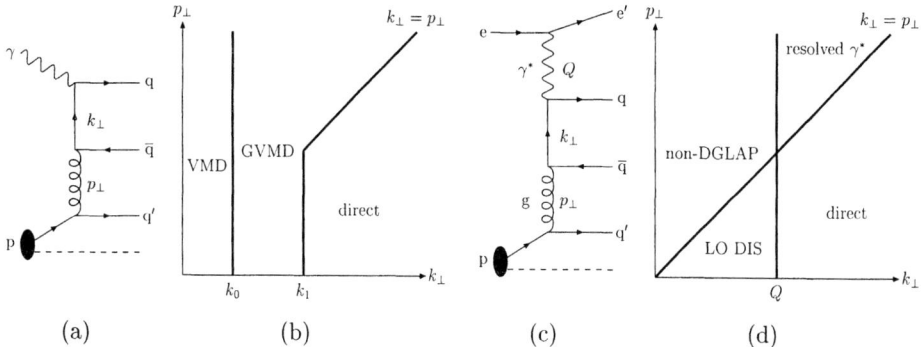

FIGURE 1. (a) Schematic graph for a hard γp process, illustrating the concept of two different scales. (b) The allowed phase space for this process, with one subdivision into event classes. (c) Schematic graph for a hard γ^*p process, illustrating the concept of three different scales. (d) Event classification in the large-Q^2 limit.

$p_\perp$ scale would not preempt it. This possibility is implicit in the standard cross section framework.

If the photon is virtual, it has a reduced probability to fluctuate into a vector meson state, and this state has a reduced interaction probability. This can be modeled by a traditional dipole factor $(m_V^2/(m_V^2 + Q^2))^2$ for a photon of virtuality Q^2, where $m_V \to 2k_\perp$ for a GVMD state. Putting it all together, the cross section of the GVMD sector then scales like

$$\int_{k_0^2}^{k_1^2} \frac{dk_\perp^2}{k_\perp^2} \frac{k_V^2}{k_\perp^2} \left(\frac{4k_\perp^2}{4k_\perp^2 + Q^2} \right)^2 . \tag{1}$$

For a virtual photon the DIS process γ^*q $\to$ q is also possible, but by gauge invariance its cross section must vanish in the limit $Q^2 \to 0$. At large Q^2, the direct processes can be considered as the $\mathcal{O}(\alpha_s)$ correction to the lowest-order DIS process, but the direct ones survive for $Q^2 \to 0$. There is no unique prescription for a proper combination at all Q^2, but we have attempted an approach that gives the proper limits and minimizes doublecounting. For large Q^2, the DIS γ^*p cross section is proportional to the structure function $F_2(x, Q^2)$ with the Bjorken $x = Q^2/(Q^2+W^2)$. Since normal parton distribution parameterizations are frozen below some Q_0 scale and therefore do not obey the gauge invariance condition, an ad hoc factor $(Q^2/(Q^2 + m_\rho^2))^2$ is introduced for the conversion from the parameterized $F_2(x, Q^2)$ to a $\sigma_{\mathrm{DIS}}^{\gamma^*\mathrm{p}}$:

$$\sigma_{\mathrm{DIS}}^{\gamma^*\mathrm{p}} \simeq \left(\frac{Q^2}{Q^2 + m_\rho^2} \right)^2 \frac{4\pi^2 \alpha_{\mathrm{em}}}{Q^2} F_2(x, Q^2) = \frac{4\pi^2 \alpha_{\mathrm{em}} Q^2}{(Q^2 + m_\rho^2)^2} \sum_{q,\bar{q}} e_q^2 \, xq(x, Q^2) . \tag{2}$$

Here m_ρ is some non-perturbative hadronic mass parameter, for simplicity identified with the ρ mass.

In order to avoid double-counting between DIS and direct events, a requirement $p_\perp > \max(k_1, Q)$ is imposed on direct events. In the remaining DIS ones, denoted lowest order (LO) DIS, thus $p_\perp < Q$. This would suggest a subdivision $\sigma_{\text{LO DIS}}^{\gamma^* p} = \sigma_{\text{DIS}}^{\gamma^* p} - \sigma_{\text{direct}}^{\gamma^* p}$, with $\sigma_{\text{DIS}}^{\gamma^* p}$ given by eq. (2) and $\sigma_{\text{direct}}^{\gamma^* p}$ by the perturbative matrix elements. In the limit $Q^2 \to 0$, the DIS cross section is now constructed to vanish while the direct is not, so this would suggest $\sigma_{\text{LO DIS}}^{\gamma^* p} < 0$. However, here we expect the correct answer not to be a negative number but an exponentially suppressed one, by a Sudakov form factor. This modifies the cross section:

$$\sigma_{\text{LO DIS}}^{\gamma^* p} = \sigma_{\text{DIS}}^{\gamma^* p} - \sigma_{\text{direct}}^{\gamma^* p} \longrightarrow \sigma_{\text{DIS}}^{\gamma^* p} \exp\left(-\frac{\sigma_{\text{direct}}^{\gamma^* p}}{\sigma_{\text{DIS}}^{\gamma^* p}}\right). \tag{3}$$

Since we here are in a region where the DIS cross section is no longer the dominant one, this change of the total DIS cross section is not essential.

The overall picture, from a DIS perspective, is illustrated in Fig. 1c-d, now with three scales to be kept track of. The traditional DIS region is the strongly ordered one, $Q^2 \gg k_\perp^2 \gg p_\perp^2$, where DGLAP-style evolution [6] is responsible for the event structure. As always, ideology wants strong ordering, while the actual classification is based on ordinary ordering $Q^2 > k_\perp^2 > p_\perp^2$. The region $k_\perp^2 > \max(Q^2, p_\perp^2)$ is also DIS, but of the $\mathcal{O}(\alpha_s)$ direct kind. The region where $k_\perp$ is the smallest scale corresponds to non-ordered emissions, that then go beyond DGLAP validity, while the region $p_\perp^2 > k_\perp^2 > Q^2$ cover the interactions of a resolved virtual photon. Comparing Figs. 1b and 1d, we conclude that the whole region $p_\perp > k_\perp$ involves no doublecounting, since we have made no attempt at a non-DGLAP DIS description but can choose to cover this region entirely by the VMD/GVMD descriptions. Actually, it is only in the corner $p_\perp < k_\perp < \min(k_1, Q)$ that an overlap can occur between the resolved and the DIS descriptions. Some further considerations show that usually either of the two is strongly suppressed in this region, except in the range of intermediate Q^2 and rather small W^2. Typically, this is the region where $x \approx Q^2/(Q^2 + W^2)$ is not close to zero, and where F_2 is dominated by the valence-quark contribution. The latter behaves roughly $\propto (1-x)^n$, with an n of the order of 3 or 4. Therefore we will introduce a corresponding damping factor to the VMD/GVMD terms.

In total, we have now arrived at our ansatz for all Q^2:

$$\sigma_{\text{tot}}^{\gamma^* p} = \sigma_{\text{DIS}}^{\gamma^* p} \exp\left(-\frac{\sigma_{\text{direct}}^{\gamma^* p}}{\sigma_{\text{DIS}}^{\gamma^* p}}\right) + \sigma_{\text{direct}}^{\gamma^* p} + \left(\frac{W^2}{Q^2 + W^2}\right)^n \left(\sigma_{\text{VMD}}^{\gamma^* p} + \sigma_{\text{GVMD}}^{\gamma^* p}\right), \tag{4}$$

with four main components. Most of these in their turn have a complicated internal structure, as we have seen. The $\gamma^* \gamma^*$ collision between two inequivalent photons contains 13 components: four when the VMD and GVMD states interact with each other (double-resolved), eight with a LO DIS or direct photon interaction on a VMD or GVMD state on either side (single-resolved, including the traditional DIS), and one where two direct photons interact by the process $\gamma^*\gamma^* \to q\bar{q}$ (direct, not to be confused with the direct process of γ^*p).

An important note is that the Q^2 dependence of the DIS and direct processes is implemented in the matrix element expressions, i.e. in processes such as $\gamma^*\gamma^* \to q\bar{q}$ or $\gamma^*q \to qg$ the photon virtuality explicitly enters. This is different from VMD/GVMD, where dipole factors are used to reduce the assumed flux of partons inside a virtual photon relative to those of a real one, but the matrix elements themselves contain no dependence on the virtuality either of the partons or of the photon itself. Typically results are obtained with the SaS 1D PDF's for the virtual (transverse) photons [5], since these are well matched to our framework, e.g. allowing a separation of the VMD and GVMD/anomalous components.

RESULTS

So far, nothing has been said about σ_L, except that gauge invariance dictates its vanishing in the limit $Q^2 \to 0$. However, we will assume that quite a similar decomposition can be made of longitudinal photon interactions as was done for the transverse one. To first approximation, this again means a separation into direct and resolved photons. In direct processes, the nature of the photon is explicitly included in the perturbative cross section formulae. Thus, for $\gamma^*q \to qg$ and $\gamma^*g \to q\bar{q}$, the differential cross sections $d\hat{\sigma}_T/d\hat{t}$ and $d\hat{\sigma}_L/d\hat{t}$ are separately available [7]. The latter is proportional to Q^2 and thus nicely vanishes in the limit $Q^2 \to 0$. Similarly the $\gamma^*\gamma^* \to q\bar{q}$ process gives four separate cross section formulae, $d\hat{\sigma}_{TT,TL,LT,LL}/d\hat{t}$ [8]. The DIS, non-direct part currently contains no explicit description of a longitudinal probing photon, only of a probed one. However, to the extent that PDF's are extracted from $F_2 \propto \sigma_T + \sigma_L$ data, effects may be implicitly included. Furthermore, perturbative calculations [9] predict $\sigma_L \ll \sigma_T$ in the large-Q^2 region, where this process dominates.

In [1,2,10] a few simple Q^2-dependent multiplicative expressions have been studied to encompass the effects of resolved longitudinal photons. The two alternative factors

$$r_1(m_V^2, Q^2) = \frac{2m_V^2 Q^2}{(m_V^2 + Q^2)^2} \qquad r_2(m_V^2, Q^2) = \frac{2Q^2}{(m_V^2 + Q^2)} \tag{5}$$

relative to the resolved transverse cross section are used for the resolved longitudinal contributions in Fig. 2, assuming a relation $m_V \simeq 2k_\perp$ for GVMD states. The differences between the r_i factors show some of the uncertainty in the modeling of resolved longitudinal photons. However, it is possible to obtain a reasonable description of all the data in both γ^*p and $\gamma^*\gamma$ with the same set of parameters (mainly constrained from γp).

More sophisticated tests of the model can be made by studying event shapes and we look forward to detailed studies by the experimental community, based on the code we now provide in the PYTHIA event generator [4].

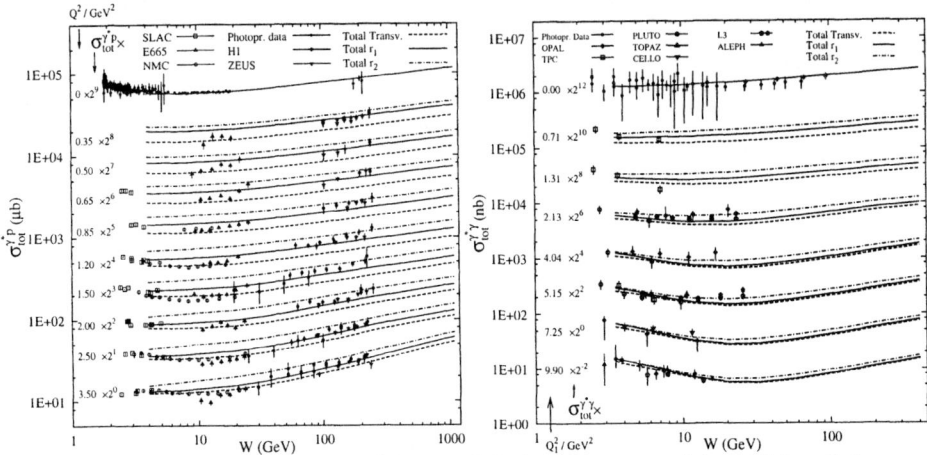

FIGURE 2. Total cross sections as a function of the invariant mass of the collision. References to the experimental measurements can be found in ref. [2].

REFERENCES

1. Friberg, C., and Sjöstrand, T., *Eur. Phys. J.* **C13**, 151 (2000) (hep-ph/9907245).
2. Friberg, C., and Sjöstrand, T., *J. High Energy Phys.* **09**, 010 (2000) (hep-ph/0007314).
3. Schuler, G.A., and Sjöstrand, T., *Phys. Lett.* **B300**, 169 (1993), *Nucl. Phys.* **B407**, 539 (1993), *Z. Phys.* **C73**, 677 (1997).
4. Sjöstrand, T., *Computer Phys. Commun.* **82**, 74 (1994); Sjöstrand, T., et al., hep-ph/0010017 (to appear in *Computer Phys. Commun.*); http://www.thep.lu.se/~torbjorn/Pythia.html.
5. Schuler, G.A., and Sjöstrand, T., *Z. Phys.* **C68**, 607 (1995), *Phys. Lett.* **B376**, 193 (1996).
6. Gribov, V.N., and Lipatov, L.N., *Sov. J. Nucl. Phys.* **15**, 438 and 675 (1972); Altarelli, G., and Parisi, G., *Nucl. Phys.* **B126**, 298 (1977); Dokshitzer, Yu.L., *Sov. Phys. JETP* **46**, 641 (1977).
7. Altarelli, G., and Martinelli, G., *Phys. Lett.* **76B** (1978) 89; Mendéz, A., *Nucl. Phys.* **B145**, 199 (1978); Peccei, P., and Rückl, R., *Nucl. Phys.* **B162**, 125 (1980); Rumpf, Ch., Kramer, G., and Willrodt, J., *Z. Phys.* **C7**, 337 (1981).
8. Budnev, V.M., Ginzburg, I.F., Meledin, G.V. and Serbo, V.G., *Phys. Rept.* **15**, 181 (1974); Baier, V.N., Kuraev, E.A., Fadin, V.S., and Khoze, V.A., *Phys. Rept.* **78**, 293 (1981).
9. Mishra, S.R., and Sciulli, F., *Phys. Lett.* **B244**, 341 (1990).
10. Friberg, C., and Sjöstrand, T., LU TP 00–31, to appear in *Phys. Lett.* **B** (hep-ph/0009003).

Jet Substructure at HERA

Claudia Glasman*

representing the ZEUS Collaboration

*Department of Physics and Astronomy, Kelvin Building,
University of Glasgow, Glasgow, G12 8QQ, UK

Abstract. Measurements of jet shapes in photoproduction performed by ZEUS are presented and compared to leading-logarithm parton-shower Monte Carlo models. The predicted differences on the size of gluon- and quark-initiated jets are used to select samples to study the dynamics of the subprocesses.

Introduction

The internal structure of a jet depends mainly on the type of primary parton (quark or gluon) from which it originated and to a lesser extent on the particular hard scattering process. QCD predicts that (a) at sufficiently high transverse energy of the jet (E_T^{jet}), where fragmentation effects become negligible, the jet structure is driven by gluon emission from the primary parton; and (b) gluon jets are broader than quark jets due to the larger colour charge of the gluon. In the first part of this article, the measurements performed by ZEUS [1] to test the QCD prediction (a) are presented and compared to leading-logarithm parton-shower Monte Carlo models. The lowest non-trivial order contribution to the jet substructure is given by $\mathcal{O}(\alpha \alpha_s^2)$ calculations and, therefore measurements of jet substructure provide a stringest test of QCD beyond leading order (LO). The results of taking prediction (b) into account to test the dynamics of the subprocesses [2] are shown in the second part of this article.

At HERA, positrons of energy $E_e = 27.5$ GeV collide with protons of energy $E_p = 820$ GeV. The main source of jets is hard scattering in γp interactions in which a quasi-real photon ($Q^2 \approx 0$, where Q^2 is the virtuality of the photon) emitted by the positron beam interacts with a parton from the proton to produce two jets in the final state. In LO QCD, there are two processes which contribute to the jet production cross section: the resolved process in which the photon interacts through its partonic content, and the direct process in which the photon interacts as a point-like particle.

In the kinematic regime studied here, the dominant subprocesses are $q_\gamma g_p \to qg$ in resolved and $\gamma g \to q\bar{q}$ in direct. The kinematics of these processes are such that the majority of the jets in the region of pseudorapitity (η^{jet}) below 0 originate from quarks and the fraction of gluon-initiated jets increases as η^{jet} increases.

CP571, *PHOTON 2000,* edited by A. J. Finch
© 2001 American Institute of Physics 0-7354-0010-5/01/$18.00

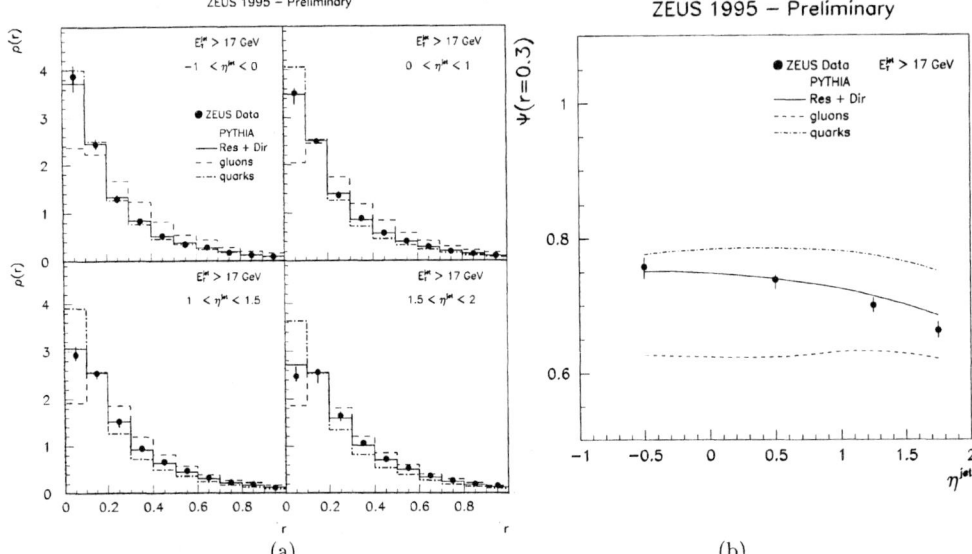

ZEUS 1995 – Preliminary

FIGURE 1. (a) Measured differential jet shape $\rho(r)$ vs. r (black dots). (b) Measured integrated jet shape $\psi(r = 0.3)$ vs. η^{jet} (black dots). PYTHIA Monte Carlo calculations are shown for comparison.

Jet substructure

The differential jet shape $\rho(r)$ is defined as the average fraction of the jet's transverse energy that lies inside an annulus in the (η) - azimuth (φ) plane of radii $r \pm \Delta r/2$, where $r = \sqrt{(\Delta\eta)^2 + (\Delta\varphi)^2}$, concentric with the jet axis. And the integrated jet shape $\psi(r)$ is the average fraction of the jet's transverse energy that lies inside a cone in the $\eta - \varphi$ plane of radius r concentric with the jet axis.

Measurements of the differential jet shape $\rho(r)$ have been performed by ZEUS [1] using an inclusive sample of jets. The jets have been identified using the longitudinally invariant k_T cluster algorithm in the inclusive mode. The jets were required to have $E_T^{jet} > 17$ GeV and $-1 < \eta^{jet} < 2$. The measurements were performed in the kinematic region $0.2 < y < 0.85$, where y is the inelasticity variable, and $Q^2 \leq 1$ GeV^2. Figure 1a shows $\rho(r)$ as a function of r in different regions of η^{jet}. The data show that the jets become broader as η^{jet} increases.

The solid histograms in figure 1a are calculations from the leading-logarithm parton-shower Monte Carlo PYTHIA. The predictions, which include initial and final state QCD radiation, give a good description of the data. The comparison of the predictions for gluon- (dashed histograms) and quark-initiated (dot-dashed histograms) jets to the data shows that the measured jets are quark-like for $-1 < \eta^{jet} < 0$ and become increasingly more gluon-like as η^{jet} increases.

The quark and gluon content of the final state has been investigated in more

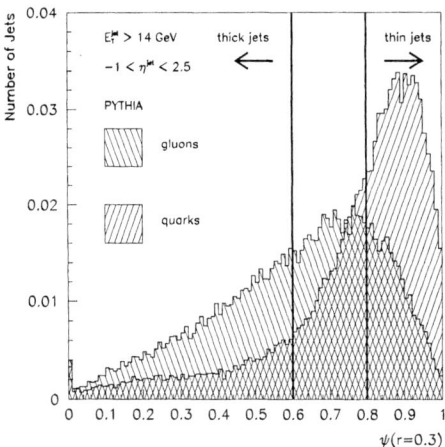

FIGURE 2. The predicted integrated jet shape $\psi(r = 0.3)$ distributions at the hadron level for samples of gluon- and quark-initiated jets simulated using the program PYTHIA.

detail by studying the η^{jet} dependence of the integrated jet shape at a fixed value of r [1]. Figure 1b shows $\psi(r)$ for $r = 0.3$ as a function of η^{jet}. The measured jet shape decreases with η^{jet}, i.e. the jets become broader as η^{jet} increases. The comparison between the predictions of PYTHIA for gluon- and quark-initiated jets and the data shows that the broadening of the jets is consistent with an increasing fraction of gluon-initiated jets as η^{jet} increases.

Substructure dependence of dijet cross sections

The predictions of the Monte Carlo models for the jet shape reproduce the data well and show the expected differences for quark- and gluon-initiated jets. Therefore, the Monte Carlo events have been used to devise a method to select samples enriched in quark- and gluon-initiated jets. The samples are selected by exploiting the QCD prediction that gluon-initiated jets should be broader than quark-initiated jets.

The predicted distributions of $\psi(r)$ for $r = 0.3$ for quark- and gluon-initiated jets are clearly different (see figure 2). A sample enriched in quark-initiated ("thin") jets has been selected by requiring $\psi(r = 0.3) > 0.8$ and a sample enriched in gluon-initiated ("thick") jets has been selected by requiring $\psi(r = 0.3) < 0.6$.

Measurements of $d\sigma/d\eta^{jet}$**.** Using the jet shape selection into "thin" and "thick" jet samples, dijet cross sections have been measured [2] as a function of η^{jet} for events with at least two jets of $E_T^{jet} > 14$ GeV and $-1 < \eta^{jet} < 2.5$ in the kinematic region given by $0.2 < y < 0.85$ and $Q^2 \leq 1$ GeV^2. Figure 3 shows $d\sigma/d\eta^{jet}$ for "thick" and "thin" jets. The cross section for "thick" jets displays a very different shape than that of the "thin" jet sample: the η^{jet} distribution for the "thin" jet sample peaks at about 0.7, whereas the "thick" jet sample distribution

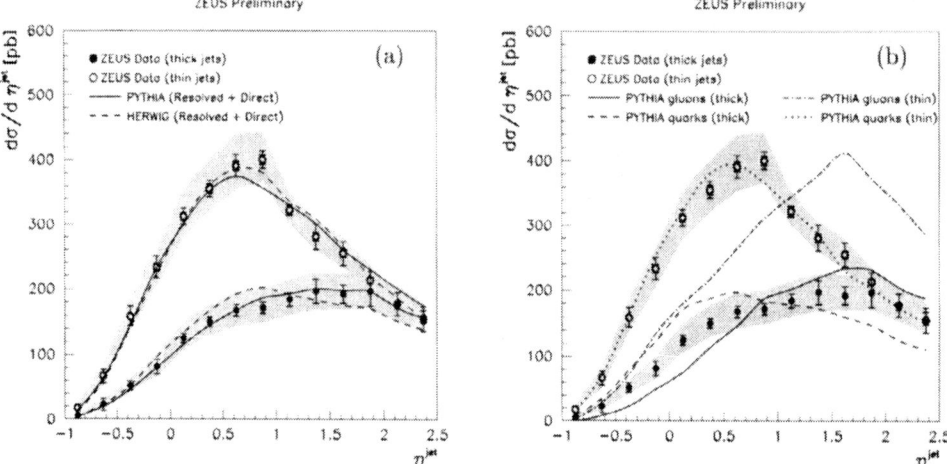

FIGURE 3. Measured $d\sigma/d\eta^{jet}$ for samples of "thick" (black dots) and "thin" (open circles) jets. The thick error bars represent the statistical errors of the data, and the thin error bars show the statistical errors and uncorrelated systematic uncertainties added in quadrature. The shaded band displays the uncertainty due to the absolute energy scale of the jets. In (a) Monte Carlo calculations using PYTHIA and HERWIG for resolved- plus direct-photon processes and in (b) the predictions of PYTHIA for quark- and gluon-initiated jets are shown for comparison.

peaks at $\eta^{jet} \approx 1.5$.

The predictions of PYTHIA and HERWIG (see figure 3a) give a good description of the data. The Monte Carlo distributions have been normalised to the total measured cross section of each type after applying the same jet shape selection as in the data. PYTHIA and HERWIG predict a similar parton content of the final state: the "thick" jet sample is composed of $15-17\%$ of gg subprocesses in the final state, $54-58\%$ of gq and $25-31\%$ of qq. The "thin" jet sample contains $54-56\%$ of qq, 41% of qg and $3-5\%$ of gg. Therefore, the "thin" jet sample is indeed dominated by quark-initiated jets in the final state and the "thick" jet sample has a high content of gluon-initiated jets coming mainly from the final-state gluon of the subprocess $q_\gamma g_p \to qg$. The measurements are compared to the predictions of quark- and gluon-initiated jets with the same jet shape selection as for the data in figure 3b. A "thin"-quark jet sample gives a good description of the "thin" jet sample, whereas the "thin"-gluon jet sample peaks one unit of pseudorapidity higher. A sample of "thick"-quark jets alone cannot describe the "thick"-jet sample and a large content of gluon-initiated jets is needed to reproduce the shape of the data.

Measurements of $d\sigma/d\cos\theta^*$. The underlying parton dynamics is reflected in the distribution of the scattering angle in the dijet centre-of-mass system, θ^*. The $\cos\theta^*$ distribution is sensitive to the spin of the exchanged particle: for gluon exchange, the cross section is proportional to $(1 - |\cos\theta^*|)^{-2}$, whereas for quark

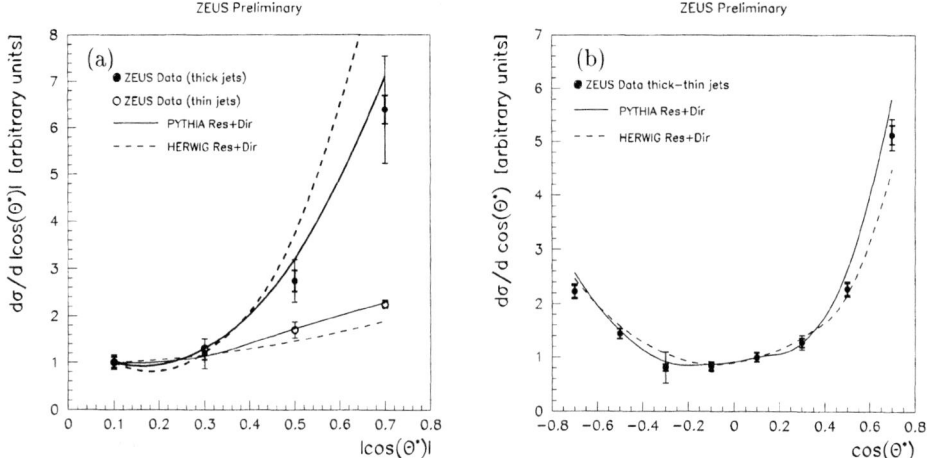

FIGURE 4. (a) Measured $d\sigma/d|\cos\theta^*|$ for two "thick"-jet events (black dots) and two "thin"-jet events (open circles). (b) Measured $d\sigma/d\cos\theta^*$ for one "thick" jet and one "thin" jet in an event (black dots). PYTHIA and HERWIG Monte Carlo calculations are shown for comparison.

exchange the cross section is proportional to $(1 - |\cos\theta^*|)^{-1}$. In a previous publication [3], the $|\cos\theta^*|$ distribution for resolved processes (which are dominated by subprocesses mediated by gluon exchange) was observed to rise more steeply than that of direct processes (which proceed via quark exchange) at high $|\cos\theta^*|$ values. In this analysis [2], samples of jets are selected according to their internal structure. This method provides a handle to better understand the dynamics of the subprocesses.

For samples of two "thick" or two "thin" jets, only the absolute value of $\cos\theta^*$ can be determined because the outgoing jets are indistinguishable. The dijet cross section as a function of $|\cos\theta^*|$ for samples of two "thick"-jet events and two "thin"-jet events has been measured for dijet invariant masses $M^{JJ} > 47$ GeV. The cross sections are presented in figure 4a and are normalised so as to have a value of unity at low $|\cos\theta^*|$ to study the difference in slope as $|\cos\theta^*| \to 1$. The $|\cos\theta^*|$ distribution for the two samples of dijet events increases as $|\cos\theta^*|$ increases, however they exhibit a different slope. For comparison, PYTHIA and HERWIG Monte Carlo predictions are shown in figure 4a. The predictions have been also normalised so as to have a value of unity at low $|\cos\theta^*|$ and give a reasonable description of the shape of the data. The different slope observed in the data can be understood in terms of the dominant subprocesses in the two samples: the two "thick"-jet sample is dominated by subprocesses mediated by gluon exchange ($gg \to gg$ and $qg \to qg$) whereas the two "thin"-jet sample is dominated by subprocesses mediated by quark exchange ($\gamma g \to q\bar{q}$).

The sample of events with a "thick" jet and a "thin" jet allows a measurement of the unfolded $d\sigma/d\cos\theta^*$ cross section, since in this case the jets can be distinguished. Figure 4b shows the measured dijet cross section as a function of $\cos\theta^*$. The dijet angular distribution was measured with respect to the "thick" jet and shows a different behaviour on the negative and positive sides: the measured cross section at $\cos\theta^* = 0.7$ is more than two times larger than the one at $\cos\theta^* = -0.7$. The Monte Carlo models give a reasonable description of the shape of the measured $d\sigma/d\cos\theta^*$ (see figure 4b). The measurements and the Monte Carlo calculations in figure 4b are normalised so as to have a value of unity at central values of $\cos\theta^*$ to study the difference in slope as $\cos\theta^* \to \pm 1$. The observed asymmetry is understood in terms of the dominant subprocess: $q_\gamma g_p \to qg$. The positive side is dominated by t-channel gluon exchange, whereas the negative side is dominated by u-channel quark exchange.

Summary and conclusions

Measurements of jet shapes have been performed for inclusive jet samples with $E_T^{jet} > 17$ GeV and $-1 < \eta^{jet} < 2$ in the kinematic regime given by $0.2 < y < 0.85$ and $Q^2 \leq 1$ GeV^2. The measured jet shape broadens as η^{jet} increases. Leading-logarithm parton-shower Monte Carlo models with initial and final state QCD radiation give a good description of the data. The observed broadening of the jet shape as η^{jet} increases is consistent with an increase of the fraction of gluon jets.

The Monte Carlo models reproduce the measurements of the jet shape and display the expected differences for quark- and gluon-initiated jets and allow the use of the jet shape to select samples enriched in quark- and gluon-initiated jets to study the dynamics of the subprocesses. Measurements of the dijet cross section as a function of η^{jet} for samples of "thick" and "thin" jets show the expected behaviour for samples enriched in gluon- and quark-initiated jets. The $|\cos\theta^*|$ distribution for a two "thick"-jet sample displays a similar behaviour to the one expected for a sample enriched in processes mediated by gluon exchange, whereas that for a two "thin"-jet sample shows a behaviour similar to a sample enriched in processes mediated by quark exchange. Finally, the $\cos\theta^*$ distribution for a sample of events with one "thick" and one "thin" jet exhibits a large asymmetry consistent with the expected dominance of t-channel gluon (u-channel quark) exchange as $\cos\theta^* \to +1(-1)$.

Acknowledgements. I would like to thank my colleagues from ZEUS for their help in preparing this report and the organisers of the conference for providing a warm atmosphere and hospitality.

REFERENCES

1. ZEUS Collaboration, "Measurements of Jet Substructure in Photoproduction at HERA", paper 530, submitted to the International Europhysics Conference on HEP, Tampere (1999).

2. ZEUS Collaboration, "Substructure Dependence of Dijet Cross Sections in Photoproduction at HERA", paper 424, submitted to the XXXth International Conference on HEP, Osaka (2000).

3. ZEUS Collaboration, M. Derrick et al, *Phys. Lett.* **B** 384 (1996) 401.

The Structure of the Virtual Photon

Claudia Glasman*

representing the ZEUS Collaboration

*Department of Physics and Astronomy, Kelvin Building,
University of Glasgow, Glasgow, G12 8QQ, UK

Abstract. Measurements of dijet cross sections for virtual photons are presented as a function of x_γ^{OBS}, the fraction of the virtual photon energy invested in the production of the dijet system, using the ZEUS detector. Comparisons to QCD predictions show that a resolved photon component is needed to describe the data up to values of the photon virtuality comparable to the scale of the interaction.

Introduction

It has been long established that real photons have a partonic structure from measurements of the photon structure function F_2^γ in $e\gamma$ interactions and observation of resolved photon processes in γp interactions and of single- and double-resolved processes in $\gamma\gamma$ interactions. Thus, it is natural to expect that virtual photons also have a partonic structure. QCD predicts that the parton densities of virtual photons become logarithmically suppressed as the virtuality of the probed photon P^2 increases for fixed μ^2, the scale of the interaction; μ^2 is usually taken as Q^2 (Q^2 is the virtuality of the probing photon) in deep inelastic scattering (DIS) $e\gamma$ and the jet transverse energy (E_T^{jet}) in jet production.

The virtual photon structure function $F_2^{\gamma^*}$ at leading order (LO) in QCD is given by

$$F_2^{\gamma^*}(x_{\gamma^*}, P^2, \mu_{F_{\gamma^*}}^2) = \sum_q x_{\gamma^*} \, e_q^2 \, [f_{q/\gamma^*}(x_{\gamma^*}, P^2, \mu_{F_{\gamma^*}}^2) + f_{\bar{q}/\gamma^*}(x_{\gamma^*}, P^2, \mu_{F_{\gamma^*}}^2)],$$

where x_{γ^*} is the fraction of the virtual photon momentum taken by the interacting parton, $\mu_{F_{\gamma^*}}^2$ is the virtual photon fragmentation scale and e_q is the quark charge. The virtual photon quark densities f_{q/γ^*} contain two terms, as in the case of the real photon,

$$f_{i/\gamma^*}(x_{\gamma^*}, P^2, \mu_{F_{\gamma^*}}^2) = f_{i/\gamma^*}^{\text{had}}(x_{\gamma^*}, P^2, \mu_{F_{\gamma^*}}^2) + f_{i/\gamma^*}^{\text{anom}}(x_{\gamma^*}, P^2, \mu_{F_{\gamma^*}}^2),$$

one term associated to the non-perturbative hadronic component (f^{had}) and a term f^{anom}, unique to the photon, which expresses the direct coupling of the photon to a $q\bar{q}$ pair, calculable in perturbative QCD (pQCD). The hadronic component is the one expected to decrease as P^2 increases. The first measurements of the

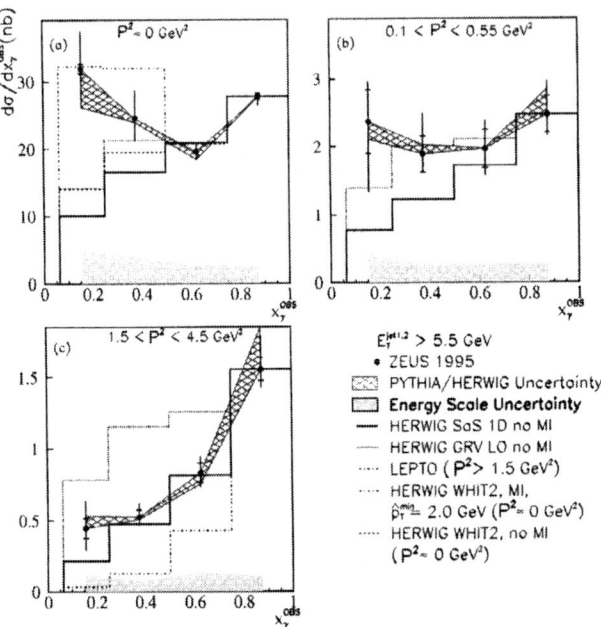

FIGURE 1. Measured dijet cross section $d\sigma/dx_\gamma^{OBS}$ (black dots) in the laboratory frame. Monte Carlo calculations using HERWIG and LEPTO are shown for comparison.

structure function of the virtual photon were done by PLUTO. Only recently new measurements from LEP have become available.

At HERA, the virtual photon structure is studied in jet production mediated by virtual photons. In LO QCD, two processes are expected to contribute to the jet cross section: resolved processes in which the virtual photon interacts with the proton via its hadronic component giving two jets in the final state, and the direct processes in which the virtual photon interacts as a point-like particle with a parton from the proton.

The dijet production cross sections in LO QCD for direct and resolved processes are given by

$$\sigma_D^{LO,ep} = \int d\Omega \; f_{\gamma^*/e}(y, P^2) \; f_{j/p}(x_p, \mu_{F_p}^2) \; d\sigma(\gamma^* j \to \text{jet jet}) \qquad \text{and}$$

$$\sigma_R^{LO,ep} = \int d\Omega \; f_{\gamma^*/e}(y, P^2) \; f_{i/\gamma^*}(x_{\gamma^*}, P^2, \mu_{F_{\gamma^*}}^2) \; f_{j/p}(x_p, \mu_{F_p}^2) \; d\sigma(ij \to \text{jet jet}),$$

where the integrals are performed over the phase space represented by "$d\Omega$"; $f_{\gamma^*/e}(y, P^2)$ is the flux of virtual photons in the positron and y is the fraction of the positron energy taken by the virtual photon; $f_{j/p}(x_p, \mu_{F_p}^2)$ are the parton densities in the proton and x_p is the fraction of the proton momentum taken by parton j; $\mu_{F_p}^2$ is the proton fragmentation scale; and $d\sigma(\gamma^*(i)j \to \text{jet jet})$ is the subprocess cross section, calculable in pQCD. In the case of resolved processes, there is an additional ingredient: the parton densities in the virtual photon $f_{i/\gamma^*}(x_{\gamma^*}, P^2, \mu_{F_{\gamma^*}}^2)$, for which up to the present there is only very little experimental information. This

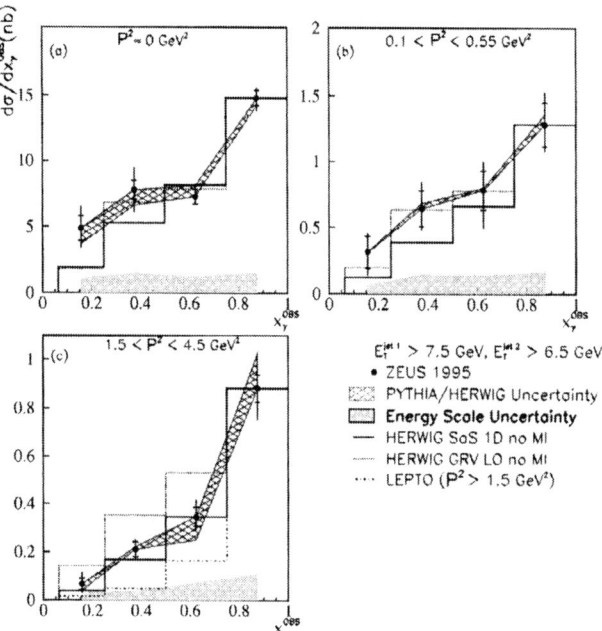

FIGURE 2. The measured dijet cross section $d\sigma/dx_\gamma^{OBS}$ (black dots) in the laboratory frame. Monte Carlo calculations using HERWIG and LEPTO are shown for comparison.

allows the study of the virtual photon structure by measuring jet cross sections.

The contribution to the dijet cross section from resolved processes is expected to decrease relative to the contribution from the direct processes as $P^2 \to \mu^2$; this means that the partonic content of the virtual photon becomes suppressed as P^2 increases. Experimentally, resolved and direct processes are separated by using the variable x_γ^{OBS},

$$x_\gamma^{OBS} = \frac{1}{2yE_e}(E_T^{jet1}e^{-\eta^{jet1}} + E_T^{jet2}e^{-\eta^{jet2}}), \qquad (1)$$

where $\eta^{jet1,2}$ is the jet pseudorapidity and E_e is the positron beam energy. x_γ^{OBS} measures the fraction of the virtual photon energy invested in the production of the dijet system and it is well defined at all orders in pQCD. For resolved (direct) processes, x_γ^{OBS} takes low (high) values.

The measurements presented here were performed with the ZEUS detector at HERA. During 1995 to 1997 HERA operated with positrons of energy $E_e = 27.5$ GeV and protons of energy $E_p = 820$ GeV.

Dijet Cross Sections in the Laboratory Frame

Dijet cross sections as a function of x_γ^{OBS} have been measured [1] for jets found using the longitudinally invariant k_T cluster algorithm in the inclusive mode. The measurements have been performed in the kinematic region given by $0.2 < y < 0.55$ and in three regions of P^2: $P^2 \approx 0$ (the photoproduction regime in which the

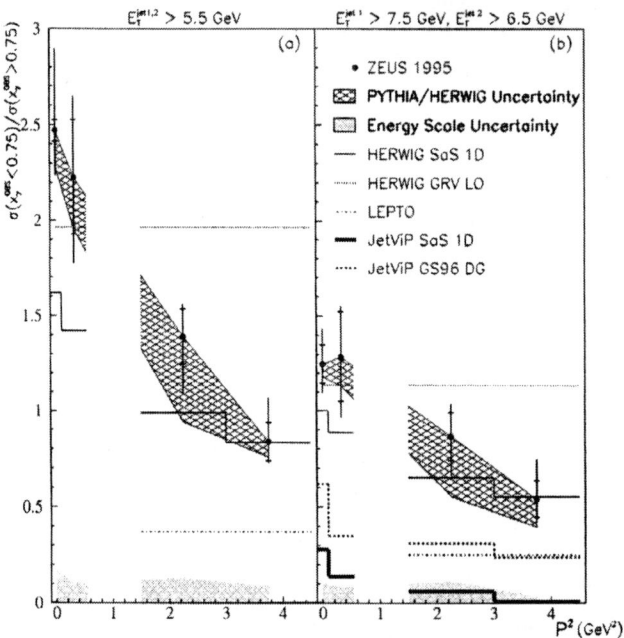

FIGURE 3. The ratio of dijet cross sections, $\sigma(x_\gamma^{OBS} < 0.75)/\sigma(x_\gamma^{OBS} > 0.75)$, as a function of P^2 (black dots). Monte Carlo (HERWIG and LEPTO) and QCD (JetVip) calculations are shown for comparison.

scattered positron is lost in the beam pipe, figure 1a), $0.1 < P^2 < 0.55$ GeV^2 (the intermediate P^2 region, in which the scattered positron is detected in the beam pipe calorimeter, figure 1b) and $1.5 < P^2 < 4.5$ GeV^2 (the low P^2 DIS regime, in which the scattered positron is detected in the uranium-scintillator calorimeter, figure 1c). The events are required to have at least two jets with $E_T^{jet} > 5.5$ GeV and $-1.125 < \eta^{jet} < 2.2$. The data show that the shape of the measured cross section changes markedly with increasing P^2: the cross section for low x_γ^{OBS} values decreases faster than at high x_γ^{OBS} values.

The measurements have been compared to several Monte Carlo models using various parametrisations of the photon parton densities (PDFs). The GRV and WHIT2 photon PDFs have been extracted for real photons and have no P^2 dependence. On the other hand, the SaS 1D photon PDFs consist of two components with different P^2 dependence: the hadronic part decreases as $\approx (m_\rho^2/(m_\rho^2 + P^2))^2$ with increasing P^2, while the P^2 dependence of the anomalous component goes like $\sim \log(\mu^2/P^2)$. The predictions of HERWIG using the GRV PDFs do not describe the data, as expected since they have no P^2 dependence. The prediction based on WHIT2 provides a reasonable description of the data in the photoproduction and in the intermediate P^2 regimes. The predictions using SaS 1D agree with the data at high x_γ^{OBS} in all regimes. The DIS Monte Carlo LEPTO is compared to the data in the DIS regime. It underestimates the data at low x_γ^{OBS}, i.e. parton shower effects are not the sole contribution to this region of x_γ^{OBS} and P^2.

The low E_T^{jet} region is affected by contributions from the presence of a possible underlying event, which may mask the P^2 evolution of the virtual photon. Therefore, measurements of the dijet cross sections have been performed at higher transverse energies. For these measurements, two jets were required with $E_T^{jet,1(2)} > 7.5(6.5)$ GeV and $-1.125 < \eta^{jet1,2} < 1.875$. The cross sections are shown in figure 2. At low x_γ^{OBS} the cross sections are also observed to decrease faster with P^2 than at high x_γ^{OBS}. The predictions from HERWIG based on the GRV PDFs are in good agreement with the data in the two lowest P^2 ranges, but fail at higher P^2. On the other hand, the predictions based on SaS 1D agree with the data at high P^2 but underestimate the data at low and intermediate P^2 at low x_γ^{OBS}. The predictions from LEPTO at high P^2 underestimate the data.

The P^2 evolution of the virtual photon structure was studied further by measuring the ratio of the dijet cross section for $x_\gamma^{OBS} < 0.75$ (enriched in resolved processes) to the dijet cross section for $x_\gamma^{OBS} > 0.75$ (enriched in direct processes). The ratio (see figure 3) falls steeply with P^2 which may be interpreted as the suppression of the resolved photon component as the virtuality of the photon increases. The prediction of HERWIG based on the GRV PDFs is constant, as expected from a photon structure without P^2 dependence. The prediction of SaS 1D decreases with P^2 but lies below the data at low P^2 and the prediction from LEPTO shows that the contribution to the ratio from parton shower effects alone is not enough to explain the P^2 dependence of the data. QCD predictions have been calculated using the program JetViP and different parametrisations of the photon PDFs. The predictions show sensitivity to the choice of PDF but lie well below the data. Hadronisation corrections ($\sim 20-30\%$) are insufficient to explain the discrepancy.

Dijet Cross Sections in the $\gamma^* p$ Frame

For $P^2 \gg 0$, the definition of x_γ^{OBS} (equation 1) is still valid. However, the photon remnant may have high transverse energy in the laboratory frame and be considered as a jet. The solution to avoid this problem is to transform to the $\gamma^* p$ frame. In such a frame the photon remnant has very low transverse energy and will not be mistaken for a jet emanating from the hard interaction.

Figure 4a shows the measurements of the dijet cross section as a function of x_γ^{OBS} in the $\gamma^* p$ frame [2]. The measurements have been performed in the kinematic region given by $0.2 < y < 0.55$ and $0.1 < P^2 < 10^4$ GeV^2. The events are required to have at least two jets with $E_{T,\gamma^* p}^{jet1(2)} > 7.5(6.5)$ GeV and $-3 < \eta_{\gamma^* p}^{jet1,2} < 0$. Also for these measurements the shape of the measured cross section changes with increasing P^2: the cross section for low x_γ^{OBS} values falls more rapidly with increasing P^2 than at high x_γ^{OBS} values.

The predictions of HERWIG based on the SaS 1D PDFs (see figure 4a) give a good description of the data in the high P^2 region but fail in the intermediate P^2 region. A resolved photon component is needed to describe the data up to

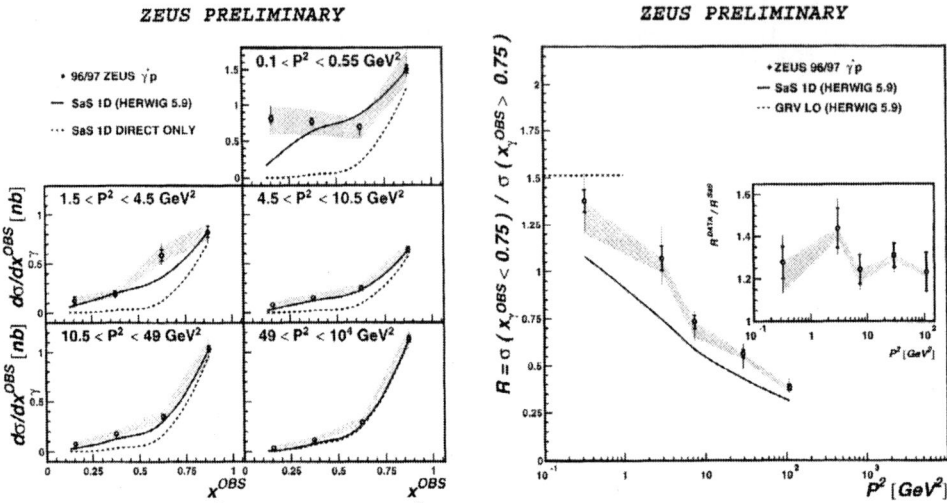

FIGURE 4. (a) The measured dijet $d\sigma/dx_\gamma^{OBS}$ (black dots) in the $\gamma^* p$ frame. (b) The ratio of dijet cross sections, $\sigma(x_\gamma^{OBS} < 0.75)/\sigma(x_\gamma^{OBS} > 0.75)$, as a function of P^2 (black dots). Monte Carlo calculations using HERWIG are shown for comparison.

$P^2 \sim 49$ GeV^2. Above this value, the HERWIG prediction is dominated by the direct component and describes the data well. Therefore, for $P^2 \geq (E_T^{jet})^2$, direct processes alone are able to describe the data.

The ratio of the cross section for $x_\gamma^{OBS} < 0.75$ and $x_\gamma^{OBS} > 0.75$ also falls steeply in the $\gamma^* p$ frame with increasing P^2 (figure 4b). The prediction of SaS 1D shows a decrease with P^2 but lies below the data in the whole P^2 range. The ratio of the data to the SaS 1D prediction (inset in figure 4b) has a constant value of ~ 1.3. This indicates that the resolved component suppression included in these PDFs is in agreement with the data, but they underestimate the fraction of resolved component by $\sim 30\%$.

Summary and Conclusions

Dijet cross sections as a function of x_γ^{OBS} have been measured in the kinematic region given by $0.2 < y < 0.55$ and $0 \lesssim P^2 < 10^4$ GeV^2 for jets found using the k_T cluster algorithm. The x_γ^{OBS} dependence of the measured cross sections changes with increasing P^2: the cross section for low x_γ^{OBS} values decreases faster than at high x_γ^{OBS} values. The predictions of HERWIG based on the SaS 1D photon PDFs describe the data for $1.5 < P^2 < 10^4$ GeV^2 but fail in the region $0.1 < P^2 < 0.55$ GeV^2. A resolved photon component is needed to describe the data up to $P^2 \sim 49$ GeV^2, i.e. where the virtual photon is probed at a scale comparable to the hard interaction scale $(E_T^{jet})^2 \sim (7$ GeV$)^2$. The ratio of the dijet cross section $\sigma(x_\gamma^{OBS} < 0.75)/\sigma(x_\gamma^{OBS} > 0.75)$ decreases with P^2; the predicted P^2 dependence of the ratio agrees with the data, but it underestimates the fraction of the resolved component by $\sim 30\%$. This result can be interpreted in terms of a resolved photon component that is suppressed as the photon virtuality increases.

Acknowledgements. I would like to thank my colleagues from ZEUS for their help in preparing this report and the organisers of the conference for providing a warm atmosphere and hospitality.

References

1. ZEUS Collaboration, J. Breitweg *et al*, *Phys. Lett.* **B** 479 (2000) 37.
2. ZEUS Collaboration, "Measurements of the Structure of Virtual Photons", paper 426, submitted to the XXXth International Conference on HEP, Osaka (2000).

Jets in DIS and the Virtual Photon Structure

Stephen Maxfield[†]

H1 Collaboration

[†]*Dept. of Physics, University of Liverpool*

Abstract.
   Single-inclusive jet cross sections in deep-inelastic scattering for photon virtualities $5 < Q^2 < 100\,\mathrm{GeV}^2$ are measured in a data sample corresponding to an integrated luminosity of 20.9 pb^{-1} as a function of the jet transverse energy E_T, of the ratio E_T^2/Q^2 and of the Bjorken scaling variable x_{Bj}. Data are compared to next-to-leading order perturbative QCD calculations using the squared four momentum transfer Q^2 and the squared jet E_T as renormalisation scale. Neither choice is able to describe the data over the full phase space region, in particular in the forward region towards the proton remnant. Possible explanations of this discrepancies are discussed.
   Dijet event rates, R_2, have also been measured for deep inelastic scattering in the of low x_{Bj} and low Q^2. R_2 is measured as a function of both x_{Bj} and Q^2 extending a single differential analysis recently published by H1. This allows a more detailed study of jet production at low x_{Bj} in a regime where both scales, Q^2 and jet E_T^2, are comparable. The data are confronted with next-to-leading order QCD calculations including both point-like and non-pointlike structure of the virtual photon.

INTRODUCTION

   Measurements of the dijet cross section in deep inelastic ep scattering [2,1] have been used to test perturbative QCD calculations in both leading and next-to-leading order. Fixed order calculations of the matrix elements in combination with proton parton densities evolved according to the DGLAP equations [3] accurately predict the fully inclusive cross section. However the jet cross sections are more demanding and in some parts of the phase space the fixed order approach is inadequate. For example, in leading order, high p_t dijets can be produced not only by direct processes in which the virtual photon interacts as a point-like particle with a parton out of the proton but also by resolved processes where a parton carrying a fraction x_γ of the photon's momentum undergoes a hard scattering with a parton from the proton (see Figure 1). In the photoproduction limit, the cross section is sensitive to the parton densities of the real photon as well as of the proton at a scale set by the

CP571, *PHOTON 2000*, edited by A. J. Finch

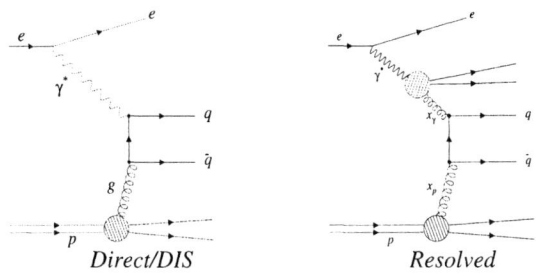

FIGURE 1. Examples of leading order diagrams contributing to the production of dijets in γp collisions. In leading order, the processes can be classified as direct (left) or resolved (right).

transverse momentum of the jets, p_t^2. In DIS, the structure of the virtual photons can still be resolved as long as $p_t^2 \gg Q^2$. Note that in photoproduction processes, divergences in the fixed order calculations *must* be absorbed into the photon parton densities but for off-shell photons the Q^2 of the photon provides a natural cut-off, the fixed order calculations are finite and it is not strictly necessary to introduce photon parton densities at all. Nevertheless, where $p_t^2 \gg Q^2$ large logarithms can still spoil the perturbation expansion and resummation into virtual photon parton densities may improve the description of the jet cross sections. Analyses of the dijet cross-section using LO matrix elements [1] have supported this concept of a resolved virtual photon with the expected Q^2 and p_t^2 evolutions. At $\mathcal{O}(\alpha_s^2)$, some of this photon structure is accounted for in the more complex matrix elements and as Q^2 increases relative to p_t^2 the resolved photon picture becomes less relevant.

Elsewhere in phase space alternative parton dynamics may be needed. Where $Q^2 \sim p_t^2$ and $x_{bj} \ll x_{jet}$ (x_{jet} is the fraction of the proton's longitudinal momentum carried by a jet), DGLAP evolution is expected to be suppressed ($\ln p_t^2/Q^2$) and large $\ln 1/x$ terms develop. This kinematic region corresponds to the forward (proton direction) part of the detector. Deviations from predictions based on DGLAP evolution here may indicate that BFKL evolution [4] is needed. In BFKL dynamics there is no strong ordering in the k_t of the parton emissions. To some extent, this is also a feature of resolved photons (see Figure 2). Deviations from the NLO predictions may simply indicate that still higher order calculations are needed. These may also help resolve other ambiguities associated with high p_t jet production in DIS: Where Q^2 and p_t are both substantial ($\gg \Lambda_{QCD}^2$) there is ambiguity in the choice of both factorisation and renormalisation scales (and in resolved photon processes there are in principle two of each!) It is important to know where these ambiguities have an important effect on the cross section.

In this talk, I will describe measurements of both single-inclusive cross sections and dijet rates in deep inelastic scattering made by H1 which explore these issues.

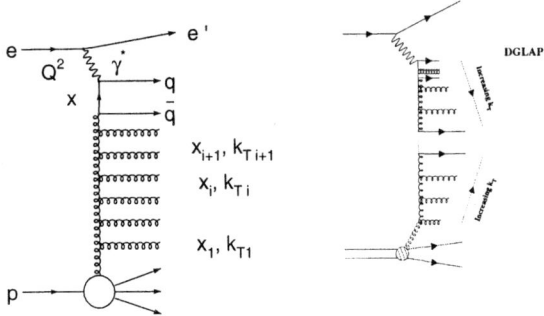

FIGURE 2. Ladder diagrams contributing to High p_t jet production in DIS. DGLAP evolution leads to a strong ordering of the k_t of the parton emissions whereas BFKL dynamics results in a strong ordering in x (and a random distribution of the k_t's (left). Resolved virtual photon processes (right) also break up the k_t ordering.

SINGLE INCLUSIVE JET CROSS SECTIONS IN DIS

A single inclusive measurement, i.e. one which counts jets rather than events, provides a fundamental test of the perturbative QCD calculation of jet production. In this analysis, the cross section is measured over a large phase space including regions where non-DGLAP dynamics and resolved virtual photon structure may play a role.

Events were selected from 21 pb^{-1} of e^+p data taken in 1996 and 1997. The scattered positron was used to measure the event kinematics. The virtuality of the photon was kept within the range $5 < Q^2 < 100$ GeV^2 and the inelasticity to $0.2 < y < 0.6$. With these cuts, $10^{-4} < x_{Bj} < 5 \cdot 10^{-3}$.

Hadronic final state objects were reconstructed from energy depositions in the calorimeter and low momentum tracks in such a way as to avoid double counting. Jets were then found by applying the k_T algorithm [5] to these objects in the Breit frame (defined by $\vec{q} + 2x_{B_j}\vec{P} = 0$). Jets were accepted if they had a minimum transverse energy of 5 GeV and pseudorapidities satisfying $-1.0 < \eta_{lab} < 2.8$, placing them well within the detector acceptance. The cross sections $d\sigma_{jet}/dE_T$ and $d\sigma_{jet}/d(E_T^2/Q^2)$ were measured for three ranges of η_{lab} and two of Q^2. $d\sigma_{jet}/dx_{B_j}$ was measured in two η_{lab} regions.

The data were corrected bin-by-bin for detector acceptance using RAPGAP [6] and ARIADNE [7] simulations which give a good overall description of distributions of the basic kinematic variables and of hadronic final state variables sensitive to the calibration of the calorimeters. The cross sections were corrected for QED radiation using HERACLES [8]. Bin migrations were small justifying the bin-by-bin procedure and the correction factors derived from RAPGAP and ARIADNE are typically close to unity and in good agreement. The difference between them is used to estimate the systematic uncertainty associated with the correction procedure.

The data are compared with $\mathcal{O}(\alpha_s^2)$ QCD calculations using the DISENT [9] program. The factorisation scale, μ_F^2, was set to Q^2. Its influence on the cross sections measured was found to be small. The renormalisation scale μ_R^2, was set to either Q^2 or E_T^2. DISENT calculates direct processes only. To investigate resolved processes, we used JETVIP [10] which incorporates photon parton densities evolved with Q^2 from parameterisations of the real photon densities. Renormalisation and factorisation scales were both set to $Q^2 + E_T^2$. JETVIP uses on-shell matrix elements and calculates the kinematics in the limit of zero quark masses ('on-shell scheme'). Although this is consistent with the use of parton density functions derived for on-shell partons, it neglects mass and k_t effects which become significant for $Q^2 \gg$ 1GeV^2. These can be approximately included by allowing the quarks to be virtual. Because the matrix elements are still massless, this 'off-shell scheme' leads to an unphysical dependence on the parameter y_{cut} which is used to isolate infrared and collinear divergences in the cross section.

Both these parton-level cross sections were corrected for hadronisation effects ($\lesssim 20\%$) using correction factors estimated from RAPGAP and ARIADNE Monte Carlos. Again the difference between ARIADNE and RAPGAP was used as an estimate of the systematic uncertainty. The upper three plots in Figure 3 (left) show the cross section as a function of jet E_T in backward, central and forward ranges of η_{lab}. Here and elsewhere, inner error bars show the statistical and outer the statistical and systematic errors added in quadrature. The data are compared with LO (dashed line) and NLO (dotted line) QCD predictions from DISENT to which hadronisation corrections have been applied.(The size of these corrections is shown in the central set of plots where the band indicates the associated uncertainty.) The QCD predictions were made with $\mu_R^2 = E_T^2$ as the scale. The shaded band in the upper plots indicates the systematic error coming from uncertainties in this renormalisation scale estimated by varying it between $4\mu_R^2$ and $1/4\mu_R^2$. The normalised difference between the NLO predictions and the data is shown in the lower section. Here, the inner and outer shaded bands show hadronisation and renormalisation scale uncertainties respectively. The NLO QCD predictions are able to describe the data for $E_T \gtrsim$ GeV^2 in all three η_{lab} ranges but for smaller values of E_T, the prediction is below the data especially in the forward region where the discrepancy is a factor of ~ 2. The NLO predictions give a better description than the LO ones. The difference between NLO and LO is at its greatest in the forward region.

If Q^2 is chosen as the renormalisation scale, a different picture emerges. Figure 3 (right) shows the normalised difference between the NLO predictions and the data with $\mu_R^2 = Q^2$. Here the NLO calculations tend to overestimate the data for $E_T \gtrsim 20$GeV^2. This is especially so for $5 < Q^2 < 20$GeV^2 shown in the upper three plots. Looking in more detail (not shown) in the forward region it is found that the discrepancy is confined to $Q^2 \lesssim 40$GeV^2. Note that the scale uncertainty is much larger for this choice of scale variable.

Distributions differential in E_T^2/Q^2 (not shown) indicate a similar trend. The

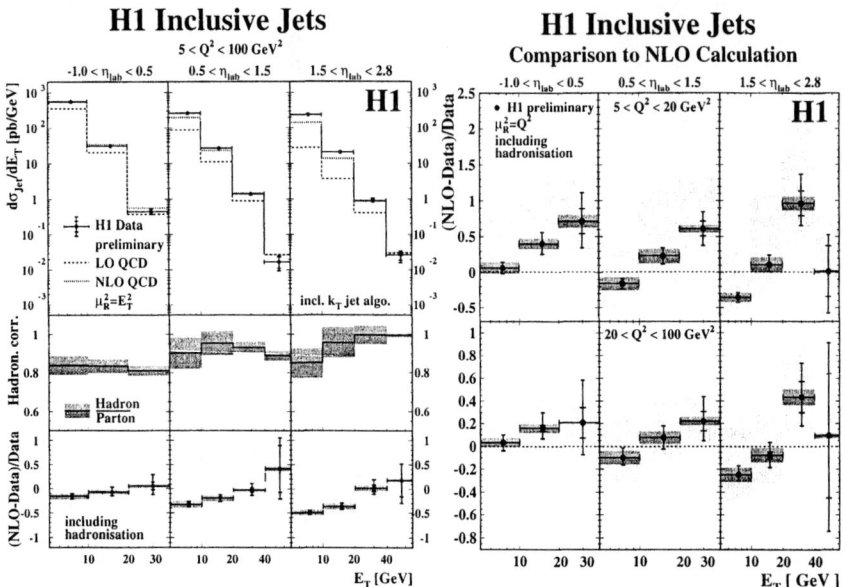

FIGURE 3. The single inclusive cross section $d\sigma_{jet}/dE_T$ shown in three regions of η_{lab}. *Left:* The top part shows the data compared to LO and NLO QCD calculations by DISENT. The shaded band shows the effect of varying the renormalisation scale between $\mu_R^2 = 4E_T^2$ and $\mu_R^2 = 1/4E_T^2$. The centre portion shows the size and uncertainty of hadronisation corrections and the lower part the difference between the NLO calculations and the data. *Right:* The difference (NLO-data)/data shown in three regions of η_{lab} and two of Q^2 with renormalisation scale set to Q^2.

NLO calculations with $\mu_R^2 = E_T^2$ successfully describe the distribution in the backward and central rapidity ranges but fail in the forward region when $2 < E_T^2/Q^2 < 50$. With $\mu_R^2 = Q^2$ the calculations only fail for $E_T^2/Q^2 > 50$.

The region of phase space where BFKL dynamics may play a rôle [11] is further isolated by requiring that $x_{jet} \equiv E_{jet}/E_p > 0.035$. Then, especially where x_{Bj} is small, large $\ln 1/x$ can develop. The discrepancy between the NLO calculations and the measured cross section $d\sigma_{jet}/dx_{Bj}$ is shown in Figure 4 *(left)*. With E_T^2 as the scale, the discrepancy is large (~ 2) in the forward region at low x_{Bj}. This is consistent with observations of π^0 production [12]. However, with $\mu_R^2 = Q^2$, the calculations become compatible with the data (given the large scale uncertainties).

The QCD predictions might be improved by including contributions with resolved virtual photons. In Figure 4 *(right)* the cross section $d\sigma_{jet}/d(E_T^2/Q^2)$ is compared with predictions made by JETVIP. The latter were made using the on-shell scheme and the off-shell scheme with two choices for y_{cut}. There is a considerable difference in the predictions from these three schemes. Unfortunately, no single scheme is

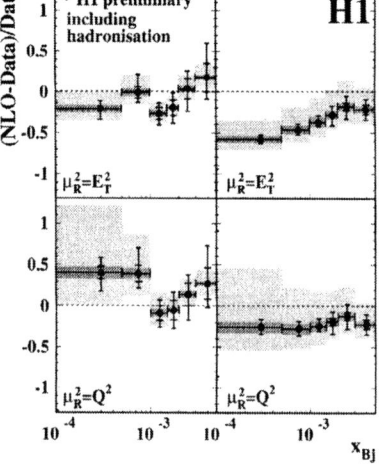

H1 Inclusive Jets
Comparison to NLO Calculation

FIGURE 4. *Left:* The difference between NLO calculations of the single inclusive cross section $d\sigma_{jet}/dx_{Bj}$ and the data shown in two regions of η_{lab} and for two choices of renormalisation parameter. *Right:* The cross section $d\sigma_{jet}/d(E_T^2/Q^2)$ in three η_{lab} ranges. JETVIP calculations employing three different schemes are superimposed.

capable of describing the data in all three η_{jet} ranges.

DIJET RATE MEASUREMENTS

The second set of measurements discussed here looks at the dijet rate, $R_2 \equiv \frac{\sigma_{2jet}}{\sigma_{DIS}}$, at low x_{Bj} and low Q^2. This rate measurement is designed to pick out the hard contributions to DIS and further investigate the limitations of the available NLO calculations.

The measurement uses data collected by H1 in 1996 and 1997 ($22pb^{-1}$). The kinematics of the events were calculated from the scattered positron and the hadronic final state was reconstructed from clusters in the calorimeters supplemented by track information. The rates were measured in the region of phase space bounded by $5 < Q^2 < 100 \text{GeV}^2$, $10^{-4} < x_{Bj} < 10^{-2}$ and $0.1 < y < 0.7$. Jet finding was by the inclusive k_T algorithm executed in the $\gamma^* p$ frame. Dijet events were defined as events containing at least two jets in the range $-1 < \eta_{lab} < 2.5$. The transverse energies of the two jets (in the $\gamma^* p$ frame) had to exceed 5GeV and $5 + \Delta\text{GeV}$ where $\Delta \geq 0$ forces an asymmetry needed for stable calculation of the NLO cross sections. The data were corrected for detector acceptance and resolution effects and

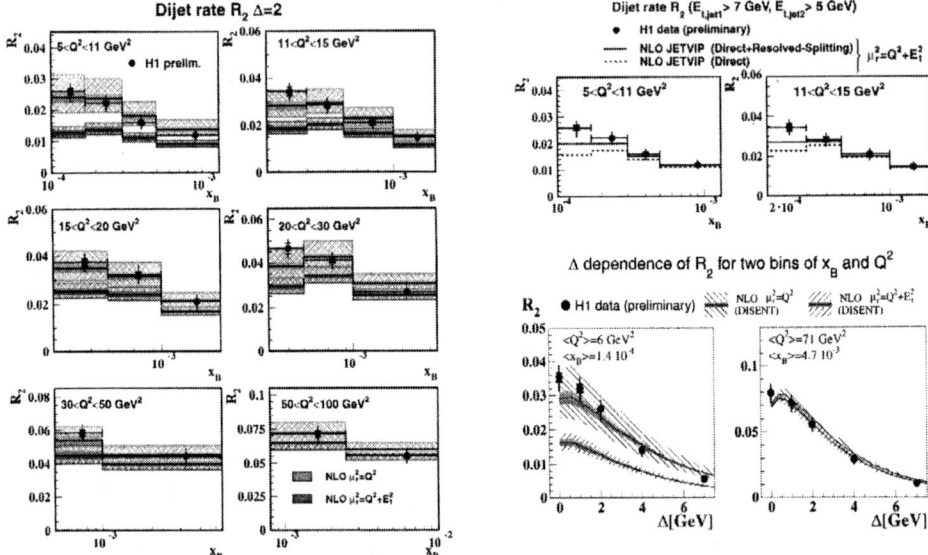

FIGURE 5. *Left:* The dijet rate R_2 as a function of x_{Bj} in ranges of Q^2. The NLO QCD predictions (DISENT) are corrected for hadronisation effects. Full bands indicate the uncertainty onthis and the hatched bands also include the scale uncertainty added in quadrature. *Upper right:* The rate compared with JETVIP NLO QCD caculations with(solid) and without(dashed) resolved photon contributions. *Lower right:* R_2 as a function of the asymmetry parameter Δ (The highest E_T jet has $E_T > 5 + \Delta$). The NLO QCD calculations are from DISENT.

for QED radiation using a combination of DJANGO/ARIADNE and RAPGAP simulations. The dominant systematic error ($\sim 10\%$) comes from a 4% uncertainty in the hadronic energy scale of the calorimetry.

The dijet rates are shown in Figure 5 (*left*) as a function of x_{Bj} for six intervals of Q^2. Δ is set to 2. The QCD calculations using DISENT were made with Q^2 and $Q^2 + E_T^2$ as scale parameters. Hadronisation corrections have been applied to the results of the calculation. They were estimated with the above monte carlos and lead to a lowering of the predictions by $\sim 10\%$. Uncertainties in these hadronisation corrections are shown as the inner shaded bands. Scale uncertainties are shown as the outer bands. Note that in this case the scale is varied between $2\mu_R^2$ and $1/2\mu_R^2$. In each q^2 range we see a rise in the rate R_2 as x_{Bj} decrreases which is simply a result of the increase in available phase space ($W^2 \sim Q^2/x_{Bj}$). We see a similar pattern to the QCD predictions as in the single inclusive case. If Q^2 is used as the renormalistation scale the predictions give a satisfactory description of the data over the whole range of Q^2 and x_{Bj} but with large uncertainties in the low Q^2 and low x_{Bj} part of the phase space. These uncertainties are significantly smaller if

$E_T^2 + Q^2$ is used as the scale. In this case the predictions are too small except for large x_{Bj} and/or Q^2.

Again we can ask if resolved virtual photon processes can provide an additional contribution. The rates were calculated using JETVIP (with off-shell partons) and SAS 1D [13] pdf's for the photon. In Figure 5 (*upper right*), the JETVIP predictions are compared with the data in the two lowest Q^2 ranges where $Q^2 < E_T^2$ and the photon structure formalism is valid. The NLO JETVIP calculations without a resolved photon contribution are shown as the dashed histograms. The full calculation, shown as the solid histograms, does indeed lead to a better description of the data if we assume that JETVIP's scale choice, $Q^2 + E_T^2$, is appropriate.

The dependence of R_2 on the asymmetry parameter Δ is shown in Figure 5 (*lower right*). Note that Δ and the E_T of the jets are closely correlated. The rate is shown in a low Q^2, low x_{Bj} range and a high Q^2, high x_{Bj} range. As $\Delta \to 0$, the NLO calculations (here DISENT) become infrared sensitive. This is responsible for the turning over of the prediction for $\Delta \lesssim 2$. We again see that in the low range there are large differences in the QCD predictions depending on the choice of scale parameter. In the high range, the predictions become insensitive to this choice and both give a good description of the rate.

CONCLUSIONS

Both the single inclusive measurements and the dijet rate measurements have established regions of phase space where NLO QCD calculations either fail to describe the data or are ambiguous.

The single inclusive jet cross section has been measured for $5 < Q^2 < 100$ GeV^2 and $0.2 < y < 0.6$. In central and backward regions of rapidity, the NLO QCD calculations give an adequate description of the data. In the forward region (i.e. in the proton direction) the predictions depend strongly on the choice of renormalisation scale. With $\mu_R^2 = E_T^2$ the predictions are too small by factors of up to two for low $E_T \lesssim 20$ GeV, low $x_{Bj} \lesssim 10^{-3}$ and for $2 < E_T^2/Q^2 < 50$. With $\mu_R^2 = Q^2$ on the other hand, the calculations give a good description of the cross section over most of the phase space, failing only for E_T between 10 and 40 GeV and for $E_T^2/Q^2 > 50$ where they give too high a cross section. Unfortunately, however, this choice of scale variable leads to scale dependencies which are up to three times larger than $\mu_R^2 = E_T^2$.

The dijet rate measurements cover a similar phase space and the corresponding NLO QCD calculations show similar tendencies. The measured rates can be described if μ_R^2 is set to Q^2 but at the cost of uncomfortably large scale uncertainties. Or, with $\mu_R^2 = E_T^2$, we get a prediction with much smaller scale uncertainty which underestimates the data especially at low x_{Bj} or low jet E_T.

The shortfall of the predictions with scale based on E_T^2 leaves room for additional contributions. The inclusion of resolved photon contributions as calculated by JETVIP does improve the description of the jet rates, however there are ambiguities in these predictions in addition to those associated with the scale choice because of problems matching virtual parton content of the photon to the massless matrix elements.

The regions where ambiguities and dependence on the scale is large tend to coincide with those where there is the largest difference between the LO and NLO predictions so it may simply be the case that there is no predictive power at NLO here and we must wait for possible improvements when NNLO calculations become available. To unambiguously identify and measure a resolved photon component or alternative parton dynamics the problematic regions must either be avoided or some means of fixing the scale adopted [14]. These precision measurements of the cross sections and the rates are a powerful resource for developing and honing the theoretical calculations.

REFERENCES

1. H1 Collab., C. Adloff et al., *Eur. Phys. J.* **C13** (2000) 609.
2. ZEUS Collab. J. Breitweg et al., *Phys. Lett.* **B479** (2000) 1, 37.
3. V.N. Gribov and L.N. Lipatov, *Sov. J. Nucl. Phys.* **15** (1972)438,675;
   Yu. L. Dokshitzer, *Sov. Phys. JETP* **46** (1977) 641;
   G. Alterelli and G. Parisi, *Nucl. Phys.* **B126** (1977) 297.
4. E. A. Kuraev, L. N. Lipatov and V. S. Fadin,*Sov. Phys. JETP***45** (1972) 199;
   Y. Y. Balitski and L. N. Lipatov,*Sov. J. Nucl. Phys.* **28** (1978) 822.
5. S. D. Ellis and D. E. Soper, *Phys. Rev.* **D48** (1993) 3160;
   S. catani, Yu. L. Dokshitzer, M. H. Seymour and B. R. Webber, *Nucl. Phys.* **B406** (1993) 187.
6. RAPGAP 2.06: H. Jung, *Comp. Phys. Commun.* **86** (1995) 147.
7. ARIADNE4.08:L. Lönnblad,*Comp. Phys. Commun.* **71** (1992) 15.
8. A. Kwiatkowski, H. Spiesberger and H.-J. Möhring,*Comp. Phys. Commun.* **69** (1992) 155.
9. S. Catani and M. H. Seymour,*Nucl. Phys.* **B485** (1997)291, Erratum-ibid. **510** (1997) 503.
10. B. Pötter,*Comp. Phys. Commun.* **119** (1999) 45.
11. A. H. Mueller,*Nucl. Phys. B (Proc. Suppl.)* **18C** (1990)125; *J. Phys.* **G 17** (1991) 1443.
12. H1 Collab., C. Adloff et al., *Phys. Lett.* **B462** (1999) 440.
13. G. A. Schuler and T. Sjöstrand, *Z. Phys.* **C 68** (1995) 607.
14. S. J. Brodsky, g. P. Lepage and P. B. Mackenzie, *Phys. Rev.* **D 28** (1983) 228;
    G. Inglemann and J. Rathsman, *Z.Phys.* **C63** (1994) 589 and references therein.

Prompt-photon production at HERA

Juan Terrón*

representing the ZEUS Collaboration

*Departamento de Física Teórica,
Universidad Autónoma de Madrid, Cantoblanco, Madrid 28049, Spain

Abstract. Measurements of inclusive prompt-photon and prompt-photon plus jet production at HERA are reported. The relevance of these measurements on constraining the parton distribution functions of the photon and on determining the $< k_T >$ of partons in the proton are discussed.

1. Introduction

Prompt photons in hadron-hadron collisions represent a probe of the parton content of the colliding hadrons as well as of the hard scattering process. The production of prompt photons at fixed target energies was considered to be an ideal tool to extract the gluon density in the proton ($qg \rightarrow q\gamma$) [1]. The advantage of measuring prompt photons instead of jets is that the former come directly from the hard scattering process whereas the latter are the result of the hadronisation of the outgoing partons. However, the extraction of the gluon density in the proton through this type of measurements at fixed-target energies has been hampered by the inadequacy of the next-to-leading order (NLO) QCD calculations to describe the data [2, 3]. In particular, the prompt-photon's transverse momentum (p_T^γ) distribution of the data is larger than that of the calculations and the discrepancy increases as p_T^γ decreases. This feature has been used to advocate values of the mean intrinsic k_T of partons in the proton ($< k_T >$) significantly larger than expected. Following this approach, the interpretation of the data for increasing values of the centre-of-mass energy ($\sqrt{s}$) reveals that the value of $< k_T >$ required to reproduce the data increases as $\sqrt{s}$ increases; thus, the interpretation of the $< k_T >$ as being *intrinsic* cannot be sustained. Recent theoretical developments go in the direction of understanding this effect in terms of soft-gluon emission [4]; i.e. to interpret $< k_T >$ as the result of initial-state gluon radiation from the colliding partons [2].

Prompt-photon production in quasi-real photon-proton collisions at HERA provides a new testing ground for theoretical developments to understand the $< k_T >$ of partons in the proton within perturbative QCD [4]. In photon-proton reactions, two types of processes contribute to prompt-photon production at lowest order (LO): either the colliding photon interacts directly with a parton in the proton (the direct process; $\gamma q_p \rightarrow \gamma q$) or the colliding photon acts as a source of partons which scatter off those in the proton (the resolved process; $q_\gamma g_p \rightarrow \gamma q$). Thus, the production of prompt photons in this reaction also represents a probe of the parton content of the photon. Measurements of the differential cross sections for inclusive

CP571, *PHOTON 2000*, edited by A. J. Finch

prompt-photon production [5] in the reaction $ep \to \gamma_{\text{prompt}} + X$ are reported in Section 2. Measurement of distributions which are sensitive to the $< k_T >$ of partons in the proton have been made for prompt-photon production in association with a jet [6], $ep \to \gamma_{\text{prompt}} + \text{jet} + X$, and are presented in Section 3. During 1996 to 1997 HERA operated with positrons of energy $E_e = 27.5$ GeV colliding with protons of energy $E_p = 820$ GeV. The data sample used in these analyses was collected with the ZEUS detector [7] at HERA and corresponds to an integrated luminosity of 38.4 pb^{-1}.

2. Inclusive prompt-photon production

Prompt-photon candidates were identified by applying an algorithm [5,8] to find electromagnetic clusters in the barrel part of the main calorimeter (BCAL). The electromagnetic section of the BCAL (BEMC) consists of cells of ~ 20 cm length azimuthally and a mean width of 5.45 cm in the Z‡ direction (see inset in Fig. 1a), at a distance of 1.3 m from the beam line. Events were retained for further analysis if the prompt-photon candidate had transverse energy $E_T^\gamma > 5$ GeV and its pseudorapidity lied in the range $-0.75 < \eta^\gamma < 1.0$. A candidate was rejected if a track pointed to it within 0.3 radians or the total E_T from other particles within a cone of unit radius in the (η, ϕ) plane exceeded $0.1 \cdot E_T^\gamma$. This isolation condition removes events in which the photon candidate is either a neutral meson (e.g. π°) produced within a jet or a photon radiated from a final state quark.

The extraction of the prompt-photon signal was performed on the basis of the shape of the cluster of BCAL cells. Two variables were used to discriminate γ, π° and η signals: (a) the mean width $< \delta Z > \equiv \sum(E_{cell} \cdot |Z_{cell} - \bar{Z}|) / \sum E_{cell}$ of the cluster in Z, and (b) the fraction f_{max} of the cluster energy found in the most energetic cell in the cluster. The $< \delta Z >$ distribution for the event sample is shown in Fig. 1a; the Monte Carlo event samples of single γ, π° and η were used to establish a cut on $< \delta Z >$ at 0.65 BEMC cell widths, such as to remove most of the η mesons. The prompt-photon signal was extracted by fitting the f_{max} distribution after the $< \delta Z >$ cut with freely-varying γ and π° contributions (see Fig. 1b). This procedure was applied bin-by-bin in every distribution studied.

Measurements of the differential cross sections $d\sigma/dE_T^\gamma$ and $d\sigma/d\eta^\gamma$ for inclusive prompt-photon production have been made in the kinematic region defined by $Q^2 < 1$ GeV^2 (where Q^2 is the virtuality of the incident photon) and photon-proton centre-of-mass energies in the range $134 < W < 285$ GeV. The main sources of systematic uncertainty are the energy scale of the calorimeter ($\pm 7\%$) and the modelling of the shower shape of electromagnetic clusters in the BEMC ($\pm 8\%$). In the figures, the inner error bars represent the statistical uncertainties and the outer error bars include the systematic uncertainties added in quadrature.

---

‡ The ZEUS coordinate system is defined as right-handed with the Z-axis pointing in the proton beam direction, hereafter referred to as forward, and the X-axis horizontal, pointing towards the centre of HERA. The pseudorapidity is defined as $\eta = -\ln(\tan\frac{\theta}{2})$, where the polar angle θ is taken with respect to the proton beam direction.

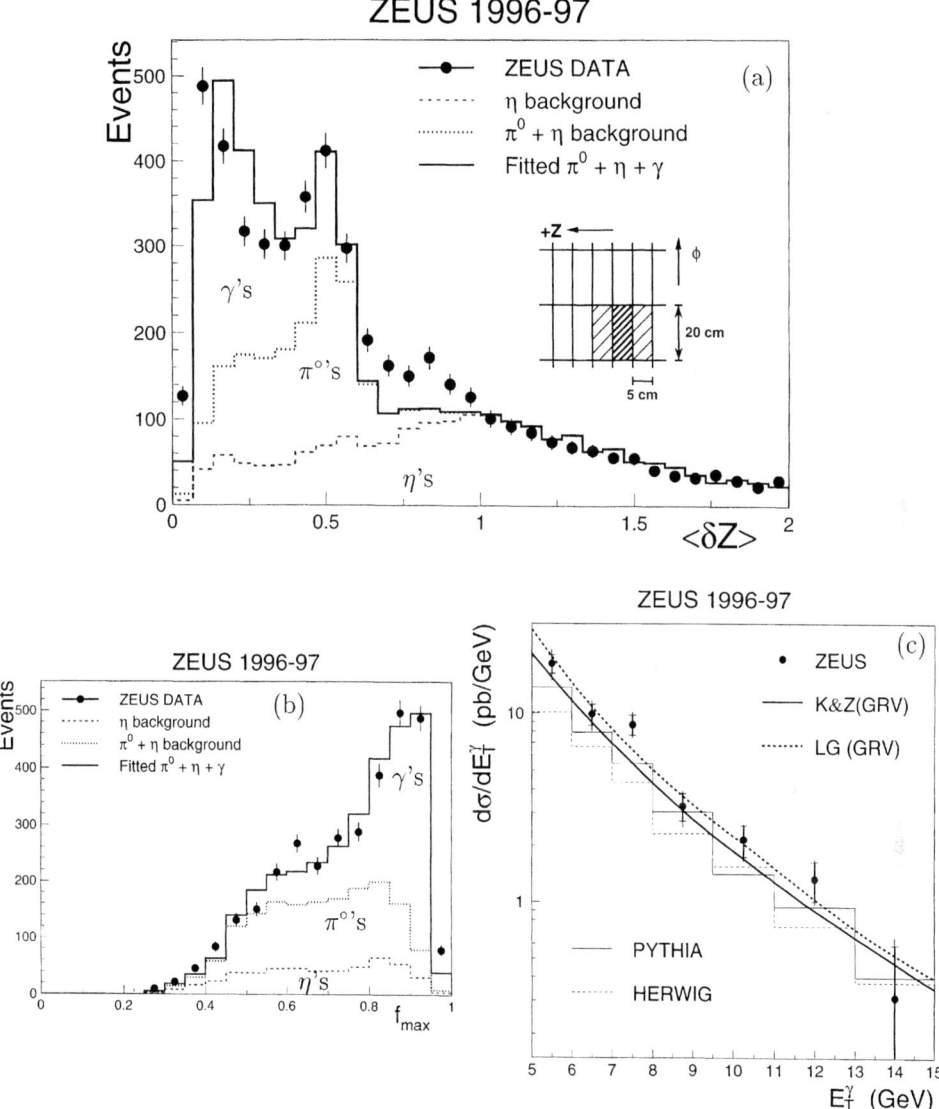

FIGURE 1. Distributions of $< \delta Z >$ (a) and f_{max} (b) for prompt-photon candidates. Differential cross section $d\sigma/dE_T^\gamma$ for inclusive prompt-photon production (c); NLO QCD calculations (curves) are shown for comparison.

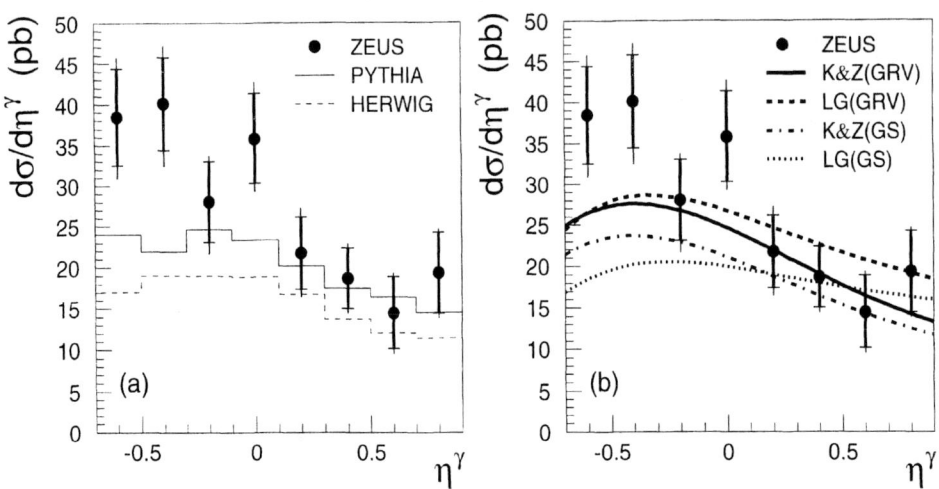

FIGURE 2. $d\sigma/d\eta^\gamma$ for inclusive prompt-photon production; in (b) NLO QCD calculations (curves) are shown for comparison.

The measured $d\sigma/dE_T^\gamma$ integrated over $-0.7 < \eta^\gamma < 0.9$ is presented in Fig. 1c and displays a steep fall-off of two orders of magnitude in the studied E_T^γ range. The measured $d\sigma/d\eta^\gamma$ integrated over $5 < E_T^\gamma < 10$ GeV, shown in Fig. 2, increases as η^γ decreases.

NLO QCD calculations are compared to the data in Figs. 1c and 2b. Two sets of calculations are shown, LG [9] and K&Z [10]. Both sets include the Born contributions to resolved and direct processes, virtual and real corrections to the direct process, and radiative contributions (e.g. $\gamma q \to gq$ in which either of the final state partons radiates a photon). In both calculations the isolation criterion was applied at the parton level and the renormalisation and factorisation scales were chosen equal to E_T^γ. The two sets differ in some respects: LG includes higher-order corrections to the resolved process whereas K&Z includes the box-diagram contribution $\gamma g \to \gamma g$. The calculations shown in Fig. 1c were performed using the GRV-HO [11] parametrisations of the photon parton distribution functions (PDFs). The calculations of $d\sigma/dE_T^\gamma$ are in reasonable agreement with the data both in shape and normalisation; within the experimental uncertainties ($\sim 20\%$), there is no need for the inclusion in the calculations of an extra $< k_T >$ of partons in the proton.

The cross section $d\sigma/d\eta^\gamma$ is sensitive to the photon PDFs. NLO QCD calculations using two sets of parametrisations of the photon PDFs (GRV-HO and GS-HO [12]) are compared to the data in Fig. 2b. The calculations describe the data well for $\eta^\gamma > 0.1$, but lie systematically below the data for lower η^γ values. It is in this region where the sensitivity to the quark PDFs in the photon is largest [13]. The observed discrepancy, together with that also seen in dijet photoproduction at HERA [14], suggests a revision of the current parametrisations of the photon PDFs.

3. Prompt-photon plus jet production

The characteristics of prompt-photon production in association with a jet have been studied [6] in the reaction $ep \rightarrow \gamma_{\text{prompt}} + \text{jet} + X$. The prompt-photon candidates were selected with similar criteria to those reported in Section 2. The prompt-photon candidate was required to have $E_T^\gamma > 5$ GeV and $-0.7 < \eta^\gamma < 0.9$. The kinematic region of the measurements is defined by $Q^2 < 1$ GeV^2 and $120 < W < 274$ GeV. The longitudinally invariant k_T-cluster algorithm [15] was used in the inclusive mode to reconstruct jets in the hadronic final state. The jet search was performed in the (η, ϕ) plane of the laboratory frame using a combination of calorimeter cells and tracks. Jets were required to have $E_T^{jet} > 5$ GeV and $-1.5 < \eta^{jet} < 1.8$. To suppress the contribution from the resolved process and thereby, that of the $< k_T >$ of partons in the photon, events were required to have a large fraction of the incident photon's energy (E_γ^{inc}) invested in the production of the prompt-photon plus jet system: $x_\gamma^{meas} > 0.9$, where $x_\gamma^{meas} = \sum_{\gamma, jet}(E - p_Z)/(2E_\gamma^{inc})$ and the sum runs over the prompt photon and the jet.

The momentum imbalance ($p_\perp = |\vec{p}_{xy}^\gamma \times \vec{p}_{xy}^{jet}|/p_T^{jet}$) and the azimuthal acollinearity ($\Delta\phi$) between the prompt photon and the jet in the transverse (X, Y) plane are sensitive to the $< k_T >$ of partons in the proton. The normalised data distributions of these quantities are shown in Fig. 3. The predictions from the Monte Carlo program PYTHIA 6.1 [16], after the generated events were passed through a full simulation of the ZEUS detector, are compared to the data in Fig. 3. The predictions of PYTHIA are shown for a variety of $< k_T >$ values of the partons in the proton: a Gaussian distribution for k_T was used and the width was varied between 0.44 GeV and 3 GeV. The observed $p_\perp$ and $\Delta\phi$ distributions favour a value of $< k_T >$ in the range 1-2 GeV. This result lies between the observed $< k_T >$ at fixed-target energies and those at the Tevatron collider.

Acknowledgements. I would like to thank the organisers for their hospitality. The help of my colleagues from the ZEUS Collaboration and, in particular, from Peter Bussey and Sungwon Lee is gratefully appreciated.

REFERENCES

1. For a recent compilation see W. Vogelsang and M. Whalley, *J. Phys.* **G23** A1 (1997).
2. J. Huston *et al*, *Phys. Rev.* **D51** (1995) 6139; L. Apanasevich *et al*, *Phys. Rev.* **D59** (1999) 074007.
3. P. Aurenche *et al*, *Eur. Phys. Jour.* **C9** (1999) 107; U. Baur *et al*, hep-ph/0005226.
4. E. Laenen, G. Oderda and G. Sterman, *Phys. Lett.* **B438** (1998) 173; S. Catani, M.L. Mangano and P. Nason, *JHEP* **9807** (1998) 024; H.-L. Lai and H.-N. Li, *Phys. Rev.* **D58** (1998) 114020; S. Catani *et al*, *JHEP* **9903** (1999) 025; N. Kidonakis and J.F. Owens, *Phys. Rev.* **D61** (2000) 094004; M.A. Kimber, A.D. Martin and M. Ryskin,

ZEUS 1996-97 Preliminary

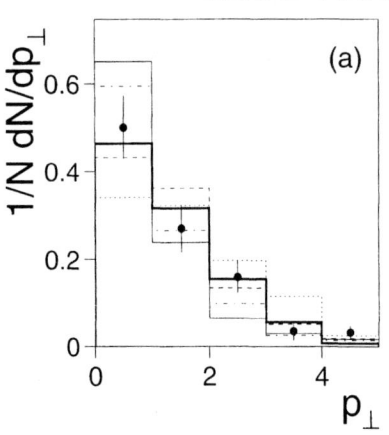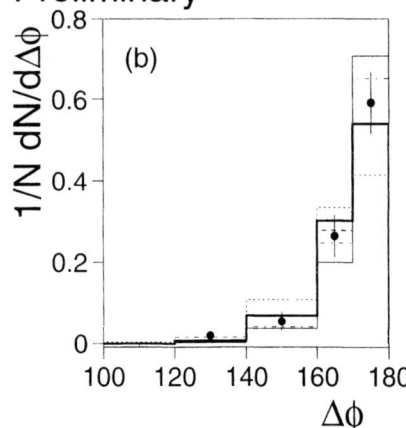

FIGURE 3. Normalised distributions of $p_\perp$ (a) and $\Delta\phi$ (b) for prompt-photon plus jet events (black dots). The predictions of PYTHIA for a variety of $< k_T >$ values are shown as histograms: 0.44 GeV (thin solid line), 1.0 GeV (dot-dashed line), 1.5 GeV (thick solid line), 2.0 GeV (dashed line) and 3.0 GeV (dotted line).

Eur. Phys. Jour. **C12** (2000) 655; E. Laenen, G. Sterman and W. Vogelsang, *Phys. Rev. Lett.* **84** (2000) 4296.

5. ZEUS Collaboration, J. Breitweg *et al*, *Phys. Lett.* **B472** (2000) 175.
6. ZEUS Collaboration, "Study of parton behaviour in the proton using prompt photon photoproduction at HERA", paper 427, submitted to the XXXth International Conference on HEP, Osaka (2000).
7. ZEUS Collaboration, M. Derrick *et al*, *Phys. Lett.* **B293** (1992) 465.
8. ZEUS Collaboration, J. Breitweg *et al*, *Phys. Lett.* **B413** (1997) 201.
9. L.E. Gordon, *Phys. Rev.* **D57** (1998) 235 and private communication; L.E. Gordon, hep-ph/9706355 and Proceedings, Photon 97, Egmond aan Zee, eds. A. Buijs and F. Erné, (World Scientific, Singapore, 1998), 173.
10. M. Krawczyk and A. Zembrzuski, hep-ph/9810253 and Proceedings, Photon 97, Egmond aan Zee, eds. A. Buijs and F. Erné, (World Scientific, Singapore, 1998), 162 and private communication.
11. M. Glück, E. Reya and A. Vogt, *Phys. Rev.* **D46** (1992) 1973.
12. L.E. Gordon and J.K. Storrow, *Nucl. Phys.* **B489** (1997) 405.
13. L.E. Gordon and W. Vogelsang, hep-ph/9606457 and Proceedings, DIS96, Rome, eds. G. D'Agostini and A. Nigro, (World Scientific, Singapore, 1997), 278.
14. ZEUS Collaboration, J. Breitweg *et al*, *Eur. Phys. Jour.* **C11** (1999) 35.
15. S. Catani *et al*, *Nucl. Phys.* **B406** (1993) 187; S.D. Ellis and D.E. Soper, *Phys. Rev.* **D48** (1993) 3160.
16. H.-U. Bengtsson and T. Sjöstrand, *Comp. Phys. Comm.* **46** (1987) 43; T. Sjöstrand, *Comp. Phys. Comm.* **82** (1994) 74.

Measurement of Dijet Cross Sections With Leading Neutrons in Photoproduction at HERA

Armen Bunyatyan

Max-Planck-Institut für Kernphysik,
Saupfercheckweg 1, 69117 Heidelberg, Germany
and Yerevan Physics Institute, Armenia
E-mail: bunar@mail.desy.de

Representing the H1 and ZEUS Collaborations

Abstract. An analysis is presented of dijets in photoproduction at the ep collider HERA where a leading neutron with an energy $E_n > 400$ GeV and angle $\theta_n < 0.8$ mrad with respect to the proton direction is detected. The average photon-proton center-of-mass energy is about 200 GeV. Mechanism of leading neutron production is studied. The differential cross sections are measured as function of jet transverse energy and pseudorapidity, the fraction of the photon momenta entering the hard scattering, x_γ, and x_π, interpreted as the fraction of pion momenta in a model where the interaction proceeds via exchange of a pion. The Monte Carlo predictions based on standard photoproduction processes and those based on the one-pion-exchange model, using different parameterizations for the pion structure function, are compared to the measurements.

INTRODUCTION

The H1 and ZEUS experiments at HERA have observed a class of events with highly energetic forward neutrons in the final state. While some leading neutron events may be explained by the hadronization of the quarks participating in the interaction, the exchange of a colour singlet particle (mainly π^+) is also expected to contribute to their production. If the one-pion-exchange process is dominant one may hope to extract the pion structure function from such measurements (see e.g. [1–3]).

The parton distributions of the pion in the region of $x \gtrsim 0.2$ have been measured by fixed target experiments. The analysis of leading neutron data in DIS at HERA [4,5] made possible for the first time a measurement of the pion structure function at small Bjorken-x ($\sim 10^{-3} - 10^{-2}$). Further tests of the pion parton distributions can be provided by studies of the high p_T jet production in hadronic interactions.

In this paper we report the recent results of H1 and ZEUS collaborations on the dijet production in photoproduction with leading neutrons [6,7]. Different predictions in the form of Monte Carlo simulations are compared to the measurements.

CP571, *PHOTON 2000,* edited by A. J. Finch

RESULTS

For this analysis the H1 collaboration used the data from 1996-97 running period. The data sample corresponds to photon virtuality of $Q^2 < 0.01$ GeV^2 and γp centre-of-mass energies of $156 < W_{\gamma p} < 242$ GeV. The events were selected which contain two jets with a transverse energy $E_T^{jet} > 7$ GeV and pseudorapidity of $-1 < \eta^{jet} < 2$ reconstructed with a cone-type jet algorithm.

ZEUS collaboration use the data collected in 1995. The data sample corresponds to the kinematical range of $Q^2 < 1$ GeV^2 and $134 < W_{\gamma p} < 269$ GeV. The k_T-algorithm was used to select jets with $E_T^{jet} > 6$ GeV and $|\eta^{jet}| < 2$.

The subsamples of dijet events with leading neutrons were selected by requiring neutron with energy $E_n > 400$ GeV and scattered angle $\theta_n < 0.8$ mrad to be detected in the Forward Neutron Calorimeters (FNC) located in the HERA tunnel at about 107 m downstream of the ep interaction point.

Neutron energy spectrum

The energy distribution of leading neutrons is sensitive to the mechanism responsible for neutron production. Figure 1 shows the neutron energy distributions measured by Forward Neutron Calorimeters of H1 and ZEUS. Distributions are not corrected for detector inefficiency and acceptance.

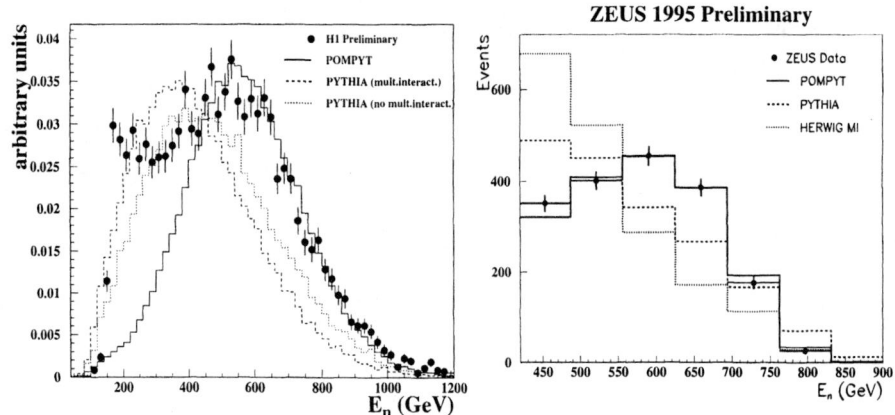

FIGURE 1. The uncorrected neutron energy spectrum for data and Monte Carlo simulations.

For comparison, the expectations from the different Monte Carlo simulations are shown which are normalized to allow for comparison of shapes. The POMPYT [8] Monte-Carlo predictions are based on the one-pion-exchange approach (using the light-cone pion flux from ref. [1]). In contrast, in the PYTHIA [9] and HERWIG [10] Monte-Carlo generators the neutrons are produced from the hadronization of proton remnant. In addition to the main hard parton scattering the PYTHIA and

HERWIG generators include the possibility for the additional (multiple) interactions between the remnant partons which were found necessary to describe the inclusive measurements of jet and energy flow.

The standard fragmentation models, both the Lund string model used by PYTHIA and the cluster fragmentation model used by HERWIG, fail to reproduce the neutron energy spectrum, predicting a distribution which is shifted towards lower energies. POMPYT, on the other hand, describes the energy spectrum well for $E_n \gtrsim 400$ GeV, thereby suggesting the necessity for a colour singlet component in the proton in the form of pion exchange and its explicit contribution to leading neutron production. A large tail at low E_n in H1 data distribution compared to Monte Carlo is due to the background from neutral electromagnetic particles (π^0, γ) which are not present in the Monte Carlo simulations used for this analysis.

Test of factorization

If the pion exchange is the dominant mechanism for production of leading neutrons, then the factorization of proton and lepton vertices is expected. Factorization can be tested by studying the ratios of jet cross sections for leading neutron events to inclusive jet cross section. Ratios have the advantage that the systematic effect are greatly reduced and that they provide a qualitative comparison of shapes of cross sections. Figure 2 shows the ratios as function of jet transverse energy E_T^{jet} and variable x_γ, which is calculated from the jets energies and pseudorapidities $(x_\gamma = \Sigma E_T^{jet} e^{-\eta^{jet}}/2E_\gamma)$ and can be interpreted as a fractions of photon momentum carried by the partons entering the hard scattering subprocess. The contribution of leading neutron events is about 4% of the inclusive jet production and practically independent of E_T^{jet} as expected from factorization. The strong increase of the ratio with x_γ may be due to phase space effects– for leading neutron events the average

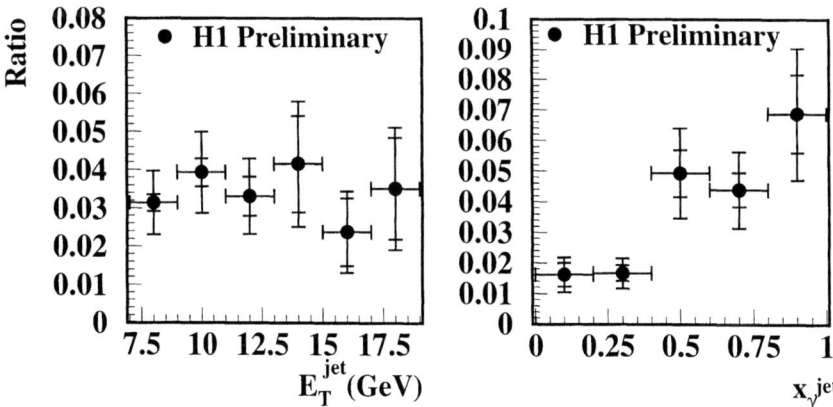

FIGURE 2. The ratios of the cross sections of dijet photoproduction with leading neutrons to inclusive dijet photoproduction.

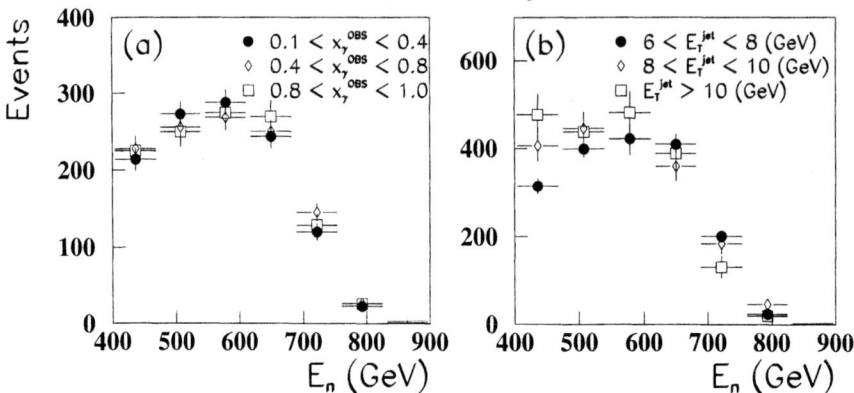

FIGURE 3. The uncorrected neutron energy spectrum for different ranges of (a) x_γ and (b) E_T^{jet}.

total energy available for jet production is smaller than in inclusive sample because significant part of proton beam energy is taken away by the neutron.

To further test the factorization properties of the dijet cross section, the neutron energy spectra are studied for different values of x_γ and E_T^{jet}. The shape of the observed neutron energy spectrum is approximately independent of x_γ and E_T^{jet}, as seen in Figure 3. The results of these studies are consistent with factorization of the proton and lepton vertices and the one-pion-exchange.

Dijet cross sections

The dijet cross sections for events with leading neutrons ($E_n > 400$ GeV and $\theta_n < 0.8$ mrad) have been measured differentially in the jet transverse energy and pseudorapidity, E_T^{jet} and η^{jet}, and the fractional momenta of the photon x_γ and the pion x_π carried by the partons involved in the hard scattering ($x_\pi = \Sigma E_T^{jet} e^{\eta^{jet}}/2(E_p^0 - E_n)$, where E_p^0 is the proton beam energy).

Figure 4 shows the measured cross sections together with Monte-Carlo predictions. The POMPYT and RAPGAP [11] Monte-Carlo models which are based on one-pion-exchange describe the cross sections well both in shapes and absolute values. PYTHIA which includes multiple interactions fails to describe the cross sections for leading neutrons although this model describes the inclusive jet measurements. The PYTHIA without multiple interactions is able to describe the cross sections, however since it failed to describe the neutron energy spectrum, the agreement in absolute values of cross section is to large extent accidental.

In the framework of the one-pion-exchange model in POMPYT, the sensitivity of the measurement to different choices for the pion structure function [12] is presented in Figure 5. It is clear that the present level of systematic uncertainty in the measurement means that no preference for any of the pion structure function

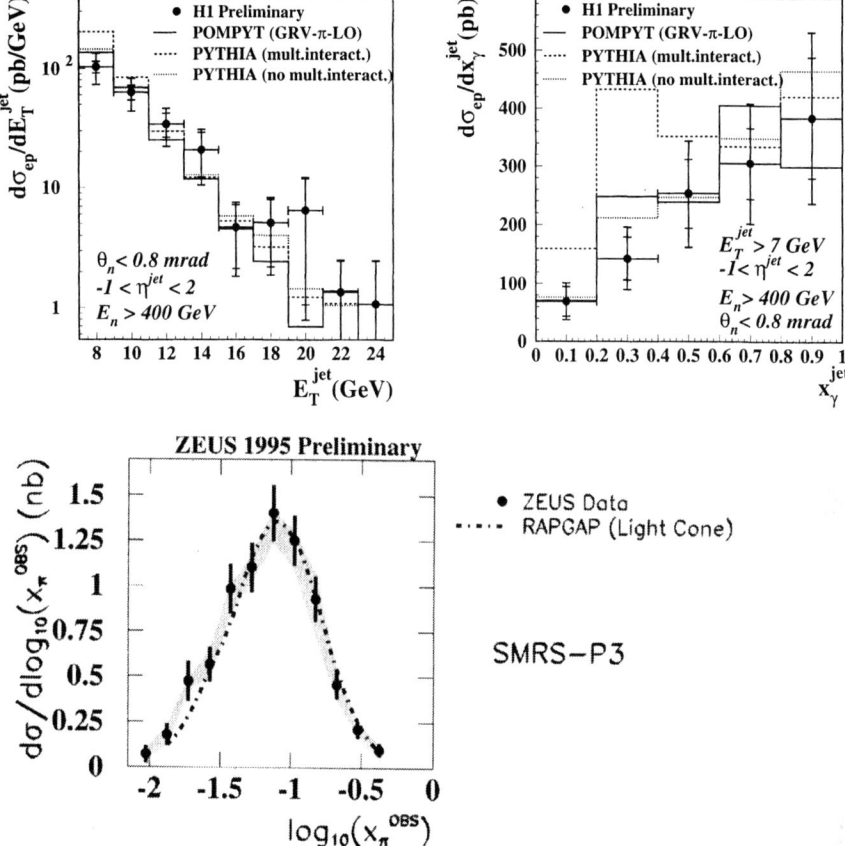

FIGURE 4. Dijet differential cross sections for events with leading neutrons $E_n > 400$ GeV and $\theta_n < 0.8$ mrad. The Monte-Carlo predictions are compared to the measurements.

parameterizations can yet be given.

SUMMARY

The photoproduction of dijets is studied in events containing an energetic forward neutron in the final state.

The contribution of events with this leading neutron is about 4% of the inclusive dijet production and independent on the jet transverse energy. The neutron energy spectrum is also independent of jet E_T and x_γ. These results are consistent with the factorization of proton and lepton vertices and the one-pion-exchange interpretation.

Models which do not include pion exchange fail to describe the neutron energy

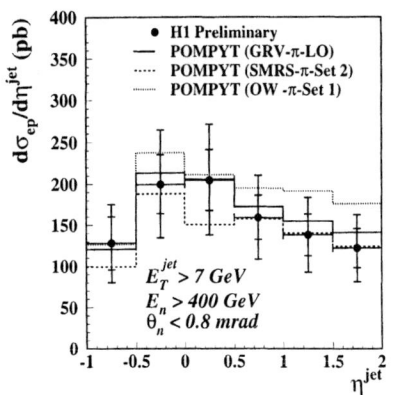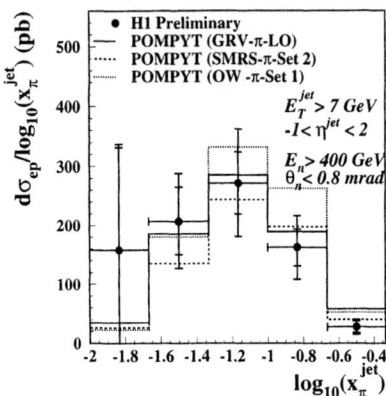

FIGURE 5. Dijet differential cross sections for events with leading neutrons $E_n > 400$ GeV and $\theta_n < 0.8$ mrad. The POMPYT Monte Carlo predictions using different parameterizations for the parton distributions in the pion are compared to the measurements.

spectrum. Models which include pion exchange describe the neutron energy spectrum well, thereby suggesting the need for a colour singlet component of proton structure in the form of pion exchange and leading neutron production. These models also reproduce the observed jet production differential cross sections. Within the present experimental uncertainties of this measurement, it is not yet possible to discriminate between different parameterizations for the pion structure function.

REFERENCES

1. H. Holtmann et al., *Phys.Lett.* **B338**, 363 (1994).
2. B. Kopeliovich, B. Povh and I. Potashnikova, *Z. Phys.* **C73**, 125 (1996).
3. M. Przybycień, A. Szczurek and G. Ingelman, *Z. Phys.* **C74**, 509 (1997).
4. H1 Collaboration, *Eur. Phys.J.* **C6**, 587 (1999).
5. ZEUS Collaboration, *Energetic neutron production in e^+p collisions at HERA*, Paper 441, XXXth ICHEP Conference, Osaka, 2000.
6. H1 Collaboration, *Measurement of dijet cross sections with leading neutrons in photoproduction at HERA*, Paper 959, XXXth ICHEP Conference, Osaka, 2000.
7. ZEUS Collaboration, *Measurement of dijet cross sections with leading neutron in photoproduction at HERA*, Paper 432, XXXth ICHEP Conference, Osaka, 2000.
8. P.Bruni, G. Ingelman, *Proceedings of the Europhysics Conference, Marseille, France, July 1993, p.595*; see also http://www3.tsl.uu.se/thcp/pompyt
9. T. Sjöstrand, *Comp. Phys. Comm.* **82**, 74 (1994).
10. G. Marchesini et al., *Comp. Phys. Comm.* **67**, 465 (1992).
11. H. Jung, *Comp. Phys. Comm.* **86**, 147 (1995).
12. M. Glück, E. Reya and A. Vogt, *Z. Phys.* **C53**, 651 (1992);
    J.F. Owens, *Phys.Rev.* **D30**, 943 (1984);
    P.J. Sutton et al., *Phys.Rev.* **D45**, 2349 (1992).

HEAVY FLAVOUR PRODUCTION

NLO calculations for heavy flavour production in two-photon collisions

Eric Laenen^a, Stefano Frixione^{b,1} and Michael Krämer^c

^{a} *NIKHEF Theory Group, Kruislaan 409, 1098 SJ Amsterdam, The Netherlands*
^{b} *CERN, Theoretical Physics Division, CH-1211 Geneva 23, Switzerland*
^{c} *Department of Physics and Astronomy, University of Edinburgh, EH9 3JZ, Scotland*

Abstract. We discuss the NLO calculations of total and diffential production rates of D* at LEP2, in tagged and untagged two-photon collisions. The sensitivity of the calculation to variations in renormalization and factorization scale, charm mass, fragmentation function etc is shown. Comparisons are made with recent experimental results, some presented at this conference.

INTRODUCTION

The production of charm and bottom quarks in two-photon collisions occurring at e^+e^- colliders allows studies of the dynamics of heavy quark production that complement those at collider and fixed-target experiments. On-shell photons are considered part of the equivalent photon cloud around the fast-moving leptons. An on-shell photon is special in that it can be either in a pointlike or hadronic state. Subprocesses in which the initial on-shell photons act pointlike are called "direct", while processes where at least one acts hadronlike are called "resolved". Resolved channels require the use of photonic parton densities. The mass of the heavy quark, $m_Q \gg \Lambda_{\mathrm{QCD}}$, sets the (minimum) scale μ for the QCD hard scattering process producing the heavy quarks, ensuring that the strong coupling coupling $\alpha_s(\mu)$ is perturbative, and that non-perturbative long distance dynamics is factorized into the photonic densities up to $\mathcal{O}(\Lambda/\mu)$.

It is thus possible to define an all-order infrared-safe cross section for open heavy flavour production. The heavy quark mass also ensures that the separation into direct and resolved production channels is unambiguous at next-to-leading order (NLO). Beyond NLO, it is important to realize that the different channels mix under evolution and that the distinction between the direct and resolved contributions becomes non-physical and scheme-dependent (see [1] for a more detailed discussion of this matter).

$^{1)}$ present address: INFN, Sez. di Genoa, Genoa, Italy.

CP571, *PHOTON 2000*, edited by A. J. Finch

By selecting two-photon collision events with either one (single-tag) or no (no-tag) scattered leptons, one imposes effectively deep-inelastic scattering ($\gamma^*\gamma$), or hadron-hadron collision kinematics ($\gamma\gamma$), respectively. We now review briefly the status of the NLO calculations for both event classes.

SINGLE TAG

For these events, occurring via the subprocess

$$\gamma^*(q) + \gamma(k) \rightarrow Q(p_1) + \bar{Q}(p_2) + X \,, \tag{1}$$

the production cross section may be expressed in terms of deep-inelastic structure functions

$$\frac{d^{2+n}\sigma}{dxdQ^2 \prod_i dV_i} = \int dz\, f_\gamma^e(z)\, \frac{2\pi\alpha^2}{x\,Q^4} \tag{2}$$

$$\times \left[(1 + (1-y)^2) \frac{d^n F_2^{\gamma,\text{charm}}(z,x,Q^2,m_Q^2,V_i)}{\prod_i dV_i} - y^2 \frac{d^n F_L^{\gamma,\text{charm}}(z,x,Q^2,m_Q^2,V_i)}{\prod_i dV_i} \right] \,,$$

where the dF_j^γ ($j = 2, L$) denote the (differential) deeply inelastic photon structure functions, α is the fine structure constant and f_γ^e is the equivalent photon momentum density (Weizsäcker-Williams function). The V_i, $i = 1, .., n$ stand for kinematic variables related to the heavy quarks, e.g. the heavy quark transverse momentum etc. The other variables are defined by

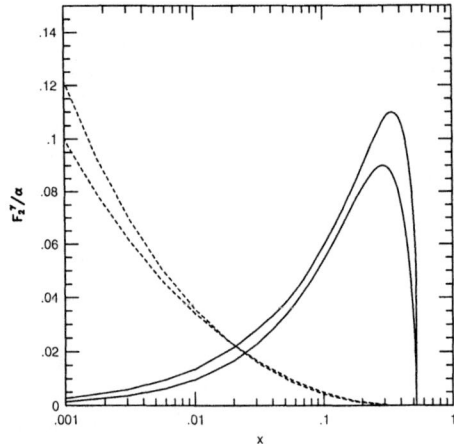

FIGURE 1. Pointlike (solid line) and hadronic (dashed) components of $F_2^{\gamma,\text{charm}}(x, Q^2)$ at $Q^2 = 10$ GeV^2. Upper solid and lower dashed curves are NLO results [2,3].

$$q = p_e - p'_e, \qquad Q^2 = -q^2, \qquad x = \frac{Q^2}{2k \cdot q}, \qquad y = \frac{k \cdot q}{k \cdot p_e}. \qquad (3)$$

with $p_e(p'_e)$ the incoming (tagged) lepton momentum. There are NLO calculations for the completely inclusive [2], the single-particle inclusive [3], and for the fully differential [4] structure functions. Recently a first measurement of the single-tag charm cross section and structure function F_2 was made by the OPAL collaboration [5], and satisfactory agreement was found with theory, within the still sizable statistical uncertainty.

The NLO corrections are all quite modest, as can be seen in Figure 1. Perhaps the most noteworthy aspect of the above observables is that an effective separation occurs in x of the pointlike (large x) and hadronic (small x) components of $F_2^{\gamma,\mathrm{charm}}(x, Q^2)$ at NLO, as shown in Fig. 1. Note that the single-tag requirement greatly suppresses the rate of this channel compared to the no-tag case. At small x, $F_2^{\gamma,\mathrm{charm}}(x, Q^2)$ is in principle sensitive to the photonic gluon density.

NO-TAG

Total cross sections and various distributions for inclusive charm and bottom quark production in $\gamma\gamma$ collisions for various e^+e^- colliders have also been calculated to NLO [6] [2]. At low energies, in the PETRA/PEP/TRISTAN range, the direct production mechanism dominates by far. At LEP2 energies, the resolved-γ contribution becomes sizeable, up to about 50% of the total cross section, depending on the photonic parton densities. The cross sections for charmed particle production are large, yielding roughly 400000 events for an integrated luminosity of $\int \mathcal{L} = 400$ pb^{-1} at LEP2; the bottom quark production cross section is suppressed by more than two orders of magnitude, due to the smaller bottom electric charge and phase space reduction by the bottom quark mass, and the fact that its higher order QCD corrections involve $\alpha_s(m_b)$, which is smaller than $\alpha_s(m_c)$. The inclusion of these corrections is important, increasing the cross section by about $30\% - 40\%$.

Total cross sections

The NLO predictions for the total charm cross section agree fairly well with experimental data [5,9,10] from PETRA energies up to LEP2 energies, provided higher-order corrections and resolved-photon contributions are included, as shown in Figure 2. The theoretical predictions appear under reasonable control, the uncertainty due to variation of the heavy quark mass and the renormalization and factorization scale being approximately $\pm 40\%$. The experimental and theoretical errors, however, do not allow to discriminate between different recent sets of photonic parton distributions, nor to constrain the value of the charm quark mass.

---

[2] The effects of summing large logarithms $\log(p_T/m_Q)$ at $p_T \gg m_Q$ have been studied in [7,8] at NLL.

However, the NLO predictions for the total bottom cross section, whose theoretical uncertainties are under somewhat better control than for charm, are considerably below the data from OPAL and L3 [11,12] shown at this conference, see Figure 2, a situation not unlike that for photoproduction at HERA [18,19]. It is not clear what the cause of this serious discrepancy might be.

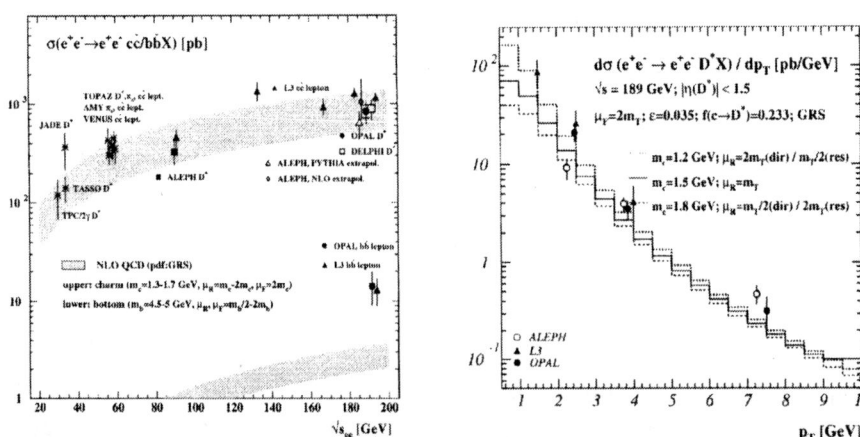

FIGURE 2. Left: comparison of NLO QCD predictions [6] and experimental results [5,9–14] for the total charm and bottom cross section as a function of the e^+e^- collider energy. GRS photonic parton densities [15]. Right: comparison between the NLO QCD prediction [1] and the OPAL/L3/ALEPH data [5,10,13] for the D^* transverse momentum distribution. The Peterson *et al.* fragmentation function [16] with $\epsilon = 0.035$ has been adopted and the probability for a charm quark to fragment into a D^* meson is set to $f(c \rightarrow D^*) = 0.270$. The L3 data [10] have been reanalyzed using an anti-tag condition for the scattered electrons [17]. To allow a common display of data sets we have scaled the L3 data to account for the different pseudorapidity range and electron anti-tag condition; numerically, the effect is $\lesssim 10\%$ and negligible on the scale of the figure.

Differential cross sections

More detailed comparisons between NLO QCD predictions and experimental data can be performed using fully differential NLO Monte Carlo programs [1], which have recently been constructed. In these codes, all final-state kinematical quantities are available on an event-by-event basis, and it is thus possible to calculate more exclusive observables at NLO and to include heavy-quark–to–heavy-meson fragmentation functions. For the direct channel, the NLO program has been constructed in [20], while for the single-resolved and double-resolved channels we use the NLO programs constructed in refs. [21,22]. To check our results we employed in addition the single-particle inclusive programs of refs. [6,23,24]. These codes allow us to compare NLO predictions with measurements in the experimentally visible region. Cross section extrapolation beyond the accepted region introduces large theoretical uncertainties [1], at least for the case at hand, and we find the visible cross section to be as important as the total cross section for the comparison of data with theory. Full dependence on the heavy quark mass is kept in the calculations. We note that simply dropping mass terms in the p_T distribution overestimates the full result by no less than $\approx 100\%$ at $m_c/p_T \approx 1$.

All four LEP collaborations have now presented [11–14], at around $\sqrt{s_{e^+e^-}} = 190$ GeV, single-particle inclusive D^* data for the differential rate with respect to the D^* transverse momentum, and pseudorapidity. Generally speaking, our NLO predictions for the D^* transverse momentum distribution compare in shape well with the OPAL and L3 measurements and slightly less so for the ALEPH data. They have a small discrepancy in absolute normalization, in particular for the L3 data, when central values for the theoretical input parameters are adopted, see Fig. 2. The pseudorapidity distribution is observed to be essentially flat in the central region, in accordance with the theoretical expectations [1,6,20]. For details see refs. [1,25]. We did not display the DELPHI D^* data in the figure, as they do not include an antitag condition and their acceptance cuts involve selections on the number of the charged tracks, which we cannot imitate in our codes.

The L3 collaboration has also presented data [12] for the differential $W_{\gamma\gamma}$ (the CM energy of the two-photon system) distribution of the charm cross section. At larger $W_{\gamma\gamma}$ values this differential cross section is completely dominated by the resolved channels, and can thus provide valuable information on the photonic gluon densities. The NLO calculations reproduce the shape of this distribution rather well, but to agree in normalization they require a rather small charm mass (about 1.3 GeV).

CONCLUSION

With so many new data on open heavy flavour production in two-photon collisions, and with theoretical calculations for these processes at NLO accuracy and well-behaved, this conference has seen the most meaningful comparisons by far of data and theory until now. While most of the charm data seem to be quite well described by NLO theory, there is a serious discrepancy with bottom data. Resolving this discrepancy remains as a challenge for the future.

E.L. would like to thank the organizers for their excellent organization of this conference, and for providing for such an enjoyable and stimulating atmosphere.

REFERENCES

1. S. Frixione, M. Krämer, and E. Laenen, Nucl. Phys. **B571**, 169 (2000), hep-ph/9908483.
2. E. Laenen, S. Riemersma, J. Smith, and W. L. van Neerven, Phys. Rev. **D49**, 5753 (1994), hep-ph/9308295.
3. E. Laenen and S. Riemersma, Phys. Lett. **B376**, 169 (1996), hep-ph/9602258.
4. B. W. Harris and J. F. Owens, Phys. Rev. **D54**, 2295 (1996), hep-ph/9603305.
5. OPAL, G. Abbiendi et al., (1999), hep-ex/9911030; R. Nisius (for the OPAL collaboration), these proceedings.
6. M. Drees, M. Krämer, J. Zunft, and P. M. Zerwas, Phys. Lett. **B306**, 371 (1993).
7. M. Cacciari et al., Nucl. Phys. **B466**, 173 (1996), hep-ph/9512246.
8. J. Binnewies, B. A. Kniehl, and G. Kramer, Phys. Rev. **D53**, 6110 (1996), hep-ph/9601278.
9. JADE, W. Bartel et al., Phys. Lett. **B184**, 288 (1987). TASSO, W. Braunschweig et al., Z. Phys. **C47**, 499 (1990). TPC/Two-Gamma, M. Alston-Garnjost et al., Phys. Lett. **B252**, 499 (1990). TOPAZ, R. Enomoto et al., Phys. Lett. **B328**, 535 (1994), hep-ex/9412001. TOPAZ, R. Enomoto et al., Phys. Rev. **D50**, 1879 (1994), hep-ex/9411002. AMY, N. Takashimizu et al., Phys. Lett. **B381**, 372 (1996). ALEPH, D. Buskulic et al., Phys. Lett. **B355**, 595 (1995).
10. M. Acciarri et al., Phys. Lett. **B467**, 137 (1999), hep-ex/9909005.
11. A. Csilling (for the OPAL collaboration), these proceedings.
12. S. Saremi (for the L3 collaboration), these proceedings.
13. U. Sieler (for the ALEPH collaboration), these proceedings.
14. A. Sokolov (for the DELPHI collaboration), these proceedings.
15. M. Gluck, E. Reya, and I. Schienbein, Phys. Rev. **D60**, 054019 (1999), hep-ph/9903337.
16. C. Peterson, D. Schlatter, I. Schmitt, and P. Zerwas, Phys. Rev. **D27**, 105 (1983).
17. V. Andreev, private communication.
18. K. Daum (for the H1 collaboration), these proceedings.
19. C. Adloff et al. [H1 Collaboration], Phys. Lett. **B467** (1999) 156 [hep-ex/9909029]; C.L. Kim (for the ZEUS collaboration), proceedings of the Int. Europhysics Conf. on High Energy Physics, Tampere, 1999.
20. M. Krämer and E. Laenen, Phys. Lett. **B371**, 303 (1996), hep-ph/9511358.
21. M.L. Mangano, P. Nason and G. Ridolfi, Nucl. Phys. **B373** (1992) 295 [hep-ph/9306337].
22. S. Frixione, M.L. Mangano, P. Nason and G. Ridolfi, Nucl. Phys. **B412** (1994) 225 [hep-ph/9306337].
23. P. Nason, S. Dawson and R.K. Ellis, Nucl. Phys. **B327** 1989) 49 (Erratum: *ibid.* **B335** 1990) 260).
24. R.K. Ellis and P. Nason, Nucl. Phys. **B312** (1989) 551.
25. S. Frixione, M. Krämer, and E. Laenen, J. Phys. G **G26**, 723 (2000), hep-ph/0002112.

Charm production in two-photon collisions measured by the ALEPH detector at LEP II energies

Uwe Sieler

Fachbereich Physik, Universität Siegen, Germany

Abstract. Charm production in two-photon collisions has been measured at LEP for center-of-mass energies of $\sqrt{s_{e^+e^-}} = 183 - 189\,\text{GeV}$ using the ALEPH detector. Charm events have been detected via muons from semi-leptonic decays and via inclusive D^* production. Preliminary results are presented.

The total cross section $\sigma(e^+e^- \to e^+e^-c\bar{c})$ has been measured. The relative fractions of the contributing processes have been determined for the acceptance range of D^* mesons. Differential D^* cross sections in the transverse momentum and the pseudorapidity of the D^*'s have been measured and compared to NLO QCD calculations.

I INTRODUCTION

Heavy flavour production in two-photon events at LEP II energies are dominated by charm production. Beauty production is suppressed because of its smaller electric charge and larger mass [1]. Direct and single resolved processes are expected to dominate the charm production and to contribute in comparable amounts [1]. Perturbative calculations become reliable as the heavy quark masses set the physical scale. Thus the study of heavy flavour production provides a good QCD test. Because the single resolved process is in LO dominated by γg fusion the measurement of the cross section can give access to the gluon content of the photon.

In this report preliminary results of the measurement of charm production by the ALEPH experiment are presented. The ALEPH detector is described in detail elsewhere [2,3]. Final states with charmed hadrons are identified by tagging muons from semi-leptonic decays [4] as well as by studying inclusive D^* production [5]. While the first approach is used to determine the total cross section $\sigma(e^+e^- \to e^+e^-c\bar{c})$ the D^* analysis gives a more detailed insight in the kinematics of the processes involved. The fractions of direct and single-resolved events are determined and differential cross sections in the transverse momentum $p_t^{D^*}$ and the pseudorapidity $|\eta^{D^*}|$ of the D^*'s are measured. The results are compared to NLO calculations which use the massive charm approach [6].

CP571, *PHOTON 2000*, edited by A. J. Finch

II CHARM IDENTIFICATION BY MUON TAG

A Event selection and muon identification

The data analyzed via muon tag were taken at $\sqrt{s} = 183\,\text{GeV}$ with an integrated luminosity of $\mathcal{L} = 52.9\,\text{pb}^{-1}$.

Hadronic two-photon events are selected by requiring the visible invariant mass W_{vis} of the event to be within $4\,\text{GeV} < W_{\text{vis}} < 65\,\text{GeV}$, the energy in charged particles E_{chrgd} to be less than $40\,\text{GeV}$, and the visible transverse momentum $p_{\text{t,vis}}$ to be less than $10\,\text{GeV}$. At least six charged tracks are required. In the analyzed data sample the number of preselected events is $N^{\text{pres}} = 112\,624$.

The muon identification is based essentially on the pattern of hits in the hadron calorimeter (HCAL) and on hits in the muon chambers surrounding HCAL. Only muons with a momentum greater than $2.4\,\text{GeV}$ are able to penetrate the hadron calorimeter and reach the muon chambers. In order to be efficient also for softer muons candidates with a momentum less than $2.4\,\text{GeV}$ are retained without requiring hits in the muon chambers. The acceptance range of muon candidates is limited to a transverse momentum greater than $700\,\text{MeV}$ and a polar angle θ in the limits $|\cos\theta| \leq 0.95$. Applying all cuts yields a number of tagged events in the preselected sample of $N^{\mu\text{tag}} = 383$.

B Measurement of the charm cross section

The total cross section $\sigma(e^+e^- \to e^+e^-c\bar{c})$ is obtained by the relation

$$\sigma(e^+e^- \to e^+e^-c\bar{c}) = \frac{N_c^{\mu\text{tag}}}{\mathcal{L} \cdot p_c}$$

where $N_c^{\mu\text{tag}}$ is the number of muon-tagged events corrected for background events, $\mathcal{L}$ is the integrated luminosity of the analyzed data sample, and p_c is the probability to identify a charm event by muon tag which was determined with the Pythia Monte Carlo [7]. This yields a total cross section of $\sigma(e^+e^- \to e^+e^-c\bar{c}) = (1300 \pm 130(stat.) \pm 370(sys.))\,\text{pb}$.

III CHARM IDENTIFICATION BY D* TAG

A Event selection and D* reconstruction

Similar to the muon-tag analysis the measurement of D* production proceeds in two steps. First, a sample of $\gamma\gamma \to$ hadrons is selected. In a second step this sample is studied for D* production. The data analyzed were taken at $\sqrt{s} = 183 - 189\,\text{GeV}$ with an integrated luminosity of $\mathcal{L} = 236.3\,\text{pb}^{-1}$.

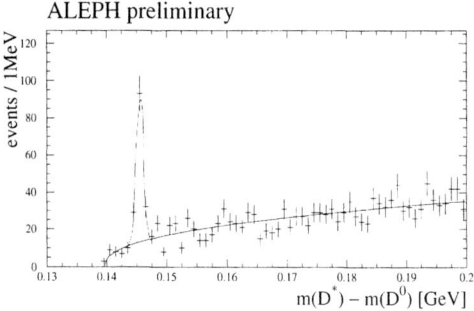

FIGURE 1. Mass difference of reconstructed D^{*+} and D^0 candidates for all considered D^0 decay modes. The solid line describes the result of the fit performed via an unbinned likelihood method.

The preselection uses the same event-shape variables as in Sec. II A with slightly different cuts. Additionally, an anti-tag condition is applied in order to reject events containing virtual photons. A tagged beam lepton is defined as an object in the ALEPH luminosity calorimeters with an energy greater than 30 GeV.

Charmed events are tagged via reconstruction of D^{*+} mesons[1] which decay via $D^{*+} \to D^0 \pi^+$. The D^0 in turn is identified in its decays to $K^- \pi^+$, $K^- \pi^+ \pi^0$, and $K^- \pi^+ \pi^- \pi^+$. The mass windows used in each decay mode to classify D^0 candidates were set according to the mass resolution determined in the Monte Carlo simulation of the detector.

Having formed a D^0 candidate each combination of this D^0 with one of the remaining π^+ candidates is considered to be a D^{*+}. In order to reduce combinatorial background from soft processes and in order to limit the kinematic range of the D^{*+} to the acceptance range of the detector with reasonable efficiency cuts were applied to the transverse momentum p_t and the pseudorapidity η of the D^{*+}:

$$2 \, \mathrm{GeV} < p_t^{D^*} < 12 \, \mathrm{GeV} , \quad |\eta^{D^*}| < 1.5 \quad . \tag{1}$$

In the following this kinematic region will be called the visible region.

Figure 1 shows the distribution of the mass difference $\Delta M = M_{D^*} - M_{D^0}$ for the analyzed data and all considered D^0 decay modes included. A clear peak is seen at 145.5 MeV. D^{*+} candidates are selected if ΔM is in the range $144 \, \mathrm{MeV} < \Delta M < 147 \, \mathrm{MeV}$ which corresponds to ± 3 standard deviations of the measured resolution. Integrating the gaussian part of the fitted parameterization over this range in the data sample analyzed yields 113 ± 15 (stat.) D^{*+} events.

---

[1] Charge conjugated particles and their decays are always included.

B Relative fractions of direct and single resolved contributions

As mentioned in the introduction open charm production in $\gamma\gamma$ collisions is dominated by contributions from *direct* and *single resolved* processes. In the present analysis we assume that only these processes contribute to the cross section which is reasonable within the uncertainties of the cross section measurement. For the simulation of D^* production in $\gamma\gamma$ events the Pythia Monte Carlo [7] was used because it allows for simulating the two considered processes separately.

While in direct processes the total invariant mass of the $\gamma\gamma$ system is contained in the two jets built by the $c\bar{c}$ pair (in LO) in the single resolved case the remnant of the resolved photon carries away a certain amount of the photons momentum in a jet near the beam pipe (*remnant jet*). Two variables sensitive to this different behaviour have been studied:

For di-jet events[2] we can define the variables $x_\gamma^\pm = \sum_{c-\text{jets}}(E \pm p_z)/\sum_{\text{hadrons}}(E \pm p_z)$ which are measures of the photons' momentum fraction taking part in the interaction. Both of them should be approximately equal to one in direct events whereas one of them should be significantly less than one for single resolved events. Thus, $x_\gamma^{\min}$, the minimum of x_γ^+ and x_γ^-, is sensitive to the different processes. A similar behaviour is seen in the fraction of transverse momentum of the D^{*+} over the visible invariant mass $p_t^{D^*}/W_{\text{vis}}$ which is defined not only for di-jet events but for all D^* events. The same variables have been used in a recent OPAL analysis [8]. The jets used for the calculation of $x_\gamma^\pm$ were reconstructed using the KTCLUS algorithm described in [9].

Figure 2 shows the distribution of $p_t^{D^*}/W_{\text{vis}}$ in data for all events found in the signal region of the mass difference spectrum. Background has been subtracted using events of the upper side band ($0.16\,\text{GeV} < \Delta M < 0.2\,\text{GeV}$) of the mass difference spectrum. Also the different components of direct and single resolved processes predicted by the Pythia Monte Carlo are shown. The contributions from direct processes have their maximum around $p_t^{D^*}/W_{\text{vis}} = 0.35$ whereas the single resolved distribution peaks at lower values of about $p_t^{D^*}/W_{\text{vis}} = 0.1$. The relative fractions are determined by fitting the sum of the two Monte Carlo distributions to data with the relative fraction as a free parameter of the fit and the total number of entries normalized to the number of entries in data.

Taking the different efficiencies for direct and single resolved events into account the fit yields a relative fraction of the direct process of $r_{\text{dir}} = (62.7 \pm 8.5)\%$ and a single resolved contribution of $r_{\text{res}} = 1 - r_{\text{dir}} = (37.3 \pm 8.5)\%$, accordingly. These fractions are valid for the visible region defined by Eq. (1). Applying the same procedure to the $x_\gamma^{\min}$ yields a consistent result (Figure 3).

The measured fraction r_{dir} is consistent within the statistical uncertainties with the fraction predicted in a NLO calculation by Frixione et al. [6] which is $r_{\text{dir}} = (70.4^{+4.6}_{-7.2})\,\%$. In this calculation the charm mass was set to 1.5 GeV

---

[2] A di-jet event in this context is an event with two jets, not counting remnant jets.

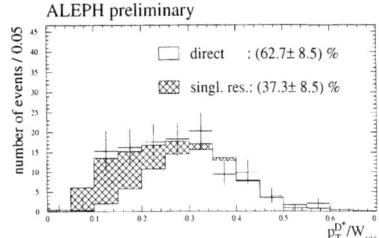

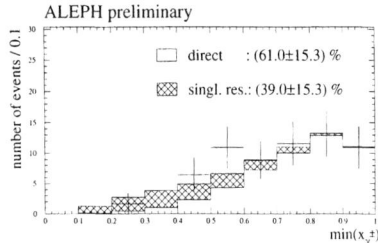

FIGURE 2. $p_t^{D^*}/W_{\text{vis}}$ distribution for reconstructed D^{*+} events in data and fitted MC contributions from direct and single resolved events.

FIGURE 3. $x_\gamma^{\min}$ distribution for reconstructed di-jet events in data containing a D^{*+} and fitted MC contributions from direct and single resolved events.

and fragmentation was modelled by the Peterson fragmentation function [10] with $\epsilon_c = 0.035$. For the resolved contribution the photonic parton densities of the GRS-HO parameterization is chosen [11].

C   Differential cross sections

The differential cross sections $d\sigma/dp_t^{D^*}$ and $d\sigma/d|\eta^{D^*}|$ for the production of D^{*+} mesons have been determined as functions of transverse momentum $p_t^{D^*}$ and of the absolute value of the pseudorapidity η^{D^*} of the D^{*+}, respectively. All considered D^0 decay modes were treated separately.

The average differential cross section $d\sigma/dp_t^{D^*}$ for a given bin $p_t^{D^*} \in [\Delta p_t^{D^*}]$ is obtained by:

$$\frac{d\sigma}{dp_t^{D^*}} = \frac{N_{\text{found}}^{D^{*+}}}{\Delta p_t^{D^*} \cdot \mathcal{L} \cdot B_* \cdot B_0 \cdot \epsilon_{p_t}} \quad , \quad p_t^{D^*} \in [\Delta p_t^{D^*}] \quad (|\eta^{D^*}| < 1.5) \quad ,$$

and analogously for $d\sigma/d|\eta^{D^*}|$. $N_{\text{found}}^{D^{*+}}$ is the number of entries found in the peak of the mass difference spectrum of events with a D^{*+} candidate in the considered acceptance range, $\Delta p_t^{D^*}$ is the width of the considered interval in $p_t^{D^*}$, $\mathcal{L}$ is the integrated luminosity, B_* is the branching ratio of the decay $D^{*+} \to D^0\pi^+$ and B_0 is the branching ratio of the considered D^0 decay, taken from [12] each. The efficiencies were determined separately for direct and single resolved processes using the Pythia Monte Carlo and were combined according to the ratio that was determined in the previous section. The results are consistent in each bin for all considered D^0 decay modes within the statistical uncertainties.

Figure 4 shows a comparison of $d\sigma/dp_t^{D^*}$ measured by ALEPH to the NLO calculation by Frixione et al. [6]. The data are consistent with the NLO calculation in

249

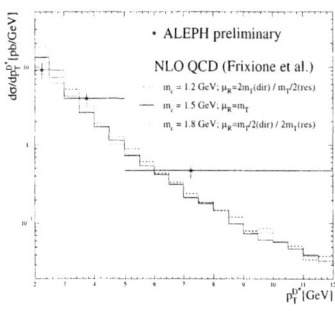

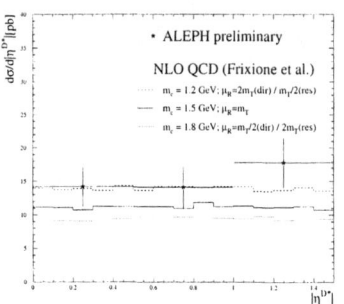

FIGURE 4. Differential cross section $d\sigma/dp_t^{D^*}$ versus the transverse momentum $p_t^{D^*}$ of the D*+.

FIGURE 5. Differential cross section $d\sigma/d|\eta^{D^*}|$ versus the absolute value of the pseudorapidity η^{D^*} of the D*+.

the first and second bin and overshoot the prediction in the third bin. Altogether, the measurement seems to favour a harder $p_t^{D^*}$ spectrum.

While the flat shape of the $d\sigma/d|\eta^{D^*}|$ distribution in data is well agreed by the calculation, the production rate is slightly underestimated by the prediction but still consistent (Figure 5).

Integrating over the visible range yields $\sigma_{vis}^{D^*}(e^+e^- \to e^+e^-D^*X) = (21.71 \pm 1.76(\text{stat.}) \pm 0.87(\text{sys.}))$pb which is consistent with the result of the NLO calculation $\sigma_{vis,\text{NLO}}^{D^*}(e^+e^- \to e^+e^-D^*X) = 17.3 + 5.1 - 2.9$ pb.

D   Total cross section

The total cross section $\sigma(e^+e^- \to e^+e^-c\bar{c})$ is obtained by extrapolating the visible D*+ cross section to the total phase space. This is performed using the separate Pythia samples for direct and single resolved processes which were mixed in such a way that their fractions in the visible region are equal to the measured fractions. The charm mass is set to $m_c = 1.5$GeV. The heavy quark fragmentation is modelled by the parameterization suggested by Peterson et al. [10] with the ϵ parameter set to $\epsilon_c = 0.031$. For the *single resolved* sample the SAS-1D parameterization [13] is used. The total D*+ cross section is then corrected by the probability of charm quarks to fragment into a D*+ taken from [14]. This yields $\sigma_{\text{Pyth.}}(e^+e^- \to e^+e^-c\bar{c}) = (653 \pm 53(\text{stat.}) \pm 169(\text{sys.}))$ pb. Alternatively, we use the NLO calculation to perform the extrapolation of the visible to the total cross section which results in $\sigma_{\text{NLO}}(e^+e^- \to e^+e^-c\bar{c}) = (1040 \pm 80(\text{stat.})^{+740}_{-340}(\text{sys.}))$ pb.

Large theoretical uncertainties are introduced to this extrapolation in each case due to poor knowledge of the charm fragmentation. Because these systematic uncertainties are strongly correlated for both of the methods the results should be compared only in their statistical uncertainty. In this respect they significantly deviate from each other. One possible reason for this discrepancy can be that NLO corrections are not negligible for this extrapolation.

IV CONCLUSIONS

Charm production in two-photon collisions has been measured at LEP energies of $\sqrt{s} = 183 - 189\,\text{GeV}$ via muon-tag and inclusive D^* production using the ALEPH detector.

The total cross section $\sigma(e^+e^- \to e^+e^- c\bar{c})$ has been extracted using the muon-tag analysis. The result is consistent with NLO predictions from [1].

Inclusive D^* production has been studied to measure relative fractions of contributions from direct and single resolved processes as well as inclusive D^* spectra. Good agreement is obtained with predictions from a NLO calculation by Frixione et al [6]. The extrapolation of the visible cross section to the total phase space is performed using a LO calculation of Pythia and the NLO calculation from [6]. Both methods significantly deviate from each other. While the Pythia result slightly undershoots the prediction from [1] the extrapolation using the NLO calculation from [6] agrees well.

REFERENCES

1. Drees M., Krämer M., Zunft J., and Zerwas P. M., *Phys. Lett.* **B306**, 371 (1993).
2. ALEPH Coll., *Nucl. Inst. Meth.* **A294**, 121 (1990).
3. ALEPH Coll., *Nucl. Inst. Meth.* **A360**, 481 (1995).
4. ALEPH Coll., *internal note CONF/2000-026*
5. ALEPH Coll., *internal note CONF/2000-048*
6. Frixione S., Krämer M., and Laenen E., *hep-ph/9908483 submitted to Nucl. Phys.*
7. Sjöstrand T., *Comp. Phys. Comm.*, **82**, 74 (1994).
8. OPAL Coll., *CERN-EP/99-157 accepted by Eur. Phys. J. C*
9. Catani S., Dokshitzer L., Seymour M. H., and Webber B. R., *Nucl. Phys.* **B406**, 187 (1993).
10. Peterson C., Schlatter D., Schmitt I., and Zerwas P., *Phys. Rev.* **D27**, 105 (1983).
11. Glück M., Reya E., and Schienbein I., *Phys. Rev.* **D60**, 54019 (1999).
12. Caso C. et al., *Eur. Phys. J.* **3**, 1 (1998).
13. Schuler G. A., Sjöstrand T., . *Z. Phys.*, **C68**, 607 (1995).
14. The LEP Collaborations ALEPH, DELPHI, L3, OPAL, the LEP Electroweak Working Group, and the SLD Heavy Flavour and Electroweak Groups, *CERN-EP/99-15.*

Inclusive D-meson and Λ_c Production in Two Photon Collisions at LEP

M. Chapkin*, V. Obraztsov*, A. Sokolov*

*Institute for High Energy Physics, 142284 Protvino, Moscow Region, Russia

Abstract. The inclusive production of D^{*+}, D^0, D^+ and Λ_c is measured by DELPHI in photon-photon collisions at LEP-II energies. The measured cross sections are compatible with the QCD calculations having the contributions from the resolved processes sensitive to the gluon density in photon. The total cross section of the charm quark production in two-photon collisions at LEP-II energies is estimated.

INTRODUCTION

Hadron production in two-photon collisions is described by the Vector meson Dominance Model (VDM) [1], the Quark Parton Model (direct process) (QPM) [2], and the hard scattering of hadronic constituents of quasi-real photons (resolved photon process)[3-7]. Charm quark production in $\gamma\gamma$ collisions has several advantages over light quark production as a test of QCD. First, the theoretical QCD calculations are less ambiguous since charm quark has a large mass compared with the typical mass scale for strong interactions. Secondly, in charm quark production the VDM contributions are small, and only the direct quark-parton model and resolved photon processes have to be considered. Studies of charm meson production might provide usefull information about the gluon density in the photon, the current charm quark mass, and the intrinsic p_T distribution of partons in resolved photons.

Fig. 1 shows some of the diagrams contributing to heavy quark production in two-photon physics. The diagram in Fig. 1a shows the direct (QPM) process in which the photon couples directly to a quark. The diagram in Fig. 1d shows the single resolved photon process.

CP571, *PHOTON 2000*, edited by A. J. Finch

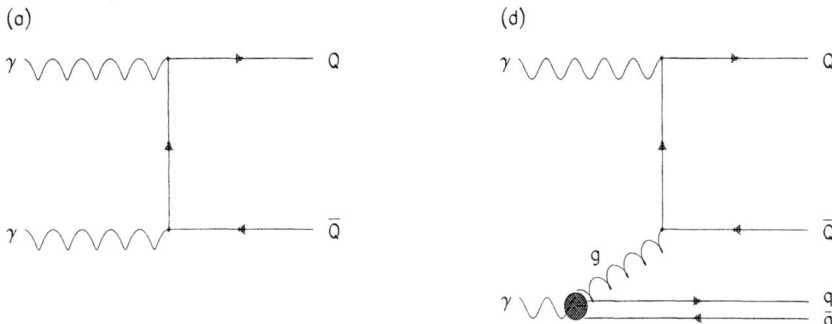

FIGURE 1. Examples of diagrams contributing to heavy quark production in $\gamma\gamma$ collisions. (a) Direct process (QPM); (d) single "resolved" process.

The direct process is dominant at low energies. However a large contribution from the resolved photon processes is predicted at high energies ($\sqrt{s_{ee}} \geq 200$ GeV) [8,9].

The charm production in $\gamma\gamma$ collisions was studied by measuring the D^{*+}, inclusive electron (muon) and K_s^0 production by JADE [10], TPC/Two-Gamma [11], TASSO [12], TOPAZ [13,14], VENUS [15] and AMY [16]. Results from LEP were recently reported by ALEPH [17], L3 [18] and OPAL [19].

In this paper we report on measurement of charm production in $\gamma\gamma$ collisions at LEP-II energies via the inclusive production of D^{*+}, D^0, D^+ mesons and Λ_c[1]. D^{*+} mesons were detected by their decay to $D^0\pi^+$, with the D^0 observed in the decay modes (1) $K^-\pi^+$, (2) $K^-\pi^+\pi^-\pi^+$ and (3) $K_s^0\pi^+\pi^-$. Inclusive D^0 and D^+ mesons were detected by their decays to $K^-\pi^+$ and $K^-\pi^+\pi^+$, respectively. Λ_c's were detected by the decays to $pK^-\pi^+$, $\Lambda\pi^+\pi^-\pi^+$, pK_s^0, $pK_s^0\pi^+\pi^-$, $pK^-\pi^+\pi^0$, $\Lambda\pi^+\pi^0$.

EXPERIMENTAL PROCEDURE

The data obtained by DELPHI detector [20,21] in 1996-1999 were used. The energies of e^+e^- collisions, and the corresponding integrated luminosities $\int Ldt$ are presented in Table 1.

TABLE 1. Integrated luminosities.

Year	96		97	98		99				Σ
$\sqrt{s_{ee}}$, GeV	161	172	183	189	192	196	200	202	204	
$\int Ldt$, pb^{-1}	10.2	10.2	54.1	157.6	25.9	76.4	83.4	40.6	0.04	458.4

Our analysis is based on the selection of a low background sample of hadronic two-photon events. Several selection criteria for charged tracks and neutral particles

---

[1] Throughout this paper charge conjugate decays are implicitly included.

in an event were applied. Charged particle tracks in the detector were accepted if the following criteria were met:

- transverse particle momentum $p_T > 100$ MeV/c;

- impact parameter of a track transverse to the beam axis $\Delta_{xy} < 3$ cm;

- impact parameter of a track along the beam axis $\Delta_z < 10$ cm;

- polar angle of a track $21° < \theta < 159°$;

- track length $l > 30$ cm;

- relative error of the track momentum $\Delta p/p < 100\%$.

To select neutral particles the calorimetric information was used. Those calorimeter clusters which are not associated to the charged particle tracks are used to form the signals from neutral particles (γ, π^0, K_L^0, n). The following thresholds were chosen for measured energy: 0.3 GeV for the showers in the electromagnetic calorimeters and 1 GeV in the hadron calorimeters.

To extract the hadronic two-photon events, the following cuts were applied to the full sample:

- visible invariant mass calculated from the four-momentum vectors of the measured charged and neutral particles is $W_{vis} < 52$ GeV/c^2;

- number of charged particles $4 \leq N_{ch} \leq 16$;

- the sum of the transverse energy components with respect to the beam direction of all charged and neutral particles is $\sum E_T^{vis} > 3$ GeV/c;

- TPC and RICH detectors were fully operational.

The last requirement reduces the integrated luminosity for the analysis to 434.1 pb^{-1}.

A comparison of the W_{vis} distributions for the data and simulated events in Fig. 2 shows that the cut $W_{vis} < 52$ GeV/c rejects the main part of non two-photon events.

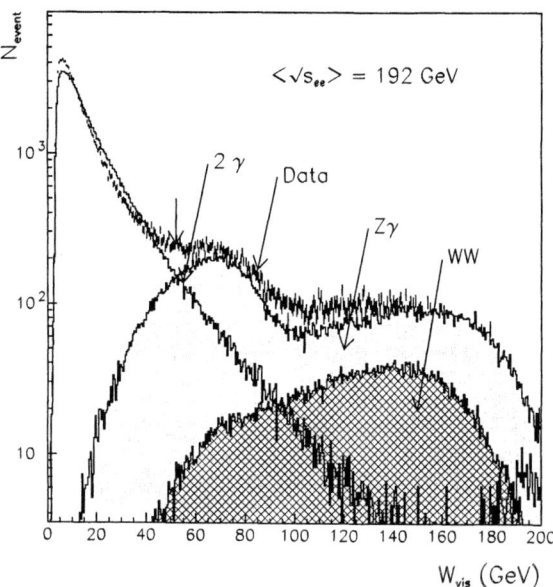

FIGURE 2. W_{vis} distributions for the data at $< \sqrt{s_{ee}} > = 192$ GeV and for the simulated $\gamma\gamma \rightarrow$ *hadrons*, $e^+e^- \rightarrow Z^0\gamma$ and $e^+e^- \rightarrow W^+W^-$ events at $\sqrt{s_{ee}} = 200$ GeV

The condition $4 \leq N_{ch} \leq 16$ decreases the combinatorial background for inclusive D mesons and Λ_c detection.

To obtain a sample of hadronic two photon events enriched in charm quark events with trigger efficiency $\geq 98\%$ [22], the additional cut was applied:

- at least one charged track from the barrel part of an event, i.e. with $45° < \theta < 135°$, has $p_T > 1.2$ GeV/c.

116323 events were selected after applying the above cuts.

The main background is from the $e^+e^- \rightarrow Z^0\gamma$ process and amounts to $\sim 3.1\%$ of the selected $\gamma\gamma$ events. The backgrounds from the $e^+e^- \rightarrow W^+W^-$ and other processes are negligible.

In order to compare the data with the theoretical predictions, samples of $\gamma\gamma \rightarrow q\bar{q}$ events simulated by TWOGAM [23] and PYTHIA 5.7 [24] programs were used. The simulated events were processed through the full chain of the DELPHI detector simulation and reconstruction programs.

As we will see below the observed distributions of charm states are well described by both generators. To compare with the other experiments we used PYTHIA.

D*+ PRODUCTION

D*+ mesons are detected by their decay into $D^0\pi^+$. The method of their reconstruction exploits the small mass difference between D*+ and D^0 mesons. The available kinetic energy of the π meson in the rest frame of the decay $D^{*+}\rightarrow D^0\pi^+$ is only 6 MeV and as a result the distribution of the variable $\Delta M = M_{D^{*+}} - M_{D^0}$ has a narrow peak. The signal is displayed by plotting ΔM for all reconstructed decay product candidates. D^0 mesons are identified via their decays to $K^-\pi^+$, $K^-\pi^+\pi^-\pi^+$ and $K_s^0\pi^+\pi^-$.

To select D^0 candidates in the $K^-\pi^+$, $K^-\pi^+\pi^-\pi^+$ decay modes, the following additional criteria were applied for the charged tracks:

- the impact parameters of the track are $\Delta_{xy} < 0.5$ cm and $\Delta_z < 2$ cm;

- the particle should not be identified as a muon;

- the particle should not be identified as a electron.

A good identification of the kaons is important for D^0 meson selection. With the Rich Imaging Cherenkov Chamber (RICH) DELPHI possesses a unique and very important hadron identification tool. The information from the liquid and gas radiators of RICH can be combined with the dE/dx information of the Time Projection Chamber (TPC) [20] to achieve separation of hadrons at different momenta.

The particle identification algorithms for the RICH and TPC detector are realised in the RIBMEAN and HADSIGN packages [25]. A package MACRIB [26] combines the benefits of RICH algorithms and dE/dx information for the high momentum region. The combination is done using a neural network, which has a much better performance than any of its input variables alone.

For separation of the signal (kaon tracks) from the background, the target of the net output x_K is set to 1 and -1 respectively. The output of the net for the data and Monte Carlo can be seen in Fig. 3. A clear separation of the kaons from background is observed. The MACRIB has a better performance than the RIBMEAN and HADSIGN packages. In addition, it is a more flexible tool as cuts can be performed as a continuous variable instead of taking discrete tags.

The charged particles satisfing the "loose" kaon criterion for $x_K > -0.5$ were used as kaon candidates for D*+ meson selection. All charged particles not identified as electron, muon or kaon were considered as pions.

For the D^0 selection, all combinations of two (for $K^-\pi^+$ decay mode) or four (for $K^-\pi^+\pi^-\pi^+$ decay mode) charged particles with the zero total charge and with one track identified as a kaon, were considered.

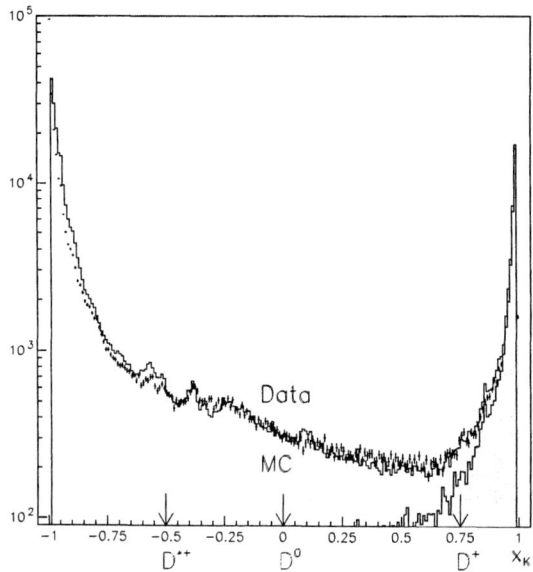

FIGURE 3. The output of the MACRIB kaon net for the charged tracks from the selected two photon events. The signal (output from kaon tracks) is accumulated at +1 (shaded area) and the background at -1.

The combination of an oppositely charged kaon and pion was considered as a candidate for the $D^0 \to K^- \pi^+$ decay if their invariant mass was within the range

$$1.83 \text{ GeV/c}^2 < M_{K^- \pi^+} < 1.90 \text{ GeV/c}^2.$$

For the $D^0 \to K^- \pi^+ \pi^- \pi^+$, the invariant mass was required to be in the range

$$1.85 \text{ GeV/c}^2 < M_{K^- \pi^+ \pi^- \pi^+} < 1.88 \text{ GeV/c}^2.$$

These ranges were determined from the resolution for $M_{K^- \pi^+}$, $M_{K^- \pi^+ \pi^- \pi^+}$ observed in the full Monte Carlo simulation.

For the K_s^0 in the $D^0 \to K_s^0 \pi^+ \pi^-$ channel, the standard DELPHI procedure of K_s^0 reconstruction [21] was applied. To form $D^0 \to K_s^0 \pi^+ \pi^-$ candidates, all the combinations of the reconstructed K_s^0 with two charged tracks with zero total charge identified as pions were used. The corresponding invariant mass was taken in the range

$$1.83 \text{ GeV/c}^2 < M_{K_s^0 \pi^+ \pi^-} < 1.90 \text{ GeV/c}^2.$$

A particle identified as a π^+ is added to the $K^- \pi^+$, $K^- \pi^+ \pi^- \pi^+$ or $K_s^0 \pi^+ \pi^-$ combinations from the above-mentioned mass ranges to form the D^{*+} candidate.

The resulting distributions of mass difference $\Delta M = (M_{\mathrm{D}^{*+}} - M_{\mathrm{D}^0})$ for different D^0 decay channels are shown in Fig. 4. A clear peak is observed around the D^{*+} and D^0 mass difference. Each histogram is fitted by the sum of a Breit-Wigner function and a background function $f(\Delta M) = \alpha \cdot (\Delta M - m_\pi)^\beta$. The distribution for the "wrong sign" combination of tracks is also shown with the pion used to form the D^{*+} from the D^0 required to have the same charge as the kaon. This produces a background spectrum from which the significance of the signal obtained from the right sign combination may be extracted. The significance of the signal for the $\mathrm{D}^0 \to \mathrm{K}_s^0 \pi^+ \pi^-$ decay was obtained from the fitting procedure. The numbers of the signal events $N_{\mathrm{D}^{*+}}$ for each D^0 decay mode are presented in the Table 2.

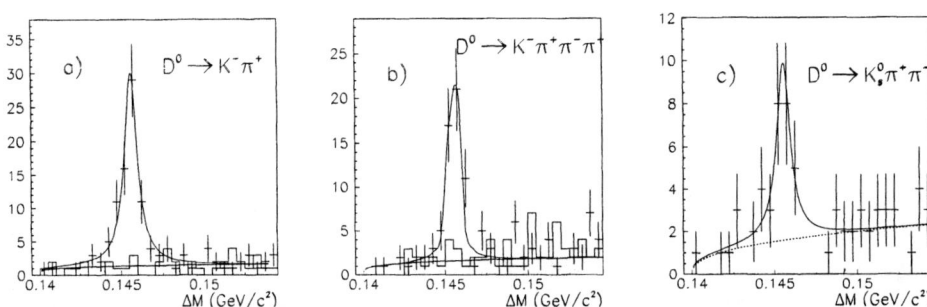

FIGURE 4. The distributions of the mass difference between the D^{*+} and D^0 for the $\mathrm{D}^0 \to \mathrm{K}^- \pi^+$(a), $\mathrm{K}^- \pi^+ \pi^- \pi^+$(b), $\mathrm{K}_s^0 \pi^+ \pi^-$(c). The histograms are fitted by the sum of a Breit-Wigner function and the background function $f(\Delta M) = \alpha \cdot (\Delta M - m_\pi)^\beta$. The distributions for "wrong sign" combinations of tracks are shown by hatched histograms.

The D^{*+} production cross section is calculated as

$$ \sigma^{\mathrm{D}^*} = \frac{N_{exp}^{\mathrm{D}^*}}{\varepsilon_{\mathrm{D}^*} \cdot L \cdot Br} $$

where $N_{exp}^{\mathrm{D}^*}$ is the observed number of D^{*+} mesons, $\varepsilon_{\mathrm{D}^*}$ is the reconstruction efficiency of D^{*+} in the corresponding mode, L is the integrated luminosity and Br is the branching ratio of D^{*+} decay to the corresponding decay chain.

The D^{*+} reconstruction efficiencies $\varepsilon_{\mathrm{D}^{*+}}$ for each D^0 decay mode calculated using PYTHIA are presented in the Table 2. Also we presented in the Table 2 the statistical uncertainties of $\varepsilon_{\mathrm{D}^{*+}}$ due to the limited number of MC events.

The main sources of systematic uncertainties are

- the branching fractions errors ($\sim 5\%$);

- uncertainties of reconstruction efficiency, determined by the changing of the PYTHIA model parameters: the p_T cut of the hard-scattering partons, the gluon density in the photon ($\sim 17\%$).

The systematic uncertainties are added in quadrature.

The calculated D^{*+} cross sections for different D^0 decay modes together with their weighted average are summarized in the Table 2.

TABLE 2. Calculation of the cross section of the process $e^+e^- \to e^+e^- D^{*+}X$.

D^0 decay mode	$K^-\pi^+$	$K^-\pi^+\pi^-\pi^+$	$K^0_s\pi^+\pi^-$
Branching fraction (%)	3.85 ± 0.09	7.6 ± 0.4	1.84 ± 0.14
$N_{D^{*+}}$	79 ± 11	54 ± 12	22 ± 7
$\varepsilon_{D^{*+}}$	$.0151 \pm .0012(stat)$	$.0054 \pm .0007(stat)$	$.0086 \pm .0019(stat)$
$\sigma_{D^{*+}}$, pb	$457 \pm 64(stat) \pm 38(MC)$	$440 \pm 98(stat) \pm 55(MC)$	$464 \pm 148(stat) \pm 96(MC)$
$< \sigma_{D^{*+}} >$, pb		$453 \pm 58(stat) \pm 80(syst)$	

To choose the Monte Carlo generator describes the data better we compare the p_T, W_{vis} and $x_E = E_{D^{*+}}/E_{vis}$ distributions (Fig. 5) for D^{*+} candidates with the PYTHIA and TWOGAM prediction.

The Monte Carlo distributions are normalized to the integrated luminosity. There is no significant disagreement between data and distributions for both MC generator programs. Although PYTHIA describes the data slightly better, we cannot fix the MC generator studying the kinematical distributions of D^{*+} candidates.

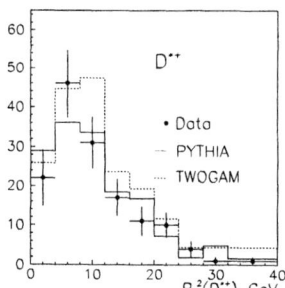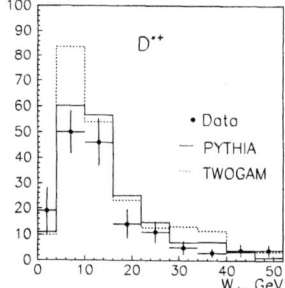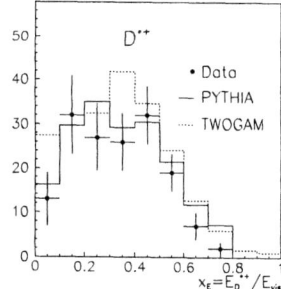

FIGURE 5. p_T momentum, W_{vis} and $x_E = E_{D^{*+}}/E_{vis}$ distributions for D^{*+} candidates are compared with the PYTHIA and TWOGAM predictions. The Monte Carlo distributions are normalized to the integrated luminosity.

The study of D^{*+} meson production in the $\gamma\gamma$ events with the requirement that at least one charged track have $p_T > 1.2$ GeV/c in the whole range of polar angle and comparison of $\sigma_{D^{*+}}$ for different cuts indicate that PYTHIA also describes better the relation of both cross sections.

D^0, D^+, AND Λ_c PRODUCTION

D^0

The D^0 were detected by their decay into $K^-\pi^+$. The large combinatorial background prohibits observation of a signal of the D^0 in the decay mode $D^0 \rightarrow K^-\pi^+\pi^-\pi^+$. For the D^0 candidates, all combinations of pairs of oppositely charged kaon and pion were used.

All charged particles satisfing the
kaon criterion with $x_K > 0$ were taken as the kaon. All charged tracks not identified as electron, muon or kaon were assumed to be pions.

The invariant $K^-\pi^+$ mass distribution is shown in Fig. 6. A clear D^0 peak is observed.

The histogram was fitted by the sum of a Gaussian function and a polynomial background. The observed number of events in the signal (N_{obs}), reconstruction efficiency (ε_{sel}) and calculated cross section (σ_{D^0}) are presented in Table 3.

The systematic uncertainties (Table 3) include the contributions from the reconstruction efficiency and branching fractions added in quadrature.

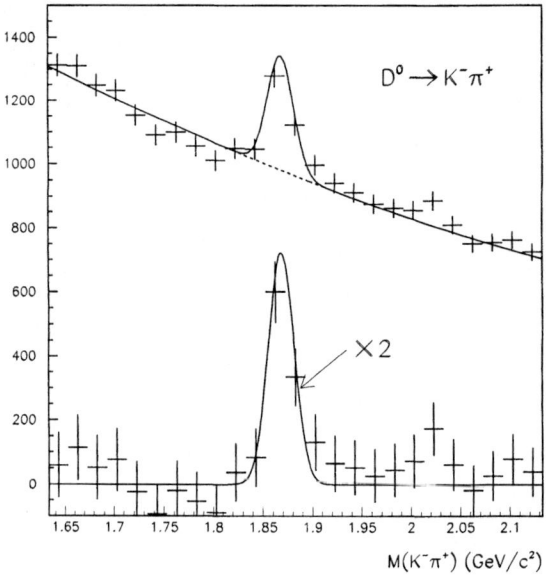

FIGURE 6. The invariant $K^-\pi^+$ mass distribution. The histogram is fitted by the sum of a Gaussian function and a polynomial background. The mass distribution scaled by a factor of 2 after background subtraction is also shown.

D⁺

The D^+ were detected by their decay into $K^-\pi^+\pi^+$. All combinations of a charged kaon and two pions with charges opposite to the kaon charge were considered.

The kaon candidates were required to satisfy the "tight" kaon selection $x_K >$ 0.75.

The invariant $K^-\pi^+\pi^+$ mass distribution is shown in Fig. 7. A clear D^+ peak is seen.

The histogram was fitted by the sum of a Gaussian function and a polynomial background. The observed number of events in the signal, the reconstruction efficiency and calculated cross section of the inclusive D^+ production are presented in Table 3.

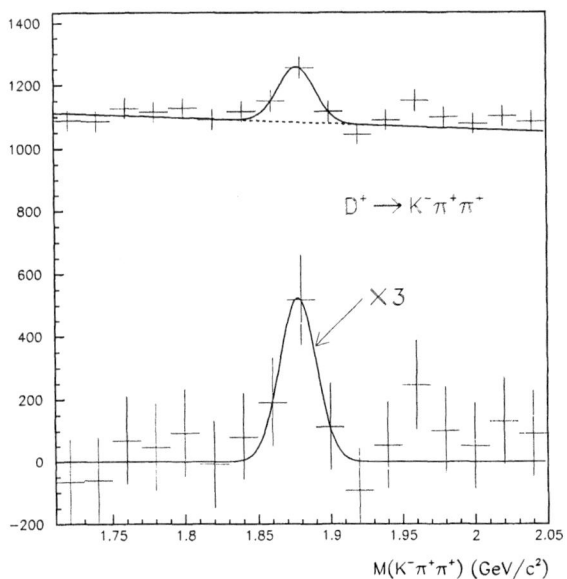

FIGURE 7. The invariant $K^-\pi^+\pi^+$ mass distribution. The histogram is fitted by the sum of a Gaussian function and a polynomial background. The mass distribution scaled by a factor of 3 after background subtraction is also shown.

Λc

The Λ_c is detected in the decay channels $pK^-\pi^+$, $\Lambda\pi^+\pi^-\pi^+$, pK_s^0, $pK_s^0\pi^+\pi^-$, $pK^-\pi^+\pi^0$ and $\Lambda\pi^+\pi^0$.

For proton identification the "proton net" from the MACRIB package was used. As proton candidates, the charged particles satisfing the "tight" proton criterion

$x_p > 0.75$ were used. For charged kaons one also required $x_K > 0.75$. The K_s^0 and Λ were reconstructed with the standard fitting procedure of the the secondary vertex [21]. For the proton from the Λ decay the "loose" proton criterion $x_p > -0.5$ was required. Two photons were considered as the π^0, if

- both photons were converted before TPC with an invariant mass of

$$0.115 \text{ GeV}/\text{c}^2 < M_{\gamma\gamma} < 0.155 \text{ GeV}/\text{c}^2;$$

- one or both photons were measured in the HPC with the invariant mass of

$$0.100 \text{ GeV}/\text{c}^2 < M_{\gamma\gamma} < 0.170 \text{ GeV}/\text{c}^2.$$

The Λ_c invariant mass distribution for the sum of all considered decay channels was shown in Fig. 8. The Λ_c signal is clearly observed.

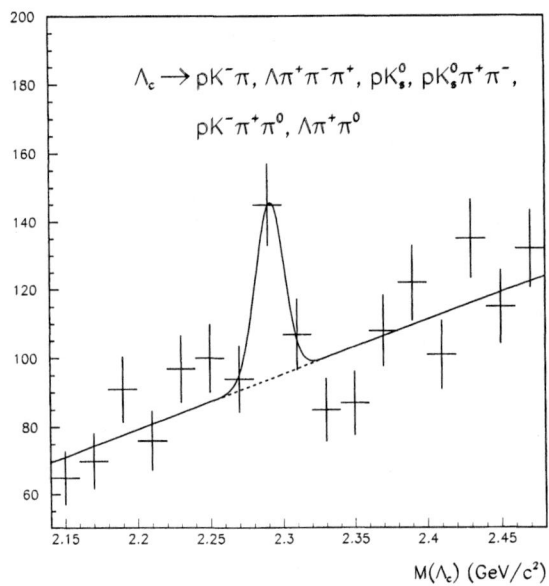

FIGURE 8. The sum of the invariant mass distributions of the $pK^-\pi^+$, $\Lambda\pi^+\pi^-\pi^+$, pK_s^0, $pK_s^0\pi^+\pi^-$, $pK^-\pi^+\pi^0$ and $\Lambda\pi^+\pi^0$ systems. The histogram was fitted by the sum of a Gaussian function and a polynomial background.

The histogram was fitted by the sum of a Gaussian function and a polynomial background. The observed number of events in signal, reconstruction efficiency and cross section of the inclusive Λ_c production are presented in Table 3.

TABLE 3. Calculation of the cross sections of the processes $e^+e^- \to e^+e^- D^0(D^+, \Lambda_c)X$.

	D^0	D^+	Λ_c
N_{obs}	498 ± 74	277 ± 66	62 ± 26
ε_{sel}	$.0325 \pm .0022(stat)$	$.0196 \pm .0024(stat)$	$.012 \pm .003(stat)$
$\sigma_{D,(\Lambda_c)}$ (pb)	$918 \pm 136(stat) \pm 62(\text{MC})$	$362 \pm 86(stat) \pm 45(\text{MC})$	$85 \pm 42(stat)$
	$\pm 157(syst)$	$\pm 62(syst)$	$\pm 14(syst)$

To compare the production of different charm mesons we estimated the inclusive cross sections of D^0 and D^+ direct production. The inclusive cross section of D^0 production can be written as

$$\sigma_{D^0} = \sigma_{D^0}^{direct} + \sigma_{D^{*0}} + Br(D^{*+} \to D^0 + X) \cdot \sigma_{D^{*+}},$$

where $\sigma_{D^0}^{direct}$ is the inclusive cross section of the direct D^0 production, $\sigma_{D^{*0(+)}}$ is the inclusive cross section of the $D^{*0(+)}$ production, $Br(D^{*+} \to D^0 + X)$ is the branching fraction of the decay $D^{*+} \to D^0 + X$. With $Br(D^{*+} \to D^0 + X) = .683$ [27] and under the assumption that $\sigma_{D^{*0}} \simeq \sigma_{D^{*+}}$ we have the following estimation: $\sigma_{D^0}^{direct} = 156 \pm 178$ pb. We took into account only statistical errors since we expect that systematic effects shift all the charm state cross sections in the same way and that for the cross section ratios it can be neglected.

For D^+ meson production we have the following relation:

$$\sigma_{D^+} = \sigma_{D^+}^{direct} + Br(D^{*+} \to D^+ + X) \cdot \sigma_{D^{*+}},$$

where $Br(D^{*+} \to D^+ + X) = .317$ [27]. The value $\sigma_{D^+}^{direct} = 218 \pm 99$ pb was obtained from the previous relation. The value of the $\sigma_{D^+}^{direct}$ and $\sigma_{D^0}^{direct}$ error is overestimated here because the correlation of $\sigma_{D^{+(0)}}$ and $\sigma_{D^{*+}}$ values is not taken into account. Within the errors $\sigma_{D^0}^{direct} \simeq \sigma_{D^+}^{direct}$.

For the D^{*+} and D^+ mesons production we have the ratio $\sigma_{D^{*+}}/\sigma_{D^+}^{direct} = 2.1^{+2.4}_{-0.9}$.

The calculated values of cross sections of the inclusive D mesons production are consistent within errors with the $(2J+1)$-relation $\sigma_{D^{*+}} = 3 \cdot \sigma_{D^0} = 3 \cdot \sigma_{D^+}$.

$C\bar{C}$ PRODUCTION CROSS SECTION

In calculating the cross section, $\sigma(\gamma\gamma \to c\bar{c})$, of charm pair production in two photon collisions we used

$$\sigma(\gamma\gamma \to D^{*+}X) = 2 \times \sigma(\gamma\gamma \to c\bar{c}) \times P_{c \to D^{*+}}.$$

Here $P_{c \to D^{*+}} = (0.255 \pm 0.017)$ is the probability of a charm quark fragmentation into D^{*+} measured by DELPHI [28] in Z^0 decays into $c\bar{c}$. Because we have no experimental measurements of the $P_{c \to D^{0(+)}(\Lambda_c)}$ and because the statistical errors of the inclusive cross sections of Λ_c and direct D^0, D^+ production are rather large, we use only D^{*+} data for the $\sigma(\gamma\gamma \to c\bar{c})$ calculation.

The measured cross section of charm quark production in two photon collisions

$$\sigma(\gamma\gamma \to c\bar{c}) = 889 \pm 128(stat) \pm 157(syst) \text{ pb}$$

agrees with theoretical predictions [8] and the OPAL [19] result and is lower than the L3 [18] value.

CONCLUSIONS

A measurement of the cross section for D^{*+}, D^0, D^+ and Λ_c inclusive production in $\gamma\gamma$ collisions at LEP-II energies has been performed.

The measured cross section values and event distributions are compatible with the QCD calculations with contributions from the resolved processes, which are sensitive to the gluon density in the photon. The measured values for the inclusive cross sections are

$$\sigma(e^+e^- \to e^+e^- D^{*+}X) = 453 \pm 58(stat) \pm 80(syst) \text{ pb},$$

$$\sigma(e^+e^- \to e^+e^- D^0X) = 918 \pm 149(stat) \pm 157(syst) \text{ pb},$$

$$\sigma(e^+e^- \to e^+e^- D^+X) = 362 \pm 97(stat) \pm 62(syst) \text{ pb},$$

$$\sigma(e^+e^- \to e^+e^- \Lambda_c X) = 85 \pm 42(stat) \pm 14(syst) \text{ pb}.$$

The estimated total cross section of the charm quark production in two-photon collisions at LEP-II energies is

$$\sigma(e^+e^- \to e^+e^- c\bar{c}) = 889 \pm 128(stat) \pm 157(syst) \text{ pb}.$$

The last result agrees with theoretical predictions.

ACKNOWLEDGEMENTS

We would like to thank V.B. Anykeyev, S. Todorova-Nova, I.A. Tyapkin for fruitful discussions on this work.

REFERENCES

1. J.J. Sakurai and D. Schildknecht, Phys. Lett. **B40** (1979) 121;
   I.F. Ginzburg and V.G. Serbo, Phys. Lett. **B109** (1982) 231.
2. S.J. Brodsky, T. Kinoshita and H. Terazawa, Phys. Rev. **D4** (1971) 1532.
3. S.J. Brodsky, T.A. DeGrand, J.F. Gunion and J.H.Weis, Phys. Rev. Lett. **41** (1978) 672; Phys. Rev. **D19** (1979) 1418.
4. D.W. Duke and J.F. Owens, Phys. Rev. **D26** (1982) 1600.
5. M. Drees and K. Grassie, Z. Phys. **C28** (1985) 451.

6. H. Abramovicz, K. Charchula and A. Levy, Phys. Lett. **B269** (1991) 458.
7. M. Glück, E.Reya and A. Vogt, Phys. Rev. **D46** (1992) 1973.
8. M. Drees, M. Krämer, J. Zunft and P.M. Zerwas, Phys. Lett. **B306** (1993) 371.
9. M. Drees and R. Godbole, Nucl. Phys. **B339** (1990) 355.
10. W. Bartel *et al.*, JADE Collab., Phys. Lett. **B184** (1987) 288.
11. M. Alston-Garnjost *et al.*, TPC/2γ Collab., Phys. Lett. **B252** (1990) 499.
12. W. Braunschweig *et al.*, TASSO Collab., Z. Phys. **C47** (1990) 499.
13. R. Enomoto *et al.*, TOPAZ Collab., Phys. Rev. **D50** (1994) 1879.
14. R. Enomoto *et al.*, TOPAZ Collab., Phys. Lett. **B328** (1994) 535.
15. S. Uehara *et al.*, VENUS Collab., Z. Phys. **C63** (1994) 213.
16. T. Aso *et al.*, AMY Collab., Phys. Lett. **B363** (1995) 249;
    Phys. Lett. **B381** (1996) 372.
17. D. Buskulic *et al.*, ALEPH Collab., Phys. Lett. **B355** (1995) 595.
18. M. Acciarri *et al.*, L3 Collab., Phys. Lett. **B453** (1999) 83;
    L3 Collab., "Measurement of Charm and Beauty Production in $\gamma\gamma$ Collisions at LEP", submitted to the Intern. Europhys. Conference HEP 99, Tampere, Finland, 15-21 July 1999.
19. G. Abbiendi *et al.*, OPAL Collab., "Inclusive Production of $D^{*\pm}$ Mesons in Photon-Photon Collisions at $\sqrt{s_{ee}}$=183 and 189 GeV and First Measurement of $F_{2,c}^{\gamma}$", preprint CERN-EP/99-157 (November 1999), submitted to Eur. Phys. J. C.
20. P. Aarnio *et al.*, DELPHI Collab., Nucl. Inst. Meth. **A303** (1991) 233.
21. P. Abreu *et al.*, DELPHI Collab., Nucl. Inst. Meth. **A378** (1996) 57.
22. V. Canale *et al.*, "The DELPHI Trigger System at LEP200", DELPHI/99-7 DAS 188 (22 February 1999).
23. S. Nova, A. Olshevski and T. Todorov, "MONTE CARLO Event Generator for Two Photon Physics", DELPHI/90-35 (November 1990).
24. T. Sjöstrand, Comput. Phys. Comm. **82** (1994) 74.
25. M. Battaglia, P.M. Kluit, "Particle identification using the RICH detectors based on the RIBMEAN package", DELPHI/96-133 (September 1996);
    W. Adam *et al.*, " Analysis techniques for the DELPHI ring imaging Cherenkov detector", DELPHI/94-112 (June 1994).
26. Z. Albrecht, M. Feindt and M. Moch, "MACRIB. High efficiency - high purity hadron identification for DELPHI", DELPHI/99-150 (October 1999).
27. Review of Particle Physics, Eur. Phys. J. **C15** (2000) 1.
28. P. Abreau *et al.*, DELPHI Collab., Eur. Phys. J. **C12** (2000) 209.

Charm and Bottom Production in Two-Photon Collisions at LEP with the L3 Detector

Sepehr Saremi

Department of Physics and Astronomy
202 Nicholson Hall
Louisiana State University
Baton Rouge, LA 70803, USA

Abstract.
Inclusive c and b quark production in $\gamma\gamma$ collisions are studied with the L3 detector at the LEP collider. Data were collected at centre-of-mass energies from 189 GeV to 202 GeV for a total integrated luminosity of 410 pb^{-1}. Hadronic final states containing c and b quarks are identified by detecting electrons or muons from their semi-leptonic decays. The cross section $\sigma(\gamma\gamma \to c\bar{c}X)$ is measured as a function of the two-photon centre-of-mass energy in the interval 5 GeV $\leq W_{\gamma\gamma} \leq$ 70 GeV. A steeper rise with energy is observed for $\sigma(\gamma\gamma \to c\bar{c}X)$ as compared to hadron-hadron cross sections and to $\sigma(\gamma\gamma \to hadrons)$. The cross section $\sigma(\gamma\gamma \to b\bar{b}X)$ is measured for the first time. This cross ection is compared to perturbative next-to-leading order QCD calculations. Inclusive production of D$^{*\pm}$ mesons in two-photon collisions is measured using the data collected at the centre-of-mass energy $\sqrt{s} =$ 189 GeV with an integrated luminosity of 176.4 pb^{-1}. Differential cross sections of the process $e^+e^- \to e^+e^- D^{*\pm}X$ are determined as functions of the transverse momentum and pseudorapidity of the D$^{*\pm}$ mesons in the kinematic region 1GeV < $p_T^{D^*}$ < 5GeV and $|\eta^{D^*}|$ < 1.4.

INTRODUCTION

The measurement of heavy flavour production in two-photon collisions provides a good test of QCD because the large physical scale set by the c or b quark masses makes the perturbative calculations reliable. Many experiments have studied c production in $\gamma\gamma$ collisions [1,2]. On the other hand b quark production in $\gamma\gamma$ collisions has not been measured before since its production is expected to be suppressed by more than two orders of magnitude relative to the production of c quark. At LEP energies, the direct and resolved processes give comparable contributions to the cross section [3] (Figure 1).

The resolved photon cross section is dominated by the photon-gluon fusion diagram. The production rate of c and b quarks in two-photon collisions depends on

CP571, *PHOTON 2000*, edited by A. J. Finch

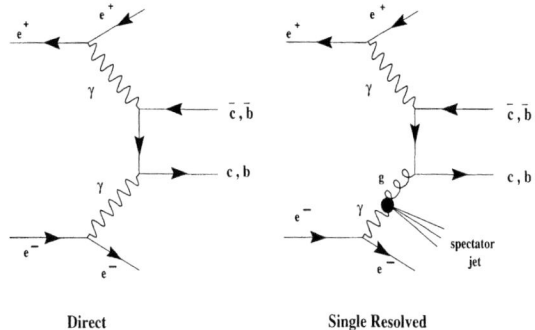

Direct **Single Resolved**

FIGURE 1. Production of heavy quarks in two-photon collisions at e^+e^- colliders.

the c and b quark masses and on the gluon density in the photon.

Here we present the results of the measurements of the $e^+e^- \rightarrow e^+e^-c\bar{c}X$ and $e^+e^- \rightarrow e^+e^-b\bar{b}X$ cross sections. Two other aspects of c production are studied as well. First the more specific case of $D^{*\pm}$ production can be studied. Second one can study the measurement of the $\gamma\gamma \rightarrow c\bar{c}X$ cross section as a function of the $\gamma\gamma$ centre-of-mass energy. In the following each of the different processes will be discussed separately.

INCLUSIVE CHARM PRODUCTION

The L3 detector has been described in detail in [4]. The event selection is done in two steps. The first one selects hadronic final states produced in two-photon collisions, the second identifies a c quark by its semi leptonic decay. Hadronic two-photon events are selected by cuts on the number of tracks, the visible energy and the visible mass. We require at least five tracks in each event; the visible energy has to be less than 0.38 $\sqrt{s}$ and the visible mass has to be greater than 3 GeV. The cut on visible energy separates the two-photon from annihilation processes which are characterized by high visible energy. The background from $e^+e^- \rightarrow e^+e^-\tau^+\tau^-$ and $e^+e^- \rightarrow \tau^+\tau^-$ events is suppressed by the five track requirement.

The analysis is limited to untagged events with small photon virtuality. Events are excluded when the most energetic cluster in the small angle calorimeters has an energy greater than 0.4 E_{beam}. Thus the interacting photons are quasi-real: $\langle Q^2 \rangle \cong 0.015$ GeV^2, where $-Q^2$ is the mass squared of the virtual photon.

To identify c quarks by their semi leptonic decay to electrons we impose the following cuts in the polar angle range of $|\cos \theta| < 0.725$:

- The distance of closest approach of the track to the average position of the e^+e^- collision point in the plane transverse to the beam must be less than 0.5 mm.

- The electromagnetic cluster is required to correspond to a track.

- To confirm that the electromagnetic shower is originated by an electron, the distribution of energies measured in the crystals of the calorimeter are compared to that of an electromagnetic cluster using a χ^2 test. The cut is $\chi^2 < 3$.

- We require $1 - 2\sigma < E_t/p_t < 1 + 2\sigma$, where σ is the detector resolution. E_t is the projection of the energy of the cluster on the plane transverse to the beam and p_t is the transverse momentum of the electron track (Figure 2).

- To remove photon conversions the invariant mass of the electron candidate and its nearest track should be greater than 0.1 GeV.

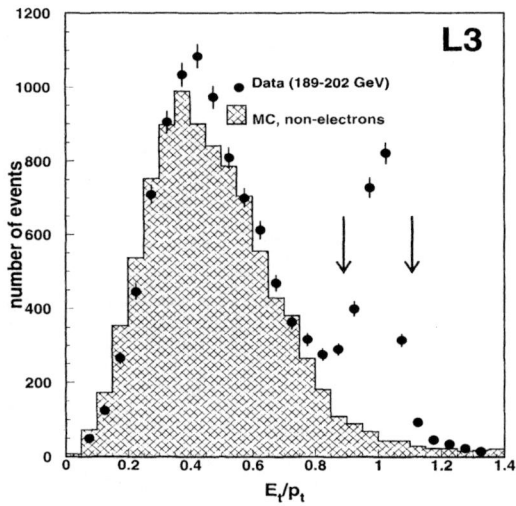

FIGURE 2. A cut on E_t/p_t clearly separates the electron signal.

The muon candidates are selected by requiring a muon in the muon chambers. Only muons with momentum greater than 2.0 GeV can reach the muon chambers. Therefore in order to select muons we require the momentum of the muons to be greater than 2.0 GeV. In order to suppress the contribution from the annihilation processes, we require the muon momentum to be less than 0.2 E_{beam}. The angular acceptance is limited to $|\cos \theta| < 0.9$.

After applying all cuts we select 2455 electron and 423 muon candidates. The purity of the electron sample is 73% and that of muon is 41%. The background in the electron sample is below 1% and in the muon sample is 4%.

The cross section for the electron channel is:

$$\sigma_{electron-tag}^{ee \to eeccX} = 1166 \pm 34 \ (\text{stat}) \ \pm 128 \ (\text{syst}) \ \text{pb}$$

and for the muon decay mode is:

$$\sigma_{muon-tag}^{ee \to eeccX} = 791 \pm 108 \ (\text{stat}) \ \pm 268 \ (\text{syst}) \ \text{pb}$$

The combined cross section for electron and muon channels is:

$$\sigma_{combined}^{ee \to eeccX} = 1136 \pm 33 \ (\text{stat}) \ \pm 128 \ (\text{syst}) \ \text{pb}$$

D* PRODUCTION

In the present study charmed vector mesons $D^*(2010)^{\pm}$ are identified by the small energy released in D^* decay, applying the mass difference technique [5] to the decay channel: $D^{*\pm} \to D^0 \pi_s^{\pm}$. We identify D^0 by its following decay modes: $D^0 \to K^{\pm}\pi^{\mp}$ or $D^0 \to K^{\pm}\pi^{\mp}\pi^0$.

The presence of a low-momentum, "soft" pion, π_s, ensures that the resolution of the mass difference $M(D^0\pi_s^+) - M(D^0)$ is superior to the resolution of the reconstructed D^0 and $D^{*\pm}$ masses themselves. The $D^{*\pm}$ signal appears as a narrow peak close to the kinematic threshold in the mass difference distributions $M(K^-\pi^+\pi_s^+) - M(K^-\pi^+)$ and $M(K^-\pi^+\pi^0\pi_s^+) - M(K^-\pi^+\pi^0)$.

This analysis was done with the data collected in 1998 at a centre-of-mass energy $\sqrt{s} = 189$ GeV and of an integrated luminosity of 176.4 pb^{-1}.

First we select hadronic two-photon events by applying cuts on the energy measured in the electromagnetic and hadron calorimeters and using tracking information. To exclude annihilation events, the total visible energy must not exceed $0.4 \sqrt{s}$, the energy deposited in the electromagnetic calorimeter must be less than 30 GeV and the energy in the hadron calorimeter less than 40 GeV. The transverse component of the missing momentum vector must be less than 10 GeV. Events are required to have at least three charged particles reconstructed in the tracking chamber.

Second we identify $D^{*\pm}$ mesons through selection of D^0 candidates, which are then combined with another track to form $D^{*\pm}$ candidates. A pair of tracks of opposite charge is considered $K^{\pm}\pi^{\mp}$ system from a D^0 decay if:

The intersection point of the tracks in the transverse plane is displaced by no more than 3 mm away from the event vertex. Also if $P_K \cdot P_{\pi} > 2 \cdot 10^{-3}$, where P_K and P_{π} are the probabilities, calculated from the measured energy loss dE/dX of each track, for kaon and pion mass hypotheses of the corresponding tracks.

The integrated cross section measured in the visible kinematic region is found to be:

$$\sigma(e^+e^- \to e^+e^-D^{*\pm}X;\ 1\text{GeV} < p_T^{D^*} < 5\text{GeV}, |\eta^{D^*}| < 1.4) = 132 \pm 22 \pm 26 \text{ pb},$$

where the first error is statistical and the second systematic. In Figure 3 the differential cross sections are compared to next-to-leading order, NLO, perturbative QCD computations and a reasonable agreement can be observed. In these calculations, three different sets of parton density parameterizations of the photon have been used: GS [6], AFG [7] and GRV-HO [8].

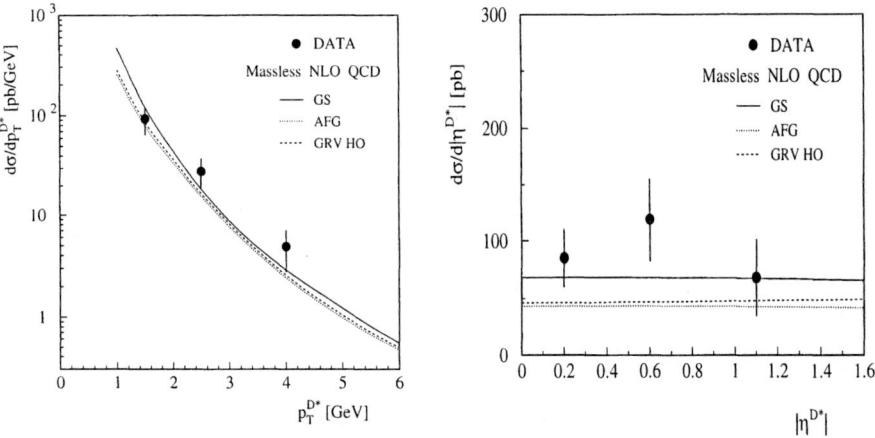

FIGURE 3. The differential cross sections of $D^{*\pm}$ production as a function of the transverse momentum of the $D^{*\pm}$ and $|\eta^{D^*}|$. The points represent the data, the error bars show the statistical and systematic errors added in quadrature. The curves represent next-to-leading order QCD calculations [9] for different parameterizations of the parton densities of the photon.

$\sigma(\gamma\gamma \to e^+e^-C\bar{C}X)$ AS A FUNCTION OF $W_{\gamma\gamma}$

We have also measured the cross section of $\gamma\gamma \to c\bar{c}X$ as a function of the photon-photon centre-of-mass energy, $W_{\gamma\gamma}$ which is determined from the hadronic final state. The visible mass, W_{vis}, of the event is calculated from the four-momentum vectors of the measured particles (tracks and calorimetric clusters including luminosity monitor). The hadronic final state is not always fully contained in the detector acceptance. Therefore an unfolding procedure is applied to go from the measured W_{vis} to the true $\gamma\gamma$ centre-of-mass energy $W_{\gamma\gamma}$. To extract the $\sigma(\gamma\gamma \to c\bar{c}X)$ cross section, the photon flux $\mathcal{L}_{\gamma\gamma}$ [10] is calculated and the hadronic two-photon processes are extrapolated to zero Q^2 [11]. For each $W_{\gamma\gamma}$ bin a numerical integration is performed over the unmeasured Q^2 of the scattered electron or positron. In Figure 4, the measured cross section $\sigma(\gamma\gamma \to c\bar{c}X)$ as a function of $W_{\gamma\gamma}$ is shown. A parametrisation by A.Donnachie and P.V. Landshoff [12] of the form $\sigma_{\text{tot}} = A\,s^{\epsilon} + B\,s^{-\eta}$ ('Pomeron + Reggeon') describes well the energy behaviour of all total hadron-hadron cross sections with a universal values $\eta = 0.34 \pm 0.02$ and $\epsilon = 0.095 \pm 0.002$ [13]. The fit in the interval $5 < W_{\gamma\gamma} < 70$ GeV by the 'Pomeron + Reggeon' form gives for the 'Pomeron' slope ϵ (η is fixed at the PDG value):

$$\epsilon = 0.400 \pm 0.062(\text{stat.}) \pm 0.096(\text{syst.})$$
$$A = 1.9 \pm 0.9(\text{stat.})$$
$$B = 44.6 \pm 11.0(\text{stat.})$$
$$\chi^2 = 9.0/2$$

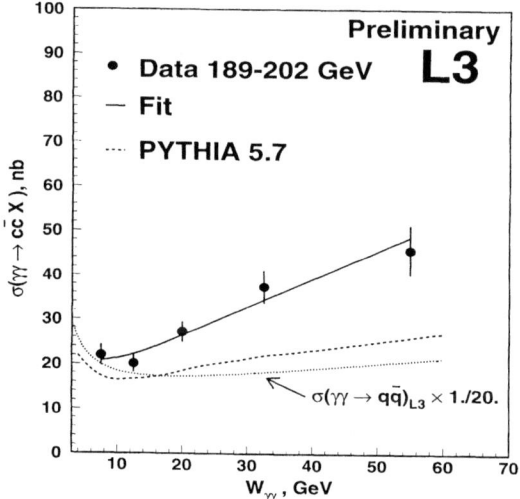

FIGURE 4. The cross section $\gamma\gamma \to c\bar{c}$ as function of $W_{\gamma\gamma}$ at $\sqrt{s} = 189 - 202$ GeV. The continuous line is the Regge fit, with fixed value of η as described in the text. The unfolding is done with PYTHIA Monte Carlo. The dashed curve shows the expectation from the PYTHIA model. The dotted curve is the total cross section $\sigma(\gamma\gamma \to \text{hadrons})$ measured by L3 scaled by an arbitrary factor of $1/20$.

The fit to the data by the form $\sigma_{\text{tot}} = A\,s^{\epsilon}$ performed in the interval $10 < W_{\gamma\gamma} < 70$ GeV gives for the 'Pomeron' slope

$$\epsilon = 0.287 \pm 0.046(\text{stat.}) \pm 0.073(\text{syst.})$$
$$A = 4.8 \pm 1.4(\text{stat.})$$
$$\chi^2 = 1.7/2$$

The value of ϵ is in agreement with the measurements of the photoproduction of J/ψ (charm) mesons at HERA, which show about two times bigger slope than the photoproduction total cross section [14,15].

INCLUSIVE BOTTOM PRODUCTION

Figure 5 shows the momentum distribution of the electron candidate at the charm selection level. As is evident from the plot, there is a discrepancy between the data and the udsc spectrum. This is undoubtedly due to the b fraction in the data.

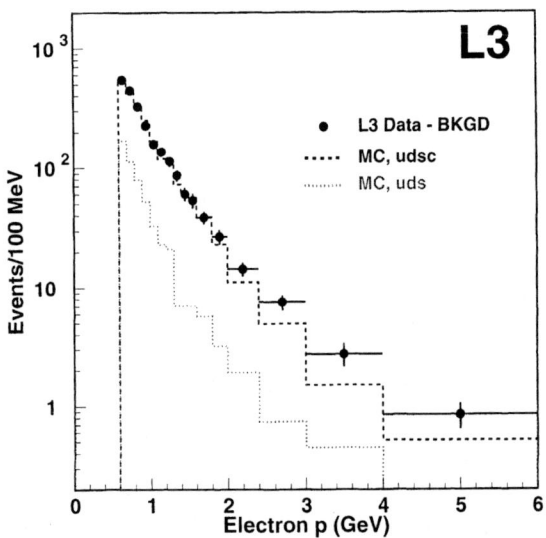

FIGURE 5. The momentum of the electron candidate. In this plot the dashed histogram is the contribution from PYTHIA Monte Carlo with only u, d, s and c quarks. The c fraction of PYTHIA is scaled to the measured $e^+e^- \to e^+e^-c\bar{c}X$ cross section.

A method for measuring b cross section is fitting the data distributions to estimate the b fraction. The transverse momentum of the lepton (from semi-leptonic decay) with respect to the nearest jet, P_t, has been chosen as the fit variable since it has the highest sensitivity to the b fraction. A three parameter fit is applied in which the fractions of b and c are free parameters. For the fit we assume the following combination of processes:

$$\text{bkgd} + \alpha_{\text{uds}}(\text{uds}) + \alpha_c(c) + \alpha_b(b)$$

Where (bkgd), (uds), (c) and (b) are the Monte Carlo distributions of the non two photon backgrounds, events generated with only the light quarks, the c quark and the b quark respectively. α_{uds}, α_c and α_b are the fractions of (uds), (c) and (b) events respectively; $\alpha_{uds} = 1\text{-}\alpha_b\text{-}\alpha_c$. The fitted distributions are shown in Figure 6. The resulting cross sections are

$$\sigma_{electron-tag}^{ee \to eebbX} = 11.10 \pm 2.60 \text{ (stat) } \pm 1.22 \text{ (syst) pb}$$

$$\sigma_{muon-tag}^{ee \to eebbX} = 18.90 \pm 3.20 \text{ (stat) } \pm 3.40 \text{ (syst) pb}$$

The combined value for the lepton tag is

$$\sigma_{combined}^{ee \to eebbX} = 14.20 \pm 2.02 \text{ (stat) } \pm 1.56 \text{ (syst) pb}$$

The systematic errors arise from the cut variation, error on charm cross section measurement and trigger efficiency. The P_t of lepton will have different shapes when using massive or massless matrix element Monte Carlos. As a result a source of systematic has been added due to the different shapes of the two Monte Carlos. This accounts for about 4% of the systematic error for both electrons and muons. The dominant systematic error is the cut variation.

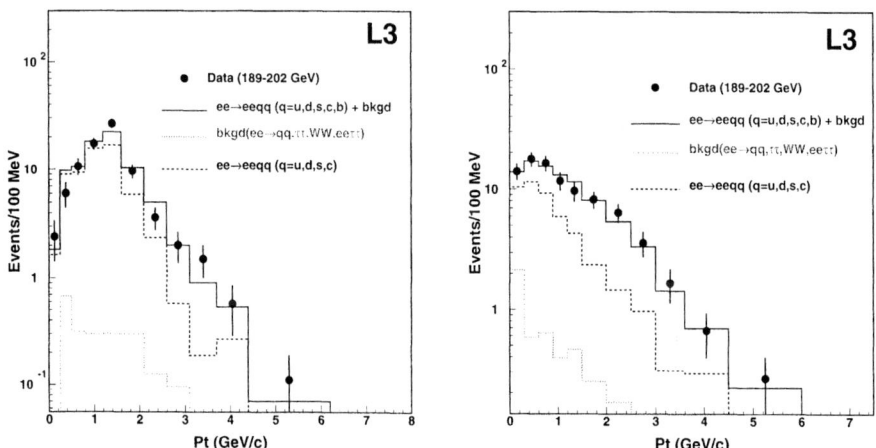

FIGURE 6. The distributions of the transverse momentum of the lepton with respect to the closest jet.

Figure 7 shows the measured value of $\sigma(e^+e^- \to e^+e^-b\bar{b}X)$ together with charm cross section measurements. The experimental results are compared with the theoretical predictions calculated to NLO accuracy. The QCD predictions for this process at $\sqrt{s} = 200$ GeV are 4.60 pb and 3.89 pb assuming mass of the b quark to be 4.5 GeV and 5.0 GeV respectively.

The $b\bar{b}$ cross section is measured for the first time in two-photon collisions and is about 4 standard deviation higher than expected.

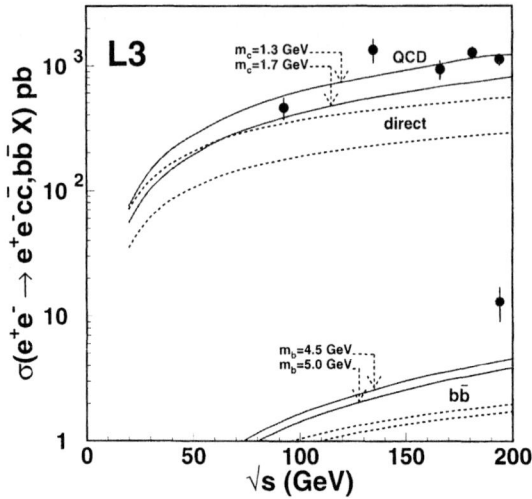

FIGURE 7. The c and b production cross sections in two-photon collisions. The L3 data from both the electron and the muon events are combined. The statistical and systematic errors are added in quadrature. The dashed line corresponds to the direct process contribution and the solid line represents the QCD prediction for the sum of the direct and resolved processes calculated to NLO.

CONCLUSIONS

The cross section for inclusive charm production in two-photon collisions is measured and the result is in good agreement with NLO QCD. The high energy behaviour of the total charm cross section as a function of the $\gamma\gamma$ centre-of-mass energy is measured for the first time. The cross section $\sigma(\gamma\gamma \to c\bar{c}X)$ is measured in the interval 5 GeV $\leq W_{\gamma\gamma} \leq$ 70 GeV. The data are compared to NLO QCD calculations and to the expectations of the PYTHIA Monte Carlo model. A steeper rise with energy is observed for $\sigma(\gamma\gamma \to c\bar{c}X)$ as compared to hadron-hadron cross sections and to $\sigma(\gamma\gamma \to hadrons)$. The inclusive production of $D^{*\pm}$ mesons in two-photon interactions at LEP is measured. The integrated and differential cross sections of inclusive $D^{*\pm}$ production are determined in the kinematic region 1 GeV $< p_T^{D^*} <$ 5 GeV, $|\eta^{D^*}| <$ 1.4. A reasonable agreement is observed between the measured differential cross sections and the predictions based on NLO QCD calculations.

The $\sigma(e^+e^- \to e^+e^-b\bar{b}X)$ is measured for the first time in two-photon collisions. This cross section is almost three times and about 4 standard deviations higher than the theoretical predictions.

ACKNOWLEDGMENTS

I would like to thank all the members of the L3 two photon group for their helpful discussions.

REFERENCES

1. JADE Collab., W. Bartel *et al.*, Phys. Lett. B **184** 288 (1987).
   TPC/Two-Gamma Collab., M. Alston-Garnjost *et al.*, Phys. Lett. B **252** 499 (1990).
   TASSO Collab., W. Braunschweig *et al.*, Z. Phys. C **47** 499 (1990).
   TOPAZ Collab., R. Enomoto *et al.*, Phys. Lett. B **328** 535 (1994), Phys. Rev. D **50** 1879 (1994), Phys. Lett. B **341** 99 (1994), Phys. Lett. B **341** 238 (1994).
   VENUS Collab., S. Uehara *et al.*, Z. Phys. C **63** 213 (1994).
   AMY Collab., T. Aso *et al.*, Phys. Lett. B **363** 249 (1995), Phys. Lett. B **381** 372 (1996).
   ALEPH Collab., D. Buskulic *et al.*, Phys. Lett. B **355** 595 (1995).
2. L3 Collab., M. Acciarri *et al.*, Phys. Lett. B **453** 83 (1999);
   L3 Collab., M. Acciarri *et al.*, Phys. Lett. B **467** 137 (1999);
   A. Stone, Thesis submitted to Louisiana State University (1999).
3. M. Drees, M. Krämer, J. Zunft and P.M. Zerwas, Phys. Lett. B **306** 371 (1993).
4. L3 Collab., B. Adeva *et al.*, Nucl. Inst. Meth. A **289** 35 (1990);
   M. Acciarri *et al.*, Nucl. Inst. Meth. A **351** 300 (1994);
   M. Chemarin *et al.*, Nucl. Inst. Meth. A **349** 345 (1994);
   I.C. Brock *et al.*, Nucl. Inst. Meth. A **381** 236 (1996);
   A. Adam *et al.*, Nucl. Inst. Meth. A **383** 342 (1996).
5. S. Nussinov, Phys. Rev. Lett. **35** 1672 (1975);
   G.J. Feldman *et al.*, Phys. Rev. Lett. **38** 1313 (1977).
6. L.E. Gordon and J.K. Storrow, Nucl. Phys. B **489** 405 (1997).
7. P. Aurenche, J.-P. Guillet and M. Fontannaz, Z. Phys. C **64** 621 (1994).
8. M. Glück, E. Reya and A. Vogt, Phys. Rev. D **46** 1973 (1992).
9. J. Binnewies, B.A. Kniehl and G. Kramer, Phys. Rev. D **53** 6110 (1996)
   J. Binnewies, B.A. Kniehl and G. Kramer, Phys. Rev. D **58** 014014 (1998).
10. V.M. Budnev, I.F. Ginzburg, G.V. Meledin and V.G. Serbo, Phys. Rep. C **15** 181 (1974).
11. L3 Coll., M. Acciari et al., Phys. Lett. B **408** 450 (1997);
    L3 Coll., paper 519 of the 29th Intern. Conf. on High-Energy Physics, Vancouver, Canada (1998).
12. A. Donnachie and P.V. Landshoff, Phys. Lett. B **296** 227 (1992).
13. Review of Particle Physics, Eur. Phys. J. C **3** 1. (1998).
14. H1 Collab., S. Aid et al., Nucl. Phys. B **472** 3. (1996).
15. ZEUS Collab., G. Breitweg et al., Z. Phys. C **75** 215 (1997).

Charm and bottom production in two-photon collisions with OPAL

Ákos Csilling[1]

University College London
Gower Street, London WC1E 6BT, UK

Abstract. A preliminary update of the previous OPAL measurement of the inclusive production of $D^{*\pm}$ mesons in anti-tagged photon-photon collisions is presented together with the first preliminary OPAL measurement of bottom production in photon-photon collisions.

INTRODUCTION

Heavy quark production in photon-photon collisions reveals the structure of a fundamental gauge boson, the photon. It can be calculated in perturbative QCD and is sensitive to the quark and gluon content of the photon. At LEP2 energies the two main processes, direct and single-resolved, have approximately the same contribution, while the double-resolved contribution is negligible.

Charm production in photon-photon collisions can be cleanly tagged e.g. by the full reconstruction of charged D^* mesons, while bottom production can be measured through a careful study of the various sources of final state leptons in hadronic two-photon events.

CHARM PRODUCTION

We present a preliminary update on the measurement of D^* production in anti-tagged photon-photon collisions [1]. The main change compared to the previous analysis is the inclusion of data collected by OPAL in 1999 at e^+e^- centre-of-mass energies $\sqrt{s_{ee}} = 192 - 202$ GeV, bringing the total integrated luminosity to 428 pb^{-1} in the centre-of-mass energy range $183 < \sqrt{s_{ee}} < 202$ GeV with a luminosity-weighted average energy of $\langle \sqrt{s_{ee}} \rangle = 193$ GeV.

---

[1] The author thanks the UK Particle Physics and Astronomy Research Council for their support. Permanent address: KFKI Research Institute for Particle and Nuclear Physics, Budapest, P.O.Box 49, H-1525 Hungary; supported by the Hungarian Foundation for Scientific Research, OTKA F-023259.

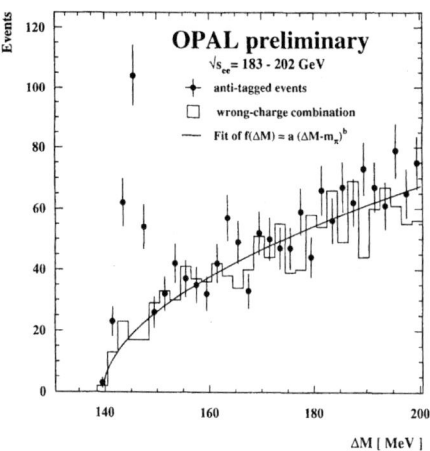

FIGURE 1. Mass difference $\Delta M \equiv M_{\mathrm{D}^*}^{\mathrm{cand}} - M_{\mathrm{D}^0}^{\mathrm{cand}}$ for the full sample. A clear peak is visible around $\Delta M \equiv M_{\mathrm{D}^*} - M_{\mathrm{D}^0} = 145.4$ MeV. The open histogram represents the wrong-charge sample which gives a good description of the combinatorial background. The result of a fit to the background is superimposed.

D^{*+} mesons are reconstructed in their decay to $\mathrm{D}^0\pi^+$, with the D^0 observed in the two decay modes $\mathrm{K}^-\pi^+$ (3-prong) and $\mathrm{K}^-\pi^+\pi^-\pi^+$ (5-prong). The charge-conjugate decay chain is used for D^{*-} mesons. The distribution of the mass difference between the D^* and the D^0, ΔM, for the combined data set is shown in Figure 1. Background is subtracted by fitting the function $f(\Delta M) = a \cdot (\Delta M - m_\pi)^b$, where a and b are free parameters and m_π is the pion mass, to the combined upper sideband of the signal and the wrong-charge distribution in the range 160.5 MeV $< \Delta M < 200.5$ MeV. The result of the fit is in good agreement with the wrong-charge combination in the signal region. In total we observe 164.3 ± 14.9 (stat) D^* mesons.

According to the PYTHIA Monte Carlo program the D^* selection efficiencies for the transverse momentum and rapidity range of $p_{\mathrm{T}}^{\mathrm{D}^*} > 2$ GeV and $|\eta^{\mathrm{D}^*}| < 1.5$ are around $31 - 34\%$ for the 3-prong and $7 - 9\%$ for the 5-prong decay modes in single-resolved and direct events. The approximately 6% lower efficiency compared to the published analysis is due to additional electron rejection cuts which significantly improve the purity of the signal.

Two methods are used to determine the relative fraction of direct and single-resolved events in the data sample. One is based on x_γ^{min}, defined in two-jet events using the cone algorithm as the minimum of $x_\gamma^\pm = \Sigma_{\mathrm{jets}}(E \pm p_z)/\Sigma_{\mathrm{hadrons}}(E \pm p_z)$, while the other uses the scaled D^* transverse momentum $x_{\mathrm{T}}^{\mathrm{D}^*} = 2p_{\mathrm{T}}^{\mathrm{D}^*}/W_{\mathrm{vis}}$, available in all events. The results of both methods are shown in Figure 2. The ratio of direct to single-resolved contributions in the dijet events determined by a fit to the x_γ^{min}

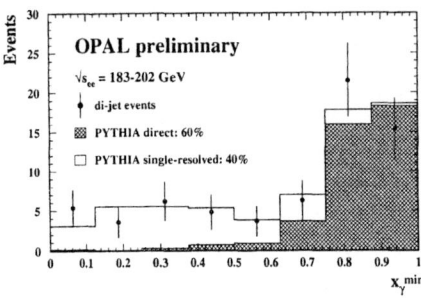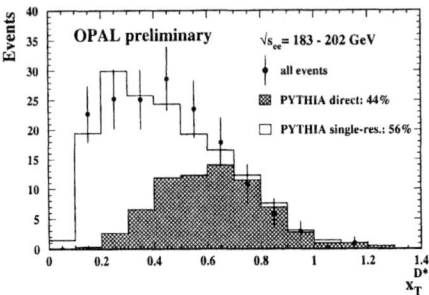

FIGURE 2. $x_\gamma^{\min}$ for dijet events (left) and $x_T^{D^*}$ for all events (right) in the signal region. The open area shows the fitted contribution from the single-resolved process, and the hatched area the one from the direct process.

distribution yields $(60\pm8)\%$ direct and $(40\pm8)\%$ single-resolved, while the fit to the $x_T^{D^*}$ distribution yields $(44\pm6)\%$ direct and $(56\pm6)\%$ single-resolved contributions, where all errors are statistical only. The requirement of two jets introduces a bias towards a larger direct component, therefore the $x_T^{D^*}$ method, where the full sample is used, is preferred.

In Figure 3 the combined differential cross-section $\mathrm{d}\sigma/\mathrm{d}p_T^{D^*}$ is compared to a next-to-leading order (NLO) calculation by Frixione et al. [2], where the matrix elements for massive charm quarks are used, and to an NLO calculation by Binnewies et al. [3], where charm is treated as a massless, active flavour in the photon parton distribution function. Despite the low transverse momenta studied, the massless calculation describes the measurement, while the massive calculation underestimates it in the region of small $p_T^{D^*}$. This is contrary to the expectation that the massive approach should be more appropriate at lower transverse momenta than the massless approach.

The integrated cross-section σ_{meas} of the anti-tagged process $e^+e^- \to e^+e^-D^*X$ in the directly observed kinematical region $2\ \mathrm{GeV} < p_T^{D^*} < 12\ \mathrm{GeV}$ and $|\eta^{D^*}| < 1.5$ is determined to be $\sigma_{\mathrm{meas}}^{D^*} = 30.7 \pm 2.8(\mathrm{stat}) \pm 3.3(\mathrm{syst})$ pb. The extrapolation of this result to the full kinematic range using Monte Carlo simulations gives $\sigma(e^+e^- \to e^+e^-c\bar{c}) = 1033 \pm 102\ (\mathrm{stat}) \pm 111\ (\mathrm{syst}) \pm 246\ (\mathrm{extr})$ pb for the total cross-section of the anti-tagged $e^+e^- \to e^+e^-c\bar{c}$ process at $\langle\sqrt{s_{ee}}\rangle = 193$ GeV. The first error is statistical, the second is systematic and the third is the extrapolation uncertainty, determined by varying the Monte Carlo parameters used in the extrapolation. The direct contribution is determined to be $\sigma(e^+e^- \to e^+e^-c\bar{c})_{\mathrm{dir}} = 362 \pm 59\ (\mathrm{stat}) \pm 42\ (\mathrm{syst}) \pm 68\ (\mathrm{extr})$ pb while the single-resolved contribution is $\sigma(e^+e^- \to e^+e^-c\bar{c})_{\mathrm{res}} = 671 \pm 110\ (\mathrm{stat}) \pm 77\ (\mathrm{syst}) \pm 178\ (\mathrm{extr})$ pb.

Figure 3 shows the total cross-section $\sigma(e^+e^- \to e^+e^-c\bar{c})$ compared to other measurements [4] and to the massive NLO calculation of [5]. Within the large band of uncertainties, the calculation is in good agreement with most measurements.

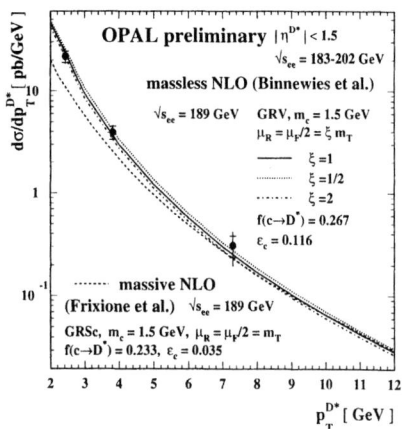
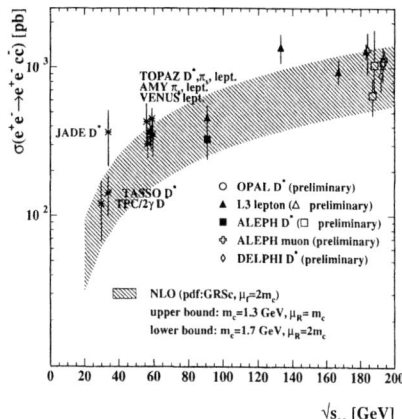

FIGURE 3. The differential D* cross-section, $d\sigma/dp_T^{D^*}$, for the process $e^+e^- \rightarrow e^+e^- D^* X$ in the range $|\eta^{D^*}| < 1.5$, compared to an NLO calculation using the massless approach [3] and another one using the massive approach [2] (left) and the total cross-section for the process $e^+e^- \rightarrow e^+e^- c\bar{c}$ also compared to an NLO calculation [5] and other measurements [4] (right).

BOTTOM PRODUCTION

A new preliminary OPAL measurement of open bottom production in photon-photon collisions is presented using data collected at e^+e^- centre-of-mass energies from $\sqrt{s_{ee}} = 189$ to 202 GeV corresponding to a total integrated luminosity of about 371 pb^{-1}.

The selection of open beauty events proceeds in three steps. First, anti-tagged photon-photon events are selected. Within these events, muon candidates are reconstructed as a signature for semileptonic beauty decays. Finally, jets are reconstructed within the selected events, and the transverse momentum of the muon candidates with respect to the axis of the closest jet is computed.

An artificial neural network trained for muon identification [6] is used to enhance the purity of the muon sample. Events are selected if they contain exactly one muon candidate with a neural net output $\mathcal{N}_\mu$ larger than 0.65. The transverse momentum of the muons, $p_{T,rel}^\mu$, is measured with respect to the axis of the closest jet reconstructed with the KTCLUS jet finding algorithm [7]. At least one jet with a transverse energy E_T^{jet} with respect to the beam axis greater than 3 GeV and at least three tracks is required. After this final cut 444 events remain in this b-enriched sample.

Four contributions to the $p_{T,rel}^\mu$ distribution have to be taken into account: hadronic photon-photon interactions with open b production, open c production, the remaining hadronic photon-photon interactions (labelled 'uds') and processes other than hadronic photon-photon interactions (labelled 'BG').

The number of b events is determined by a fit to the $p_{T,rel}^\mu$ distribution after

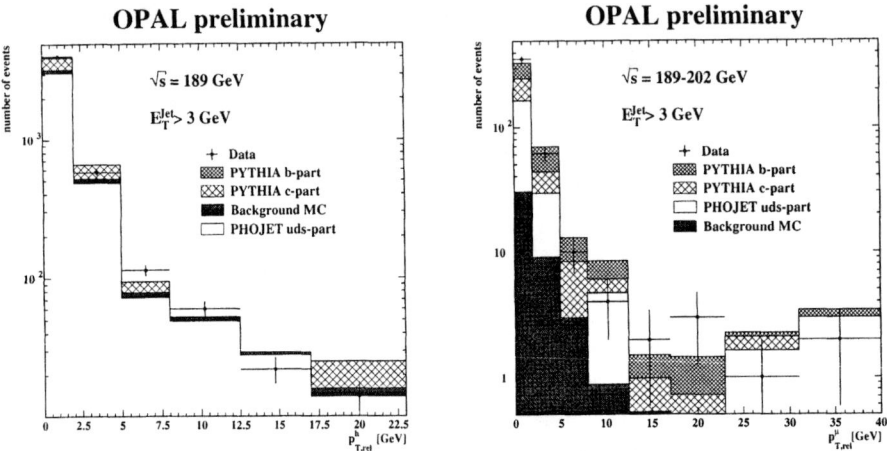

FIGURE 4. Distribution of the transverse momentum with respect to the associated jet for pions and kaons in the b-depleted sample (left) and for muons in the b-enriched sample (right).

all the other contributions have been fixed. The previous OPAL measurement of open charm production [1] is used to fix the absolute contributions of open c production via the direct and single-resolved processes, where the shape is given by the PYTHIA Monte Carlo program. Most of the muons found in the uds background are due to decays of pions and kaons into muons, therefore the rate of pion and kaon production is measured separately. A b-depleted sample of photon-photon events is selected with $\mathcal{N}_\mu < 0.65$ and with pion and kaon identification using $\mathrm{d}E/\mathrm{d}x$ measurements. The transverse momentum relative to the jet axis for the charged hadrons in this b-depleted sample, $p^h_{T,\mathrm{rel}}$, shown in Figure 4 (left), was used to find the scaling factor of 1.35 ± 0.03 for the PHOJET Monte Carlo cross-section.

The predictions for processes other than anti-tagged hadronic photon-photon interactions are taken entirely from Monte Carlo simulations. The fraction f_{BG} of background events is about 10%. The main contributions to this background are e^+e^- annihilation events into hadrons (54%), deep inelastic electron-photon scattering events (29%), production of four fermions from other reactions than photon-photon scattering processes (9%) and τ pairs produced in photon-photon interactions (7%).

The final result of the one-parameter fit to the b-enriched sample, shown in Figure 4 (right), yields a beauty fraction $f_{\mathrm{b}} = (27.2 \pm 4.8 \,(\mathrm{stat}))\%$ with $\chi^2 = 5.8$ (ndf=7), corresponding to 121 ± 21 events. The fraction of the other processes are approximately $f_{\mathrm{C}} = 24\%$, $f_{\mathrm{uds}} = 37\%$ and $f_{\mathrm{BG}} = 10\%$.

The cross-section $\sigma(e^+e^- \to e^+e^-b\bar{b})$ is determined from the measured number of events using the branching ratios of b hadrons into muons taken from [8]. The total open b cross-section is determined to be $\sigma(e^+e^- \to e^+e^-b\bar{b}) =$

OPAL preliminary

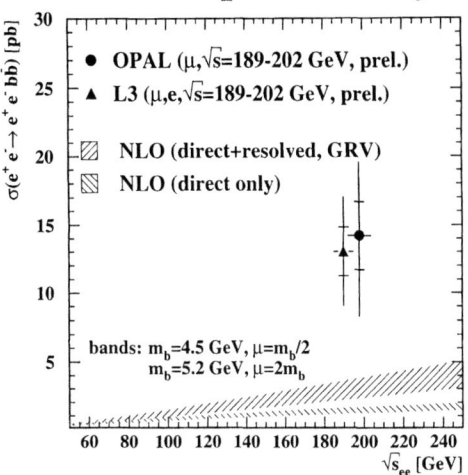

FIGURE 5. The total beauty production cross-section $\sigma(e^+e^- \rightarrow e^+e^-b\bar{b})$ measured in the range $\sqrt{s_{ee}} = 189 - 202$ GeV compared to the preliminary L3 result and the prediction of an NLO calculation.

14.2 ± 2.5 (stat) $^{-4.8}_{+5.3}$ (sys) pb. The precision of the measurement is currently limited by the systematic errors associated with the background from charm production, but the improved charm measurement presented above will help to reduce this uncertainty.

Figure 5 shows the measured open b cross-section compared to the preliminary L3 result performed in the same $\sqrt{s_{ee}}$ range using semileptonic decays of b hadrons into muons and electrons [9], and an NLO calculation that uses matrix elements for massive b quarks [5], with the direct contribution shown separately. The prediction of the NLO calculation for the total cross-section at $\sqrt{s_{ee}} = 200$ GeV is 3.88 pb and 2.34 pb for a b quark mass of 4.5 GeV and 5.2 GeV, respectively, significantly lower than both measurements. A recent measurement of the open beauty cross-section in photon-proton collisions is also about 2.5 standard deviations higher than expected from an NLO calculation [10].

CONCLUSION

The inclusive production of $D^{*\pm}$ mesons has been measured in anti-tagged photon-photon collisions using the OPAL detector at LEP. The contribution of the direct process in the kinematical region $p_T^{D^*} > 2$ GeV and $|\eta^{D^*}| < 1.5$ is determined from the $x_T^{D^*}$ distribution to be $r_{\text{dir}} = (44 \pm 6)\%$.

The measured differential cross-section as a function of the D^* transverse momentum and pseudorapidity is compared to NLO calculations using the massless and the massive approaches, and despite the low values of $p_T^{D^*}$ studied the massless calculation is in good agreement with the data, while the massive calculation underestimates them for lower values of $p_T^{D^*}$.

The cross-section of the anti-tagged $e^+e^- \rightarrow e^+e^- D^* X$ process is measured in the restricted kinematical range of $p_T^{D^*} > 2$ GeV and $|\eta^{D^*}| < 1.5$ to be $\sigma_{\text{meas}}^{D^*} = 30.7 \pm 2.8(\text{stat}) \pm 3.3(\text{syst})$ pb. The extrapolation of this result to the total charm cross-section introduces large uncertainties both on the theoretical and experimental side, therefore at present a comparison in the restricted kinematic range is preferred.

The open b cross-section in photon-photon events has been measured using the semi-leptonic decays of b hadrons into muons. The spectrum of the transverse momentum of the muons with respect to the closest jet axis is fitted after all background contributions have been fixed independently, leading to a measured total cross-section of $\sigma(e^+e^- \rightarrow e^+e^- b\bar{b}) = 14.2 \pm 2.5$ (stat) $^{-4.8}_{+5.3}$ (sys) pb, in agreement with the preliminary L3 measurement. However, the NLO QCD calculation underestimates the measured total cross-section by about 2 standard deviations.

REFERENCES

1. OPAL Collaboration, G. Abbiendi et al., *Inclusive Production of $D^{*\pm}$ Mesons in Photon-Photon Collisions at $\sqrt{s}_{ee} = 183$ and 189 GeV and a First Measurement of $F_{2,c}^{\gamma}$*, CERN-EP/99-157, accepted by Eur. Phys. J. C (2000).
2. S. Frixione, M. Krämer and E. Laenen, Nucl. Phys. B571 (2000) 169.
3. J. Binnewies, B.A. Kniehl and G. Kramer, Phys. Rev. D53 (1996) 6110;
   J. Binnewies, B.A. Kniehl and G. Kramer, Phys. Rev. D58 (1998) 014014.
4. G. Altarelli, T. Sjöstrand and F. Zwirner, *Physics at LEP2*, CERN 96-01 (1996);
   ALEPH Collaboration, D. Buskulic et al., Phys. Lett. B355 (1995) 595;
   U. Sieler, ALEPH Collaboration, *Charm production in two-photon collisions measured by ALEPH at LEP II*, in these proceedings;
   L3 Collaboration, M. Acciarri et al., Phys. Lett. B453 (1999) 83;
   S. Saremi, L3 Collaboration, *Charm and bottom production in two photon collisions at LEP* , in these proceedings;
   A. Sokolov, DELPHI Collaboration, *Inclusive D-meson and Λ_c production in two photon collisions at LEP*, in these proceedings.
5. M. Drees, M. Krämer, J. Zunft and P.M. Zerwas, Phys. Lett. B306 (1993) 371.
6. OPAL Collaboration, G. Abbiendi et al., Eur. Phys. J. C16 (2000) 41.
7. S. Catani et al., Nucl. Phys. B406 (1993) 187;
   S.D. Ellis and D.E. Soper, Phys. Rev. D48 (1993) 3160;
   ZEUS Collaboration, J. Breitweg et al., Eur. Phys. J. C1 (1998) 109.
8. D.E. Groom et al., Review of Particle Physics, Eur. Phys. J. C15 (2000) 1.
9. L3 Collaboration, *Beauty Production in $\gamma\gamma$ Collisions at LEP*, L3 note 2565, submitted to XXXth Int. Conf. on High Energy Physics, Osaka, Japan, 2000.
10. H1 Collaboration, C. Adloff et al., Phys. Lett. B467 (1999) 156.

Open Charm and Beauty Production at HERA

Karin Daum[*][1]

Rechenzentrum, Universität Wuppertal
Gaußstraße 20, D-42097 Wuppertal, Germany

Abstract.
Measurements on open charm and beauty production in *ep* collisions at a center-of-mass energy of $\sqrt{s} = 300$ GeV performed by the H1 experiment at HERA are presented. Final states containing charm are identified by the reconstruction of $D^{*\pm}$ meson. Events containing muons and at least two jets were used to select data samples enriched with beauty. The results cover the region of negative four-momentum transfer squared, Q^2, from photoproduction ($Q^2 \approx 0$) to deep inelastic scattering. The experimental results are compared with QCD predictions.

INTRODUCTION

The study of heavy flavour production in lepton-proton scattering provides an important tool for testing the standard model of strong interactions. At the *ep* collider HERA, which was operated at a center-of-mass energy of $\sqrt{s} = 300$ GeV, heavy quarks are almost exclusively produced by the *photon gluon fusion (PGF)* process, $\gamma g \rightarrow Q\overline{Q}$ ($Q = c, b$), where a real or virtual photon emitted by the electron[2] interacts with a gluon in the proton producing a heavy quark pair $Q\overline{Q}$.

The dominant contribution to heavy flavour production is due to the exchange of an almost real photon *(photoproduction)*, where the negative four-momentum transfer square carried by the photon is ($Q^2 \approx 0$). The heavy quarks hadronize and may be detected as *"open charm (beauty)"*, i.e. charmed/beauty hadrons visible in the final state.

The kinematics of the *ep* interaction is described by three independent variables, the center-of-mass energy $\sqrt{s}$, the four-momentum transfer squared of the photon $q^2 = -Q^2$ and either one of the scaling variables $y = (q \cdot P)/(l \cdot P)$, the inelasticity of the *ep* interaction, or Bjorken-*x* $x = Q^2/(2P \cdot l)$. Here P and l denote the four-momentum of the proton and the electron, respectively. The γp center-of-mass energy squared is given by $W^2_{\gamma p} = W^2 \approx y \cdot s - Q^2$.

---

[1] permanent at DESY, Notkestraße 85, D-22607 Hamburg, Germany, email: daum@mail.desy.de
[2] Hereafter, a reference to electrons implies a reference to either electrons or positrons.

CP571, *PHOTON 2000*, edited by A. J. Finch
© 2001 American Institute of Physics 0-7354-0010-5/01/$18.00

Open heavy flavour production at HERA is dominated by charm production. The increasing statistical precission of these data allows detailed test of perturbative QCD ($pQCD$) because of the high luminosity delivered by HERA during the last years. Due to the large mass m_b and the charge of beauty quarks the cross section for $\sigma(ep \rightarrow eb\overline{b}X)$ is expected to be roughly two orders of magnitude smaller than $\sigma(ep \rightarrow ec\overline{c}X)$. However, the study of beauty production in lepton nucleon scattering is of special interest because the calculations in pQCD are expected to be more reliable due to the large scale, i.e. m_b, involved in this process.

The results presented here [1–3] are based on roughly 19 pb^{-1} of data recorded by H1 [4] in 1996 and 1997 at HERA, when positrons with an energy of 27.5 GeV were collided with protons of 820 GeV.

OPEN HEAVY FLAVOUR PRODUCTION

The description of open heavy flavour production is based on pQCD. In leading order (LO) the *direct* photon gluon fusion process ($\gamma g \rightarrow Q\overline{Q}$) is the dominant contribution. In photoproduction (γp) sizeable contributions from *resolved* photon interactions, i.e. $gg \rightarrow Q\overline{Q}$, are expected due to the partonic structure of the photon. In higer orders, however, the resolved photon processes are part of the higher order contributions and the distinction between direct and resolved processes becomes impossible.

NLO Calculation in the DGLAP Scheme

Several schemes are used to perform NLO calculations. All approaches assume the scale to be hard enough to apply pQCD and to guarantee the validity of the factorization theorem.

Here, the massive approach is adopted which is a fixed order calculation (in α_s) with massive quarks, i.e. $m_Q \neq 0$, assuming three active flavours in the proton. The densities of the three light quarks and the gluon in the proton and the photon are obtained by the DGLAP evolution. Heavy quarks are produced perturbatively [5,6] via boson gluon fusion. These calculations are reliable near threshold, where for the renormalization scale μ the relation $\mu^2 \approx m_Q^2$ is valid. However, they break down for $\mu^2 \gg m_Q^2$ due to large logarithms $\ln(\mu^2/m_Q^2)$. Based on the NLO calculations of order α_s^2 in the coefficient functions [7,8] the Monte Carlo integration program HVQDIS [9] and the FMNR code [10] provide the four-momenta of the outgoing partons in the DIS and the γp regime, respectively. Thus the calculation of visible differential inclusive heavy meson production cross sections becomes possible.

The heavy flavoured quarks are fragmented to heavy flavoured mesons according to the longitudinal Peterson fragmentation function [11], which is controlled by a single parameter ϵ. In the analysis of open charm production the D^{*+} meson is given a transverse momentum p_t with respect to the charm quark in addition, according to the function $\exp(-\alpha p_t^2)$, to account for the experimentally observed p_t smearing of hadrons with respect to the quark direction. The parameter α is chosen such that an average transverse momentum $\langle p_t \rangle \approx 350$ MeV is obtained as observed in e^+e^- data. In the analysis open beauty production both, the longitudinal fragmentation model and a more complex modeling of the fragmentation process by picking up light quark pairs from the vacuum have been used. Finally, the weak decay of beauty hadrons into muons is included.

CCFM Evolution

The results on open charm production in DIS will also be compared with predictions based on the CCFM evolution equation [12]. This evolution scheme may be most appropriate to describe the parton evolution at small x. In the parton cascade, gluons are emitted in an angular ordered region to account for coherence effects. Due to this ordering, the unintegrated gluon distribution in CCFM depends on the maximum allowed angle in addition to the momentum fraction x and the transverse momentum of the propagator gluon. The cross section is then calculated according to the k_t-factorization theorem by convoluting the unintegrated gluon density with the off- shell boson gluon fusion matrix element with massive quarks for the hard scattering process, well suited for heavy flavour production.

Based on a recent solution of the CCFM equation [13] a full hadron level Monte Carlo generator CASCADE has been developed [14] in which the full generation of charm events, including the initial state gluon radiation according to CCFM and Lund string fragmentation (JETSET) is also possible. The fragmentation of charmed quarks to D^{*+} mesons is performed using the Peterson fragmentation function$\epsilon = 0.078$. A refined version of CASCADE with the unintegrated gluon density [15] as extracted from the H1 F_2 data [16] has been used.

OPEN CHARM PRODUCTION IN DIS

Open charm production is tagged by the observation of $D^{*\pm}$ mesons in the final state. $D^{*\pm}$ mesons are identified by the decay chain $D^{*\pm} \rightarrow D^0 \pi^{\pm}$, $D^0 \rightarrow K^{\mp} \pi^{\pm}$ in the visible range of the transverse momentum $p_\perp(D^*) > 1.5$ GeV and pseudorapidity $|\eta(D^*)| < 1.5$ in the laboratory frame, where the pseudorapidity is defined as $\eta(D^{*\pm}) = -\ln\tan(\Theta_{D^{*\pm}}/2)$. For the DIS selection the event kinematics is restricted to $0.05 < y < 0.7$ and 1 GeV$^2 < Q^2 < 100$ GeV^2.

In figure 1 the inclusive single differential $D^{*\pm}$ cross sections in the visible region are shown as a function of the kinematic quantities W, x_{Bj} and Q^2 and as a function of the D^{*+} observables $p_t(D^*)$, $\eta(D^*)$ and $z_D(D^*) = P \cdot p_{D^*}/P \cdot q$, where P, q and p_{D^*} denote the four-momenta of the incoming proton, the exchanged photon and the observed $D^{*\pm}$. Also shown in Fig.1 are the expectations from the NLO calculations of the HVQDIS program using the GRV 98 HO parton density parameterization [17]. The dark shaded band indicates the uncertainties in this calculation by varying

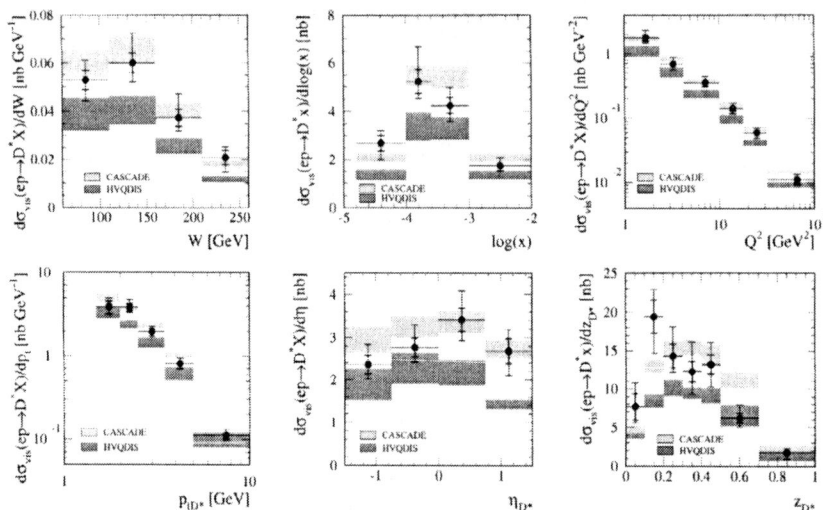

FIGURE 1. Single differential inclusive cross section $\sigma(ep \to eD^{*+}X)$ versus W, x_{Bj}, Q^2 and $p_{T\,D^*}$, η_{D^*}, z_{D^*}. The NLO DGLAP expectation (HVQDIS) is indicated by the lower shaded band (upper bounds: $m_c = 1.3$ GeV, $\epsilon = 0.035$, lower bounds: $m_c = 1.5$ GeV, $\epsilon = 0.10$). The upper shaded band is the CCFM expectation (CASCADE) (1.3 GeV $< m_c <$ 1.5 GeV, $\epsilon = 0.078$).

m_c and ϵ from $m_c = 1.3$ GeV and $\epsilon = 0.035$ (upper limit) to $m_c = 1.5$ GeV and $\epsilon = 0.10$ (lower limit). The renormalization scale and the factorization scale are set to $\mu_r^2 = \mu_f^2 = Q^2 + 4m_c^2$. Although the total visible cross section prediction of HVQDIS is smaller than experimentally observed, the agreement with the data in the shapes of the different single differential cross sections is reasonable. A difference in shape is observed only for the case of the pseudorapidity η. In the forward direction, the observed D^{*+} meson production cross section is considerably larger than predicted by this calculation. This excess is also manifested in the larger cross section observed in the data at small z_D.

Figure 1 also includes the predictions of the CCFM calculations using the CAS-CADE program (light shaded band) with m_c varying between 1.3 GeV and 1.5 GeV and using $\epsilon = 0.078$. The expectations from the CASCADE program are found to agree better with data in general and especially in the forward η region, where the HVQDIS program fails to describe the data. It is interesting to note that the CCFM calculation, which starts from completely different principles and aims specifically to describe low x phenomena, is able to describe open charm production at HERA better than the DGLAP based NLO calculations with the chosen settings.

The charm contribution $F_2^c(x, Q^2)$ to the proton structure function is obtained by using the expression for the one photon exchange cross section for charm production

$$\frac{d^2\sigma^c}{dxdQ^2} = \frac{2\pi\alpha^2}{Q^4x}\left(1 + (1-y)^2\right)F_2^c(x, Q^2). \tag{1}$$

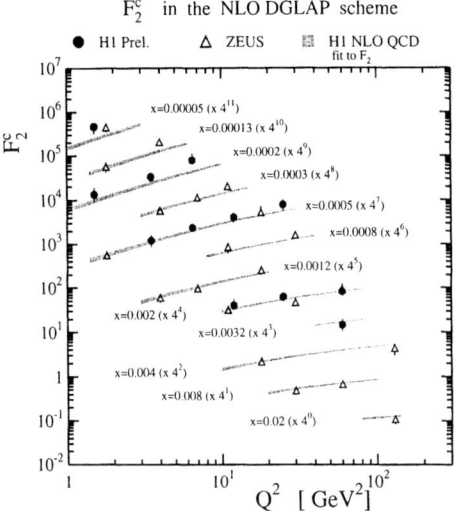

FIGURE 2. The charm contribution to the proton structure function F_2^c as derived from the inclusive D^{*+} meson analysis as a function of Q^2 for different bins in x. The error bars refer to the statistical (inner) and the total error (outer), respectively. The shaded bands represent the predictions of the NLO DGLAP evolution based on the parton densities in the proton obtained by the NLO fit to the inclusive F_2.

The range of y of the measurements is restricted so that contribution from the second structure function F_1 is everywhere negligible.

The visible inclusive D^{*+} cross sections $\sigma_{\text{vis}}^{\text{exp}}(x, Q^2)$ in bins of x and Q^2 are converted to a bin center corrected $F_2^{c\ \text{exp}}(\langle x \rangle, \langle Q^2 \rangle)$ by the relation:

$$F_2^{c\ \text{exp}}(\langle x \rangle, \langle Q^2 \rangle) = \frac{\sigma_{\text{vis}}^{\text{exp}}(x, Q^2)}{\sigma_{\text{vis}}^{\text{theo}}(x, Q^2)} \cdot F_2^{c\ \text{theo}}(\langle x \rangle, \langle Q^2 \rangle) , \qquad (2)$$

where $\sigma_{\text{vis}}^{\text{theo}}$ and $F_2^{c\ \text{theo}}$ are the theoretical predictions of HVQDIS.

In Fig. 2 the charm contribution to the proton structure function F_2^c is shown as a function of Q^2 for different bins in x in the NLO DGLAP picture together with the results from ZEUS [19]. Also shown is the expectation using the result on the gluon density from the H1 NLO DGLAP fit to the inclusive F_2 measurement [20]. The band indicates the uncertainty in this extraction introduced by varying the charm quark mass between 1.3 GeV and 1.5 GeV. Other sources of uncertainties in the determination of the gluon density in the proton are not yet taken into account. The charm data show large scaling violations visible in the variation of F_2^c with Q^2 for constant x_{Bj}. The amount of scaling violation present in the charm data is well predicted by the NLO DGLAP evolution on the gluon density extracted from the inclusive F_2 measurement.

OPEN BEAUTY IN PHOTOPRODUCTION

Two independent analyses of open beauty in photoproduction [2,3] have been performed by H1. Based on data from different years they make use of different features of beauty hadron decays for tagging. Both analyses are based on the semileptonic decay of beauty hadrons resulting in muons identified in the final state. The muon has to be observed in the central region of the detector, i.e. $35^\circ < \theta^\mu < 130^\circ$, and its transverse momentum $p_\perp^\mu$ has to exceed 2 GeV. In addition, at least two high $E_\perp^{jet}$ jets have to be found in the events. At least one of the jets has to contain a muon candidate fulfilling the requirements above. The γp-regime is defined by requiring that no electron candidate with $\theta_e < 177.8^\circ$ is observed ($Q^2 < 1$ GeV^2).

Large Mass Tagging

The analysis of the 1996 data [2] makes use of the fact that b quarks are much heavier than c quarks for a statistical separation of b and c events. In contrast to charmed hadrons the decay products of beauty hadrons are expected to show large transverse momenta $p_{T,rel}$ with respect to the direction flight of the decaying beauty hadron. Since this analysis does not attempt to reconstruct the beauty hadron directly, its direction is approximated by the thrust axis of the jet containing the muon.

Apart from semileptonic charm and beauty decays other hadrons do contribute to the muon sample because of misidentification. The misidentification probabilities for pions, kaons and protons have been parameterized using Monte Carlo simulations. They are verified by studying K_s^0 and ϕ decays in data, which present unambiguous sources of pions and kaons, respectively.

In figure 3 the measured $p_{T,rel}^\mu$ distributions is shown together with the fitted contributions of beauty and charm, using the shapes from the AROMA Monte Carlo simulations, and the hadronic background determined from data. The data exhibit a significant tail to large $p_{T,rel}^\mu$ values. As expected, the $p_{T,rel}^\mu$ distribution in b events is significantly harder than in c events, while the distribution of the hadronic background is similar in shape to the expectation from the semileptonic charm decay.

The visible electroproduction cross section of b quarks, determined from the number of muons N_b^μ attributed to b quark decays, is measured to be:

$$\sigma_{vis}(ep \to eb\bar{b}X \to \mu X) = 176 \pm 16(stat.)^{+26}_{-17}(syst.)\text{pb}$$

in the kinematic range $Q^2 < 1$ GeV^2, $0.1 < y < 0.8$, $p_\perp^\mu > 2$ GeV and $35^\circ < \theta^\mu < 130^\circ$.

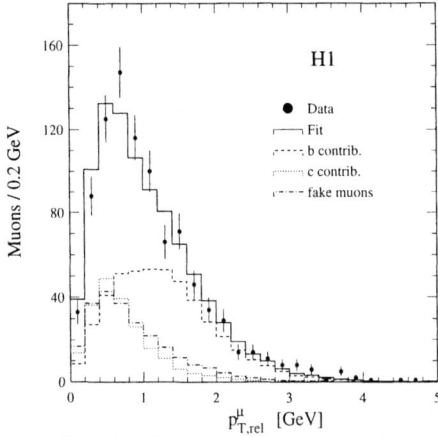

FIGURE 3. The measured $p^{\mu}_{T,rel}$ distribution in the data and the fitted sum (solid line) of the contribution of beauty (dashed line), charm (dotted line) and the fixed fake muon background(dashed dotted line).

Lifetime Tagging

The second analysis [3] makes use of the longevity of b mesons. This only became possible since 1997 when the central silicon tracker (CST) [4] of H1 was fully commissioned. For each muon candidate the impact parameter δ is calculated in the plane perpendicular to the beam axis. Its magnitude is given by the distance of closest approach of the track. Its sign is positive if the intercept of the track with the jet axis is downstream of the primary vertex, and negative otherwise. Decays of long-living particles are expected to have large positive impact parameters, whereas particles from the primary vertex should yield a symmetric distribution around zero due to resolution effects.

Figure 4a shows the observed impact parameter distribution for the data together with histograms indicating the contribution from b production and from the backgrounds. The decomposition is obtained from a likelihood fit. The fit uses the shapes of the δ distribution of b and c events from the AROMA Monte Carlo simulation and the fake muon events from real data. The observed δ distribution clearly shows a tail to large positive values, which indicates a sizeable contribution of b events in the data.

The 1997 data also enable the measurement of both, δ and $p^{\mu}_{T,rel}$ simultaneously for the events. By imposing a cut on one of these two variables it is possible to further enrich the b component while studying the distribution of the other quantity. Figures 4b and 4c show the observed δ distribution when cutting in $p^{\mu}_{T,rel}$ and vice versa, together with the predictions obtained by the fit to the distribution of Figure 4a. A good description of the data is observed in these restricted areas of phase space where the b contribution clearly dominates. Therefore, a fit in the δ-$p^{\mu}_{T,rel}$

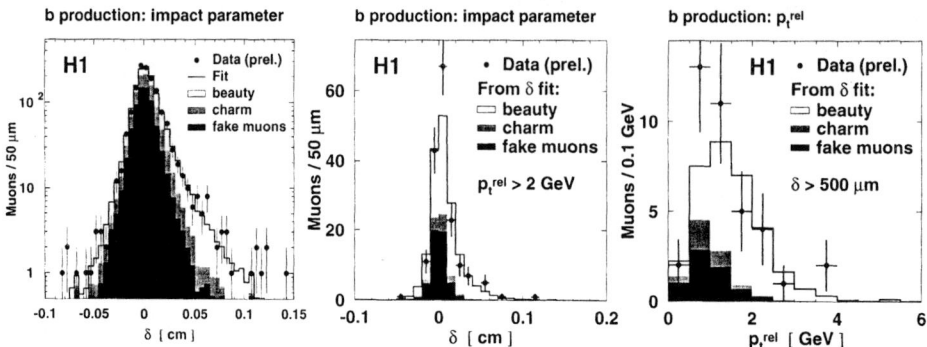

FIGURE 4. Results from the impact parameter analysis: the measured δ distribution (a) for all selected muon events and its decomposition from the likelihood fit, and (b) for those events having a muon with $p^{\mu}_{T,rel} > 2$ GeV, and the observed $p^{\mu}_{T,rel}$ distribution for events having muons with $\delta > 500\mu$m.

plane is performed, which leads to a cross section of

$$\sigma_{vis}(ep \to eb\bar{b}X \to \mu X) = 160 \pm 16(stat.) \pm 29(syst.) \text{ pb}$$

in the kinematic range $Q^2 < 1$ GeV^2, $0.1 < y < 0.8$, $p^{\mu}_{\perp} > 2$ GeV and $35^o < \theta^{\mu} < 130^o$, agreeing well with the measurement based on the 1996 data. Taking into account correlated systematic uncertainties the combination of both measurements yields to

$$\sigma_{vis}(ep \to eb\bar{b}X \to \mu X) = 170 \pm 22 \text{ pb}.$$

NLO pQCD predicts (104 ± 17)pb [2] , using the FMNR program [10] together with a Peterson type fragmentation, where the error reflects the uncertainties due to variations in the renormalization or factorization scale, and due to fragmentation. This value is smaller than the H1 result. Such a discrepancy is now established in both ep and $p\bar{p}$ interactions [21].

CONCLUSIONS

New results on differential inclusive D^{*+} meson production cross sections in deep inelastic ep scattering from the 1996 and 1997 H1 data have been presented. These measurements favor the predictions of the CCFM based CASCADE program over those of the NLO DGLAP based HVQDIS program in the parameter space explored in the current analysis. By extrapolating the visible D^{*+} meson production cross section to full phase space in $p_t(D^{*+})$ and $\eta(D^{*+})$, the charm contribution to the proton structure F_2^c has been extracted. The data show large scaling violations.

Two independent analyses using data from different years and different tagging techniques have shown that the open beauty photoproduction cross section in ep scattering at HERA is above NLO pQCD expectation.

REFERENCES

1. C. Adloff et al.(H1. Collab.), Contributed paper to the 30th International Conference on High-Energy Physics ICHEP2000, Osaka, Japan, July 2000, abstract, 984.
2. C. Adloff et al.(H1. Collab.), Phys. Lett.B **467**, 156 (1999).
3. C. Adloff et al.(H1. Collab.), Contributed paper to the 30th International Conference on High-Energy Physics ICHEP2000, Osaka, Japan, July 2000, abstract, 979,982.
4. I. Abt et al.(H1. Collab.), Nucl. Inst. Meth. **A 386**, 310 and 348 (1997).
5. E. Laenen et al., Nucl. Phys.B **392** 162,229 (1993), Nucl. Phys.B **291** 325 (1992); S. Riemersma, J. Smith, and W.L. van Neerven: Phys. Lett. **B 347** 142 (1995).
6. S. Frixione et al., Phys. Lett. **B 348** 633 (1995).
7. B.W. Harris and J. Smith, Nucl. Phys. **B 452** 109 (1995), Phys. Lett. **B 353** 535 (1995).
8. R.K. Ellis and P. Nason, Nucl. Phys. **B 312** 551 (1989); P. Nason, S. Dawson and R.K. Ellis, Nucl. Phys. **B 303** 607 (1988), Nucl. Phys. **B 327** 49 (1989).
9. B.W. Harris and J. Smith, Phys. Rev. **D 57** 2806 (1998).
10. M.L. Mangano, P. Nason and G. Ridolfi, Nucl. Phys. **B 373** 295 (1992); S. Frixione et al. Nucl. Phys. **B 412** 225 (1994).
11. C. Peterson et al.,Phys. Rev. **D 27** 105 (1983).
12. M. Ciafaloni,Nucl. Phys. **B 296** 49 (1988); S. Catani, F. Fiorani and G. Marchesini,Phys. Lett. **B 234** 339 (1990), Nucl. Phys. **B 336** 18 (1990); G. Marchesini,Nucl. Phys. **B 445** 45 (1995).
13. H. Jung, Proc. 7th DIS Workshop, Nucl. Phys. **B 79** 429 (1999) (Proc. Suppl.), hep-ph/9905554.
14. H. Jung, Proc. Workshop on Monte Carlo Generators for HERA Physics, 1999, DESY-PROC-1999-02, p.75, hep-ph/9908497; H. Jung and G. Salam, to be published.
15. S.P. Baranov, H. Jung and N.P. Zotov, Proc. Workshop on Monte Carlo Generators for HERA Physics, 1999, DESY-PROC-1999-02, p.484.
16. S. Aid et al. (H1 Collab.), Nucl. Phys. **B 470** 3 (1996).
17. M. Glück, E. Reya and A. Vogt, Eur. Phys. J. **C 5** 461 (1998).
18. C. Adloff et al. (H1 Collab.), Z. Phys.C **72** 593 (1996).
19. J. Breitweg et al. (ZEUS Collab.), Phys. Lett.B **407** 402 (1997) 402, Euro. Phys. J.C **12** 1 (2000).
20. C. Adloff et al. (H1 Collab.), Contributed paper to the 30th International Conference on High-Energy Physics ICHEP2000, Osaka, Japan, July 2000 abstract 945
21. F. Abe et al. (CDF Collab.), Phys. Rev. Lett. **71** 2396 (1993), Phys. Rev. **D 53** 1051 (1996); S. Abachi et al. (D0 Collab.) Phys. Rev. Lett. **74** 3548 (1995), Phys. Lett. **B 370** 239 (1996).

Recent Results on Charm Photoproduction

Don Hochman*

On behalf of the ZEUS Collaboration

*Weizmann Institute of Science,Rehovot Israel 76100

Abstract. Photoproduction of $D_s^{\pm}$ mesons has been measured in the ZEUS detector at HERA and compared with predictions of NLO pQCD calculations. The ratio of $D_s^{*\pm}$ to $D^{*\pm}$ cross sections has been compared to results from e^+e^- experiments. Orbitally excited P-wave charm mesons have been observed in the $D^{*\pm}\pi^{\mp}$ final state. The fraction of $D^{*\pm}$'s originating from these mesons has been calculated and compared with that from e^+e^- interactions. No evidence for radially excited mesons decaying to $D^{*\pm}\pi^+\pi^-$ was found. The inelastic production of J/ψ mesons has been measured and compared to LO and NLO pQCD predictions.

INTRODUCTION

Charm photoproduction measurements have been performed at the HERA ep collider in the ZEUS detector from data taken during 1995-2000. Electrons or positrons with energy $E_e = 27.5\,GeV$ collided with protons of energy $E_p = 820\,GeV$ (1995-1997) or $E_p = 920\,GeV$ (1998-2000). The ZEUS detector description can be found elsewhere [1].

The decay chain $D_s^{\pm} \to \phi\pi^{\pm} \to K^+K^-\pi^{\pm}$ (38pb^{-1} integrated luminosity) was studied [2] as a continuation of a previous analysis of charm photoproduction [3]. The study of $D_s^{\pm}$ photoproduction provides another test of next-to-leading order (NLO) perturbative quantum chromodynamics (pQCD) calculations.

Orbitally excited P-wave D mesons can decay to a D^* by pion emission. Two of these states ($D_1(2420)$ and $D_2^*(2460)$) have been found to decay into narrow states [4] with properties predicted by Heavy Quark Effective Theory (HQET) [5] and a third broad state has been seen by the CLEO collaboration [6]. A radial excitation of the $D^{*\pm}$ with a mass of about 2.6 GeV decaying into $D^{*\pm}\pi^+\pi^-$ has been reported by DELPHI [7] but not seen by OPAL and CLEO [8,9].

Inelastic J/ψ photoproduction proceeds via direct (resolved) processes, where the virtual photon (parton from the photon) interacts with a parton from the incoming proton. In the dominant process, boson gluon fusion (BGF), the latter parton is a gluon. Photon diffraction to J/ψ also contributes. The inelasticity variable,

$z = \frac{P \cdot p_\psi}{P \cdot q}$, can be used to distinguish these processes. Here P, p_ψ and q are the four-momenta of the incoming proton, J/ψ and exchanged photon, respectively. From previous ZEUS data [10] the diffractive process dominates at $z > 0.9$, the direct photon process dominates at $0.4 < z < 0.9$. The resolved photon contribution is expected to dominate at $z \lesssim 0.2$ [11].

Color singlet and color octet models have been used to calculate the above non-diffractive production processes in pQCD. For the former, the charm-anticharm pair ($c\bar{c}$) from the hard process is identified with the physical J/ψ state. In this model in leading order (LO) only the BGF diagram contributes to the direct channel. In the color octet model the $c\bar{c}$ pair from the hard process emits one or more soft gluons to evolve into the physical J/ψ state. The free parameters of the model can be extracted from J/ψ cross-section measurements and used in other inelastic J/ψ production experiments.

$D_S^\pm$ PHOTOPRODUCTION

$D_s^\pm$ production was studied for: $Q^2 < 1.0\,GeV^2, 130 < W_{\gamma p} < 280\,GeV, 3 < p_\perp^{D_s} < 12\,GeV, |\eta^{D_s}| < 1.5$, where Q^2 is the photon virtuality, $W_{\gamma p}$ is the virtual photon proton center of mass energy, $p_\perp^{D_s}$ is the transverse momentum of the $D_s^\pm$ and η^{D_s} is the pseudorapidity of the $D_s^\pm$. The effective mass of two opposite charge track combinations, assumed to be kaons, was calculated and plotted in Fig. 1a. A clear enhancement at the ϕ mass is seen. The effective mass of the combinations in this enhancement region and another track assumed to be a pion was then obtained. The peak in the $D_s^\pm$ mass region contained 339 ± 48 $D_s^\pm$ mesons (Fig. 1b), corresponding to a cross section of $\sigma_{ep \to D_s X} = 3.79 \pm 0.59(stat)^{+0.26}_{-0.46}(syst) \pm 0.94(br)\,nb$.

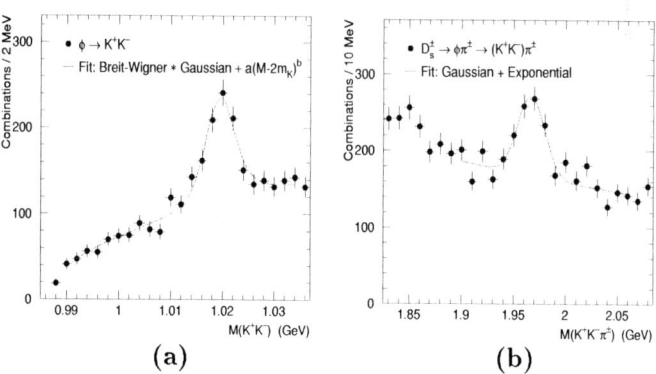

FIGURE 1. (a) $M(K^+K^-)$ distribution for events inside the $D_s^\pm$ mass range, $(1.94 < M(K^+K^-\pi^\pm) < 2.00\,GeV)$. The solid curve is a fit to a Breit-Wigner convoluted with a Gaussian-shaped resonance and a background parameterization, $a[M(K^+K^-) - 2m_K]^b$. (b) $M(K^+K^-\pi^\pm)$ distribution for events in the ϕ mass range, $(1.0115 < M(K^+K^-) < 1.0275\,GeV)$. The solid curve is a fit to a Gaussian plus an exponential background.

Distributions in $p_\perp^{D_s}$ and η^{D_s} were compared with those for $D^{*\pm}$ production [3] and with a fixed order NLO calculation [12] in which charm was produced by the BGF process. The signal is above the prediction (Fig. 2), particularly for η along the proton beam direction, as was the case for $D^{*\pm}$ production.

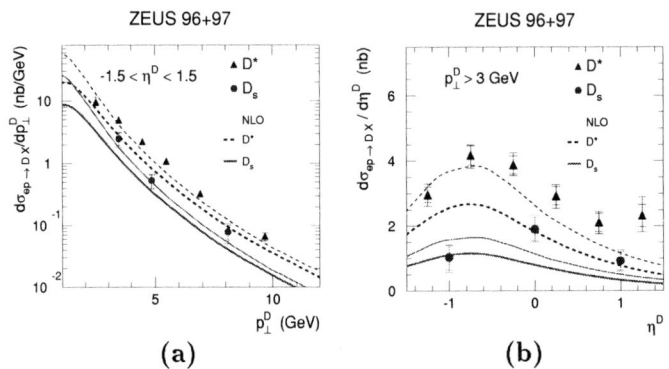

FIGURE 2. Differential cross sections for the photoproduction reaction $ep \to DX$: (a) $d\sigma/dp_\perp^D$ and (b) $d\sigma/d\eta^D$, where D stands for D^* or D_s. Inner (outer) error bars show statistical (statistical and systematic added in quadrature) errors. The D_s (dots) and D^* (triangles) data are compared with NLO predictions for D_s (full curves) and D^* (dashed curves) with two parameter settings: $m_c = 1.5\,GeV$, $\mu_R = m_\perp$ (thick curves) and $m_c = 1.2\,GeV$, $\mu_R = 0.5 m_\perp$ (thin curves).

The ratio of the cross section for $D_s^\pm$ to $D^{*\pm}$ production at HERA has been compared to that from $e^+ e^-$ experiments, where the latter result is taken from a recent compilation [13] of fragmentation fractions to charm mesons ($f(c \to D)$). The results from the two types of interactions are:
$\sigma_{ep \to D_s X}/\sigma_{ep \to D^* X} = 0.41 \pm 0.07^{+0.03}_{-0.05}$ and $f(c \to D_s^+)/f(c \to D^{*+}) = 0.43 \pm 0.04$.
The strangeness suppression factor, γ_s, (the ratio of the probability to produce a strange quark to that to produce a non-strange quark), has also been compared to that from $e^+ e^-$ experiments. From HERA the value of the above cross section ratio and the PYTHIA Monte Carlo was used and for $e^+ e^-$ the quantity $2f(c \to D_s^+)/[f(c \to D^0) + f(c \to D^+)]$ served as an estimator for γ_s. The values of γ_s for HERA and $e^+ e^-$, respectively, were $0.27 \pm 0.04^{+0.02}_{-0.03}$ and 0.26 ± 0.03, implying consistency with universal charm fragmentation.

EXCITED CHARM MESONS

As a basis for the study of higher excitatations of charm mesons, an enlarged sample of data (integrated luminosity of 110pb^{-1}), containing a clean signal of $D^{*\pm}$ mesons from both photoproduction and deep inelastic scattering, was used [14]. Events in the mass range $1.83 < M(K\pi) < 1.90\,GeV$, $0.144 < M(K\pi\pi_s) - M(K\pi) < 0.147\,GeV$ were chosen (π_s is the low momentum pion in the D^* decay).

The background (estimated from events in which the K and π in the D^0 mass range have the same charge) has been subtracted, yielding 27286 ± 232 $D^{*\pm}$.

For orbital excitations an extra track, π_4, was added to the $D^{*\pm}$ candidate and the effective mass combination, $M(K\pi\pi_s\pi_4) - M(K\pi\pi_s) + M(D^*)(2.010\,GeV)$, was evaluated. An enhancement in the mass distribution with total charge zero is seen in Fig. 3a. This spectrum was fitted to D_1^0 and D_2^{*0} Breit-Wigner shapes with masses and widths fixed [4], and convoluted with a Gaussian function with a width as in the Monte Carlo simulation. The background was described by $x^\alpha e^{-\beta x + \gamma x^2}$, where $x = M(K\pi\pi_s\pi_4) - M(K\pi\pi_s) - m_\pi$, with α, β and γ constant. Helicity angle distributions for D_1^0 and D_2^{*0} proportional to $1 + 3\cos^2\theta$ and $1 - \cos^2\theta$, respectively, were folded in for the fit. Here, θ is the angle between π_4 and π_s in

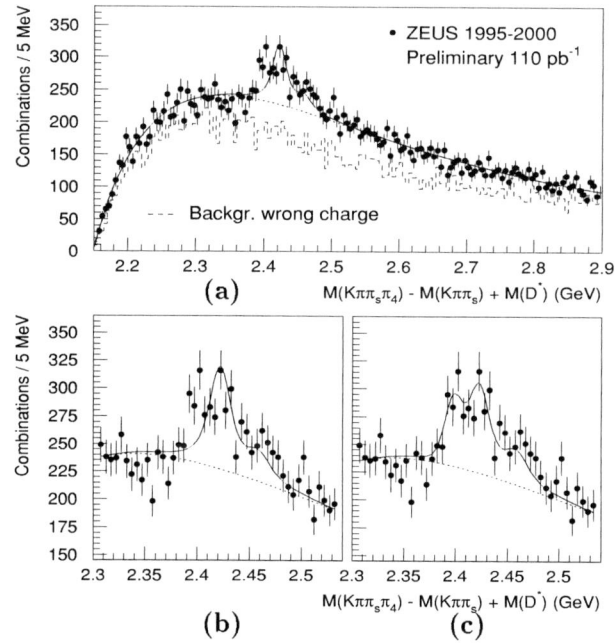

FIGURE 3. Distribution of $M(K\pi\pi_s\pi_4) - M(K\pi\pi_s) + M(D^*)(2.010\,GeV)$ for D^* and π_4 with opposite charges. (a)-(b) Fit to two Breit-Wigner shapes convoluted with a Gaussian (full curve). The dashed histogram is for D^* and π_4 with the same charge. (c) Fit included extra Gaussian with free mass and width. The dotted curves are the fitted combinatorial background.

the $D^{*\pm}$ rest frame. A closer look, (Fig. 3b), indicates an excess of events near $2.4\,GeV$. An extra Gaussian was included to better fit the data. The fit (Fig. 3c) yielded 526 ± 65, 203 ± 60, and 211 ± 49 entries for the number of D_1^0, D_2^{*0} and the additional Gaussian combinations. The mass of the extra Gaussian was

$2398.1 \pm 2.1^{+1.6}_{-0.8} MeV$ with a width consistent with the detector resolution.

The ratios of the rates of the $D^{*\pm}\pi^{\mp}$ decay channel of D^0_1 and D^{*0}_2 to $D^{*\pm}$ are $3.40 \pm 0.42^{+0.78}_{-0.63}\%$ and $1.37 \pm 0.40^{+0.96}_{-0.33}\%$, respectively. Extrapolating to the full kinematic region and using [13] along with known branching ratios [4] as well as isospin conservation, the result is given in table 1. This result is consistent with those from e^+e^- experiments.

TABLE 1. Comparison of D^0_1 and D^{*0}_2 production rates. For ZEUS errors are statistical, systematic and extrapolation errors. For CLEO, OPAL and ALEPH the statistical and systematic errors have been added in quadrature. The DELPHI results are without systematic errors.

Experiment	$f(c \to D^0_1)$ [%]	$f(c \to D^{*0}_2)$ [%]
ZEUS Prelim.	$1.46 \pm 0.18^{+0.33}_{-0.27} \pm 0.06$	$2.00 \pm 0.58^{+1.40}_{-0.48} \pm 0.41$
CLEO [15]	1.8 ± 0.3	1.9 ± 0.3
OPAL [16]	2.1 ± 0.8	5.2 ± 2.6
ALEPH Prelim. [17]	1.6 ± 0.5	4.7 ± 1.0
DELPHI Prelim. [18]	1.9 ± 0.4	4.7 ± 1.3

In order to search for a radially excited D^* meson, $D^{*'}$, the combination $M(K\pi\pi_s\pi_4\pi_5) - M(K\pi\pi_s) + M(D^*)$, where π_4 and π_5 are oppositely charged pions with $p_\perp > 0.125 GeV$, was fitted along with a background distribution of $x^\alpha e^{-\beta x}$, where $x = M(K\pi\pi_s\pi_4\pi_5) - M(K\pi\pi_s) - 2m_\pi$. No peak was seen in the expected mass range, $(2.59 - 2.67\, GeV)$ (Fig. 4). An upper limit was obtained by fitting

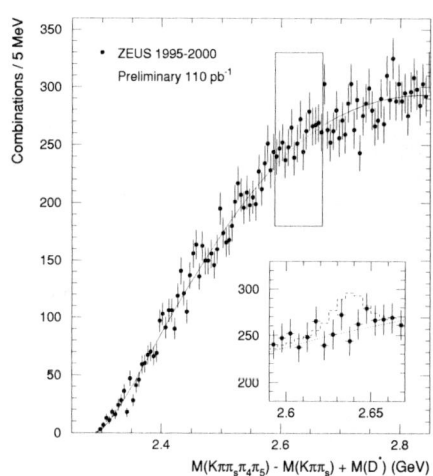

FIGURE 4. Distribution of $M(K\pi\pi_s\pi_4\pi_5) - M(K\pi\pi_s) + M(D^*)$ for the $D^{*\pm}$ candidates with the $D^{*'}$ window within the rectangle. Inset: Dashed histogram is the Monte Carlo signal normalized to the upper limit and added to the fit interpolation (dotted curve) in the $D^{*'}$ window.

the background outside this range, interpolating within the range and subtracting this from the data in the mass range. The 95% confidence level upper limit for the $D^{*\pm}\pi^+\pi^-$ decay relative to $D^{*\pm}$ was found to be 2.3%. Extrapolating to the full kinematic range and using [13], $f(c \to D^{*'+}) \cdot B_{D^{*'+} \to D^{*+}\pi^+\pi^-} < 0.7\%$ at 95% confidence level was obtained. The equivalent OPAL limit is 1.2% [8].

INELASTIC J/ψ PRODUCTION

The $\mu^+\mu^-$ decay channel of J/ψ for $0.4 < z < 0.9$, $50 < W_{\gamma p} < 180\,GeV$ and $Q^2 < 1\,GeV^2$, using the 1996-1997 data, has been studied [19]. Distributions in z, rapidity and transverse momentum squared of the J/ψ and a comparison with theoretical expectations are shown in Fig. 5. The z dependence of the data is not described in magnitude by the LO color singlet and octet model with octet matrix elements calculated from the CDF data [20,23]. On the other hand, the NLO color singlet model [21] roughly fits the spectrum for $p_{\perp\psi} > 1\,GeV$. For $p_{\perp\psi} > 2\,GeV$ the LO model with octet matrix elements from CLEO data agrees with the data for high z only [22,24]. Currently there is no calculation in NLO for the rapidity distribution. The NLO calculation agrees with the $p^2_{\perp\psi}$ data.

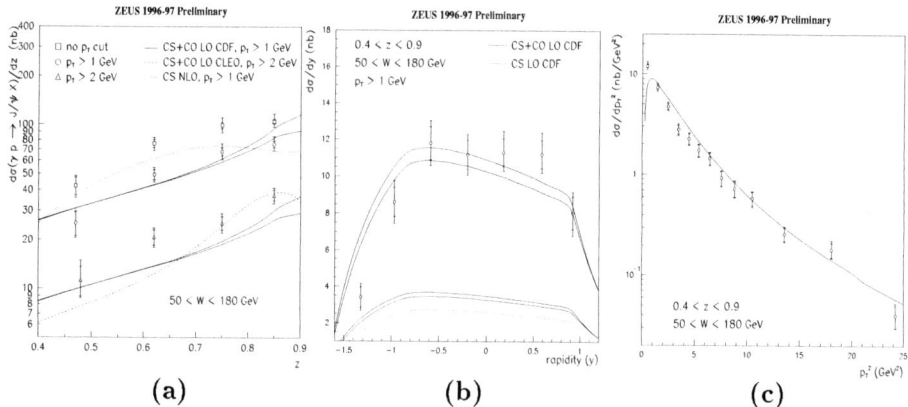

FIGURE 5. (a) z distribution for various $p_{\perp\psi}$ cuts: no cut (squares), $p_{\perp\psi} > 1\,GeV$ (circles) and $p_{\perp\psi} > 2\,GeV$ (triangles). Inner (outer) error bars are statistical (quadratic sum of statistical and systematic) errors. Lower pair of solid curves are a prediction of color singlet and octet model [20] for $p_{\perp\psi} > 1\,GeV$. Separation of the curves indicates the uncertainty in the color octet matrix elements. Upper pair of solid curves includes a scale factor of ~ 3. Dotted curve is the color singlet NLO prediction for the direct photon process and $p_{\perp\psi} > 1\,GeV$ [21]. Dashed curve is prediction of the color singlet and octet models for $p_{\perp\psi} > 2\,GeV$ [22]. (b) Rapidity distribution for $p_{\perp\psi} > 1\,GeV$. Solid curves as in (a). Dotted curve is the LO contribution of the direct photon color singlet component. (c) $p^2_{\perp\psi}$ distribution: Solid curve is the prediction of the NLO calculation [21].

REFERENCES

1. ZEUS Collaboration, Derrick M. et al., *Phys. Lett.* **B293**, 465 (1992). *The ZEUS Detector: Status Report 1993*,DESY 1993.
2. ZEUS Collaboration, Breitweg J. et al., *Phys. Lett.* **B481**, 213 (2000).
3. ZEUS Collaboration, Breitweg J. et al., *Eur. Phys. J.* **C6**, 67 (1999).
4. Caso C. et al., Particle Data Group, *Eur. Phys. J.* **C3**, 1 (1998).
5. Isgur N. and Wise M. B., *Phys. Lett.* **B232**, 113 (1989); Neubert N., *Phys. Reports.* **A245**, 259 (1994).
6. CLEO Collaboration, Anderson S. et al., *Nucl. Phys.* **A663**, 647 (2000).
7. DELPHI Collaboration, Abreu P. et al., *Phys. Lett.* **B426**, 231 (1998).
8. OPAL Collaboration, submitted to the XXIX International Conference on High Energy Physics, ICHEP 98, Vancouver Canada,July 1998; *OPAL PN*, 352.
9. CLEO Collaboration, Rodriquez J. L. *hep-ex.* **9901008**.
10. ZEUS Collaboration, Breitweg J. et al., *Z. Phys.* **C76**, 599 (1997); ZEUS Collaboration, contributed paper 814 to the XXIX ICHEP Conference, Vancouver, Canada (1998).
11. Jung H., Schuler A. and Terron J., ,*Int. Jour. of Mod. Phys.* **A7**, 7955 (1992).
12. Frixione S. et al., *Nucl. Phys.* **B454**, 3 (1995); *Phys. Lett* **B348**, 633 (1995).
13. Gladilin L., *hep-ex.* **9912064**.
14. ZEUS Collaboration, contributed paper 448 to the XXX ICHEP Conference, Osaka, Japan (2000).
15. CLEO Collaboration, Avery P. et al., *Phys. Lett.* **B331**, 236 (1994).
16. OPAL Collaboration, Ackerstaff K. et al., *Z. Phys.* **C76**, 425 (1997).
17. ALEPH Collaboration, *Production of D_1 and D_2^* mesons in hadronic Z decays*, Contributed paper to HEP99, Tampere, July 15-21, Abstract 5_411.
18. DELPHI Collaboration, *Narrow D^{**} production in c and b jets*, Contributed paper to ICHEP98, Vancouver, July 23-29, Paper 240.
19. ZEUS Collaboration, contributed paper 446 to the XXX ICHEP Conference, Osaka, Japan (2000).
20. Kniehl B. A. et al., *Eur. Phys. J.* **C6**, 493 (1999); Kniehl B. A., *Proceedings of the Workshop 1998-1999 on Monte Carlo Generators for HERA Physics*, ed. by Doyle A. T., Grindhammer G., Ingelman G. and Jung H., DESY-PROC-1999-02,p.427.
21. Kramer M. et al., *Phys. Lett.* **B348**, 657 (1995); Kramer M., *Nucl. Phys.* **B459**, 3 (1996).
22. Beneke M. et al., *hep-ex.* **0001062**, subitted to *Phys. Rev. D.*
23. CDF Collaboration, Abe F. et al., *Phys. Rev. Lett.* **79**, 578 (1997).
24. CLEO Collaboration, Balest R. et al., *Phys. Rev.* **D52**, 2661 (1995).

Exclusive Photoproduction of J/ψ Mesons at HERA

Alexander A. Savin

DESY, Hamburg, Germany and Moscow State University, Moscow, Russia [1]

Abstract. The exclusive photoproduction of J/ψ mesons, $\gamma p \to J/\psi p$, has been studied in ep collisions with the ZEUS and H1 detectors at HERA using the e^+e^- and $\mu^+\mu^-$ decay modes. The total cross section has been measured as a function of the photon-proton centre-of-mass energy, W, in the kinematic range $20 \le W \le 290$ GeV for ZEUS and $26 \le W \le 285$ GeV for H1. The differential cross section $d\sigma/dt$, where t is the squared four-momentum at the proton vertex, has been measured. The Pomeron trajectory has been determined.

INTRODUCTION

The study of exclusive diffractive vector mesons production at HERA, $\gamma p \to V p$, where V=(ρ, ω, ϕ, J/ψ), has shown that when Q^2 (photon virtuality) is large the cross section increases with energy faster than expected for soft processes [1,2], in good agreement with the expectations of perturbative QCD (pQCD). The same effect was observed also at low photon virtualities for heavy meson production [3,4].

Assuming that the mass of the vector meson plays a role similar to Q^2, it is natural to expect that J/ψ photoproduction can be described in the framework of pQCD, with the charm mass used as a perturbative scale. The reaction $\gamma p \to J/\psi$ can be viewed in this case as a three-step process: the photon fluctuates into a $c\bar{c}$ state, the $c\bar{c}$ system scatters on the proton via (colourless) two-gluon exchange, and the J/ψ is formed from the outgoing $c\bar{c}$ pair. Within pQCD the two-gluon exchange amplitude can be shown to be directly proportional to the gluon density of the proton squared [5] leading to a fast rise of the total cross section with W as a consequence of the fast increase of the gluon density in the proton at small values of Bjorken-x, since $x \propto (Q^2 + M_{J/\psi}^2)/W^2$. In this framework, the t-dependence of the cross section should not vary with W, i.e. no shrinkage of diffractive peak is expected.

The J/ψ photoproduction can also be described in the framework of Regge phenomenology [6]. In this approach $d\sigma/dt \propto (W^2)^{2\alpha_{I\!P}(t)-2}$, where $\alpha_{I\!P}(t) = \alpha_{I\!P}(0)+\alpha'_{I\!P}t$

---

[1] On behalf of the H1 and ZEUS collaborations

is the Pomeron trajectory, which then can be determined by studying the W dependence of $d\sigma/dt$ at fixed t.

The *soft pomeron*, parametrized by Donnachie and Landshoff as $\alpha_{I\!P}(t) = 1.08 + 0.25t$, alone is not sufficient to describe the cross section behaviour when a large scale is involved [7].

ZEUS 99-00 Preliminary

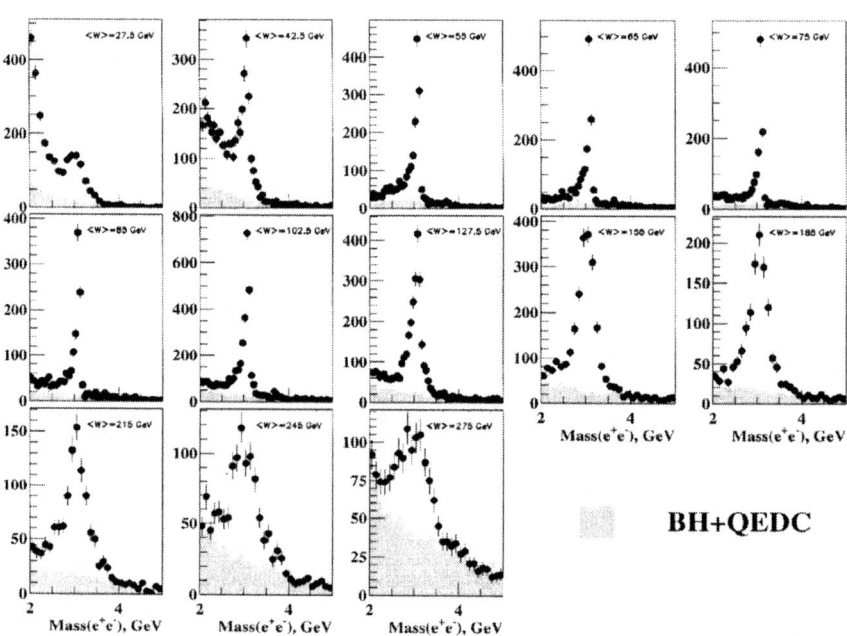

FIGURE 1. Invariant-mass distributions of the e^+e^- pairs in the different W regions measured by ZEUS. The histograms represent Monte Carlo distributions of the non-resonant background explained in the text.

DATA SELECTION

In this analysis the J/ψ mesons were identified via their decays into e^+e^- and $\mu^+\mu^-$ with a branching ratio of 6% each. The data were taken with the ZEUS and H1 detectors while HERA was operating with positrons (electrons) of 27.5 GeV and protons of 820 or 920 GeV.

Photoproduction of J/ψ mesons was studied using the reaction $ep \to eJ/\psi p$ for very low values of Q^2 ($< Q^2 > \approx 10^{-2} - 10^{-5}$ GeV^2) .

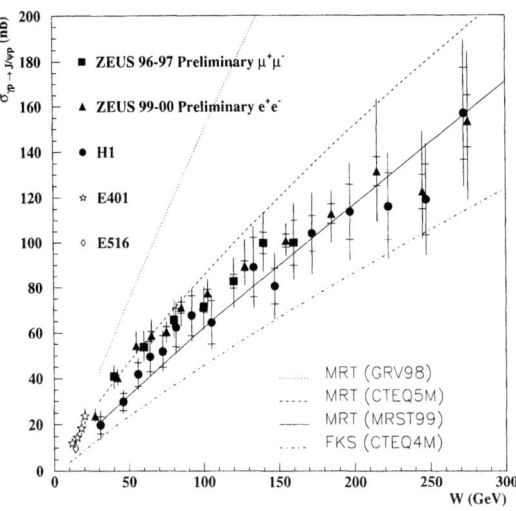

FIGURE 2. The elastic J/ψ photoproduction cross section as a function of W from HERA [4] and from fixed-targed experiments [8,9]. The QCD predictions [10,11] using various parametrizations of the gluon density in the proton [12–14] are shown.

Compared to earlier HERA data, the present results cover a wider W range. Since W is correlated to the J/ψ polar angle, the wider W coverage was obtained by using not only events in which the e^+e^- or $\mu^+\mu^-$ tracks were reconstructed in the central tracking detector, but also events in which one or both of the e^+e^- were detected by calorimeter, outside of the coverage of the central trackers. Also Forward Muon Detector was used in H1 case to detect one of the muons in the very forward direction outside of the central trackers region. In the H1 case the $\mu^+\mu^-$ data cover the region $26 \leq W \leq 150$ GeV and the e^+e^- data the region $135 \leq W \leq 285$ GeV . In the ZEUS case, $30 < W < 170$ GeV and $20 < W < 290$ GeV for the $\mu^+\mu^-$ and e^+e^- data respectively.

Both experiments selected only exclusive events where the scattered lepton and the scattered proton escaped in the beampipe and were not detected. The data were corrected for the proton-dissociative reaction, $\gamma p \rightarrow J/\psi N$, where N is a low mass state, which escapes undetected in the beam pipe. In ZEUS case the size of the proton-dissociative background was measured to be $\approx 20\%$ for $\mu^+\mu^-$ and $\approx 13\%$ for e^+e^- case independently of W. In H1 case the correction is typically $10 - 13$ % in the central and backward regions and 30% in the forward region.

The observed J/ψ mass distributions for the e^+e^- decay mode are shown in Fig. 1 for ZEUS. The expected background from the Bethe-Heitler and QED Compton scattering processes are also shown.

301

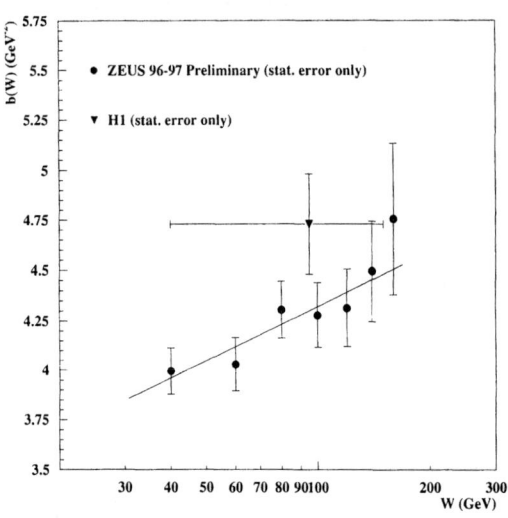

FIGURE 3. The b-slope measurement in bins of W. The result of the fit explained in the text is shown. The horizontal error bar for H1 point indicates the W-range entering the determination.

RESULTS

W dependence of the cross section

The cross section for elastic photoproduction of J/ψ mesons, as a function of W, is shown in Fig. 2. While overall the agreement between the H1 and ZEUS measurements is good, at low W the measurements disagree substantially. A fit of the form W^δ to the data yields a value of $\delta = 0.83 \pm 0.07(tot.)$ for H1 and $\delta = 0.65 \pm 0.03(stat.) \pm 0.09(syst.)$ for the ZEUS $\mu^+\mu^-$ sample and $\delta = 0.66 \pm 0.02(stat.)^{+0.05}_{-0.08}(syst.)$ for the ZEUS e^+e^- sample.

These results confirm the steep rise of the cross section observed previously [3,4]. The data are compared to predictions from pQCD calculations [10,11] using various gluon-density parametrizations [12–14]. The theoretical predictions exhibit a strong sensitivity to the input parametrization of the gluon density, both in shape and in normalization.

$d\sigma/dt$

The differential cross section $d\sigma/dt$ has been measured. Fits of the form $d\sigma/dt \propto \exp(bt)$ were performed and the corresponding results for b are plotted in Fig. 3.

An estimate of $\alpha'_{I\!P}$ was obtained by ZEUS by studying the W dependence of b, which was fitted to the function $b = b_0 + 2\alpha'_{I\!P} \cdot ln(W/90\ GeV)$ yielding
$\alpha'_{I\!P} = 0.098 \pm 0.035(stat) \pm 0.050(syst)$.

The Pomeron Trajectory

A more direct determination of the Pomeron trajectory can be obtained by measuring the energy dependence of the elastic cross section at fixed values of t. In Fig. 4, $d\sigma/dt$ is shown versus W in six bins of t. The lines are results of the fit of the form $d\sigma/dt \propto (W^2)^{2\alpha_{I\!P}(t)-2}$. The resulting values of $\alpha_{I\!P}(t)$ are plotted in Fig. 5 as a function of t.

A linear fit to the data yields $\alpha_{I\!P}(t)$ equal:
$(1.193 \pm 0.011(stat)^{+0.015}_{-0.010}(syst)) + (0.105 \pm 0.024(stat)^{+0.022}_{-0.020}(syst)) \cdot t/GeV^2$ and
$(1.27 \pm 0.05) + (0.08 \pm 0.17) \cdot t/GeV^2$ for ZEUS and H1 respectively.

The fits are also shown in Fig. 5 , including lines which reflect the one standard deviation uncertainty. The resulting trajectories are inconsistent with the DL parametrization of the soft Pomeron and also with the so called hard Pomeron trajectory, also introduced by the same authors [7]. The data are in agreement with the NLO BFKL calculation [15].

The slope, $\alpha'_{I\!P}$, is close to but non-zero and suggests the presence of small shrinkage we observed already in Fig. 3.

REFERENCES

1. ZEUS Coll., Breitweg J. et al., *Phys. Lett.* **B 487**, 273 (2000).
2. H1 Coll., Adloff C. et al., *Phys. Lett.* **B 483**, 360 (2000).
3. ZEUS Coll., Breitweg J. et al., *Z. Phys.* **C 75**, 215 (1997).
4. H1 Coll., Adloff C. et al., *Phys. Lett.* **B 483**, 23 (2000).
5. Ryskin M.G., Roberts R.G., Martin A.D., Levin E.M., *Z. Phys.* **C 76**, 231 (1997).
6. Collins P.D.B., *An Introduction to Regge Theory and High Energy Physics,* Cambridge University Press, Cambridge (1977).
7. Donnachie A., Landshoff P.V., *Phys. Lett.* **B 437**, 408 (1998).
8. E516 Coll, Denby B.H. et al., *Phys. Rev. Lett.* **52**, 795 (1984).
9. E401 Coll, Binkley M. et al., *Phys. Rev. Lett.* **48**, 73 (1982).
10. Frankfurt L., Koepf W., Strikman M,, *Phys. Rev.* **D 57**, 512 (1998).
11. Martin A.D., Ryskin M.G., Teubner T., *Phys. Lett.* **B 454**, 339 (1999).
12. Lai H.L. et al., *Phys. Rev.* **D 55**, 1280 (1997).
13. Glück M., Reya E., Vogt A., *Eur. Phys. J.* **C 5**, 461 (1998).
14. Martin A.D. et al., *Eur. Phys. J.* **C 4**, 463 (1998).
15. Brodsky S.J. et al., *JETP Lett.* **70**, 155 (1999).

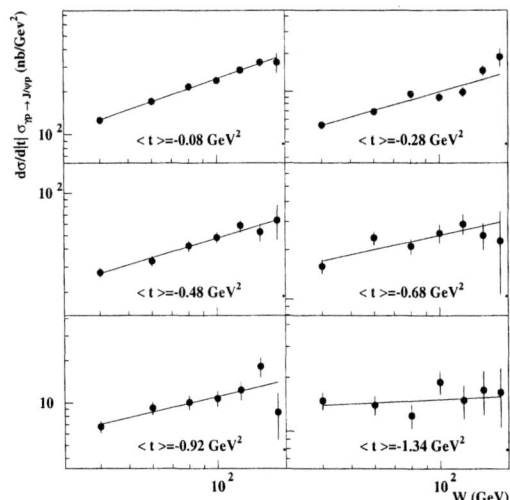

FIGURE 4. The differential cross section $d\sigma/dt$ as a function of W at fixed values of t for ZEUS. The lines represent the result of the fit explained in the text.

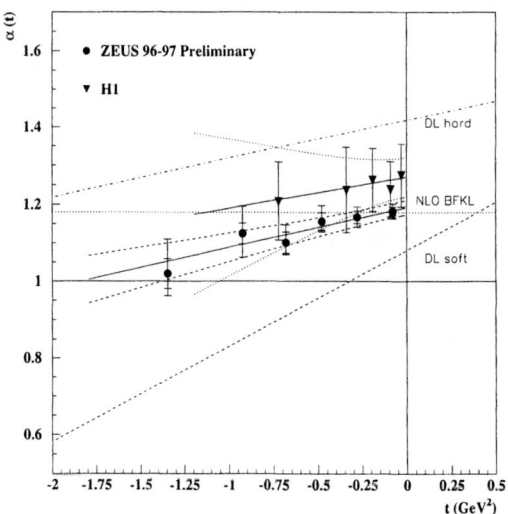

FIGURE 5. The measured Pomeron trajectory. The solid lines represent the results of the fit explained in the text. The one standard deviation contour is indicated for H1 and ZEUS measurements. Also shown are the soft and hard Donnachie-Lanshoff pomeron trajectories [7] and a result based on a NLO BFKL calculation [15] .

Exclusive Electroproduction of Charmonium at HERA

Tetsuo Abe[a]

on behalf of the H1 and ZEUS collaborations

[a] University of Tokyo, 7-3-1 Hongo, Bunkyo-ku, Tokyo 113-8654, Japan

October 24, 2000

Abstract. Exclusive electroproduction of charmonium (J/ψ) in ep collisions at HERA has been studied by both H1 and ZEUS using integrated luminosities of $27\,\mathrm{pb}^{-1}$ and $75\,\mathrm{pb}^{-1}$, respectively. The J/ψ state is identified via its leptonic decay channels. Cross sections have been measured as functions of Q^2 and W. Angular distributions of the decay leptons have been measured and the ratio $R = \sigma_L/\sigma_T$ is extracted as a function of Q^2.

INTRODUCTION

Vector Dominance Model (VDM) in conjunction with Regge phenomenology has described vector meson (VM) production in the measurements of elastic photoproduction of the light VMs (ρ, ω and ϕ) performed both in fixed target experiments [1–4] and at HERA [5–10]. At sufficiently high energies ($W > 10\,\mathrm{GeV}$, where W is the center-of-mass energy of the γp system.), the reaction displays the characteristics of a diffractive soft process, most notably a slow rise of the cross section ($\propto W^\delta$, $\delta \simeq 0.22$). However, a different cross-section behavior has been observed in J/ψ photoproduction [11]; the cross section rises more rapidly with W ($\delta \simeq 0.6 - 0.9$). This is inconsistent with such a soft production mechanism. A similar steep W-dependence is observed in the exclusive electroproduction of the light VMs. At Q^2 values close to the J/ψ mass squared ($Q^2 \approx M_{J/\psi}^2$), the observed W-dependence [12,13] yields values of δ ($\simeq 0.4 - 0.7$) similar to those obtained in J/ψ photopro-

CP571, *PHOTON 2000*, edited by A. J. Finch

duction. Here, Q^2 indicates the negative four-momentum transfer squared at the lepton vertex or virtuality of the exchanged photon (γ^*).

An alternative approach to VM production based on perturbative QCD (pQCD) has been proposed when a hard scale is available. The reaction is understood as a scattering between the proton and a color dipole ($q\bar{q}$) into which the virtual photon fluctuates, mediated by two gluons in a color singlet state. Long after this interaction, the dipole forms a J/ψ state. The cross section is, in the leading-log approximation, proportional to the square of the gluon density in the proton, $[xg(x, \mu^2)]^2$. It is interesting to study the interplay of the two hard scales, Q^2 and $M_{J/\psi}^2$, because the hard scale μ can be given by a combination of them. The observed steep rise in the cross section is explained by the rapid increase of the gluon density towards low x, since $W^2 \propto 1/x$ at fixed Q^2.

The study of J/ψ electroproduction has been performed with the recent high-luminosity data sets from HERA. This talk is based on the H1 published results [15] using 27 pb^{-1} taken in 1995-97 and the ZEUS preliminary results [16] using 75 pb^{-1} taken in 1996-99. In both analyses, a scattered e is tagged with the calorimeter, and the J/ψ state is identified via its leptonic decay channels, requiring two oppositely-charged tracks in the central tracker not associated with the scattered e. Measured ep cross sections are translated into $\gamma^* p$ ones as functions of Q^2 and W.

W-DEPENDENCE

Figure 1 shows the measured cross sections as a function of W at three Q^2 values, 3.1, 6.8 and 16 GeV^2. For the H1 data, the cross sections at the Q^2 values, 3.5, 10.1 and 33.6 GeV^2 have been rescaled according to the Q^2-dependence from the H1 measurement described in the next section. The cross sections at each Q^2 value are fitted with the function W^δ. The solid lines in Figure 1 are the results of the fit to the ZEUS data. The resulting δ values for the H1 and ZEUS data are shown in the inset of Figure 1. The obtained W-slopes in electroproduction are consistent with those of photoproduction by H1 [17] and ZEUS [18]. The cross sections are also compared with the pQCD-based calculations from Frankfurt et al. (FKS) [19] using CTEQ4M and from Martin et al. (MRT) [20] using CTEQ5M. The theories and the data are consistent within the uncertainty of the measurements although the rise with W predicted by these calculations tends to be steeper.

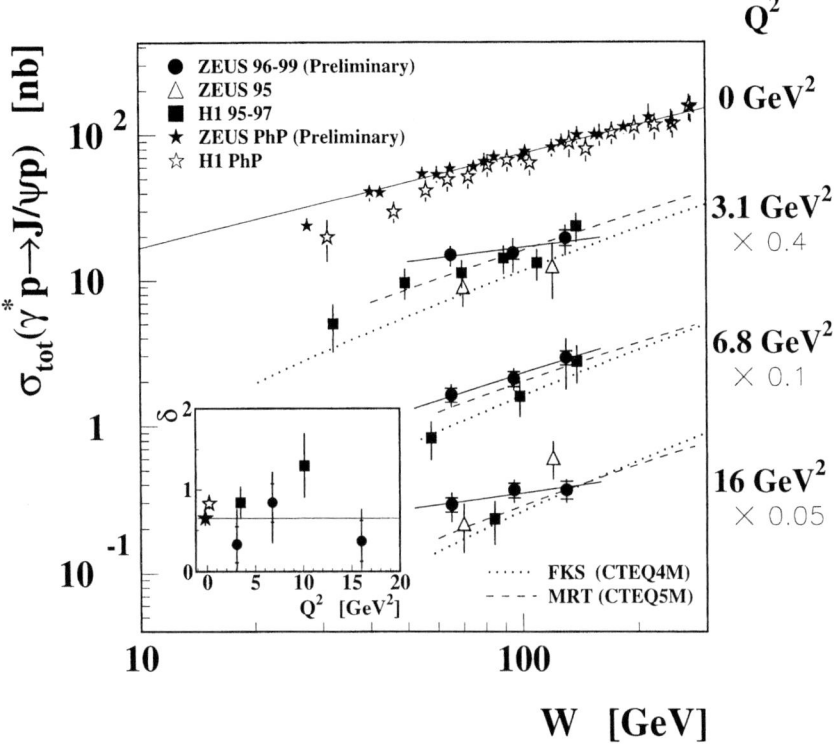

FIGURE 1. W-dependence of the cross section. The H1 cross sections at the Q^2 values, 3.5, 10.1 and 33.6 GeV^2 have been rescaled according to the Q^2-dependence from the H1 measurement. The solid lines are the results of the fit to the ZEUS data using the function W^δ. The resulting δ values for the H1 and ZEUS data are shown in the inset. The cross sections are compared with the pQCD-based calculations from Frankfurt et al. (FKS) using CTEQ4M and from Martin et al. (MRT) using CTEQ5M.

Q^2-DEPENDENCE

The Q^2-dependence of the cross section at $W = 90\,\text{GeV}$ is shown in Figure 2. The cross sections are fitted in the range $Q^2 > 2\,\text{GeV}^2$ with a simple parameterisation of $1/(Q^2 + M_{J/\psi}^2)^n$, yielding $n = 2.38 \pm 0.11$ for the H1 data and $n = 2.60 \pm 0.11(stat.)^{+0.09}_{-0.08}(syst.)$ (the solid line in Figure 2) for the ZEUS data. The extrapolation of the ZEUS fit, however, overshoots the photoproduction measurement by ZEUS [18]. The photoproduction data of H1, however, is consistent with the value extrapolated from the fit to the H1 data for $Q^2 > 2\,\text{GeV}^2$. The data are also compared with the predictions from FKS using CTEQ4M and MRT using CTEQ5M and MRST99. Again the theories describe the data fairly well.

RATIO TO ρ ELECTROPRODUCTION

In VDM, the ratios of the different VM cross sections are given by the couplings of the photon to the VMs and by the elastic VM-proton cross sections. Under the simplified assumption that the photon-VM coupling is proportional to the quark-current decomposition of the photon (SU(4) flavor symmetry), and that the VM-proton cross sections are universal, the cross-section ratios are given by $\rho : \omega : \phi : J/\psi = 9 : 1 : 2 : 8$. This prediction is expected to hold in the region of $Q^2 \gg M_{VM}^2$. In the pQCD approach, this hypothesis also holds if the coupling strengths between the dipole and the two gluons, and between the dipole and the VM, are assumed to be universal.

The measured cross sections are compared with the ρ cross sections [21]. The ratio $r = \sigma_{\text{tot}}^{\gamma^* p \to J/\psi p}/\sigma_{\text{tot}}^{\gamma^* p \to \rho p}$ for both the H1 and ZEUS data is shown as a function of Q^2 in Figure 3. The previous ZEUS measurements in photoproduction [22] and electroproduction [12] are also shown. The value of r rises rapidly as Q^2 becomes larger. The ratio at $Q^2 = 53\,\text{GeV}^2$ is consistent with the expectation from SU(4) flavor symmetry 8/9. This value of Q^2 is about five times higher than $M_{J/\psi}^2$.

POLARIZATION OF THE J/ψ

In VM production, it is possible to determine the polarization of the VM using the spin density matrix elements obtained from the angular distributions of the decay particles [24] [23]. The previous analyses of the helicity angles in both photoproduction [5–7] and electroproduction [12–14] suggest that the s-channel helicity conservation hypothesis holds to a good approximation, allowing one of the

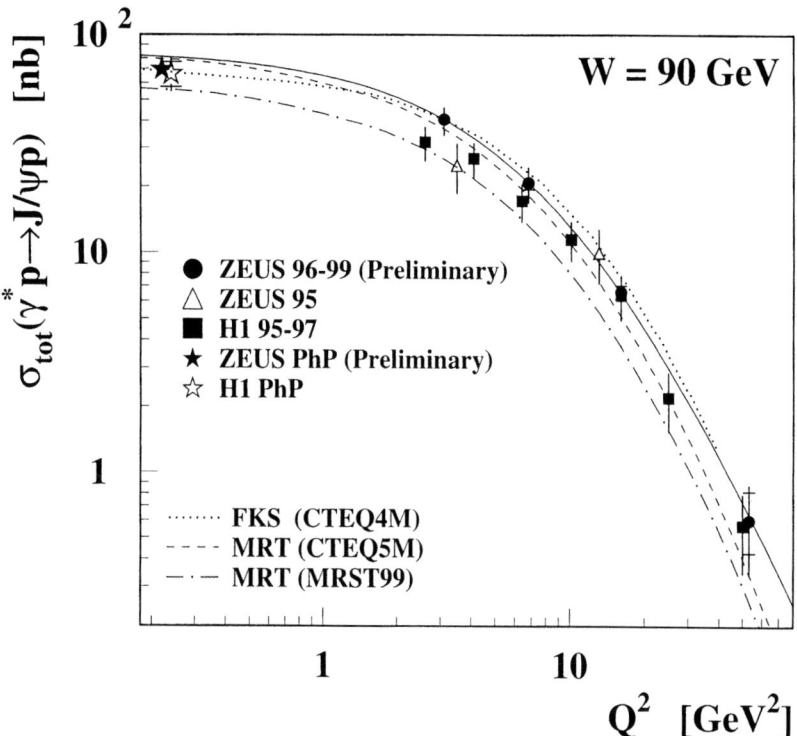

FIGURE 2. Q^2-dependence of the cross section. The solid line is a result of the fit to the ZEUS data for $Q^2 > 3\,\text{GeV}^2$ using the form $1/(Q^2 + M_{J/\psi}^2)^n$. The data are compared with the predictions from FKS using CTEQ4M and MRT using CTEQ5M and MRST99.

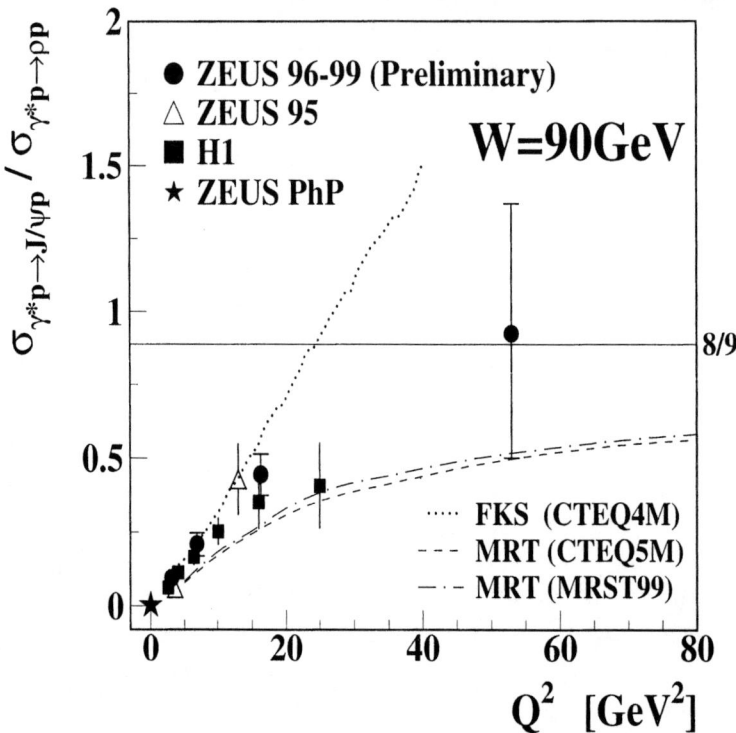

FIGURE 3. The ratio of the J/ψ to ρ cross sections as a function of Q^2. The previous ZEUS measurements in photoproduction and electroproduction are also shown.

density matrix elements (r_{00}^{04}) to be related directly to the ratio of the longitudinal to transverse cross sections ($R = \sigma_L / \sigma_T$);

$$R = \frac{r_{00}^{04}}{\epsilon (1 - r_{00}^{04})}$$

where ϵ is the ratio of the longitudinal to transverse fluxes of γ^*, $\epsilon = \Gamma_L / \Gamma_T \sim 1$. The matrix element r_{00}^{04} is determined from the distribution of the helicity angle (θ_h);

$$\frac{d\sigma}{d\cos\theta_h} \propto 1 + r_{00}^{04} + (1 - 3r_{00}^{04})\cos^2\theta_h.$$

The ratio R has been measured by fitting the $\cos\theta_h$ distribution. H1 obtained $R = 0.18^{+0.18}_{-0.14}$, $0.94^{+0.79}_{-0.43}$ at $Q^2 = 4$, $16\,\mathrm{GeV}^2$, respectively, and ZEUS $R = -0.06^{+0.23}_{-0.09}$, $0.08^{+0.17}_{-0.14}$, $1.49^{+1.54}_{-0.83}$ at $Q^2 = 3.1$, 6.8, $16\,\mathrm{GeV}^2$, respectively. The value of R increases with Q^2. The values are about one order of magnitude smaller than those measured in ρ production at the same Q^2 [12,13,21]. However, the measured values are consistent with the fitted parameterisation of R for ρ [21] replacing the ρ mass by the J/ψ one.

CONCLUSIONS

Exclusive J/ψ electroproduction has been studied at HERA by both H1 and ZEUS. W-slopes of the cross sections have been measured and found to be consistent with that of photoproduction and also with the pQCD-based calculations. The Q^2-dependence is fairly well-described by pQCD. The ratio to ρ production rises with Q^2 and is consistent with the SU(4) expectation (8/9) at $Q^2 = 53\,\mathrm{GeV}^2$. The ratio $R = \sigma_L / \sigma_T$ also rises with Q^2. Although the measured values of R are about one order of magnitude smaller than those for ρ production, they are consistent with the fitted parameterisation of R for ρ when the ρ mass is replaced by that of the J/ψ.

REFERENCES

1. R.M. Egloff et al., Phys. Rev. Lett. 43 (1979) 657.

2. R.M. Egloff et al., Phys. Rev. Lett. 43 (1979) 1545.

3. D. Aston et al., Nucl. Phys. B 209 (1982) 56.

4. J. Busenitz et al., Phys. Rev. D 40 (1989) 1.

5. ZEUS Collaboration, M. Derrick et al., Z. Phys. C 69 (1995) 39.

6. H1 Collaboration, S. Aid et al., Nucl. Phys. B 463 (1996) 3.

7. ZEUS Collaboration, M. Derrick et al., Z. Phys. C 73 (1997) 253.

8. ZEUS Collaboration, J. Breitweg et al., Eur. Phys. J. C 2 (1998) 247.

9. ZEUS Collaboration, M. Derrick et al., Z. Phys. C 73 (1996) 73.

10. ZEUS Collaboration, M. Derrick et al., Phys. Lett. B 377 (1996) 259.

11. A. Savin, Study of exclusive photoproduction of J/ψ mesons in ep interactions, these proceedings.

12. ZEUS Collaboration, J. Breitweg et al., Eur. Phys. J. C 6 (1999) 603.

13. H1 Collaboration, C. Adloff et al., Eur. Phys. J. C 13 (2000) 371.

14. ZEUS Collaboration, J. Breitweg et al., Eur. Phys. J. C 12 (2000) 393.

15. H1 Collaboration, C. Adloff et al., Eur. Phys. J. C 10 (1999) 373.

16. ZEUS Collaboration, Paper contribution to XXXth International Conference on High Energy Physics ICHEP2000, Osaka, Japan, Paper 438.

17. H1 Collaboration, C. Adloff et al., Phys. Lett. B 483 (2000) 23.

18. ZEUS Collaboration, Paper contribution to XXXth International Conference on High Energy Physics ICHEP2000, Osaka, Japan, Paper 437.

19. L. Frankfurt, W. Koepf and M. Strikman, Phys. Rev. D 57 (1998) 512.

20. A.D. Martin, M.G. Ryskin and T. Teubner, Phys. Rev. D 62 (2000) 014022.

21. ZEUS Collaboration, Paper contribution to XXXth International Conference on High Energy Physics ICHEP2000, Osaka, Japan, Paper 439.

22. ZEUS Collaboration, J. Breitweg et al., Z. Phys. C 75 (1997) 215.

23. K. Schilling and G. Wolf, Nucl. Phys. B61 (1973) 381-413.

24. H1 Collaboration, C. Adloff et al., Phys. Lett. B 483 (2000) 360-372.

EXCLUSIVES

Two-Photon Exclusive Processes in QCD*

Stanley J. Brodsky

Stanford Linear Accelerator Center
Stanford University, Stanford, California 94309
sjbth@slac.stanford.edu

Abstract. Exclusive two-photon reactions such as Compton scattering at large angles, deeply virtual Compton scattering, and hadron production in photon-photon collisions provide important tests of QCD at the amplitude level, particularly as measures of hadron distribution amplitudes and skewed parton distributions.

A central focus of study in QCD are the wavefunctions which describe hadrons in terms of their quark and gluon degrees of freedom at the amplitude level. Of particular interest are the gauge- and process-independent meson and baryon valence-quark distribution amplitudes $\phi_M(x, Q)$, and $\phi_B(x_i, Q)$ which control exclusive processes involving a hard scale Q; for example, meson distribution amplitudes play a key role in the analysis of exclusive semi-leptonic and two-body hadronic B-decays [1–6]. There has recently been considerable progress both in calculating hadron wavefunctions from first principles in QCD and in measuring them using diffractive di-jet dissociation.

Two-photon processes such as $\gamma^*\gamma \to$ hadrons, Compton scattering $\gamma p \to \gamma p$ at large momentum transfer, and $\gamma\gamma \to$ hadron pairs at high momentum transfer and fixed θ_{cm}, can play a crucial role in understanding the perturbative and non-perturbative structure of QCD, first by testing the validity and empirical applicability of leading-twist factorization theorems, second by verifying the structure of the underlying perturbative QCD subprocesses, and third, through measurements of angular distributions and ratios which are sensitive to the shape of the distribution amplitudes. In effect, Compton scattering and photon-photon collisions are microscopes for testing fundamental scaling laws of PQCD and for measuring distribution amplitudes. In addition, as I shall discuss in the next section, deeply virtual Compton scattering $\gamma^*p \to \gamma p$ for far off-shell initial photons has emerged as one of the most important and interesting exclusive QCD reactions.

---

*) Work supported by the Department of Energy under contract number DE-AC03-76SF00515.

CP571, *PHOTON 2000*, edited by A. J. Finch
2001 American Institute of Physics 0-7354-0010-5

DEEPLY VIRTUAL COMPTON SCATTERING

The virtual Compton scattering amplitude $\frac{d\sigma}{dt}(\gamma^*p \to \gamma p)$ has extraordinary sensitivity to fundamental features of proton structure [7–14]. Even though the final state photon is on-shell, the deeply virtual Compton process probes the elementary quark structure of the proton near the light cone as an effective local current. In contrast to deep inelastic scattering, which measures only the absorptive part of the $t = 0$ forward virtual Compton amplitude, deeply virtual Compton scattering allows the measurement of the phase and spin structure of proton matrix elements for general momentum transfer t. The scaling, Regge behavior, and phase structure of deeply virtual Compton scattering have been discussed in the context of the co-variant parton model in Ref. [15]. The interference of Compton and bremsstrahlung amplitudes gives an electron-positron asymmetry in the $e^\pm p \to e^\pm \gamma p$ cross section which is proportional to the real part of the Compton amplitude [15].

To leading order in $1/Q$, the deeply virtual Compton scattering amplitude factorizes as the convolution in x of the amplitude $t^{\mu\nu}$ for hard Compton scattering on a quark line with the generalized Compton form factors $H(x, t, \zeta)$, $E(x, t, \zeta)$, $\tilde{H}(x, t, \zeta)$, and $\tilde{E}(x, t, \zeta)$ of the target proton. Here x is the light-cone momentum fraction of the struck quark, and $\zeta = Q^2/2P \cdot q$ plays the role of the Bjorken variable. The form factor $H(x, t, \zeta)$ describes the proton response when the helicity of the proton is unchanged, and $E(x, t, \zeta)$ is for the case when the proton helicity is flipped. Two additional functions $\tilde{H}(x, t, \zeta)$, and $\tilde{E}(x, t, \zeta)$ appear, corresponding to the dependence of the Compton amplitude on quark helicity. These "skewed" parton distributions involve non-zero momentum transfer, so that a probabalistic interpretation is not possible. However, there are remarkable sum rules connecting the chiral-conserving and chiral-flip form factors $H(x, t, \zeta)$ and $E(x, t, \zeta)$ with the corresponding spin-conserving and spin-flip electromagnetic form factors $F_1(t)$ and $F_2(t)$ and gravitational form factors $A_q(t)$ and $B_q(t)$ for each quark and anti-quark constituent [7]. Thus deeply virtual Compton scattering is related to the quark contribution to the form factors of a proton scattering in a gravitational field.

One can construct space-like electromagnetic, electroweak, gravitational couplings, or any local operator product matrix element from the diagonal overlap of the LC wavefunctions [16]. In the case of the generalized form factors of deeply virtual Compton scattering, the computation [17,18] requires not only the diagonal matrix element $n \to n$ for $\zeta < x < 1$, where parton number is conserved, but also an off-diagonal $n+1 \to n-1$ convolution for $0 < x < \zeta$. This second domain occurs since the current operator of the final-state photon with positive light-cone momentum fraction ζ can annihilate a $q\overline{q'}$ pair in the initial proton wavefunction. The off-diagonal terms are referred to in the literature as the "ERBL" contributions, since they resemble virtual Compton scattering on an exchanged mesonic system $\gamma^* q\overline{q'} \to \gamma$ and thus obey the same evolution equations in $\log q^2$ as the meson distribution amplitudes [19–22]. In fact, the light cone Fock representation shows that there are underlying relations between the Fock states of different particle number which interrelate the two domains.

NON-PERTURBATIVE CALCULATIONS OF THE PION DISTRIBUTION AMPLITUDE

The distribution amplitude $\phi(x, \tilde{Q})$ can be computed from the integral over transverse momenta of the renormalized hadron valence wavefunction in the light-cone gauge at fixed light-cone time [23]:

$$\phi(x, \tilde{Q}) = \int d^2\vec{k}_\perp \, \theta\left(\tilde{Q}^2 - \frac{\vec{k}_\perp^{\,2}}{x(1-x)}\right) \psi^{(\tilde{Q})}(x, \vec{k}_\perp), \tag{1}$$

where a global cutoff in invariant mass is identified with the resolution $\tilde{Q}$. The distribution amplitude $\phi(x, \tilde{Q})$ is boost and gauge invariant and evolves in $\ln \tilde{Q}$ through an evolution equation [24,19,21]. Since it is formed from the same product of operators as the non-singlet structure function, the anomalous dimensions controlling $\phi(x, Q)$ dependence in the ultraviolet $\log Q$ scale are the same as those which appear in the DGLAP evolution of structure functions [30]. The decay $\pi \to \mu\nu$ normalizes the wave function at the origin: $a_0/6 = \int_0^1 dx \phi(x, Q) = f_\pi/(2\sqrt{3})$. One can also compute the distribution amplitude from the gauge invariant Bethe-Salpeter wavefunction at equal light-cone time. This also allows contact with both QCD sum rules [26] and lattice gauge theory; for example, moments of the pion distribution amplitudes have been computed in lattice gauge theory [27–29]. Conformal symmetry can be used as a template to organize the renormalization scales and evolution of QCD predictions [30,31]. For example, Braun and collaborators have shown how one can use conformal symmetry to classify the eigensolutions of the baryon distribution amplitude [32].

Dalley [33] has recently calculated the pion distribution amplitude from QCD using a combination of the discretized light-cone quantization [34] method for the x^- and x^+ light-cone coordinates with the transverse lattice method [35,36] in the transverse directions, A finite lattice spacing a can be used by choosing the parameters of the effective theory in a region of renormalization group stability to respect the required gauge, Poincaré, chiral, and continuum symmetries. The overall normalization gives $f_\pi = 101$ MeV compared with the experimental value of 93 MeV. Figure 1 (a) compares the resulting DLCQ/transverse lattice pion wavefunction with the best fit to the diffractive di-jet data (see the next section) after corrections for hadronization and experimental acceptance [37]. The theoretical curve is somewhat broader than the experimental result. However, there are experimental uncertainties from hadronization and theoretical errors introduced from finite DLCQ resolution, using a nearly massless pion, ambiguities in setting the factorization scale Q^2, as well as errors in the evolution of the distribution amplitude from 1 to 10 GeV^2. Instanton models also predict a pion distribution amplitude close to the asymptotic form [38]. In contrast, recent lattice results from Del Debbio et al. [29] predict a much narrower shape for the pion distribution amplitude than the distribution predicted by the transverse lattice. A new result for the proton distribution amplitude treating nucleons as chiral solitons has recently been

317

derived by Diakonov and Petrov [39]. Dyson-Schwinger models [40] of hadronic Bethe-Salpeter wavefunctions can also be used to predict light-cone wavefunctions and hadron distribution amplitudes by integrating over the relative k^- momentum. There is also the possibility of deriving Bethe-Salpeter wavefunctions within light-cone gauge quantized QCD [41] in order to properly match to the light-cone gauge Fock state decomposition.

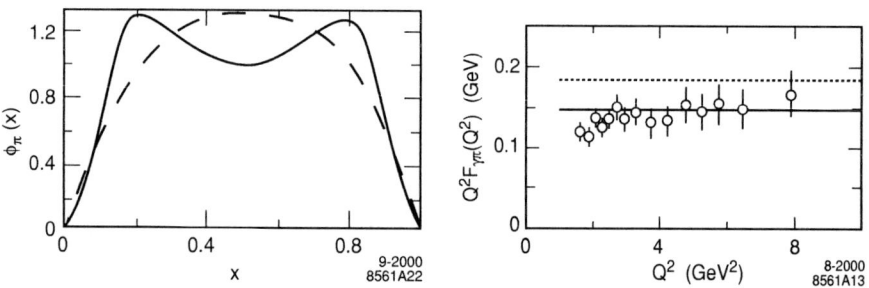

FIGURE 1. (a) Preliminary transverse lattice results for the pion distribution amplitude at $Q^2 \sim 10\text{GeV}^2$. The solid curve is the theoretical prediction from the combined DLCQ/transverse lattice method [33]; the chain line is the experimental result obtained from jet diffractive disso-ciation [37]. Both are normalized to the same area for comparison. (b) Scaling of the transition photon to pion transition form factor $Q^2 F_{\gamma\pi^0}(Q^2)$. The dotted and solid theoretical curves are the perturbative QCD prediction at leading and next-to-leading order, respectively, assuming the asymptotic pion distribution The data are from the CLEO collaboration [42].

MEASUREMENTS OF THE PION DISTRIBUTION AMPLITUDE BY DI-JET DIFFRACTIVE DISSOCIATION

The shape of hadron distribution amplitudes can be measured in the diffractive dissociation of high energy hadrons into jets on a nucleus. For example, consider the reaction [43–45] $\pi A \rightarrow \text{Jet}_1 + \text{Jet}_2 + A'$ at high energy where the nucleus A' is left intact in its ground state. The transverse momenta of the jets balance so that $\vec{k}_{\perp i} + \vec{k}_{\perp 2} = \vec{q}_\perp < R^{-1}{}_A$. The light-cone longitudinal momentum fractions also need to add to $x_1 + x_2 \sim 1$ so that $\Delta p_L < R_A^{-1}$. The process can then occur coherently in the nucleus. Because of color transparency and the long coherence length, a valence $q\bar{q}$ fluctuation of the pion with small impact separation will penetrate the nucleus with minimal interactions, diffracting into jet pairs [43]. The $x_1 = x$, $x_2 = 1 - x$ dependence of the di-jet distributions will thus reflect the shape of the pion valence light-cone wavefunction in x; similarly, the $\vec{k}_{\perp 1} - \vec{k}_{\perp 2}$ relative transverse momenta of the jets gives key information on the second derivative of the underlying shape of the valence pion wavefunction [44–46]. The diffractive nuclear amplitude extrapolated to $t = 0$ should be linear in nuclear number A if color transparency is correct. The integrated diffractive rate should then scale as $A^2/R_A^2 \sim A^{4/3}$.

The E791 collaboration at Fermilab has recently measured the diffractive di-jet dissociation of 500 GeV incident pions on nuclear targets [37]. The results are consistent with color transparency, and the momentum partition of the jets conforms closely with the shape of the asymptotic distribution amplitude, $\phi_\pi^{\text{asympt}}(x) = \sqrt{3}f_\pi x(1-x)$, corresponding to the leading anomalous dimension solution [19,21] to the perturbative QCD evolution equation.

THE PHOTON-TO-PION TRANSITION FORM FACTOR AND THE PION DISTRIBUTION AMPLITUDE

The simplest and perhaps most elegant illustration of an exclusive reaction in QCD is the evaluation of the photon-to-pion transition form factor $F_{\gamma \to \pi}(Q^2)$ which is measurable in single-tagged two-photon $ee \to ee\pi^0$ reactions. The form factor is defined via the invariant amplitude $\Gamma^\mu = -ie^2 F_{\pi\gamma}(Q^2)\epsilon^{\mu\nu\rho\sigma}p_\nu^\pi\epsilon_\rho q_\sigma$. As in inclusive reactions, one must specify a factorization scheme which divides the integration regions of the loop integrals into hard and soft momenta, compared to the resolution scale $\tilde{Q}$. At leading twist, the transition form factor then factorizes as a convolution of the $\gamma^*\gamma \to q\bar{q}$ amplitude (where the quarks are collinear with the final state pion) with the valence light-cone wavefunction of the pion:

$$F_{\gamma M}(Q^2) = \frac{4}{\sqrt{3}} \int_0^1 dx \phi_M(x, \tilde{Q}) T_{\gamma \to M}^H(x, Q^2). \qquad (2)$$

The hard scattering amplitude for $\gamma\gamma^* \to q\bar{q}$ is $T_{\gamma M}^H(x, Q^2) = [(1-x)Q^2]^{-1}(1 + \mathcal{O}(\alpha_s))$. The leading QCD corrections have been computed by Braaten [47]. The evaluation of the next-to-leading corrections in the physical α_V scheme is given in Ref. [48]. For the asymptotic distribution amplitude $\phi_\pi^{\text{asympt}}(x) = \sqrt{3}f_\pi x(1-x)$ one predicts $Q^2 F_{\gamma\pi}(Q^2) = 2f_\pi \left(1 - \frac{5}{3}\frac{\alpha_V(Q^*)}{\pi}\right)$ where $Q^* = e^{-3/2}Q$ is the BLM scale for the pion form factor. The PQCD predictions have been tested in measurements of $e\gamma \to e\pi^0$ by the CLEO collaboration [42]. See Fig. 1 (b). The flat scaling of the $Q^2 F_{\gamma\pi}(Q^2)$ data from $Q^2 = 2$ to $Q^2 = 8$ GeV^2 provides an important confirmation of the applicability of leading twist QCD to this process. The magnitude of $Q^2 F_{\gamma\pi}(Q^2)$ is remarkably consistent with the predicted form, assuming the asymptotic distribution amplitude and including the LO QCD radiative correction with $\alpha_V(e^{-3/2}Q)/\pi \simeq 0.12$. One could allow for some broadening of the distribution amplitude with a corresponding increase in the value of α_V at small scales. Radyushkin [49], Ong [50] and Kroll [51] have also noted that the scaling and normalization of the photon-to-pion transition form factor tends to favor the asymptotic form for the pion distribution amplitude and rules out broader distributions such as the two-humped form suggested by QCD sum rules [52].

The two-photon annihilation process $\gamma^*\gamma \to$ hadrons, which is measurable in single-tagged $e^+e^- \to e^+e^-$ hadrons events, provides a semi-local probe of $C = +$ hadron systems $\pi^0, \eta^0, \eta', \eta_c, \pi^+\pi^-$, etc. The $\gamma^*\gamma \to \pi^+\pi^-$ hadron pair process is related to virtual Compton scattering on a pion target by crossing. The leading twist amplitude is sensitive to the $1/x - 1/(1 - x)$ moment of the two-pion distribution amplitude coupled to two valence quarks [53,54].

I EXCLUSIVE TWO-PHOTON ANNIHILATION INTO HADRON PAIRS

Two-photon reactions, $\gamma\gamma \to H\overline{H}$ at large s $= (k_1 + k_2)^2$ and fixed θ_{cm}, provide a particularly important laboratory for testing QCD since these cross-channel "Compton" processes are the simplest calculable large-angle exclusive hadronic scattering reactions. The helicity structure, and often even the absolute normalization can be rigorously computed for each two-photon channel [55]. In the case of meson pairs, dimensional counting predicts that for large s, $s^4 d\sigma/dt(\gamma\gamma \to M\overline{M})$ scales at fixed t/s or $\theta_{c.m.}$ up to factors of $\ln s/\Lambda^2$. The angular dependence of the $\gamma\gamma \to H\overline{H}$ amplitudes can be used to determine the shape of the process-independent distribution amplitudes, $\phi_H(x, Q)$. An important feature of the $\gamma\gamma \to M\overline{M}$ amplitude for meson pairs is that the contributions of Landshoff pitch singularities are power-law suppressed at the Born level – even before taking into account Sudakov form factor suppression. There are also no anomalous contributions from the $x \to 1$ endpoint integration region. Thus, as in the calculation of the meson form factors, each fixed-angle helicity amplitude can be written to leading order in $1/Q$ in the factorized form $[Q^2 = p_T^2 = tu/s; \tilde{Q}_x = \min(xQ, (l - x)Q)]$:

$$\mathcal{M}_{\gamma\gamma \to M\overline{M}} = \int_0^1 dx \int_0^1 dy \phi_{\overline{M}}(y, \tilde{Q}_y) T_H(x, y, s, \theta_{c.m.}) \phi_M(x, \tilde{Q}_x), \qquad (3)$$

where T_H is the hard-scattering amplitude $\gamma\gamma \to (q\bar{q})(q\bar{q})$ for the production of the valence quarks collinear with each meson, and $\phi_M(x, \tilde{Q})$ is the amplitude for finding the valence q and $\bar{q}$ with light-cone fractions of the meson's momentum, integrated over transverse momenta $k_\perp < \tilde{Q}$. The contribution of non-valence Fock states are power-law suppressed. Furthermore, the helicity-selection rules [56] of perturbative QCD predict that vector mesons are produced with opposite helicities to leading order in $1/Q$ and all orders in α_s. The dependence in x and y of several terms in $T_{\lambda,\lambda'}$ is quite similar to that appearing in the meson's electromagnetic form factor. Thus much of the dependence on $\phi_M(x, Q)$ can be eliminated by expressing it in terms of the meson form factor. In fact, the ratio of the $\gamma\gamma \to \pi^+\pi^-$ and $e^+e^- \to \mu^+\mu^-$ amplitudes at large s and fixed θ_{CM} is nearly insensitive to the running coupling and the shape of the pion distribution amplitude:

$$\frac{\frac{d\sigma}{dt}(\gamma\gamma \to \pi^+\pi^-)}{\frac{d\sigma}{dt}(\gamma\gamma \to \mu^+\mu^-)} \sim \frac{4|F_\pi(s)|^2}{1 - \cos^2\theta_{\text{c.m.}}}. \tag{4}$$

The comparison of the PQCD prediction for the sum of $\pi^+\pi^-$ plus K^+K^- channels with recent CLEO data [57] is shown in Fig. 2. The CLEO data for charged pion and kaon pairs show a clear transition to the scaling and angular distribution predicted by PQCD [55] for $W = \sqrt{(s_{\gamma\gamma}} > 2$ GeV. It is clearly important to measure the magnitude and angular dependence of the two-photon production of neutral pions and $\rho^+\rho^-$ cross sections in view of the strong sensitivity of these channels to the shape of meson distribution amplitudes. QCD also predicts that the production cross section for charged ρ-pairs (with any helicity) is much larger that for that of neutral ρ pairs, particularly at large $\theta_{\text{c.m.}}$ angles. Similar predictions are possible for other helicity-zero mesons.

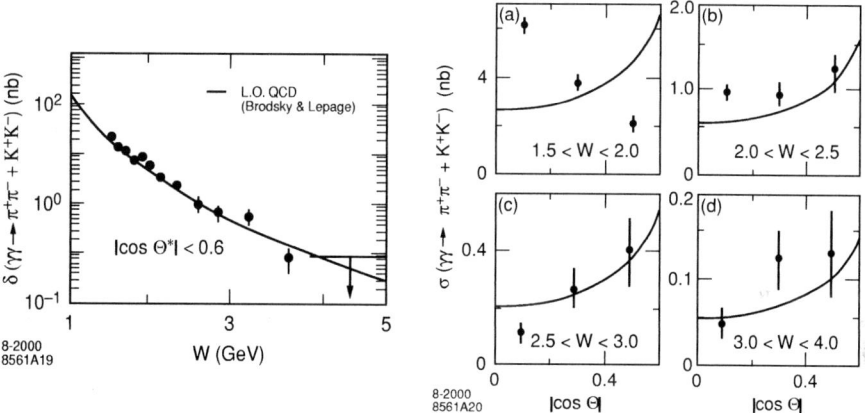

FIGURE 2. Comparison of the sum of $\gamma\gamma \to \pi^+\pi^-$ and $\gamma\gamma \to K^+K^-$ meson pair production cross sections with the scaling and angular distribution of the perturbative QCD prediction [55]. The data are from the CLEO collaboration [57].

Baryon pair production in two-photon annihilation is also an important testing ground for QCD. The only available data is the cross channel reaction, $\gamma p \to \gamma p$. The calculation of T_H for Compton scattering requires the evaluation of 368 helicity-conserving tree diagrams which contribute to $\gamma(qqq) \to \gamma'(qqq)'$ at the Born level and a careful integration over singular intermediate energy denominators [58,59,9]. Brooks and Dixon [60] have recently completed a recalculation of the Compton process at leading order in PQCD, extending and correcting earlier work. It is useful to consider the ratio $s^6 d\sigma/dt(\gamma p \to \gamma p)/t^4 F_1^2(ep \to ep)$ where $F_1(t)$ is the elastic helicity-conserving Dirac form factor since the power-law fall-off, the normalization of the valence wavefunctions, and much of the uncertainty from the scale of the

QCD coupling cancel. The scaling and angular dependence of this ratio is sensitive to the shape of the proton distribution amplitudes and appears to be consistent with the distribution amplitudes motivated by QCD sum rules. The normalization of the ratio at leading order is not predicted correctly by perturbative QCD. However, it is conceivable that the QCD loop corrections to the hard scattering amplitude are significantly larger than those of the elastic form factors in view of the much greater number of Feynman diagrams contributing to the Compton amplitude relative to the proton form factor. The perturbative QCD predictions for the phase of the Compton amplitude phase can be tested in virtual Compton scattering by interference with Bethe-Heitler processes [61].

A debate has continued [62–65] on whether processes such as the pion and proton form factors and elastic Compton scattering $\gamma p \to \gamma p$ might be dominated by higher-twist mechanisms until very large momentum transfer. If one assumes that the light-cone wavefunction of the pion has the form $\psi_{\text{soft}}(x, k_\perp) = A \exp(-b\frac{k_\perp^2}{x(1-x)})$, then the Feynman endpoint contribution to the overlap integral at small $k_\perp$ and $x \simeq 1$ will dominate the form factor compared to the hard-scattering contribution until very large Q^2. However, this ansatz for $\psi_{\text{soft}}(x, k_\perp)$ has no suppression at $k_\perp = 0$ for any x; $i.e.$, the wavefunction in the hadron rest frame does not fall-off at all for $k_\perp = 0$ and $k_z \to -\infty$. Thus such wavefunctions do not represent well soft QCD contributions. Endpoint contributions are also suppressed by the QCD Sudakov form factor, reflecting the fact that a near-on-shell quark must radiate if it absorbs large momentum. One can show [21] that the leading power dependence of the two-particle light-cone Fock wavefunction in the endpoint region is $1 - x$, giving a meson structure function which falls as $(1 - x)^2$ and thus by duality a non-leading contribution to the meson form factor $F(Q^2) \propto 1/Q^3$. Thus the dominant contribution to meson form factors comes from the hard-scattering regime. Radyushkin [63] has argued that the Compton amplitude is dominated by soft end-point contributions of the proton wavefunctions where the two photons both interact on a quark line carrying nearly all of the proton's momentum. This description appears to agree with the Compton data at least at forward angles where $-t < 10$ GeV^2. From this viewpoint, the dominance of the factorizable PQCD leading twist contributions requires momentum transfers much higher than those currently available. However, the endpoint model cannot explain the empirical success of the perturbative QCD scaling $s^7 d\sigma/dt(\gamma p \to \pi^+ n) \sim$ const at relatively low momentum transfer in pion photoproduction [66].

CONCLUSIONS

The leading-twist QCD predictions for exclusive two-photon processes such as the photon-to-pion transition form factor and $\gamma\gamma \to$ hadron pairs are based on rigorous factorization theorems. The recent data from the CLEO collaboration on $F_{\gamma\pi}(Q^2)$ and the sum of $\gamma\gamma \to \pi^+\pi^-$ and $\gamma\gamma \to K^+K^-$ channels are in excellent agreement with the QCD predictions. It is particularly compelling to see a transition in angular dependence between the low energy chiral and PQCD regimes. The success of leading-twist perturbative QCD scaling for exclusive processes at presently experimentally accessible momentum transfer can be understood if the effective coupling $\alpha_V(Q^*)$ is approximately constant at the relatively small scales Q^* relevant to the hard scattering amplitudes [48]. The evolution of the quark distribution amplitudes in the low-Q^* domain at also needs to be minimal. Sudakov suppression of the endpoint contributions is also strengthened if the coupling is frozen because of the exponentiation of a double logarithmic series.

One of the formidable challenges in QCD is the calculation of non-perturbative wavefunctions of hadrons from first principles. The recent calculation of the pion distribution amplitude by Dalley [33] using light-cone and transverse lattice methods is particularly encouraging. The predicted form of $\phi_\pi(x, Q)$ is somewhat broader than but not inconsistent with the asymptotic form favored by the measured normalization of $Q^2 F_{\gamma\pi^0}(Q^2)$ and the pion wavefunction inferred from diffractive di-jet production.

Clearly much more experimental input on hadron wavefunctions is needed, particularly from measurements of two-photon exclusive reactions into meson and baryon pairs at the high luminosity B factories. For example, the ratio $\frac{d\sigma}{dt}(\gamma\gamma \to \pi^0\pi^0)/\frac{d\sigma}{dt}(\gamma\gamma \to \pi^+\pi^-)$ is particularly sensitive to the shape of pion distribution amplitude. Baryon pair production in two-photon reactions at threshold may reveal physics associated with the soliton structure of baryons in QCD [67]. In addition, fixed target experiments can provide much more information on fundamental QCD processes such as deeply virtual Compton scattering and large angle Compton scattering.

ACKNOWLEDGMENTS

Work supported by the Department of Energy under contract number DE-AC03-76SF00515. I am grateful to Markus Diehl and Hans Paar for their input to this talk and their help as session organizers at Photon 2000.

REFERENCES

1. Szczepaniak, A, Henley, E. M. and Brodsky, S. J.,*Phys. Lett.* **B243**, 287 (1990).
2. Szczepaniak, A. *Phys. Rev.* **D54**, 1167 (1996).
3. Ball, P. and Braun, N. V., *Phys. Rev.* **D58**, 094016 (1998); hep-ph/9805422.
4. Beneke, M., Buchalla, G., Neubert, M. and Sachrajda, C. T., hep-ph/9905312.
5. Keum, Y., Li, H. and Sanda, A. I., hep-ph/0004004.
6. Keum, Y. Y., Li, H. and Sanda, A. I., hep-ph/0004173.
7. Ji, X., Talk presented at 12th Int. Symp. on High-Energy Spin Physics (SPIN96), Amsterdam, Sep. 1996, hep-ph/9610369; *Phys. Rev. Lett.* **78** (1997) 610; Ji, X., *Phys. Rev.* **D55**, 7114 (1997), hep-ph/9609381. Ji, X., *J. Phys.* **G 24**, 1181 (1998),

323

hep-ph/9807358; Ji, X. and Osborne, J., *Phys. Rev.* **D 58**, 094018 (1998), hep-ph/9801260.

8. Radyushkin, A. V. *Phys. Rev.* **D56**, 5524 (1997), hep-ph/9704207; Radyushkin, A. V., *Phys. Rev.* **D 59**, 014030 (1999), hep-ph/9805342.

9. Guichon, P. A. and Vanderhaeghen, M. *Prog. Part. Nucl. Phys.* **41**, 125 (1998), hep-ph/9806305.

10. Vanderhaeghen, M., Guichon, P. A., and Guidal, M., *Phys. Rev. Lett.* **80**, 5064 (1998).

11. Collins, J. C. and Freund, A., *Phys. Rev.* **D 59**, 074009 (1999), hep-ph/9801262.

12. Diehl, M., Feldmann, T., Jakob, R. and Kroll, P., *Phys. Lett.* **B460**, 204 (1999) hep-ph/9903268; Diehl, M., Feldmann, T., Jakob, R. and Kroll, P., *Eur. Phys. J.* **C8**, 409 (1999), hep-ph/9811253.

13. Blumlein, J. and Robaschik, D., hep-ph/0002071.

14. Penttinen, M., Polyakov, M. V., Shuvaev, A. G., and Strikman, M., hep-ph/0006321.

15. Brodsky, S. J., Close, F. E. and Gunion, J. F., *Phys. Rev.* **D 5**, 1384 (1972); *Phys. Rev.* **D 6**, 177 (1972); *Phys. Rev.* **D 8**, 3678 (1973).

16. Brodsky, S. J. and Drell, S. D., *Phys. Rev.* **D22**, 2236 (1980).

17. Brodsky, S. J., Diehl, M., and Hwang, D. S., hep-ph/0009254.

18. Diehl, M., Feldmann, T., Jakob, R. and Kroll, P., hep-ph/0009255.

19. Lepage, G. P. and Brodsky, S. J., *Phys. Lett.* **B 87**, 359 (1979).

20. Brodsky, S. J. and Lepage, G. P., SLAC-PUB-2294 *Presented at the Workshop on Current Topics in High Energy Physics*, Cal Tech., Pasadena, Calif., Feb 13-17, 1979.

21. Lepage, G. P. and Brodsky, S. J., *Phys. Rev.* **D22**, 2157 (1980).

22. Efremov, A. V. and Radyushkin, A. V., *Theor. Math. Phys.* **42**, 97 (1980).

23. Brodsky, S. J. and Lepage, G. P., *Phys. Rev. Lett.* **53**, 545 (1979); *Phys. Lett.* **87B**, 359 (1979); Lepage, G. P. and Brodsky, S. J., *Phys. Rev.* **D22**, 2157 (1980).

24. Lepage, G. P. and Brodsky, S. J., *Phys. Rev. Lett.* **43**, 545 (1979).

25. Brodsky, S. J., Frishman, Y., Lepage, G. P. and Sachrajda, C., *Phys. Lett.* **91B**, 239 (1980).

26. Shifman, M. A., Vainshtein, A. I., and Zakharov, V. I., *Nucl. Phys.* **B147**, 385 (1979).

27. Martinelli, G. and Sachrajda, C. T., *Phys. Lett.* **B190**, 151 (1987).

28. Daniel, D., Gupta, R., and Richards, D. G., *Phys. Rev.* **D43**, 3715 (1991).

29. Del Debbio, L., Di Pierro, M., Dougall, A., and Sachrajda, C., [UKQCD collaboration], *Nucl. Phys. Proc. Suppl.* **83-84**, 235 (2000), hep-lat/9909147.

30. Brodsky, S. J., Frishman,Y., Lepage, G. P., and Sachrajda, C., *Phys. Lett.* **91B**, 239 (1980).

31. Brodsky, S. J., Gardi, E., Grunberg, G., and Rathsman, J., hep-ph/0002065.

32. Braun, V. M., Derkachov, S. E., Korchemsky, G. P., and Manashov, A. N., *Nucl. Phys.* **B553**, 355 (1999) hep-ph/9902375.

33. Dalley, S., hep-ph/0007081.

34. Pauli, H.-C. and Brodsky, S. J., *Phys. Rev.* **D32**, 1993 and 2001 (1985).

35. Bardeen, W. A. and Pearson, R. B., *Phys. Rev.* **D14**, 547 (1976).

36. Burkardt, M., *Phys. Rev.* **D54**, 2913 (1996).

37. Ashery, D., [E791 Collaboration], hep-ex/9910024.

38. Petrov, V. Y., Polyakov, M. V., Ruskov, R., Weiss, C., and Goeke, K., *Phys. Rev.* **D59**, 114018 (1999), hep-ph/9807229.
39. Diakonov, D. and Petrov, V. Y., hep-ph/0009006.
40. Hecht, M. B., Roberts, C. D., and Schmidt, S. M., nucl-th/0008049.
41. Srivastava, P. P. and Brodsky, S. J., *Phys. Rev.* **D61**, 025013 (2000), hep-ph/9906423, and in preparation.
42. Gronberg, J., *et al.* [CLEO Collaboration], *Phys. Rev.* **D57**, 33 (1998), hep-ex/9707031.
43. Bertsch, G., Brodsky, S. J., Goldhaber, A. S., and Gunion, J. F., *Phys. Rev. Lett.* **47**, 297 (1981).
44. Frankfurt, L., Miller, G. A,., and Strikman, M., *Phys. Lett.* **B304**, 1 (1993), hep-ph/9305228.
45. Frankfurt, L., Miller, G. A., and Strikman, M., hep-ph/9907214.
46. Brodsky, S., Diehl, M., Hoyer, P., and Peigne, S., in preparation.
47. Braaten, E. and Tse, S.-M., *Phys. Rev.* **D35**, 2255 (1987).
48. Brodsky, S. J., Ji, C., Pang, A., and Robertson, D. G., hep-ph/9705221.
49. Radyushkin, A. V., *Acta Phys. Polon.* **B26**, 2067 (1995).
50. Ong, S., *Phys. Rev.* **D52**, 3111 (1995).
51. Kroll, P. and Raulfs, M., *Phys. Lett.* **387B**, 848 (1996).
52. Chernyak, V. L. and Zhitnitsky, A. R., *Phys. Rep.* **112**, 173 (1984).
53. Muller, D., Robaschik, D., Geyer, B., Dittes, F. M., and Horejsi, J., *Fortsch. Phys.* **42**, 101 (1994), hep-ph/9812448.
54. Diehl, M., Gousset, T., and Pire, B., hep-ph/0003233.
55. Brodsky, S. J. and Lepage, G.P., *Phys. Rev.* **D24**, 1808 (1981).
56. Brodsky, S. J. and Lepage, G.P., *Phys. Rev.* **D24**, 2848 (1981).
57. Paar, H., *et al.* CLEO collaboration (to be published); See also Boyer, J. *et al.*, *Phys. Rev. Lett.* **56**, 207 (1980); TPC/Two Gamma Collaboration (H. Aihara *et al.*), *Phys. Rev. Lett.* **57**, 404 (1986).
58. Farrar, G. R. and Zhang, H., *Phys. Rev. Lett.* **65**, 1721 (1990).
59. Kronfeld, A. S. and Nizic, B., *Phys. Rev.* **D44**, 3445 (1991).
60. Brooks, T. and Dixon, L., hep-ph/0004143.
61. Brodsky, S. J., Close, F. E., and Gunion, J. F., *Phys. Rev.* **D6**, 177 (1972).
62. Isgur, N. and Llewellyn Smith, C. H., *Phys. Lett.* **B217**, 535 (1989).
63. Radyushkin, A. V., *Phys. Rev.* **D58**, 114008 (1998), hep-ph/9803316.
64. Bolz, J. and Kroll, P., *Z. Phys.* **A356**, 327 (1996), hep-ph/9603289.
65. Vogt, C., this conference, hep-ph/0010040.
66. Anderson, R. L., *et al.*, *Phys. Rev. Lett.* **30**, 627 (1973).
67. Sommermann, H., Seki, M. R., Larson. S. and Koonin, S. E., *Phys. Rev.* **D45**, 4303 (1992); Brodsky, S. J., and Karliner, M., in preparation.

Interference of the Two-Photon and Bremsstrahlung Production for the $\pi^+\pi^-$ System at B- and ϕ-Factories

I.F. Ginzburg[1,2], A. Schiller[3], and V.G. Serbo[2,1]

[1] *Institute of Mathematics, Novosibirsk, 630090 Russia*
[2] *Novosibirsk State University, Novosibirsk, 630090 Russia*[1]
[3] *Institut für Theoretische Physik, Universität Leipzig, D-04109 Leipzig, Germany*

Abstract. In the process $e^-e^+ \to e^-e^+\pi^+\pi^-$ the two-pion system is produced via two mechanisms — two–photon and bremsstrahlung ones. The first of them gives a C-even system, the second one — C-odd. We study the charge asymmetry of pions in the differential pion momenta cross section obliged by an interference between these two mechanisms. At low effective mass of dipion this asymmetry is directly related to the s- and p-phases of elastic $\pi\pi$ scattering. At higher energies it can give new information about $f_0(400 - 1300)$ meson (former σ), $f_0(980)$ meson, etc. We discuss the magnitude of the charge asymmetry and how this effect can be seen at modern colliders.

I   INTRODUCTION

We consider the process $e^-e^+ \to e^-e^+\pi^+\pi^-$ in which the $\pi^+\pi^-$ system can be produced either via the two-photon mechanism (the amplitude $\mathcal{M}_1$ of Fig. 1) or via bremsstrahlung (the amplitudes $\mathcal{M}_2$ and $\mathcal{M}_3$ of Fig. 2). In the differential cross section the interference of these mechanisms results in terms which are antisymmetric under $\pi^+ \leftrightarrow \pi^-$ exchange. They could give information about the relative phases of production amplitudes.

At low effective masses of dipions this interference is directly related to the difference of s- and p-phase shifts of the elastic $\pi\pi$ scattering. These phase shifts are of primary importance for low energy hadron physics, in particular, for the chiral dynamics which pretends to be the "low energy QCD" [1], [2]. At higher energies the interference can give new information about the $f_0(400 - 1200)$ meson (former σ), the $f_0(980)$ meson, etc. Their nature is now the subject of wide discussions.

Such C-odd effects were firstly studied almost three decades ago in Ref. [3] for small total transverse momenta of the produced pion pair, $\mathbf{k}_\perp^2 \ll m_\pi^2$. However,

[1] This work was partially supported by grants RFBR 00-02-17592, 00-15-96691, "Universities of Russia" 015.0201.16 and Sankt-Petersburg Center of Higher Education.

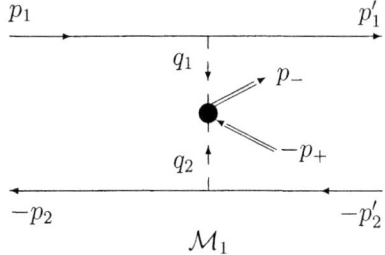

$$\mathcal{M}_1$$

FIGURE 1. Amplitude $\mathcal{M}_1$ for the two-photon production of pions. The colliding particles with initial 4-momenta (energies) p_1 (E_1) and p_2 (E_2) and final momenta (energies) p'_1 (E'_1) and p'_2 (E'_2) emit virtual photons with $q_i = p_i - p'_i$ $(\omega_i = E_i - E'_i)$. These photons produce the C-even $\pi^+\pi^-$ system with total 4-momentum $k = p_+ + p_- = q_1 + q_2$ and effective mass $W = \sqrt{k^2}$, furthermore $s = (p_1 + p_2)^2 = 4E_1E_2$.

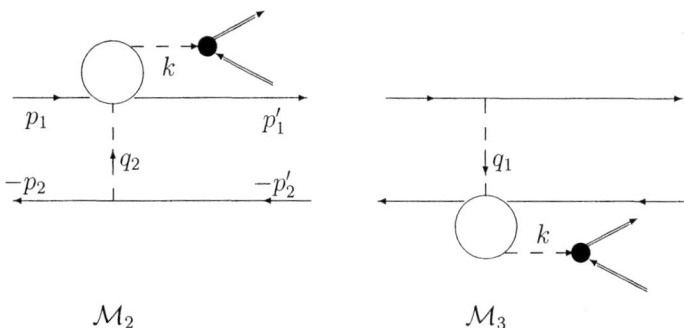

$$\mathcal{M}_2 \qquad\qquad \mathcal{M}_3$$

FIGURE 2. Amplitudes $\mathcal{M}_2$ and $\mathcal{M}_3$ for the bremsstrahlung production of pions. The pion pair in C-odd state is produced by one virtual photon with 4-momentum $k = p_+ + p_-$ emitted by the electron ($\mathcal{M}_2$) or by the positron ($\mathcal{M}_3$). The open circles represent the virtual Compton scattering.

this region gives only a small fraction of the entire charge asymmetry discussed. In the present paper we obtain formulae which allow to study the charge asymmetry in the main region.

A similar problem was discussed in the work [5] for the ep-scattering with respect to HERA experiments. Unfortunately, the results obtained in that paper are incorrect. The same process (including the charge asymmetry of pions) was considered in Ref. [6], however for the region of large virtualities of one photon, $-q_1^2 \gg W^2$, where the number of events is considerably smaller than that in the main region discussed in the present paper.

Note, that recently a charge asymmetry was observed at CLEO in the $e^+e^- \rightarrow e^+e^-\pi^+\pi^-$ reaction detecting additionally an electron scattered at large angle [4].

Based on the collider parameters the expected number of events **per year** will be about 10^6 pion pairs at the collider DAΦNE and 10^8 pion pairs at PEP-II. We will show that the charge asymmetry is of the order of 1% and that the signal to background ratio can be significantly improved choosing appropriate cuts.

We consider an experimental set-up when only pion momenta $\mathbf{p}_+$ and $\mathbf{p}_-$ are measured (so called *no tag* experiments). It corresponds to the cross section $d\sigma/(d^3p_+\, d^3p_-)$ for the $e^-e^+ \to e^-e^+\pi^+\pi^-$ process.

II   QUALITATIVE DESCRIPTION OF DIFFERENT CONTRIBUTIONS

The cross section of the process can be written via the amplitudes $\mathcal{M}_j$ shown in Figs. 1—2

$$d\sigma = d\sigma_{C=+1} + d\sigma_{C=-1} + d\sigma_{\text{interf}}$$

where

$$d\sigma_{C=+1} \propto |\mathcal{M}_1|^2, \quad d\sigma_{C=-1} = d\sigma_2 + d\sigma_3 \propto |\mathcal{M}_2|^2 + |\mathcal{M}_3|^2,$$

$$d\sigma_{\text{interf}} = d\sigma_{12} + d\sigma_{13}, \quad d\sigma_{12} \propto 2\,\text{Re}(\mathcal{M}_2^*\mathcal{M}_1), \quad d\sigma_{13} \propto 2\,\text{Re}(\mathcal{M}_3^*\mathcal{M}_1).$$

Let us discuss these contributions qualitatively.

The two-photon mechanism (Fig. 1) produces C-even dipions (see review [7]). It provides the main contribution to the total cross section. The corresponding part of the cross section $d\sigma_{C=+1}$ can be expressed via the amplitudes M_{ab} describing the collisions of virtual photons with helicities a and b ($a, b = \pm 1,\ 0$). Its dominant part is given by the region $q_1^2 \approx 0$ and $q_2^2 \approx 0$. The produced pairs are distributed almost uniformly over their total rapidity and they are peaked at small values of their total transverse momentum $k_\perp$.

The bremsstrahlung mechanism (Fig. 2) produces pion pairs in C-odd state. Its contribution to the cross section $d\sigma_{C=-1}/(d^3p_+\, d^3p_-)$ was calculated in Ref. [8]. It is proportional to $|F_\pi(k^2)|^2$ where F_π is the formfactor of the pion. The main contribution to $d\sigma_2$ is given by the region where q_2 is almost real. In the kinematic region being essential for both two-photon production and interference, the $k_\perp$ distribution of the pions is relatively wide.

Note that both contributions $d\sigma_1$ and $d\sigma_2$ are **charge symmetric**, they do not change under pion exchange $\pi^+ \leftrightarrow \pi^-$.

The interference of C-even and C-odd contributions $d\sigma_{\text{interf}} = d\sigma_{12} + d\sigma_{13}$ where $d\sigma_{12} \propto \text{Re}(\mathcal{M}_2^*\mathcal{M}_1)$ and $d\sigma_{13} \propto \text{Re}(\mathcal{M}_3^*\mathcal{M}_1)$. This contribution $d\sigma_{\text{interf}}$ is **antisymmetric** under pion exchange $\pi^+ \leftrightarrow \pi^-$. Therefore, this interference determines the **charge asymmetry of pions**.

The main contribution to $d\sigma_{12}$ is given by an almost real photon q_2. The produced pions fly mainly along the electron, $k_z = p_{+z} + p_{-z} > 0$, and the dipion transverse momentum distribution is not peaked at small $k_\perp$. Therefore the transverse momentum of the electron is not small, $\mathbf{q}_{1\perp} = \mathbf{k}_\perp$.

III RESULTS

After integrating over transverse momenta of the scattered leptons, we obtain

$$d\sigma_{12} = [G_{++} \operatorname{Re}(F_\pi^* M_{++}) + G_{+-} \operatorname{Re}(F_\pi^* M_{+-}) + G_{0+} \operatorname{Re}(F_\pi^* M_{0+})] \frac{d^3 p_+ d^3 p_-}{\varepsilon_+ \varepsilon_-}$$

where G_{ab} are simple analytical functions given in Ref. [9]. In the region of small transverse momentum of the produced pair $k_\perp \to 0$ our result for $d\sigma_{12}$ coincides with that of Ref. [3].

Variables. We consider the symmetric and antisymmetric combinations of pion 4-momenta:

$$k = p_+ + p_-, \qquad \Delta = p_+ - p_-.$$

The discussed charge asymmetry is proportional to components of the 4-vector Δ. To describe that asymmetry in our calculations, we use the variables

$$K_- = \frac{(p_2 - p_1)\Delta}{(p_2 + p_1)k} = \left\{ \frac{p_{+z} - p_{-z}}{\varepsilon_+ + \varepsilon_-} \right\}_{e^- e^+ \text{ c.m.s.}}, \quad v = \mathbf{p}_{+\perp}^2 - \mathbf{p}_{-\perp}^2 = \mathbf{k}_\perp \mathbf{\Delta}_\perp.$$

The "longitudinal" variable is K_- and the "transverse" variable is v. Note that K_- is proportional to the difference of the longitudinal momenta of π^+ and π^- in the $e^- e^+$ center-of-mass.

Studied Quantities. To take into account the symmetry properties and to summarize different contributions, it is natural to introduce the following quantities related to the charge asymmetry of pions

$$\Delta \sigma_K = \int_{\mathcal{D}} \epsilon(K_-) d\sigma, \quad \Delta \sigma_v = \int_{\mathcal{D}} \epsilon(v) \epsilon(x - y) d\sigma.$$

where $x = (\varepsilon + k_z)/(2E_1)$, $y = (\varepsilon - k_z)/(2E_2)$ with $\varepsilon = \varepsilon_+ + \varepsilon_-$. The function $\epsilon(z)$ is given by the standard step function $\theta(z)$: $\epsilon(z) = \theta(z) - \theta(-z)$.

The Model and Results. For the pion pair production the **strong interaction effects** are of primary interest. Nevertheless, the pure **Born QED model** (point-like pions) gives a reasonable but rough description of the physical picture near the threshold.

First, we present values of the entire effect, i.e. $\Delta \sigma_K$ and $\Delta \sigma_v$, and of the background $\sigma_B = \sigma_{C=+1} + \sigma_{C=-1}$ at different c.m.s. energies $\sqrt{s}$ for the production of pion pairs (see Table 1).

For the collider DAΦNE with an expected annual luminosity $\mathcal{L} = 5$ fb^{-1} it corresponds to $N_B = 2 \cdot 10^6$ pion pairs with a signal of about $N_S = 30\,000$ events, i.e. the **signal to background ratio (S/B)** and the **statistical significance (SS)** are equal to

$$\frac{S}{B} = \frac{N_S}{N_B} = 1.5\%, \quad SS = \frac{N_S}{\sqrt{N_B}} = 21.$$

TABLE 1. Effects and background

$\sqrt{s}$, GeV	1	4	10	200
$\Delta\sigma_K$, pb	-6.1	-26	-35	-56
$\Delta\sigma_v$, pb	3.3	17	27	51
σ_B, pb	420	2900	6200	27000

For the collider PEP-II with annual luminosity $\mathcal{L} = 30$ fb^{-1} it corresponds to $1.9 \cdot 10^8$ pion pairs which leads to about 10^6 events, i.e. $S/B = 0.6\%$ and $SS = 77$.

In Fig. 3 we present the distribution of the effect and background over $W = \sqrt{k^2}$ — the invariant mass of the $\pi^+\pi^-$ system. It can be seen that both signal and background are concentrated near the threshold.

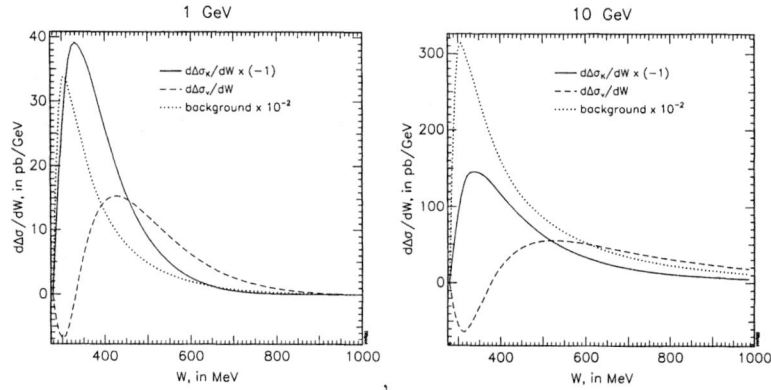

FIGURE 3. Distribution over W of the contributions $\Delta\sigma_K$ and $\Delta\sigma_v$ and of the background at $\sqrt{s} = 1$ GeV and 10 GeV.

Next we consider two different invariant mass regions:
1) $W = 300 \div 350$ MeV — near the threshold
2) $W = 475 \div 525$ MeV — far from the threshold but below the f_2 resonance and average over an interval $\Delta W = 50$ MeV.
By this choice we would like to demonstrate that signal and background have different distributions over the total pair transverse momentum $k_\perp$. As a result, both **S/B** ratio and **SS** can be considerably improved by applying some cut in $k_\perp$.

In Fig. 4 we show curves for signal and background integrated over $k_\perp > k_0$, where k_0 is a cut-off from below. We can see that the background drops considerably faster than the signal.

A similar behavior can be demonstrated for signal and background contributions integrated over $x > x_0$ where $x = (\varepsilon + k_z)/(2E_1)$ (for ultra-relativistic pions flying along the initial electron or positron momentum, the quantity x is the fraction of energy transferred from the electron to the $\pi^+\pi^-$ system).

Some numerical examples:

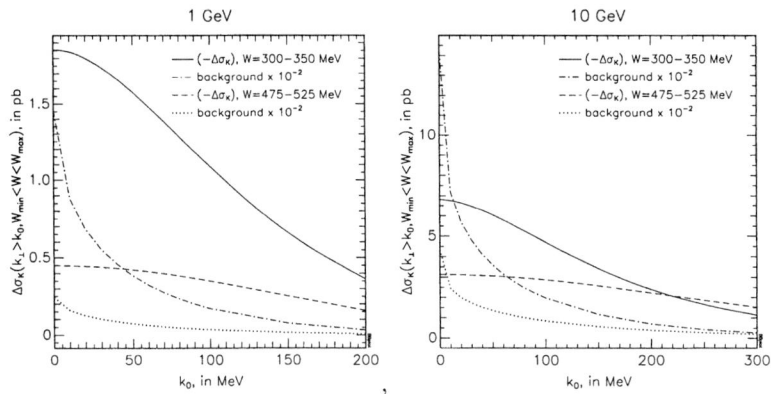

FIGURE 4. The contribution $\Delta\sigma_K$ and the background at $\sqrt{s} = 1$ GeV and 10 GeV for the intervals $W = 300 \div 350$ MeV and $W = 475 \div 525$ MeV and $k_\perp > k_0$ is function of k_0.

DAΦNE (region $W = 300 \div 350$ MeV):
without cuts $\Delta\sigma_K = -1.85$ pb, $\sigma_B = 145$ pb, $S/B = 1.3\%$;
with cuts $k_\perp > k_0 = 0.1$ GeV and $0.3 < x < 0.9$
$\Delta\sigma_K = -0.71$ pb, $\sigma_B = 10.4$ pb, $S/B = 6.8\%$, $SS = 16$.
PEP-II (region $W = 475 \div 525$ MeV): with cuts $k_\perp > k_0 = 0.15$ GeV and $0.3 < x < 0.95$
$\Delta\sigma_K = -0.81$ pb, $\sigma_B = 8.6$ pb, $S/B = 9.5\%$, $SS = 48$.

We have also analyzed the charge asymmetry of **muons** in the process $e^-e^+ \to e^-e^+\mu^+\mu^-$.

IV   CONCLUSION

1. The measurement of the charge asymmetry of pions in the $e^+e^- \to e^+e^-\pi^+\pi^-$ process can give an essential information about phase shifts of the elastic $\pi\pi$ scattering near the threshold.

2. The analytical expression for this charge asymmetry, i.e. for the contribution $d\sigma_{\text{interf}}/(d^3p_+d^3p_-)$, is obtained.

3. The numerical analysis of the pion charge asymmetry is performed in the framework of a QED model. The corresponding number of events was found large enough to be studied at B- and ϕ-factories.

4. The **S/B** ratio is about 1 % and can be considerably improved by applying appropriate cuts in the $k_\perp$ and x distributions.

REFERENCES

1. The Second DAΦNE Physics Handbook. Eds. L. Maiani, G. Pancheri, N. Paver (INFN, Frascati, 1995)

2. Int. Workshop "e^+e^- collisions from ϕ to J/Ψ" (Novosibirsk, March 1-5, 1999).

3. V.L. Chernyak, V.G. Serbo. Nucl. Phys. **B 67** (1973) 464.

4. V. Savinov (CLEO), private communication.

5. M. Galynsky, E.A. Kuraev, P.G. Ratcliffe, B.G. Shaikhatdenov, hep-ph/000361

6. M. Diehl, T. Gousset, B. Pire, hep-ph/0003233.

7. V.M. Budnev, I.F. Ginzburg, G.V. Meledin, V.G. Serbo, *Phys. Rept.* **15C** (1975) 575.

8. G.L. Kotkin, V.G. Serbo. *Yad. Fiz. [Sov. Jour. of Nucl. Phys.]* **21** (1975) 785

9. I.F. Ginzburg, A. Schiller, V.G. Serbo. Charge asymmetry of pions in the process $e^-e^+ \rightarrow e^-e^+\pi^+\pi^-$, (in preparation).

$\gamma^*\gamma \to \pi\pi$ at large Q^2

Markus Diehl[*1], Thierry Gousset[†], and Bernard Pire[¶]

Stanford Linear Accelerator Center, Stanford University, Stanford, CA 94309, U.S.A.
†*SUBATECH, B.P. 20722, 44307 Nantes, France* [2]
¶*CPhT, Ecole Polytechnique, 91128 Palaiseau, France* [3]

Abstract. The QCD analysis of the process $\gamma^*\gamma \to \pi\pi$ at large Q^2 and small center-of-mass energy allows one to access a new hadronic observable describing the exclusive transition from a $q\bar{q}$ or gg state to a pair of mesons. A fruitful study may be envisaged at existing machines.

I INTRODUCTION

Exclusive hadron production in two-photon collisions provides a tool to study a variety of fundamental aspects of QCD and has long been a subject of great interest (cf. [1–3] and references therein). Recently a new aspect of this has been pointed out, namely the physics of the process $\gamma^*\gamma \to \pi\pi$ in the region where Q^2 is large but W^2 small [4,5]. This process factorizes [6,7] into a perturbatively calculable, short-distance dominated scattering $\gamma^*\gamma \to q\bar{q}$ or $\gamma^*\gamma \to gg$, and non-perturbative matrix elements measuring the transitions $q\bar{q} \to \pi\pi$ and $gg \to \pi\pi$. We call these matrix elements generalized distribution amplitudes (GDAs) to emphasize their close connection to the distribution amplitudes introduced long ago in the QCD description of exclusive hard processes [8].

II FACTORIZATION

We are interested in $e + \gamma \to e + \pi\pi$. To lowest order in QED the interaction between the lepton and the hadron side is mediated by one-photon exchange. The contribution we focus on is the one where the $\gamma^*\gamma \to \pi\pi$ subprocess appears, as depicted in Fig. 1. The two diagrams where both photons attach to the lepton line are referred to as the bremsstrahlung contribution. We notice that in this case the pions are in a C-odd state, in contrast to the former one where they emerge in a

---

[1] Supported by the Feodor Lynen Program of the Alexander von Humboldt Foundation.
[2] Unité mixte 6457 de l'Université de Nantes, de l'École des Mines de Nantes et de l'IN2P3/CNRS
[3] Unité mixte 7644 du CNRS

C-even state. The phenomenological interest of this observation will be discussed in Sect. IV.

The kinematical regime in which we study the reaction is that of large $Q^2 = -q^2$, where $q = k - k'$ is the momentum tranferred from the electron (see Fig. 1), and small $W^2 = (p+p')^2$. To be specific we shall explore the domain $Q^2 \geq 4$ GeV^2 and $W^2 \leq 1$ GeV^2. The Q^2 at which the leading contribution to be discussed starts to drive the cross section is, however, a matter of experimental determination because present theory can at best estimate the size of power corrections [9].

Let us now discuss the factorization of the process at large Q^2. To do this it is useful to visualize the process in the Breit frame, where the incoming real photon of momentum $-\frac{1}{2}(Q+W^2/Q)\,\hat{z}$ collides with a static virtual photon of momentum $Q\hat{z}$ and the pion pair emerges with momentum $\frac{1}{2}(Q - W^2/Q)\,\hat{z}$. Neglecting W^2 compared with Q^2, a simple spacetime cartoon can be drawn (see Fig. 2), which in addition shows the momentum sharing between the two pions on one hand, and between the two partons on the other hand.

Although a complete proof of factorization relies on a very detailed study of loop corrections, we may motivate it by emphasizing the similarities between the one-pion and the two-pion channels. The result of factorization is that the amplitude of the $\gamma^*\gamma \to \pi\pi$ reaction is a combination of a short distance transition from $\gamma^*\gamma$ to

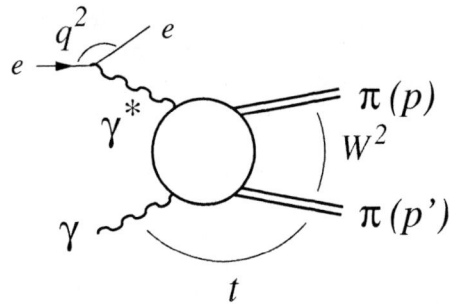

FIGURE 1. The subprocess $\gamma^*\gamma \to \pi\pi$ in the reaction $e + \gamma \to e + \pi + \pi$.

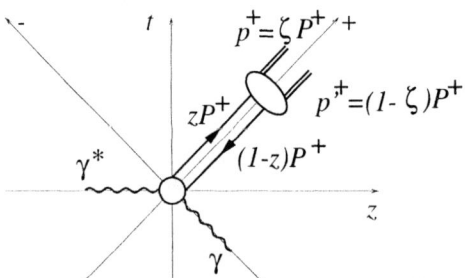

FIGURE 2. Spacetime cartoon for $\gamma^*\gamma \to \pi\pi$ in the Breit frame.

two partons (two quarks or two gluons) and a long distance process of hadronization of two partons into two hadrons.

At leading order in α_S, the two photons couple to a $q\bar{q}$ pair. The relevant diagrams are displayed in Fig. 3. The matrix element of the time ordered product of the two electromagnetic current reads

$$T^{\mu\nu} = -g_T^{\mu\nu} \sum_q \frac{e_q^2}{2} \int_0^1 dz \, \frac{2z-1}{z(1-z)} \, \Phi_q^+(z,\zeta,W^2), \tag{1}$$

where Φ_q^+ is the quark component of the two-pion distribution amplitude. The index + indicates that the process selects the C-even component of the quark GDA. We notice that the amplitude (1) is independent of Q^2. This is understood to be true up to logarithmic scaling violation. We also remark that both photons have the same helicity. This property is specific to the $q\bar{q}$ channel. In the gluon channel, i.e., for the $\gamma^*\gamma \to gg$ subprocess, the photons can have opposite helicities [10], but still the virtual photon has to be transversely polarized. The phenomenological consequences of all these facts will be discussed in Sect. IV.

III   THE GENERALIZED DISTRIBUTION AMPLITUDE

The operator definition of the two-pion distribution amplitude is, in $A^+ = 0$ gauge,

$$\Phi_q(z,\zeta,W^2) = \int \frac{dx^-}{2\pi} e^{-iz(P^+ x^-)} \langle \pi(p)\pi(p')|\bar{q}(x^-)\gamma^+ q(0)|0\rangle,$$

$$\Phi_g(z,\zeta,W^2) = \frac{1}{P^+} \int \frac{dx^-}{2\pi} e^{-iz(P^+ x^-)} \langle \pi(p)\pi(p')|F^{+\mu}(x^-)F_\mu^+(0)|0\rangle.$$

The standard analysis of QCD radiative corrections shows that the amplitude (1) is modified by logarithms of Q^2. The leading logarithmic corrections may be factorized into the distribution amplitude, which thus depends on a factorization scale μ^2 through a linear combination of terms $(\log \mu^2)^{-d_n}$ with anomalous dimensions d_n. We refer the reader to Ref. [5] for a review in the present context. We only

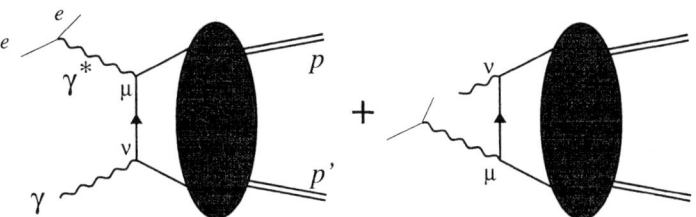

FIGURE 3. Leading order amplitude.

mention here that in the considered channel quarks and gluons mix under evolution and that all d_n's are positive, except for one which is zero. This implies that the GDA tends to a non vanishing asymptotic form when $\mu^2 \to \infty$. This asymptotic form reads:

$$\sum_{q=1}^{n_f} \Phi_q^+(z, \zeta, W^2) = 18 n_f z(1-z)(2z-1)$$

$$\times \left[B_{10}(W^2) + B_{12}(W^2) P_2(2\zeta - 1) \right],$$

$$\Phi_g(z, \zeta, W^2) = 48 z^2 (1-z)^2$$

$$\times \left[B_{10}(W^2) + B_{12}(W^2) P_2(2\zeta - 1) \right].$$

One finds that the first subasymptotic term has the same polynomials dependence in z and ζ, whereas solutions with yet higher d_n involve polynomials of higher degree in both z and ζ.

One interesting phenomenological consequence of the interrelation of z and ζ is that, since the ζ dependence may be rewritten as a partial wave expansion, the $\cos \theta$ dependence of the cross section provides information on the z behavior of Φ. The latter would be otherwise difficult to extract since the amplitude is given as an integral over this variable (see Eq. (1)).

Let us now construct a simple model for the GDA. To perform this task, we take advantage of the energy-momentum sum-rules:

$$\int_0^1 dz\, (2z-1)\, \Phi_q(z, \zeta, W^2) = \frac{2}{(P^+)^2} \langle \pi^+(p)\pi^-(p')|T_q^{++}(0)|0\rangle,$$

$$\int_0^1 dz\, \Phi_g(z, \zeta, W^2) = \frac{1}{(P^+)^2} \langle \pi^+(p)\pi^-(p')|T_g^{++}(0)|0\rangle,$$

and of the form factor decomposition of $T^{\mu\nu}$:

$$\langle \pi(p)\pi(p')|T_{q,g}^{\mu\nu}(0)|0\rangle = \frac{1}{2} T_{q,g}^{(1)}(W^2) \left[(p+p')^\mu (p+p')^\nu - W^2 g^{\mu\nu} \right]$$

$$+ \frac{1}{2} T_{q,g}^{(2)}(W^2) (p-p')^\mu (p-p')^\nu.$$

We have to perform an analytic continuation from W^2 to 0, which leads to

$$\frac{9 n_f}{10} B_{12}(0) = \sum_q T_q^{(2)}(0) = R_\pi$$

where $R_\pi \approx 50\%$ is the total momentum fraction carried by quarks and antiquarks in a pion.

We simplify the discussion on energy dependence by using first that below the inelastic threshold the partial wave phases are related to the phase shifts $\delta_0(W^2)$

and $\delta_2(W^2)$ of $\pi\pi$ elastic scattering as a consequence of Watson's theorem [11]. We then assume $|B_{12}(W^2)|$ to be constant and thus equal to $B_{12}(0)$. Finally, we estimate B_{10} through a soft pion theorem that gives $B_{10}(0) = -B_{12}(0)$ [11].

Our model two-pion distribution amplitude thus reads

$$\Phi_{u/d} = 10z(1-z)(2z-1)R_\pi \left[\frac{\beta^2-3}{2} e^{i\delta_0(W^2)} + \beta^2 e^{i\delta_2(W^2)} P_2(\cos\theta) \right]$$

IV PHENOMENOLOGY

Let us now briefly outline some useful phenomenological features of the $e\gamma \rightarrow e\pi\pi$ process (see Ref. [5] for more details).

There are three independent helicity amplitudes A_{++}, A_{0+} and A_{-+}, but the first one dominates at large Q^2 and may be written as

$$A_{++} = \sum_q \frac{e_q^2}{2} \int_0^1 dz \frac{2z-1}{z(1-z)} \Phi_q^{\pi\pi}(z,\zeta,W^2),$$

at leading order in α_S. The two other amplitudes are non-leading:

$$A_{0+}/A_{++} \propto 1/Q, \qquad A_{-+}/A_{++} \propto \alpha_S(Q^2).$$

In the case of $\pi^+\pi^-$ production, the cross section gets a contribution from the Bremsstrahlung process and can be decomposed as

$$d\sigma = d\sigma_B + d\sigma_I + d\sigma_G.$$

The interference with the bremsstrahlung process allows us to access the $\gamma^*\gamma$ process at the amplitude level, thanks to the different C-conjugation properties of the two processes. This contribution is selected by charge asymmetries such as

$$d\sigma(\pi^+(p)\pi^-(p')) - d\sigma(\pi^-(p)\pi^+(p'))$$

and by angular distributions derived from this. This interference part may be written as

$$d\sigma_I(Q^2, W^2, \cos\theta, \varphi) \propto e_l \left(C_0 + C_1 \cos\varphi + C_2 \cos 2\varphi + C_3 \cos 3\varphi \right)$$

with the dominant term at large Q^2 given by

$$C_1 = \mathrm{Re}\left\{ F_\pi^* A_{++} \right\} [1 - (1-x)(1+\epsilon)] \sin\theta$$
$$- \mathrm{Re}\left\{ F_\pi^* A_{0+} \right\} \sqrt{2x(1-x)} \, 2\epsilon \cos\theta$$
$$+ \mathrm{Re}\left\{ F_\pi^* A_{-+} \right\} (1-x) \sin\theta$$

With our model distribution presented above we estimate counting rates around 10^4 events for an integrated luminosity of 20–30 fb^{-1} at an e^+e^- c.m. energy of 10 GeV. This would allow one to measure the interference term with some $O(10\%)$ statistical errors.

V SUMMARY

The study of $\gamma^*\gamma \rightarrow \pi^+\pi^-$, $\pi^0\pi^0$ at large photon virtuality and small c.m. energy is a powerful new tool for investigating the confining mechanism which takes place in the transformation of quarks or gluons into two mesons. The simplest model for the two-pion distribution amplitude uses the asymptotic solution to the evolution equations, $\pi\pi$ elastic phase shifts, and R_π.

Encouraging rates are predicted for the kinematics and luminosity of existing B factories, and one may expect data to come soon from the BABAR, BELLE, and CLEO experiments.

The same GDA appears in deep electroproduction processes too [12], where it opens the possibility of extracting skewed parton distributions from 2π electroproduction without careful separation of resonant ρ vs. non-resonant 2π production.

ACKNOWLEDGMENTS

We acknowledge useful discussions with O.V. Teryaev.

REFERENCES

1. H. Terazawa, Rev. Mod. Phys. **45**, 615 (1973).
2. V.M. Budnev *et al.*, Phys. Rept. **15C**, 181 (1975).
3. S.J. Brodsky, hep-ph/9708345, proc. of PHOTON 97, Egmond aan Zee, Netherlands, May 1997, World Scientific;
   M.R. Pennington, Nucl. Phys. **B** (Proc. Suppl.) **82**, 291 (2000), hep-ph/9907353.
4. M. Diehl, T. Gousset, B. Pire and O.V. Teryaev, Phys. Rev. Lett. **81**, 1782 (1998), hep-ph/9805380.
5. M. Diehl, T. Gousset, B. Pire , Phys. Rev. **D62**, 73014 (2000), hep-ph/0003203.
6. D. Müller *et al.*, Fortschr. Phys. **42**, 101 (1994), hep-ph/9812448.
7. A. Freund, Phys. Rev. **D61**, 074010 (2000), hep-ph/9903489.
8. G.P. Lepage and S.J. Brodsky, Phys. Rev. **D22**, 2157 (1980).
9. N. Kivel and L. Mankiewicz, hep-ph/0008168.
10. N. Kivel, L. Mankiewicz and M. V. Polyakov, Phys. Lett. **B467**, 263 (1999), hep-ph/9908334.
11. M.V. Polyakov, Nucl. Phys. **B555**, 231 (1999), hep-ph/9809483.
12. M. Diehl, T. Gousset and B. Pire, to appear in the Procs. of the Workshop on Exclusive and Semiexclusive Processes at High Momentum Transfer, Jefferson Lab, Newport News, VA, USA, May 1999, hep-ph/9909445;
   B. Lehmann-Dronke *et al.*, Phys. Lett. **B475**, 147 (2000), hep-ph/9910310.

Cross Section Measurement of τ Pairs in Two-Photon Collisions with the L3 detector at LEP 2

Daniel Haas[†]

University of Basle
Institute for Physics
Klingelbergstrasse 82
CH – 4056 Basel
[†] *email:Daniel.Haas@cern.ch*

Abstract. The production of τ pairs in $\gamma\gamma$ collisions is studied with the L3 detector at LEP. Data were collected at $\sqrt{s}$ = 189 - 202 GeV for a total integrated luminosity of 393.0 pb^{-1}. An exclusive decay channel is considered, with $\tau^{\pm} \rightarrow e^{\pm}\nu_{\tau}\nu_{e}$ and $\tau^{\mp} \rightarrow \rho^{\mp}\nu_{\tau}$, with $\rho^{\mp} \rightarrow \pi^{\mp}\pi^{0}$. The cross section $\sigma(e^{+}e^{-} \rightarrow e^{+}e^{-}\tau^{+}\tau^{-})$ is compared to $\mathcal{O}(\alpha^{4})$ QED calculations. The cross section of τ prduction is measured in $\gamma\gamma$ collisions for the first time at LEP 2.

INTRODUCTION

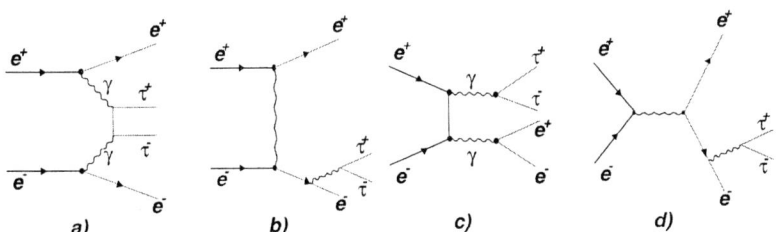

FIGURE 1. Principal Feynman diagrams of $\mathcal{O}(\alpha^{4})$ of the two-photon process $e^{+}e^{-} \rightarrow \tau^{+}\tau^{-}$: a) Multiperipheral, b) Bremsstrahlung, c) Conversion, d) Annihilation

Two-photon physics offers a wide field of research at LEP 1 and 2. The analysis presented here measures the cross section of τ pairs in two-photon collisions at energies of 189 GeV and 192 – 202 GeV. The four principal feynman diagrams for the QED process are shown in figure 1. This process has been observed and measured for the first time by L3 at 91 GeV [1] but not yet at LEP 2 energies

CP571, *PHOTON 2000,* edited by A. J. Finch
© 2001 American Institute of Physics 0-7354-0010-5/01/$18.00

where the cross section is expected to be large and high integrated luminosities are available.

The data analysed here has been taken by the L3 detector [2] at LEP in 1998 and 1999. L3 collected 172.1 pb^{-1} at 189 GeV in 1998 and 220.9 pb^{-1} at 192 - 202 GeV in 1999. The events were mainly triggered by the charged particle trigger of the time-expansion chamber (TEC) [3] and the newly implemented inner TEC trigger [4], added in 1997 specially to improve acceptance.

MONTE CARLO SIMULATION

For the calculation of efficiencies and for the comparison of data with QED predictions, the Vermaseren Monte Carlo [6] is used. For the cross section calculation, the DIAG36 generator [7] is used. It takes into account the full set of QED diagrams up to $\mathcal{O}(\alpha^4)$ and their interference terms.

For background studies mainly resonances and $q\bar{q}$-production, the EGPC [8] Monte Carlo is used. The events were fully simulated in the detector, including detector and trigger inefficiencies. Data and Monte Carlo were treated with the same programs.

EVENT SELECTION AND DATA SAMPLES

a) Decay mode	Fraction Γ_i/Γ in %	b)	$\pi^-\pi^0\nu_\tau$	$e^-\bar{\nu}_e\nu_\tau$	$\mu^-\bar{\nu}_\mu\nu_\tau$	$\pi^-\nu_\tau$
$\pi^-\pi^0\nu_\tau$	25.32 ± 0.15	$\pi^-\pi^0\nu_\tau$	6.41	**9.00**	**8.80**	5.62
$e^-\bar{\nu}_e\nu_\tau$	17.81 ± 0.07	$e^-\bar{\nu}_e\nu_\tau$		3.17	**6.18**	3.95
$\mu^-\bar{\nu}_\mu\nu_\tau$	17.37 ± 0.09	$\mu^-\bar{\nu}_\mu\nu_\tau$			3.00	3.85
$\pi^-\nu_\tau$	11.08 ± 0.13	$\pi^-\nu_\tau$				1.23

TABLE 1. a) The main decays of the τ^- (The τ^+ modes are charge conjugates) and their branching ratios. b) The combined branching ratios for two τ's.

Table 1 a) shows the most important τ decays and their resulting branching-ratios as found in the PDG [5] while table 1 b) shows the branching ratios for the combination of two τ decays. The symmetric decays[1] are not suited for analysis because they cannot be distinguished very well from background processes mainly $\gamma\gamma \rightarrow ee, \mu\mu$ and $q\bar{q}$. The three most interesting channels are highlighted in the table. This analysis concentrates on the channel, where one τ decays into an electron while the other decays into a $\rho^\pm$, decaying further into $\pi^\pm\pi^0$.

The events are selected by using the following criteria:

- 2 tracks of opposite charge requiring:

---

[1] with both τ's going to the same decay channel

340

- at least 12 hits in the tracking chamber,
- the distance of closest approach to the interaction point in the transverse plane smaller than 10 mm,
- a transverse momentum greater than 0.3 GeV and less than 10 GeV,
- a matching signal in the electromagnetic calorimeter.

- 2 neutral particles identified as photons requiring:

  - an energy greater than 0.1 GeV with at least 2 crystals firing,
  - a shower shape compatible with an electromagnetic particle
  - a separation of 150 mrad in the $r\phi$ plane to the nearest track,
  - that the two photons form a π^0 with $0.115 < m_{\pi^0} < 0.155$ GeV.

- The higher energy charged particle must be identified as an electron by:

  - a positive neural network identification[2],
  - a strict cut on the shower shape in the electromagnetic calorimeter,
  - a momentum greater than 0.6 GeV.

- The least energetic charged particle is assumed to be a pion.

- To reject exclusive $\gamma\gamma$ collison events, the sum of the transverse momenta of all observed particles must be greater than 0.2 GeV.

After these cuts the selection efficiencies are in the order of 7.0 % with purities of about 70.0 %. The data have been corrected for trigger efficiencies as a function of the momentum of the higher energy particle (the electron), the opening angle between the two charged particles in the ϕ-plane and the applied trigger conditions[3], using a loose preselection sample. This function was then applied as a bin-by-bin correction and reweighting of each individual event. The trigger efficiencies for all trigger levels can be found in table 2.

Figure 2 shows the good agreement between Monte Carlo and data after the final selection and corrections.

RESULTS

The observed number of events are given in table 3, together with the predictions of the DIAG36 Monte Carlo generator. The effects of detector acceptance and trigger efficiencies are already included in the numbers. Good agreement is found between the data and the MC predictions.

---

[2] The neural network has been trained with SNNS [9] to distinguish between electrons, muons and pions by using ten input variables: momentum, dE/dx, E_9, E_t/p_t, no. of crystals, E_1/E_9, E_9/E_{25}, the energy in the hadron calorimeter, the energy in a cone of 7 degrees in the hadron

	1998	1999
Level 1	$\approx 77.0 \pm 1.3$ %	$\approx 66.0 \pm 1.8$ %
Level 2	95.2 ± 0.3 %	94.6 ± 0.3 %
Level 3	98.0 ± 0.2 %	96.3 ± 0.3 %

TABLE 2. The trigger efficiencies for the different trigger levels in 1998 and 1999. The shown level 1 triggerefficiency is a mean value over the full spectrum. The real correction is applied bin-by-bin in two dimensions (electron momentum and charged particle opening angle).

	189 GeV	192 - 202 GeV
N_{obs}	85 ± 9.2	97 ± 9.8
N_{exp}	82.9	96.8
$N_{\mathrm{obs}} - N_{\mathrm{bkg}}$	60.3 ± 6.5	66 ± 6.6
$N_{\mathrm{exp}} - N_{\mathrm{bkg}}$	58.2	65.8

TABLE 3. Number of observed events for 189 GeV and 192-202 GeV in comparison with the MC-expectations, before and after background-subtraction

In order to compare the cross section with QED calculations, the data are then corrected for the detection efficiency and normalized to the integrated luminosity. The comparison is given in the range $10° \leq \theta \leq 170°$ and for the invariant mass $W_{\gamma\gamma} \geq 3.6$ GeV. The combined τ branching ratio of 9.0 % from table 1 is then used to calculate the total cross section. The results can be seen in figure 3 and are:

189GeV : $\sigma_{Data} = 458.5 \pm 49.4_{\mathrm{stat}} \pm 34.4_{\mathrm{sys}}$ pb   ($\sigma_{\mathrm{QED}} = 442.6$ pb)

196GeV : $\sigma_{Data} = 453.7 \pm 45.3_{\mathrm{stat}} \pm 42.1_{\mathrm{sys}}$ pb   ($\sigma_{\mathrm{QED}} = 452.3$ pb)

The main systematics come from the variation of the cuts. Other important factors are the uncertainty on the level 1 trigger efficiency and the uncertainty from the Monte Carlo statistics. The error on the trigger efficiencies of level 2 and level 3 is small.

Tau pair production in two-photon collisions has been measured for the first

---

calorimeter, and the number of crystals for a minimum ionising particle
[3] The trigger conditions have been changed from 1998 to 1999. Because of an increased luminosity delivered by LEP, the charged particle triggers had to be restricted, resulting in lower efficiencies in 1999.

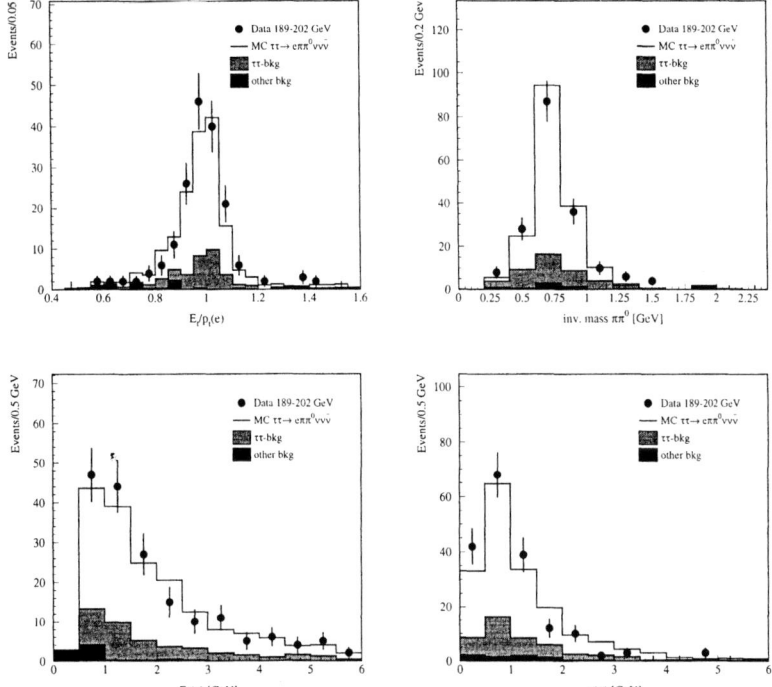

FIGURE 2. Final distributions after all applied cuts: The plots show the combined 1998 and 1999 dataset. Upper left shows the E_t/p_t distribution of the electron, that has to peak at 1 for electromagnetic particles. Upper right shows the distribution of the combined $\pi\pi^0$ mass coming from one of the τ's. Lower left shows the energy distribution of the electron and the lower right the momentum of the charged π.

time at LEP 2. The obtained results are in good agreement with $\mathcal{O}(\alpha^4)$ QED calculations.

REFERENCES

1. Production of e, mu and tau Pairs in Untagged Two-Photon Collisions at LEP, Phys. Lett. **B 403** (1997) 168 - 176.
2. L3 Collab., B. Adeva *et al.*, Nucl. Inst. Meth. **A 289** (1990) 35;
   L3 Collab., O. Adriani *et al.*, Phys. Rev. **236** (1993) 1.
3. P. Bene *et al.*, Nucl. Inst. Meth. **A 306** (1991) 150.
4. D. Haas *et al.*, Nucl. Inst. Meth. **A 420** (1999) 101 - 116.
5. The European Physical Journal C (1998) Volume 3, 1-4.
6. J.A.M. Vermaseren, Nucl. Phys. **B 229** (1983) 347.

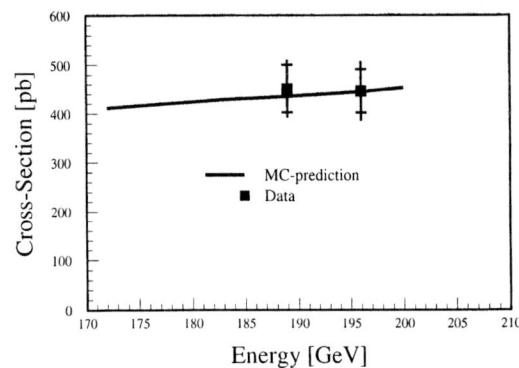

FIGURE 3. The cross section of $e^+e^- \rightarrow e^+e^- \gamma\gamma \rightarrow e^+e^- \tau^+\tau^-$. The points are the 1998 and 1999 data, the line shows the $\mathcal{O}(\alpha^4)$ QED calculations from DIAG36

7. F.A. Berends, P.H. Daverveldt and R. Kleiss. Comp. Phys. Com. **40** (1986) 285.
8. F. Linde, "Workshop on detector and event simulation in high energy physics Monte Carlo", eds. K. Bos and B. van Eijl, Amsterdam, 1991.
9. Stuttgart Neural Network Simulator,
   http://www-ra.informatik.uni-tuebingen.de/SNNS/

Two-photon annihilation into pion pairs

C. Vogt [1]

Fachbereich Physik
Universität Wuppertal
42097 Wuppertal, Germany

Abstract. We discuss pion pair production in two-photon collisions in two different kinematical regimes. When both photons are real and at moderately large center-of-mass energy $\sqrt{s}$ we elaborate on partonic transverse momentum and Sudakov corrections within the hard scattering approach. We also point out the difference between our approach and that of other authors. When one of the photons is highly virtual the produced pion pair can be described in terms of a two-pion distribution amplitude, for which we derive the perturbative limit at large s.

Due to the pointlike structure of the photon exclusive hadron production in two-photon collisions provides a very useful field for the test of perturbative QCD. In the limit of large $\sqrt{s}$, the amplitude of $\gamma^{(*)}\gamma \to \pi\pi$ factorises into a perturbatively calculable hard photon-parton scattering, which in lowest order can simply be obtained from one-gluon exchange diagrams, and soft parts that are expressed in terms of distribution amplitudes describing the transition of partons to pions [1].

At large c.m. energies $\sqrt{s}$, transverse momenta of the partons relative to the pion are negligible and the conventional collinear hard scattering formula can be applied [2]. At moderately large $\sqrt{s}$ of a few GeV, however, the collinear approach is known to suffer severely from substantial endpoint contributions where the strong coupling α_s becomes large, such that perturbation theory is not applicable [3]. These problems can be overcome by including transverse momenta and Sudakov corrections [4,5]. The correspondingly modified hard scattering approach leads to perturbative predictions, which in most cases are not sufficient to account for the experimental data [6–8]. Hadronic form factors and Compton scattering, for example, are dominated by soft contributions at presently accessible c.m. energies [9–11]. In this talk we will discuss Sudakov suppressions in $\gamma\gamma \to \pi^+\pi^-$ in the few GeV region and point out the difference between our approach and that of Ref. [12].

Another interesting, more theoretically motivated application of the hard scattering approach is the process $\gamma^*\gamma \to \pi^+\pi^-$ at large photon virtuality Q^2 and large s. In this kinematical regime, we will briefly outline the calculation of the

---

[1] Supported by the Deutsche Forschungsgemeinschaft

perturbative limit of the two-pion distribution amplitude (2π-DA) [13].

SUDAKOV SUPPRESSION IN $\gamma\gamma \to \pi^+\pi^-$

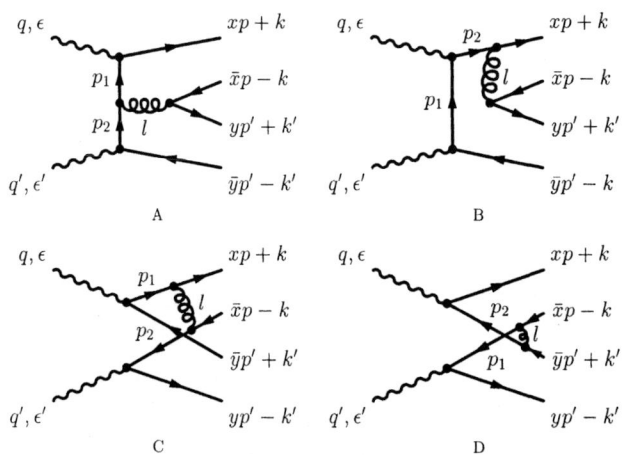

FIGURE 1. The four basic diagrams for $\gamma\gamma \to \pi\pi$. We use the notation $\bar{x} \equiv 1 - x$.

In close analogy to the calculation of hadronic form factors [5–7] in the modified perturbative approach we can express the helicity amplitude $\mathcal{M}_{\lambda\lambda'}$ of the process $\gamma\gamma \to \pi^+\pi^-$ in transverse configuration space as the convolution

$$\mathcal{M}_{\lambda\lambda'}(s, \Theta) = \int dx\,dy\,\frac{d^2\mathbf{b}_\perp}{4\pi}\,\frac{d^2\mathbf{b}'_\perp}{4\pi}\,\hat{\Psi}_\pi(x, \mathbf{b}_\perp)\,\hat{\Psi}_\pi(y, \mathbf{b}'_\perp)$$
$$\times \hat{T}_{H,\lambda\lambda'}(x, y, \mathbf{b}_\perp, \mathbf{b}'_\perp; s, \Theta, \mu_R)\,\exp[-S(x, y, b_\perp, b'_\perp; \mu_R)], \qquad (1)$$

where Θ is the scattering angle in the center-of-mass system of the produced pions and λ, λ' are the photon helicities. The hat denotes the Fourier transform of a function w.r.t. the transverse momenta $\mathbf{k}_\perp$, $\mathbf{k}'_\perp$ of the partons relative to the pions. The Fourier conjugated variables $\mathbf{b}_\perp$, $\mathbf{b}'_\perp$ are the transverse separations of the quark-antiquark pairs and x, y describe how they share the pions' longitudinal momenta.

Using a phenomenological ansatz for the wave function of the pion's valence Fock state we write

$$\Psi_\pi(x, \mathbf{k}_\perp) = \frac{\sqrt{6}\,\pi}{f_\pi}\,\exp\left[-\frac{k_\perp^2}{8\pi f_\pi^2\,x\,(1 - x)}\right], \qquad (2)$$

with $f_\pi = 131$ MeV being the pion decay constant. Integrating Eq. (2) over transverse momenta leads to the asymptotic form of the pion distribution amplitude

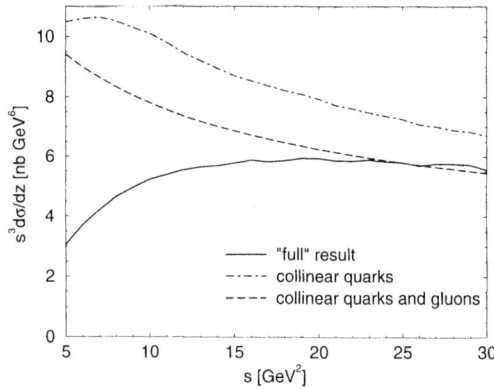

FIGURE 2. Differential cross section at c.m. angle $\Theta = 90°$ with different approximations.

ϕ_π. The use of the asymptotic form is justified through the phenomenology of the π-γ-transition form factor [15] and the parameters of expression (2) are fixed by various pion decay processes [14]. The Gaussian $k_\perp$-dependence describes well soft contributions [6,10,11].

The Sudakov corrections are incorporated in the factor e^{-S}, where S is the Sudakov function [4] (see also [16]). Since it suppresses large quark-antiquark separations it serves as a natural infrared cut-off and thus no external regulator is needed to avoid the singularity of α_s.

In leading order QCD, the hard photon-parton scattering amplitude T_H is to be calculated from 20 one-gluon exchange diagrams, four representatives of which are shown in Fig.1. Following the authors of Ref. [5] we choose the renormalisation scale μ_R to be the largest mass scale appearing in the gluon virtualities. Owing to the structure of the hard scattering amplitude its analytical Fourier transform cannot be calculated exactly and we have to resort to approximations. As the longitudinal momentum fractions occur quadratically in the gluon propagators we keep transverse momentum corrections there if not otherwise stated.

Our results for the differential cross section at $\Theta = 90°$ using different approximations for the quark propagators are shown in Fig.2. The dot-dashed line shows the result obtained by replacing the quark propagators by their collinear limits. In the solid curve we take into account transverse momenta in quark propagators in those integration regions where they have singularities. We see that the effect in the few GeV region is dramatic, which means that one can generally not ignore $k_\perp$-corrections in quark propagators, as has been done in Ref. [17], for instance. For comparison we also show the result of the collinear hard scattering approach, i.e. completely neglecting $k_\perp$-corrections in quark as well as in gluon propagators, where we have frozen α_s below 1 GeV (dashed line). We note that our main result, given by the solid curve, approaches the collinear approximation for $s \gtrsim 20$ GeV^2,

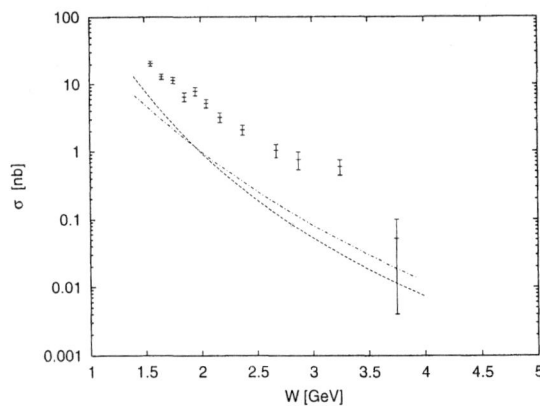

FIGURE 3. The combined cross section $\sigma(\gamma\gamma \to \pi^+\pi^-, K^+K^-)$ as a function of the c.m. energy $W = \sqrt{s}$.

i.e. for c.m. energies above 4-5 GeV. In brief, the $k_\perp$-corrections of the quark propagators effect the transition amplitude such that it reduces its absolute magnitude while receiving a large phase.

In Fig. 3 we therefore only compare our upper estimate, obtained by ignoring transverse momenta in the quark propagators and given by the dot-dashed line, with the data of the combined cross section $\sigma(\gamma\gamma \to \pi^+\pi^-, K^+K^-)$ of Ref. [18], where we have accounted for the contributions from kaons by a relative factor $(f_K/f_\pi)^4 \simeq 2.2$. For comparison we again show the collinear approximation. As we can see, the curves are already far below the data, so that the inclusion of $k_\perp$-corrections in the quark propagators would further increase the discrepancy.

In Ref. [2] it was shown that, in the collinear approximation, essential parts of the hard scattering amplitudes are accidentally proportional to the pion form factor. Using this relation and a phenomenological value for F_π the authors of [12] obtained reasonable agreement with the data. However, we would like to emphasise that the assumed value for the pion form factor, $Q^2 F_\pi(Q^2) = 0.3$ GeV^2, is rather large for a perturbative calculation. With a renormalisation scale of the order of the typical virtuality of the exchanged gluon and using the asymptotic form of the pion distribution amplitude, the pion form factor in the collinear approach reads [1] $Q^2 F_\pi(Q^2) = 8\pi f_\pi \alpha_s(Q^2)$ and ranges between 0.17 and 0.1 GeV^2 for 1 GeV $\lesssim Q \lesssim 4$ GeV. Since the pion form factor enters the cross section for $\gamma\gamma \to \pi^+\pi^-$ quadratically that accounts for the difference between our result for the collinear approximation and that of Ref. [12].

Finally, we would like to point out that with the inclusion of $k_\perp$-corrections in the hard scattering amplitude the simple relation between the cross section and the pion form factor does not longer hold. In particular, our predictions are independent of any phenomenological value for the pion form factor.

THE PERTURBATIVE LIMIT OF THE 2π-DA

We now turn to the kinematical regime where one of the photons has a large virtuality Q^2. In Refs. [19] it was shown that for $s \ll Q^2$ the helicity amplitude of $\gamma^* \gamma \to \pi\pi$ factorises in a hard part and a generalised distribution amplitude $\Phi_{2\pi}$:

$$\mathcal{M}_{\lambda\lambda'}(\zeta, s) = \frac{1}{2}\delta_{\lambda\lambda'} \sum_q e_0^2 e_q^2 \int_0^1 dz \frac{2z-1}{z(1-z)} \Phi_{2\pi}^q(z, \zeta, s), \tag{3}$$

where the light-cone fractions $z = k^+/P^+$ and $\zeta = p^+/P^+$ respectively describe how the partons and the pions share the light-cone plus component of the total momentum $P = p + p'$ of the pions and the sum runs over all quark flavours q. The 2π-DA, first discussed in [20], represents the collinear hadronisation of two partons into a pion pair. The helicity selection rule, expressed through the Kronecker delta, immediately follows from the collinear scattering of massless quarks. Note that apart from logarithmic corrections the leading order expression (3) is completely independent of Q.

If we demand that $s, -t, -u \gg \Lambda^2$, where Λ is a typical hadronic scale of the order of 1 GeV, while keeping the constraint that the photon virtuality is the dominant scale, $s \ll Q^2$, we can use the conventional hard scattering approach [1] to calculate the helicity amplitude (3) in terms of the hard scattering amplitude T_H and two single pion DAs ϕ_π:

$$\mathcal{M}_{\lambda\lambda'}(s, t, u) = \frac{f_\pi^2}{24} \int_0^1 dx\, dy\, \phi_\pi(\bar{y})\, \phi_\pi(x)\, T_{H,\lambda\lambda'}(x, y, s, t, u). \tag{4}$$

Using light-cone gauge and organising the result in powers of $\sqrt{s}/Q$ one can show [13] that the leading contributions are independent of Q, reflecting the correct scaling behaviour, and come from the diagrams of the group B in Fig. 1. Moreover, the helicity selection rule of Eq. (3) is reproduced. Reexpressing Eq. (4) through the light-cone fractions z and ζ for each diagram, we can then read off the large-s limit of the 2π-DA for a flavour $q = u$ by comparison of Eqs. (3) and (4):

$$\Phi_{2\pi}^u(z, \zeta, s) = \frac{8\pi f_\pi^2}{9} \left\{ \Theta(\zeta - z) \frac{\zeta}{\zeta - z} \phi_\pi\left(\frac{z}{\zeta}\right) I(\bar{z}, \bar{\zeta}, s; \phi_\pi) \right.$$

$$\left. -\Theta(z - \zeta) \frac{\bar{\zeta}}{z - \zeta} \phi_\pi\left(\frac{\bar{z}}{\bar{\zeta}}\right) I(z, \zeta, s; \phi_\pi) \right\}, \tag{5}$$

where the integral I is given by $I(z, \zeta, s; \phi_\pi) = \int_0^1 dx \frac{\alpha_s}{s} \frac{z + \bar{x}\zeta}{z - x\zeta} \frac{\phi_\pi(x)}{\bar{x}}$. The 2π-DAs for u- and d-quarks are related by $\Phi_{2\pi}^u(z, \zeta, s) = -\Phi_{2\pi}^d(\bar{z}, \zeta, s)$ and since higher Fock states are suppressed by powers of α_s/s there is no s-quark contribution. The $1/s$ scaling of Eq. (5) is a characteristic feature of the hard scattering approach [1,2].

Our result manifestly fulfills the charge conjugation relation $\Phi_{2\pi}^q(z, \zeta, s) = -\Phi_{2\pi}^q(\bar{z}, \bar{\zeta}, s)$ and it can be shown to comply with a general polynomiality condition [21]. It possesses integrable logarithmic singularities at $z = \zeta$, which reflect the above mentioned endpoint problems of the collinear hard scattering approach when the exchanged gluon becomes soft.

CONCLUSIONS

Two-photon annihilation into pion pairs allows for a sensitive test of perturbative QCD. Using a self-consistent approach, where there are no large endpoint contributions spoiling the applicability of perturbation theory, we have shown that the hard contributions are not sufficient to explain the experimental data of $\gamma\gamma \to \pi^+\pi^-$. Therefore considerable soft contributions have to be expected. New data are desireable to determine the onset of the perturbative regime, which seems not to start below c.m. energies of 4-5 GeV. When one of the photons is far off-shell and at large s, where transverse momenta become irrelevant, the collinear hard scattering approach can be applied to calculate the perturbative limit of the 2π-DA in terms of the conventional pion distribution amplitudes.

REFERENCES

1. Lepage, G. P. and Brodsky, S. J., Phys. Rev. D **22**, 2157 (1980).
2. Brodsky, S. J. and Lepage, G. P., Phys. Rev. D **24**, 1808 (1981).
3. Isgur, N. and Llewellyn Smith, C. H., Phys. Rev. Lett. **52**, 1080 (1984); Phys. Lett. B **217**, 535 (1989); Radyushkin, A. V., Nucl. Phys. A **527**, 153c (1991).
4. Botts, J. and Sterman, G., Nucl. Phys. B **325**, 62 (1989).
5. Li, H. and Sterman, G., Nucl. Phys. B **381**, 129 (1992).
6. Jakob, R. and Kroll, P., Phys. Lett. D **315**, 463 (1993).
7. Bolz, J. et al., Z. Phys. C **66**, 267 (1995).
8. Stefanis, N. G., Schroers, W. and Kim, H.-Ch., Phys. Lett. B **449**, 299 (1999).
9. Jakob, R., Kroll, P. and Raulfs, M., J. Phys. G **22**, 45 (1996); Radyushkin, A. V., Phys. Rev. D **58**, 114008 (1998); Afanasev, A. V., hep-ph/9808291.
10. Diehl, M., Feldmann, Th., Jakob, R. and Kroll, P., Eur. Phys. J. C **8**, 409 (1999).
11. Vogt, C., hep-ph/0007277.
12. Brodsky, S. J. et al., Phys. Rev. D **57**, 245 (1998).
13. Diehl, M., Feldmann, Th., Kroll, P. and Vogt, C., Phys. Rev. D **61**, 074029 (2000).
14. Brodsky, S. J., Huang, T. and Lepage, G. P., in *Banff Summer Institute, Particles and Fields 2*, Capri, A. Z. and Kamal, A. N. (Eds.), 143 (1983).
15. Kroll, P. and Raulfs, M., Phys. Lett. B **387**, 848, (1996);
    Musatov, I. V. and Radyushkin, A. V., Phys. Rev. D **56**, 2713 (1997).
16. Stefanis, N. G., Schroers, W. and Kim, H.-Ch., hep-ph/0005218.
17. Hyer, T., Phys. Rev. D **47**, 3875 (1993).
18. CLEO Collaboration, Dominick, J. et al., Phys. Rev. D **50**, 3027 (1994).
19. Müller, D. et al., Fortschr. Phys. **42**, 101 (1994); Diehl, M. et al. , Phys. Rev. Lett. **81**, 1782 (1998); Freund, A., Phys. Rev. D **61**, 074010 (2000).
20. V. N. Baier and A. G. Grozin, Sov. J. Nucl. Phys. **35**, 899 (1982).
21. Polyakov, M. V., Nucl. Phys. B **555**, 231 (1999).

Measurement of the Mass, Width, and Two-Photon Width of the $\eta_c(1S)$

Hans P. Paar

Physics Department 0319
University of California, San Diego
La Jolla, CA 92037-0319, U.S.A.
[e-mail: hpaar@ucsd.edu]

Abstract. I present recent measurements of the mass, width, and two-photon width of the $\eta_c(1S)$ obtained with the CLEO detector at the Cornell e^+e^- collider CESR. The $\eta_c(1S)$ has been detected in its decay $\eta_c(1S)\rightarrow K^0_S K^-\pi^+$ and its charge conjugate. The results are based upon an integrated luminosity of $13.4\,\mathrm{fb}^{-1}$ at a centre-of-mass energy near the $\Upsilon(4S)$. The $\eta_c(1S)$ mass was measured to be $2980.4 \pm 2.3(\mathrm{stat}) \pm 0.6(\mathrm{syst})\,\mathrm{MeV}/c^2$ and its width $27.0\pm5.8(\mathrm{stat})\pm1.4(\mathrm{syst})\,\mathrm{MeV}$. The two-photon width of the $\eta_c(1S)$ was measured to be $7.6 \pm 0.8(\mathrm{stat}) \pm 0.4(\mathrm{syst}) \pm 2.3(\mathrm{Br})\,\mathrm{keV}$ with the last uncertainty associated with the branching fraction $\mathcal{B}[\eta_c(1S)\rightarrow K\overline{K}\pi]$. The measured mass and two-photon width agree with the world average while the measured width differs from the world average by approximately two standard deviations. The measured width is now agrees with theory for somewhat larger values of α_s.

INTRODUCTION

The measurement of the width and the two-photon width of the $\eta_c(1S)$ are of interest because there are QCD based next-to-leading order (NLO) calculations [1] that relate these to one another and to the partial width of $J/\psi\rightarrow e^+e^-$. The world average experimental values are [2] $\Gamma[\eta_c(1S)] = 13.2^{+3.8}_{-3.2}\,\mathrm{MeV}$ and $\Gamma_{\gamma\gamma}[\eta_c(1S)\rightarrow\gamma\gamma] = 7.4\pm1.4\,\mathrm{keV}$. The experimental value for the two-photon width of the $\eta_c(1S)$ is in good agreement with the NLO theoretical expectation. However, the experimental value for the width of the $\eta_c(1S)$ is barely consistent with the theoretical expectation. Thus a measurement of the width of the $\eta_c(1S)$ is of interest to see if theory and experiment can be reconciled. The precision of the two-photon width is limited by the precision of the branching fraction of the $\eta_c(1S)$ into the hadronic final state under consideration. Unfortunately, these are known only to approximately 30 % as their values all rely upon a 1986 measurement of the inclusive rate [3] of $\eta_c(1S)\rightarrow\gamma X$ which has a 28 % error. Thus detecting $\eta_c(1S)$ decay in several decay modes and averaging does not help since the errors in their respective branching fractions are correlated. The NLO calculations have sizable

CP571, *PHOTON 2000*, edited by A. J. Finch
© 2001 American Institute of Physics 0-7354-0010-5/01/$18.00

corrections of order α_s making them less precise too. A calculation to order α_s^2 is very desirable.

CLEO has observed the $\eta_c(1S)$ in its decay $\eta_c(1S) \to K_S^0 K^- \pi^+$ (charge conjugate state are implied throughout). After a brief review of theoretical estimates for the width and the two-photon width and a discussion of the CLEO detector, the event selection, and experimentally results on the mass, width, and two-photon width are presented.

THEORETICAL ESTIMATES

The interpretation of the width and the two-photon width of the $\eta_c(1S)$ in terms of QCD based next-to-leading-order (NLO) calculations can be done using the results of Ref. [1]. The ratio of the width to the two-photon width of the $\eta_c(1S)$ is

$$\frac{\Gamma[\eta_c(1S)]}{\Gamma(J/\psi \to \gamma\gamma)} = \frac{9}{8} \left(\frac{\alpha_s}{\alpha}\right)^2 \frac{1 + 4.8\,\alpha_S/\pi}{1 - 3.4\,\alpha_s/\pi} \tag{1}$$

The ratio of the two-photon width of the $\eta_c(1S)$ to the leptonic width of the J/ψ is given in Ref. [1] as

$$\frac{\Gamma(\eta_c(1S) \to \gamma\gamma)}{\Gamma(J/\psi \to e^+ e^-)} = \frac{4}{3} \left(1 + 1.96 \frac{\alpha_S}{\pi}\right) \frac{|\psi(0)|^2_{\eta_c(1S)}}{|\psi(0)|^2_{J/\psi}} \tag{2}$$

The factor 1.96 is obtained from the ratio of $(1 - 3.38\alpha_s/\pi)$ and $(1 - 5.33\,\alpha_s/\pi)$ by the approximation $(1 - 5.33\,\alpha_s/\pi)^{-1} = (1 + 5.33\,\alpha/\pi$, good to order $(\alpha_s/\pi)^2$. The NLO calculation neglects terms of order $(\alpha_s/\pi)^2$ so the previous approximation is consistent with that. We prefer not to make that particular approximation and replace (2) by

$$\frac{\Gamma(\eta_c(1S) \to \gamma\gamma)}{\Gamma(J/\psi \to e^+ e^-)} = \frac{4}{3} \frac{1 - 3.38\,\alpha_S/\pi}{1 - 5.33\,\alpha_s/\pi} \frac{|\psi(0)|^2_{\eta_c(1S)}}{|\psi(0)|^2_{J/\psi}} \tag{3}$$

Depending upon the value of α_s used, the ratio in (3) is at least 15% larger than in (2). We show in Fig. 1 the relation between the width and the two-photon width of the $\eta_c(1S)$ as α_s varies from 0.15 to 0.35 in steps of 0.05. We have used the world average value [2] $\Gamma(J/\psi \to e^+ e^-) = 5.26 \pm 0.37$ keV and assumed equal wavefunctions at the origin for the $\eta_c(1S)$ and J/ψ. The datapoint shown with a circle is the world average $\Gamma(\eta_c(1S)) = 13.2^{+3.8}_{-3.2}$ MeV and $\Gamma[\eta_c(1S) \to \gamma\gamma] = 7.5^{+1.6}_{-1.4}$ where the error on the two-photon width does not include the uncertainty from the hadronic branching fraction. It is seen that the world average favors an unreasonably low value $\alpha_s = 0.21$.

The NLO calculations have sizable corrections of order α_s. The term $5.33\,\alpha_s/\pi$ is about 0.5, uncomfortably large. A calculation to order $(\alpha_s/\pi)^2$ is obviously desirable.

DETECTOR, DATASET, AND EVENT SIMULATION

The CLEO II detector [4] is a general purpose detector that provides charged particle tracking, precision electro-magnetic calorimetry, charged particle identification, and muon detection. Charged particle detection over 95% of the solid angle is achieved by tracking devices in two different configurations, situated in a magnetic field of 1.5 T. In the first configuration (CLEO II) tracking is provided by three concentric wire chambers while in the second configuration (CLEO II.V) the innermost wire chamber is replaced by a precision three-layer silicon vertex detector [5]. The momentum resolution is 0.5% at $p = 1$ GeV. The drift chambers are surrounded by a time-of-flight (TOF) system. Energy loss (dE/dx) in the outer drift chamber and the TOF system provide pion-kaon separation. A CsI based electro-magnetic calorimeter giving an energy resolution of 4% for 100 MeV electro-magnetic showers provides π^0 detection. A super-conducting coil and muon detectors surround the calorimeter. Redundant triggers provide efficient registration of mutiparticle final states.

The Cornell Electron Storage Ring CESR operates at a center-of-mass energy of approximately 10.6 GeV. The results in this report are based upon on-4S data collected on the $\Upsilon(4S)$ resonance with an integrated luminosity of 9.10 fb^{-1} and on 4.29 fb^{-1} off-4S data collected 60 MeV below the $\Upsilon(4S)$ for the study of non-$B\overline{B}$

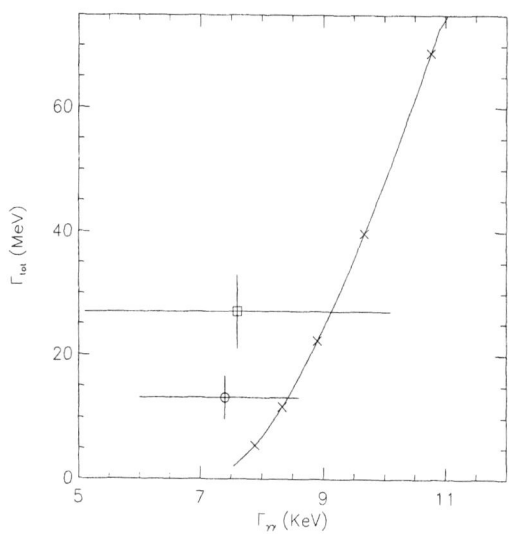

FIGURE 1. Relation of the two-photon width and the width of the $\eta_c(1S)$ when vaying α_s. The crosses correspond to $\alpha_s = 0.15, 0.20, 0.25, 0.30, 0.35$ from smaller to larger widths. The datapoint shown with a circle represents world averages while the square represents CLEO's measurements.

backgrounds. The number of $B\overline{B}$ pairs is $(9.63 \pm 0.19) \times 10^6$.

The event simulation uses uses the BGMS [6] formalism for the event generation. Photon form-factors based upon vector-meson dominance with the J/ψ mass are used. The line shape is parameterized by a relativistic spin zero Breit-Wigner. The Monte Carlo simulation of the CLEO II detector is based upon GEANT. The detector is simulated down to the component level. The trigger simulation and the event reconstruction use this information to determine the detector response. Simulated events are processed in the same manner as the data.

EVENT SELECTION

The $\eta_c(1S)$ is detected through its decay $\eta_c(1S) \to K_S^0 K^- \pi^+$ with $K_S^0 \to \pi^+ \pi^-$. The event selection requires exactly four good quality tracks whose sum of charges is zero. A K_S^0 vertex fit is made using pairs of oppositely charged tracks. A vertex is accepted if its corresponding decay path is at least three standard deviations for CLEOII or five for CLEOII.V. The standard deviations are determined on an event-to-event basis and the decay path requirement corresponds to approximately 1.5 mm. We further require that the K_S^0 momentum points back to the e^+e^- interaction point and that the $\pi^+\pi^-$ invariant mass is within four standard deviations of the nominal K^0 mass. The mass resolution is also determined on an event-by-event basis and the mass requirement correspond to an approximately 24 MeV/c^2 wide window. If there is more than one K_S^0 candidate passing these selection criteria, the event is rejected. Of the remaining two tracks the one with lowest momentum is identified as a pion or a kaon using dE/dx and TOF information which work best at low momentum. The higher momentum particle is identified as a pion or kaon depending upon the assignment of the lower momentum particle. The transverse component of the vector sum of the tracks p_T has to be less than 0.6 GeV and the total energy measured in the calorimeter, not associated with one of the four tracks has to be less then 0.6 GeV. We show in Fig. 2 the p_T distribution of the $K_S^0 K^- \pi^+$. The dots with error bars are the sideband subtracted data while the histogram is the result of the simulation. The peak at zero p_T corresponds to signal events. There is good agreement between data and simulation over a wide range of p_T. Because incorrectly evaluated background can cause a disagreement this represents a very stringent check on our understanding of the event selection and its efficiency in the presence of considerable background. The visible energy of the event (the sum of the four track's energies) has to be less then 6 GeV. For the $\eta_c(1S)$ mass measurement only we require that the pion and kaon be in the central region ($|\cos\theta| < 0.71$) and that each has a transverse momentum greater than 225 MeV. This ensures that both tracks traverse the entire tracking volume for optimal momentum resolution and therefore optimal invariant $K_S^0 K^- \pi^+$ mass resolution.

To keep the systematics related to the trigger efficiency small, only events satisfying the two-track trigger are used. It's efficiency was measured using the simulation which in turn is checked using data taken with redundant triggers.

RESULTS

We show in Fig. 3 the $K_S^0 K^- \pi^+$ invariant mass distribution, separately for CLEOII and CLEOII.V. Clear peaks are seen just below 3.0 GeV. The solid line is the result of a simultaneous fit to the CLEOII and CLEOII.V mass plots. A spin 0 relativistic Breit-Wigner convolved with two Gaussian's is used to fit the signal. Its parameters are taken from the Monte Carlo simulation. The background is fit with a form $A[m(K_S^0 K^- \pi^+)]^{-n}$ where A and n are constants and are determined by the fit. There are 300 ± 32 events in the mass peaks. To measure the mass of the $\eta_c(1S)$, tighter cuts are used as discussed in the Section on event selection and there are now 195 ± 24 events in the mass peaks. With these tighter cuts the resolution of the $K_S^0 K^- \pi^+$ invariant mass is approximately $8 \, \text{MeV}/c^2$. There are now 195 ± 24 events in the mass peaks. The fitted mass $M[\eta_c(1S)] = 2980.4 \pm 2.3(\text{stat}) \pm 0.6(\text{syst}) \, \text{MeV}/c^2$. The mass scale is calibrated using the measured masses of the K_S^0, ϕ, D^0, and J/ψ. The 0.6 MeV systematic error is estimated from these calibrations. The world average value [2] is $2979.8 \pm 1.8 \, \text{MeV}/c^2$, there is excellent agreement.

A fit to the invariant mass distribution of Fig. 3 in the manner discussed above gives a width $\Gamma[\eta_c(1S)] = 27.0 \pm 5.8(\text{stat}) \pm 1.4(\text{syst}) \, \text{MeV}$. The fit has a $\chi^2 = 226$

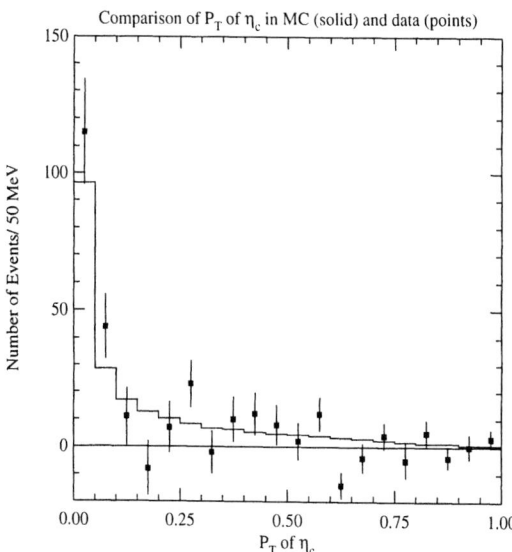

FIGURE 2. Transverse momentum distribution p_T of $K_S^0 K^- \pi^+$. The dots with error bars are the sideband subtracted data while the histogram is the simulation.

for 243 degrees of freedom. To determine the systematic error the mass resolution needs to be calibrated. This is done using the D^+ in its decay $D^+ \rightarrow K_S^0 K^- \pi^+$ for a comparison of the mass resolution in data and simulation. The agreement is better than 0.1 MeV leading to a negligible contribution to the systematic error. Particle identification errors in pion-kaon separation (misidentification contributes to tails in the invariant mass distribution) and imperfect understanding of detector resolution are the main contributers to the 1.4 MeV systematic error. The CLEO measurement may be compared with the world average [2] value $\Gamma[\eta_c(1S)] = 13.2^{+3.8}_{-3.2}$ MeV. CLEO's measurement. The world average differs by approximately two standard deviations.

The number of signal events in the mass peaks in the invariant mass plot can be related to the product of $\Gamma[\eta_c(1S) \rightarrow \gamma\gamma]\mathcal{B}[\eta_c(1S) \rightarrow K_S^0 K^- \pi^+]$. The branching fraction $\mathcal{B}[\eta_c(1S) \rightarrow K_S^0 K^- \pi^+] = \frac{1}{3}\mathcal{B}[\eta_c(1S) \rightarrow K\overline{K}\pi] = \frac{1}{3}(5.5 \pm 1.7) \times 10^{-2}$. This 30 % error is large and dominates the determination of the two-photon width of the $\eta_c(1S)$. Using the acceptance from the simulation and the integrated luminosity we find $\Gamma[\eta_c(1S) \rightarrow \gamma\gamma] = 7.6 \pm 0.8(\text{stat}) \pm 0.4(\text{syst}) \pm 2.3(\text{Br})$ keV. The systematic error is dominated by uncertainties in the trigger efficiency (0.2 keV), tracking efficiency (0.2 keV), and the possible presence of $K^*(892)K$ substructure in the final state (0.2 keV). The last error due to uncertainty in the branching fraction. The CLEO measurement (without the error from the uncertainty in the $\eta_c(1S) \rightarrow K\overline{K}\pi$ branching fraction) may be compared with the world average [2]

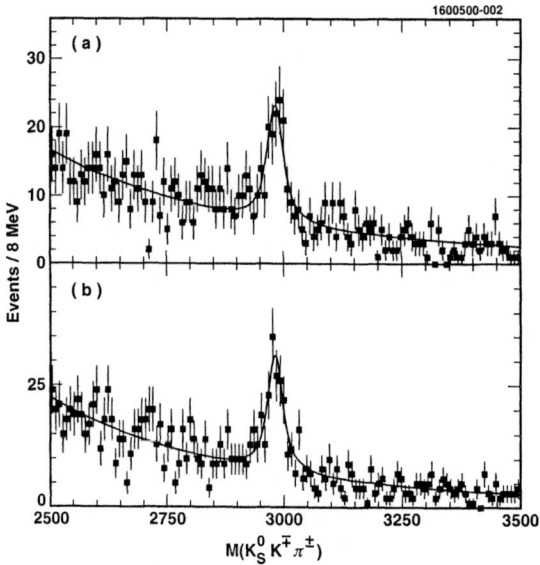

FIGURE 3. Invariant mass of $K_S^0 K^- \pi^+$ for (a) CLEOII and (b) for CLEOII.V. The data are the dots with error bars, the fit is a simultaneous fit to CLEOII and CLEOII.V data, see text.

value $\Gamma[\eta_c(1S){\to}\gamma\gamma] = 7.5^{+1.6}_{-1.4}$. There is excellent agreement between the two in part due to the fact that the same central values for the branching fractions are used.

INTERPRETATION

We show in Fig. 1 the CLEO measurements of the width and the two-photon width of the $\eta_c(1S)$. Also shown are their world averages. The curve is discussed in the Section on Theoretical Estimates. It is seen that the CLEO measurements are consistent with the NLO calculations, as are the world averages, but with a larger but still low value of $\alpha_s = 0.26 \pm 0.02$.

The largest α_s/π correction in (1) and (3) is the term with $5.33\alpha_s/\pi$ and or approximately 0.5. It is clearly desirable that a calculation to order $(\alpha_s/\pi)^2$ be done to improve the accuracy of the calculation. Perhaps this is the reason for the still unreasonably low value of α_s.

On the experimental side, the hadronic branching fractions of the $\eta_c(1S)$ need to be determined with much better precision. They are all dependent upon a (1986) measurement of the branching fraction of the inclusive decay $\eta_c(1S){\to}\gamma X$ [3] which has a 28 % error. Even worse, the uncertainties in the various hadronic decay modes are correlated because of their common dependence upon the inclusive decay rate so averaging measurements of the two-photon widths from different hadronic decay modes will not help. A new measurement of the branching fraction of the inclusive decay $\eta_c(1S){\to}\gamma X$ is desirable.

REFERENCES

1. W. Kwong, P.B. Mackenzie, Rogerio Rosenfeld, and J.L. Rosner, *Phys. Rev.* D **37**, 3210 (1988).
2. D.E. Groom *et al.* (Particle Data Group), *Eur. Phys. Jour.* C **15**, 1 (2000).
3. J. Gaiser *et al.*, (Crystal Ball Collaboration), *Phys. Rev.* D **34**, 711 (1986).
4. Y. Kubota *et al.* (CLEO Collaboration), *Nucl. Instrum. Methods Phys. Res.*, Sec. A **320**, 66 (1992).
5. T. Hill *et al.*, *Nucl. Instrum. Methods Phys. Res.*, Sec. A **418**, 32 (1998).
6. V.M. Budnev, I.F. Ginzburg, G.V. Meledi, and V.G. Serbo, *Phys. Rep.* C **15**, 181 (1975).

Study of the two photons decay of charmonium states formed in $\bar{p}p$ annihilations

Presented by Wander Baldini[a,1] for the E835 Collaboration[2]

[a] *Università degli studi di Ferrara and INFN - Dipartimento di Fisica,*
Via Paradiso 12, 44100 Ferrara Italy.

Abstract. Experiment E835 at the Fermilab Antiproton Accumulator studies the charmonium states directly formed in $\bar{p}p$ annihilations. In this paper we report the preliminary measurements of the mass, total with and partial width to $\gamma\gamma$ of the foundamental state η_c ($^1S_0$). Results on the two photons partial widths $\Gamma_{\gamma\gamma}(\chi_2)$ and $\Gamma_{\gamma\gamma}(\chi_0)$, and a measurement of α_s at the quark charm mass are also reported.

INTRODUCTION

The charmonium family of $\bar{c}c$ states is a rich source of experimental data for the study of the spin dependence of the QCD interactions and of the decay of $\bar{q}q$ bounded states. Since it's discovery in 1974, a large amount of data has been collected, mostly in e^+e^- annihilation experiments. With this technique, since the annihilation proceeds through an exchange of an intermediate virtual photon, only the states with the same quantum number of the photon can be directly formed. All others states are studied through radiative decays.

A technique exploiting $\bar{p}p$ annihilations was pioneered by experiment R704 at the CERN Intersecting Storage Ring and then systematically used by E760 in the Fermilab Antiproton Accumulator. This technique allows the direct formation of all the charmonium states and provides a more precise way of determining the

---

[1] e-mail: baldini@fe.infn.it

[2] G. Garzoglio, K. E. Gollwitzer, A. Hahn, W. Marsh, J. Peoples Jr., S. Pordes, G. Stancari, J. Streets, S. Werkema (*Fermi National Accelerator Laboratory*), M. Ambrogiani, W. Baldini, D. Bettoni, R. Calabrese, P. Dalpiaz, E. Luppi, M. Martini, R. Mussa, M. Savriè (*INFN and University of Ferrara*), A. Buzzo, M. Lo Vetere, M. Macrì, M. Marinelli, M. Pallavicini, C. Patrignani, E. Robutti, A. Santroni (*INFN and University of Genova*), G. Lasio, M. Mandelkern, J. Schultz, M. Stancari, G. Zioulas (*University of California, Irvine*), X. Fan, S. Jin, J. Kasper, P. Maas, T. K. Pedlar, J. Rosen, K. K. Seth, A. Tomaradze (*Northwestern University*), S. Argirò, S. Bagnasco, G. Borreani, R. Cester, F. Marchetto, E. Menichetti, M. M. Obertino, N. Pastrone, P. Rumerio (*INFN and University of Torino*) and R. McTaggart (*Pennsylvania State University*).

CP571, *PHOTON 2000,* edited by A. J. Finch

© 2001 American Institute of Physics 0-7354-0010-5/01/$18.00

resonance parameters.

Although the charmonium production cross section $\bar{p}p \to \bar{c}c$ is very small ($\simeq 1\,\mu b$) compared to the total $\bar{p}p$ cross section ($\simeq 70\,mb$), we can extract a clear signal selecting the electromagnetic final states.

THE EXPERIMENT E835

The E835 detector is a non magnetic spectrometer that covers the whole azimuthal angle and the polar angles θ from $2°$ to $70°$.

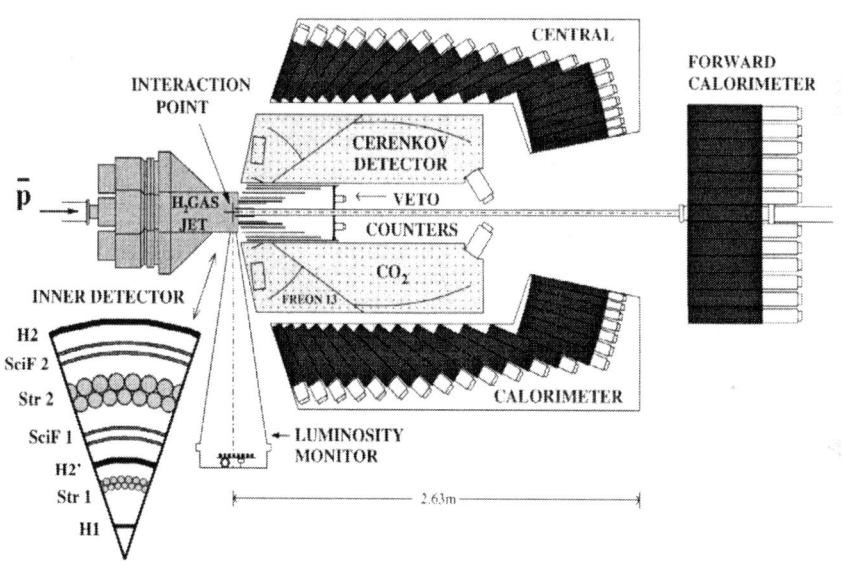

FIGURE 1. Schematic view of the E835 detector.

A schematic view can bee seen in fig. 1. The main components are: a *molecular hydrogen jet target*, an *inner charged tracking system*, a *threshold Čerenkov counter*, two electromagnetic calorimeters: the *central (CCAL)* and the *forward (FCAL)*, and a *luminosity monitor*.

The $\bar{p}p$ annihilations are obtained by intersecting the stocastically cooled antiproton beam ($\sigma_E \simeq 0.4 MeV$) with the *molecular hydrogen jet target* [1] that provides a stream of hydrogen clusters with a density up to $5.0 \times 10^{14} atoms/cm^3$. The typical instantaneous luminosity during the data taking is $2.0 \div 3.0 \times 10^{31}\ cm^{-2} sec^{-1}$.

The *inner charged tracking system* gives a precise measurement of the polar and azimuthal angles of charged particles. It contains: four azimuthally segmented scintillator hodoscopes, used for triggering purposes and for dE/dx measurement, a four layers scintillating fiber detector for the measurement of the polar angle, and a four layers Straw Chamber to measure the azimuthal angle.

The *threshold Čerenkov counter* is used to distinguish between electron and other charged particles (mainly pions) in the first level trigger and during the offline selection of the events.

The *central calorimeter* plays an important role in the study of two photons events. It is composed of 1280 lead glass detectors, pointing to the interaction point and arranged in 20 "rings" and 64 "wedges". It covers the whole azimuthal angle and the polar angles from $10.6°$ to $70.0°$. It measures the energy and the angles of electrons and photons with the following resolutions:

$$\frac{\sigma_E}{E} = \frac{6\%}{\sqrt{E(GeV)}} + 1.4\% \; ; \qquad \sigma_\theta = 6 \, mrad \; ; \qquad \sigma_\varphi = 11 \, mrad$$

The $CCAL$ provides a first level trigger for $\gamma\gamma$ and e^+e^- events (in conjunction with the the inner tracking system). The *forward calorimeter* extends up to $2°$ the angular coverage in the forward region.

The *luminosity monitor* gives a precise measurement of the integrated luminosity by measuring the rate of the recoiling protons near $90°$ in the lab and comparing it with the well known elastic $\bar{p}p$ cross section.

DATA SELECTION

E835 studies the two photon decay of charmonium states through the process: $\bar{p}p \rightarrow \bar{c}c \rightarrow \gamma\gamma$. The selection of two photon events is very difficult due to the presence of a large background, coming mainly from the reactions:

$$\bar{p}p \rightarrow \pi^0\gamma \rightarrow 3\gamma \qquad\qquad \bar{p}p \rightarrow \pi^0\pi^0 \rightarrow 4\gamma$$

where one or two photons, respectively, are not detected because of either, out of the detector's acceptance or below the $CCAL$ energy threshold (20 MeV).

The $\gamma\gamma$ event selection begins with the hardware trigger and the online filter (second level trigger). It requires essentially two back to back, high energy deposits in the $CCAL$, with an invariant mass above $2.5 \, GeV$ and no hits in the inner hodoscopes. In the offline analysis the data sample is further purified by requiring the invariant mass of the energy deposits to be within 20% of the center of mass energy, and reducing the angular acceptance in the forward region (since the background is forward peaked). A 4C kinematical fit to the $\gamma\gamma$ hypothesis is then applied and only events with $P(\chi^2) > 5\%$ are kept.

RESULTS

Experiment E835 collected $143\,pb^{-1}$ of data from Oct. 1996 to Sept. 1997 (RUN I) and about $100\,pb^{-1}$ from Jan. 2000 until now (Aug. 2000, the experiment will take data until Nov. 2000).

A good fraction of the data taking has been dedicated to the study of the two photons decay of the charmonium states. In particular, E835 measured the line paramenters of the η_c (1^1S_0) state, the two photons partial width of the P-wave states χ_2 (1^3P_2) and χ_0 (1^3P_0) and performed a search of the radially excited η'_c (2^1S_0) resonance in the energy range $3570 < E_{cm}(MeV) < 3660$. At the present time E835 is improving its measurements of the χ_0 resonance.

η_c results

E835 collected about $18pb^{-1}$ of data in the η_c region. In fig. 2 is shown the measured cross section (solid dots) compared with the level of background predicted by a Monte Carlo calculation. This level has been obtained by first measuring the $\bar{p}p \to \pi^0\gamma$ and $\bar{p}p \to \pi^0\pi^0$ cross sections and then estimating how many events will pass the cuts applied for the $\gamma\gamma$ selection .

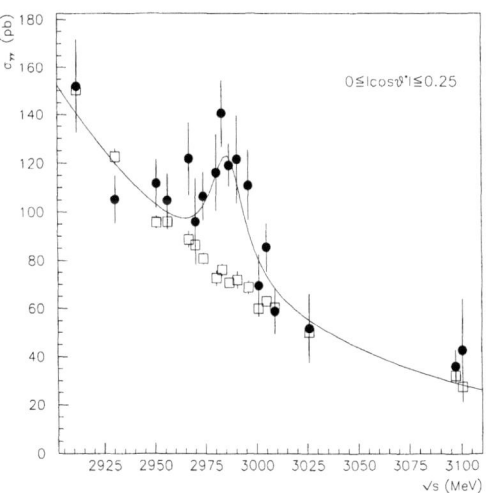

FIGURE 2. $\bar{p}p \to \gamma\gamma$ measured cross section (solid dots), in the η_c region, compared with the calculated background (open squares). The line represents the fit to the data of a Breit-Wigner distribution plus a power law background.

The resonance paramenters, M_{η_c}, Γ_{η_c} and the product of the branching ratios

$B(\eta_c \to \bar{p}p) \times B(\eta_c \to \gamma\gamma)$ have been calculated by fitting the data with the maximum likelihood method to a Breit-Wigner function plus a power law background:

$$\sigma_{\gamma\gamma}(E) = \underbrace{const \times \frac{B(\eta_c \to \bar{p}p) \times B(\eta_c \to \gamma\gamma)}{(E - M_{\eta_c}c^2)^2/\Gamma_{\eta_c}^2 + 1}}_{Breit-Wigner\ term} + \underbrace{A\,(2.990/E)^B}_{background\ parametrization}$$

Where the known constant factor depends on the spin and on the momentum of the incoming particles. The preliminary results are the following:

$$M_{\eta_c} = 2985.4^{+2.1}_{-2.0}\ MeV;\ \ \Gamma_{\eta_c} = 21.1^{+7.5}_{-6.2}\ MeV;\ \ B(\eta_c \to \bar{p}p) \times B(\eta_c \to \gamma\gamma) = 21.8^{+3.4}_{-3.3} \times 10^{-8}$$

Taking from the PDG '98 [2] the value $B(\eta_c \to \bar{p}p) = (1.2 \pm 0.4) \times 10^{-3}$ we obtain, for the partial width to $\gamma\gamma$, $\Gamma_{\gamma\gamma} = 3.85^{+1.5}_{-1.2}\ keV$.

In fig. 3 the above results are compared with those obtained from other experiments and with theoretical predictions.

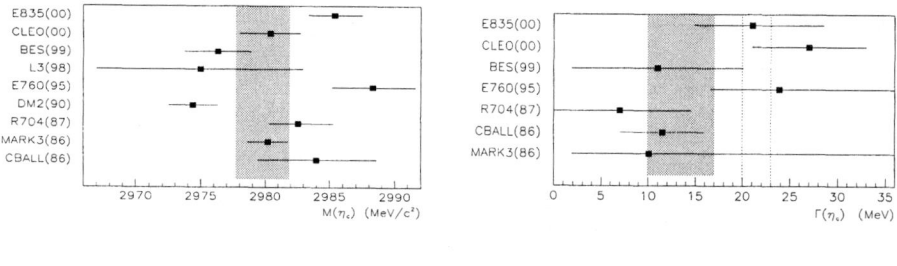

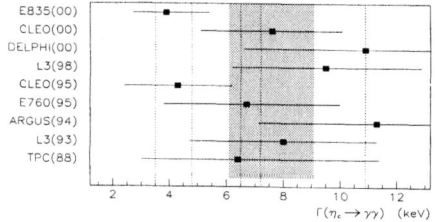

FIGURE 3. Comparison of the various measurements of the mass, the total width and the partial width to $\gamma\gamma$ of the η_c resonance. The shaded area and the dashed lines show the world average (from PDG '98) and theoretical predictions, respectively.

χ_{c0} and χ_{c2} partial width measurement

The χ_{c0} and χ_{c2} resonances can decay into channels $\gamma\gamma$ and $J/\psi\gamma$. If we define R the ratio between the two branching ratios:

$$R \equiv \frac{B(\chi_{0,2} \to \gamma\gamma)}{B(\chi_{0,2} \to J/\psi\gamma \to e^+e^-\gamma)} = \frac{\sigma(\bar{p}p \to \chi_{0,2} \to \gamma\gamma)}{\sigma(\bar{p}p \to \chi_{0,2} \to J/\psi\gamma \to e^+e^-\gamma)}$$

we can then obtain $\Gamma(\chi_{0,2} \to \gamma\gamma)$ by measuring R from our data and taking the well measured $\Gamma(\chi_{0,2} \to J/\psi\gamma \to e^+e^-\gamma)$ and Γ_{χ_2} from the PDG '98. For the χ_0 total width we use our measurement [3] obtained studying the decay: $\chi_0 \to J/\psi\gamma \to e^+e^-\gamma$. Our results from the 1997 run are [4]:

$$\Gamma(\chi_2 \to \gamma\gamma) = 0.27 \pm 0.05 \pm 0.03 \; keV \; ; \qquad \Gamma(\chi_0 \to \gamma\gamma) = 1.61 \pm 0.8 \pm 0.65 \; keV$$

Where the first error include counting statistics and errors from efficiencies, acceptance and backgrounds and the second is due to uncertainties on $B(\chi_{0,2} \to J/\psi\gamma \to e^+e^-\gamma)$ and on the total widths. The measured cross sections and the comparison of the partial width measurements are shown in fig.4.

α_s measurement

Perturbative QCD with first order radiative corrections predicts [5] that in the ratio between the two photon and the gluonic width of the same resonance unknown factors involving m_c and the wavefunctions cancel:

$$\frac{\Gamma(\eta_c \to \gamma\gamma)}{\Gamma(\eta_c \to gg)} = \frac{8\alpha^2 \left(1 - \frac{3.4}{\pi}\alpha_s\right)}{9\alpha_s^2 \left(1 - \frac{4.8}{\pi}\alpha_s\right)} \; ; \qquad \frac{\Gamma(\chi_2 \to \gamma\gamma)}{\Gamma(\chi_2 \to gg)} = \frac{8\alpha^2 \left(1 - \frac{16}{3\pi}\alpha_s\right)}{9\alpha_s^2 \left(1 - \frac{2.2}{\pi}\alpha_s\right)}$$

If we assume that

$$\Gamma_{gg}(\eta_c) \sim \Gamma_{TOT}(\eta_c) \qquad and \qquad \Gamma_{gg}(\chi_2) \sim \Gamma_{TOT}(\chi_2) - \Gamma(\chi_2 \to J\psi\gamma)$$

from our measurements of $\Gamma(\eta_c \to \gamma\gamma)$ and $\Gamma(\chi_2 \to \gamma\gamma)$ we obtain:

$$\alpha_s(m_c) = 0.33^{+0.06}_{-0.03} \; (\eta_c) \; ; \qquad \alpha_s(m_c) = 0.38 \pm 0.02 \; (\chi_2)$$

The combined result is shown in fig. 5, compared with the PDG '98 compilation of α_s measurements [2]. The poor determination of $\Gamma(\chi_0 \to \gamma\gamma)$, (due to the small amount of data taken in that energy region during the 1997 run: $\sim 3.5\,pb^{-1}$) does not allow us to calculate α_s from χ_0 data. More data have recently been taken ($\sim 30\,pb^{-1}$) and the analysis is in progress.

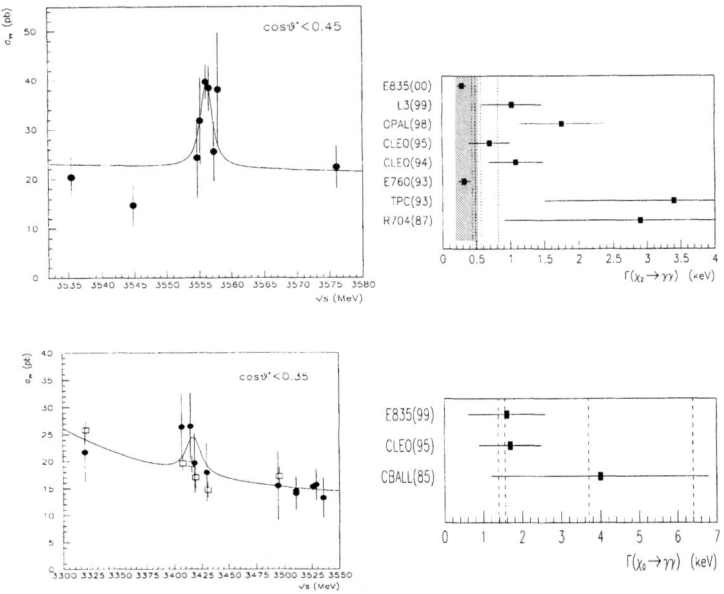

FIGURE 4. In the top-left and bottom-left plot are shown the measured $\gamma\gamma$ cross section as a function of the c.m. energy, in the χ_2 and χ_0 region. The curve represent the fit to the data with a Breit-Wigner plus a power law background. The comparisons of our partial width measurements with those of other experiments are shown in the two plots on the right. The shaded area and the dashed lines show the world average (from PDG '98) and theoretical predictions.

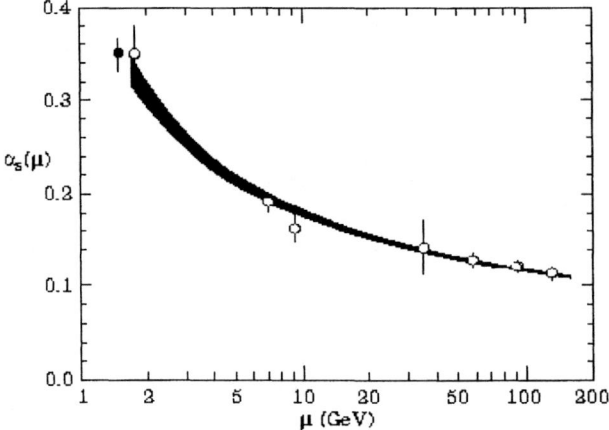

FIGURE 5. E835 estimate for α_s (solid circle) superimposed to the PDG '98 compilation of α_s measurements (open circles)

CONCLUSIONS

Experiment E835 has studied the two photons decay of the η_c, χ_0 and χ_2 resonances. Preliminary results for the mass, the total width and the partial width to $\gamma\gamma$ of the η_c resonance have been presented. The two photon partial widths $\Gamma_{\gamma\gamma}(\chi_0)$ and $\Gamma_{\gamma\gamma}(\chi_2)$ and the value of α_s that can be calculated from $\Gamma_{\gamma\gamma}(\eta_c)$ and $\Gamma_{\gamma\gamma}(\chi_2)$ have also been reported.

More data in the χ_0 region are being taken at the present time. This will significantly improve the $\Gamma_{\gamma\gamma}(\chi_0)$ measurement and, in turn, will provide another independent measurement of the strong coupling constant in the charm quark energy region.

REFERENCES

1. D. Allspach *et al.*, Nucl. Instr. and Meth. **A410**(1998),195. **A 424**(1999),304.
2. Particle Data Group, C. Caso *et al.*,, Eur. Phys. Journ. **C3**,(1998).
3. M.Ambrogiani *et al.*, Phys. Rev. Lett. **83**(2000),2902.
4. M.Ambrogiani *et al.*, Phys. Rev. **D62**(2000),52002.
5. W. Kwong *et al.*, Phys. Rev. **D37**(1988),3210.

$K_s^0 K_s^0$ Final State and Glueball Searches and $\Lambda\bar{\Lambda}$ Production in Two-Photon Collisions in L3 at LEP

Saverio Braccini[*]

[*] *University of Geneva,*
24, Quai Ernest Ansermet, CH-1211 Geneve 4, Switzerland.
E-mail : Saverio.Braccini@cern.ch

Abstract.
The $K_S^0 K_S^0$ final state in two-photon collisions is studied with the L3 detector at LEP using data collected at centre-of-mass energies from 91 GeV to 202 GeV. The mass spectrum is dominated by the formation of the $f_2'(1525)$ tensor meson in the helicity two state. The two-photon width times the branching ratio is measured to be $\Gamma_{\gamma\gamma}(f_2') \times \mathrm{Br}(f_2' \to K\bar{K}) = 0.076 \pm 0.006 \pm 0.011$ keV. Clear evidence for destructive $f_2(1270)$-$a_2(1320)$ interference is observed. In addition a clear signal for $f_J(1750)$ is observed. The study of the decay angular distribution in the 1750 MeV mass region shows that the spin two helicity two wave is dominant. No signal is observed in the region around 2.2 GeV. The upper limit for the two-photon partial width of the $\xi(2230)$ tensor glueball candidate of $\Gamma_{\gamma\gamma}(\xi(2230)) \times \mathrm{Br}(\xi(2230) \to K_S^0 K_S^0) < 1.4$ eV at 95% C.L. is derived. The production of $\Lambda\bar{\Lambda}$ pairs in two-photon collisions is also studied. The cross section is compared to quark-diquark model predictions.

INTRODUCTION

Electron-positron storage rings are a good laboratory to investigate the behaviour of two-photon interactions via the process $e^+e^- \to e^+e^-\gamma^*\gamma^* \to e^+e^- X$, where γ^* is a virtual photon. The outgoing electron and positron carry nearly the full beam energy and their transverse momenta are usually so small that they are not detected. In this case the two photons are quasi-real. This kind of event is characterized by an initial state $e^+e^-\gamma^*\gamma^*$, calculable by QED, and a low multiplicity final state. This process is particularly useful in the study of the formation of hadron resonances and baryon anti-baryon pairs.

The cross section is given by the convolution of the QED calculable luminosity function $\mathcal{L}$, giving the flux of the photons, with the two-photon cross section $\sigma(\gamma\gamma \to X)$. In the case of the formation of a resonance R, $\sigma(\gamma\gamma \to R)$ can be expressed by a Breit-Wigner function. This leads to the proportionality relation

CP571, *PHOTON 2000*, edited by A. J. Finch
© 2001 American Institute of Physics 0-7354-0010-5/01/$18.00

$\sigma(e^+e^- \to e^+e^-R) = \mathcal{K}\,\Gamma_{\gamma\gamma}(R)$ that allows to extract the two-photon width from the cross section. The quantum numbers of the resonance must be compatible with the initial state of the two quasi-real photons. A neutral, unflavoured meson with even charge conjugation, J$\neq$1 and helicity-zero (λ=0) or two (λ=2) can be formed. In order to decay into $K_S^0 K_S^0$, a resonance must have $J^{PC} = $ (even)$^{++}$. For the 2^{++}, 1^3P_2 tensor meson nonet, the $f_2(1270)$, the $a_2^0(1320)$ and the $f_2'(1525)$ can be formed. However, since these three states are close in mass, interferences must be taken into account. According to SU(3), the $f_2(1270)$ interferes constructively with the $a_2^0(1320)$ in the K^+K^- final state but destructively in the $K^0 \bar{K}^0$ final state [1].

Since gluons do not couple directly to photons, the two photon width of a glueball is expected to be very small. A state that can be formed in a gluon rich environment but not in two photon fusion has the typical signature of a glueball. According to lattice QCD predictions [2], the lowest lying glueball has $J^{PC}= 0^{++}$ and a mass between 1400 and 1800 MeV. The 2^{++} tensor glueball is expected in the mass region around 2200 MeV. Since several 0^{++} states have been observed in the 1400-1800 MeV mass region, the scalar ground state glueball can mix with nearby quarkonia, making the search for the scalar glueball and the interpretation of the scalar meson nonet a complex problem [3] [4] [5].

Using the Brodsky-Lepage hard scattering approach [6], predictions have been made for the production of baryon anti-baryon pairs in two-photon interactions. The cross section for the process $\gamma\gamma \to p\bar{p}$ is measured by CLEO [7]. Data are found to be inconsistent with the prediction of a pure quark model [8] and agreement is found with the prediction of a quark di-quark model [9], which takes into account non-perturbative quark-quark correlations inside the baryon. The cross section $\sigma(\gamma\gamma \to \Lambda\bar{\Lambda})$ is found by CLEO [10] to be higher than the value predicted by the quark di-quark model.

A study of the reactions $e^+e^- \to e^+e^-K_S^0 K_S^0$ and $e^+e^- \to e^+e^-\Lambda\bar{\Lambda}X$ is presented here. The data correspond to an integrated luminosity of 143 pb^{-1}collected by the L3 detector at LEP at $\sqrt{s} = 91$ GeV, 52 pb^{-1}at $\sqrt{s} = 183$ GeV and 393 pb^{-1}at $\sqrt{s} = 189 - 202$ GeV. The L3 detector is described in detail elsewhere [11]. These analyses are mainly based on the central tracking system and the high resolution electromagnetic calorimeter. The events are collected predominantly by the charged particle track triggers [12].

$K_S^0 K_S^0$ FINAL STATE AND GLUEBALL SEARCHES

A study of the reaction $\gamma\gamma \to K_S^0 K_S^0$ is performed by selecting four charged track events with a net charge of zero and $|\vec{p_T}(\pi^+\pi^-\pi^+\pi^-)|^2 < 0.1$ GeV^2. Within these events two K_S^0 candidates are searched for by secondary vertex reconstruction. Events with photons are rejected. Fig. 1(a) shows the distribution of the mass of one K_S^0 candidate versus the mass of the other candidate. There is a strong enhancement corresponding to the $K_S^0 K_S^0$ exclusive formation over a small background. We require that the invariant masses of the two K_S^0 candidates must be

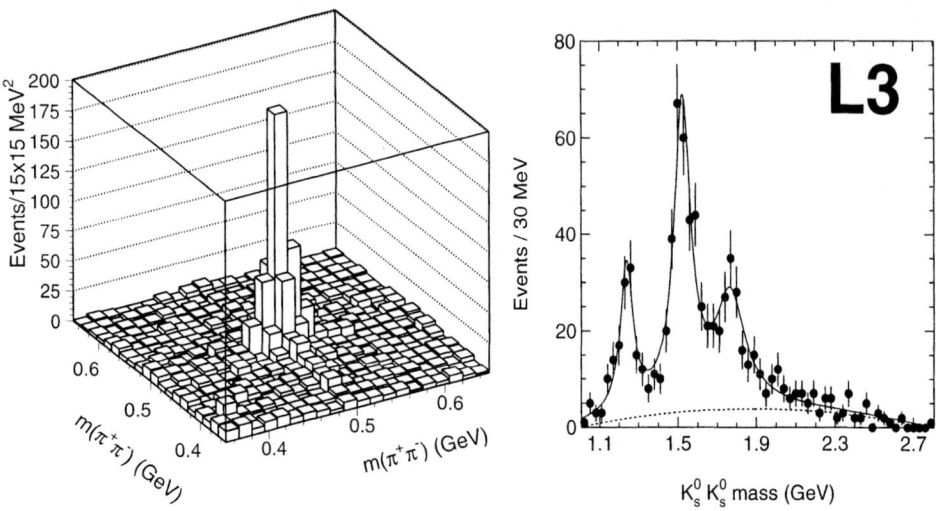

FIGURE 1. (a) The mass of a K_S^0 candidate versus the mass of the other candidate. (b) The $K_S^0 K_S^0$ mass spectrum.

inside a circle of 40 MeV radius centred on the peak of the $K_S^0 K_S^0$ signal. With these selection criteria, 802 events are found in the data sample. The resulting $K_S^0 K_S^0$ mass spectrum is shown in fig. 1(b). The spectrum is dominated by the formation of the $f_2'(1525)$ tensor meson. The mass region between 1100 and 1400 MeV shows destructive $f_2(1270) - a_2^0(1320)$ interference in the $K_S^0 K_S^0$ final state [1]. A clear signal is visible for the $f_J(1750)$. No excess is observed around 2230 MeV. A maximum likelihood fit using three Breit-Wigner functions plus a second order polynomial for the background is performed on the full $K_S^0 K_S^0$ mass spectrum. The fit is shown in fig. 1(b) and the results are summarized in Table 1.

In order to correct the data for the detector acceptance and efficiency, a Monte Carlo procedure is used [13]. The nominal $f_2'(1525)$ parameters [14] are used for the generation. The angular distribution of the two K_S^0's in the two-photon center-of-mass system is generated according to phase space (J=0) i.e. uniform in $\cos \theta^*$

TABLE 1. Results from the maximum likelihood fit on the full $K_S^0 K_S^0$ mass spectrum.

	$f_2(1270)$-$a_2(1320)$	$f_2'(1525)$	$f_J(1750)$
Mass (MeV)	1239 ± 6	1523 ± 6	1767 ± 14
Width (MeV)	78 ± 19	100 ± 15	187 ± 60
Area	123 ± 22	331 ± 37	220 ± 55

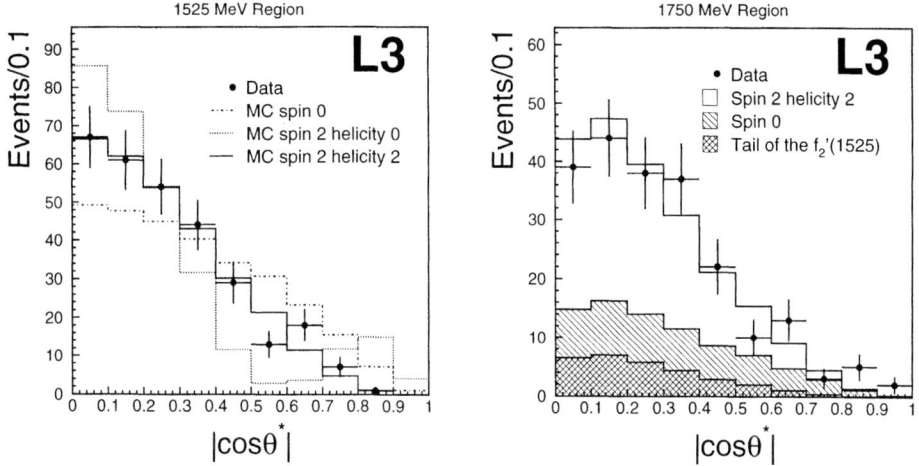

FIGURE 2. (a) The experimental angular distribution is compared with different spin-helicity Monte Carlo predictions for the $f_2'(1525)$. (b) The result of the fit of the angular distribution in the 1750 MeV mass region.

and in ϕ^*, where θ^* and ϕ^* are the polar and azimuthal angles taking the z direction parallel to the electron beam. In order to take into account the helicity of a spin-two resonance, a weight is assigned to each generated event according to the weight functions: $w = (\cos^2 \theta^* - \frac{1}{3})^2$ for the spin-two helicity-zero (J=2, λ=0) contribution and $w = \sin^4 \theta^*$ for the spin-two helicity-two (J=2, λ=2) contribution. All the events are passed through the full detector simulation program and are reconstructed following the same procedure used for the data.

To determine the spin and the helicity state in the mass region of the $f_2'(1525)$, a study of the angular distribution of the two K_S^0's in the mass region between 1400 and 1640 MeV in the two-photon center of mass is performed. The experimental polar angle distribution is compared with the Monte Carlo in Fig. 2(a) for pure (J=0), (J=2, λ=0) and (J=2, λ=2) cases. A χ^2 is calculated normalizing the Monte Carlo distributions to the same number of events as in the data. Bins are grouped in order to have at least 10 entries both in the data and in the Monte Carlo. The confidence level values for the (J=0) and (J=2, λ=0) hypotheses are less then 10^{-6}. For the (J=2, λ=2) hypothesis a confidence level of 99.9% is obtained. The contributions of (J=0) and (J=2, λ=0) are found to be compatible with zero when fitting the three waves simultaneously. Thus data are in agreement with a pure spin two, helicity two wave contribution. According to the theoretical predictions [15], the (J=2, λ=2) contribution largely dominates over (J=2, λ=0). From the cross section the value $\Gamma_{\gamma\gamma}(f_2'(1525)) \times Br(f_2'(1525) \to K\bar{K}) = 0.076 \pm 0.006 \pm 0.011$ keV is obtained for the two-photon width. The main source of systematic error is due to the background parameterization in the fitting procedure. This result is in good agreement with the value previously published by L3 [16].

To investigate the spin composition in the mass region between 1640 and 2000 MeV the angular distribution of the two K_S^0's in the two-photon center of mass is studied. A resonance with a mass of 1750 MeV and a total width of 200 MeV is generated with the same Monte Carlo procedure adopted for the $f_2'(1525)$. The effect of the tail of the $f_2'(1525)$ in this mass region is taken into account by using the Monte Carlo distribution for the $f_2'(1525)$ with (J=2, λ=2). The fraction of the events belonging to the $f_2'(1525)$ in the 1750 MeV mass region is found to be 14%. A fit of the angular distribution is performed using a combination of the two waves (J=0) and (J=2, λ=2) for the signal plus the distribution of the tail of the $f_2'(1525)$. The contribution due to the (J=2, λ=0) wave is considered negligible with respect to (J=2, λ=2) according to the theoretical predictions [15] and to our experimental results for the $f_2'(1525)$. The normalization is fixed to the same number of events as in the data and bins are grouped in order to have at least 10 entries both in the data and in the Monte Carlo. The fit is shown in Fig. 2(b). The fraction of (J=0) is found to be 24±16%. The confidence level is 68%. Thus the (J=2, λ=2) wave is found to be dominant also in the 1750 MeV mass region. If the (J=2, λ=0) wave is also considered, its contribution is found to be compatible with zero.

The BES Collaboration [17] reported the presence of both 2^{++} and 0^{++} waves in the 1750 MeV mass region in K^+K^- in the reaction $e^+e^- \to J/\psi \to K^+K^-\gamma$. The (J=0) fraction is estimated to be 30±10%, in good agreement with our measurement.

For the two-photon width of the (J=2, λ=2) state, we measure $\Gamma_{\gamma\gamma}(f_2(1750)) \times$ Br($f_2(1750) \to K\bar{K}$)= 0.049 ± 0.011 ± 0.013 keV. The systematic error takes into account the selection criteria, the trigger, the fitting procedure, the uncertainty on the total width and on the (J=2, λ=2) fraction.

The (J=2, λ=2) signal around 1750 MeV may be due to the formation of a radially excited tensor meson state. The f_2'' is in fact predicted to have a mass of 1740 MeV and a two-photon width of 1.040 keV [18]. The 1750 MeV region is also very interesting for glueball searches and for the understanding of the 0^{++} scalar meson nonet. In the scenario presented in [3], the $f_0(1500)$ is identified with the scalar 0^{++} ground state glueball. This hypothesis includes the prediction of a further scalar state, the $f_0'(1500-1800)$, mainly composed by $s\bar{s}$. This state will couple strongly to $K\bar{K}$ and its observation is essential to support the $f_0(1500)$ glueball nature. Our and BES measurements support this hypothesis.

The BES Collaboration confirmed the observation by the Mark III Collaboration of a resonance, the $\xi(2230)$ [19], produced in radiative decays of the J/ψ. Due to its narrow width and its production in gluon rich environment, this state is considered a glueball candidate. Its mass is consistent with the lattice QCD prediction for the ground state tensor glueball.

Since we do not observe any signal in the 2230 MeV mass region, a Monte Carlo simulation is used to determine the detection efficiency for the $\xi(2230)$ under the hypothesis of a pure (J=2, λ=2) state. For the simulation we use a mass of 2230 MeV and a total width of 20 MeV. A mass resolution of σ= 60 MeV is found. The

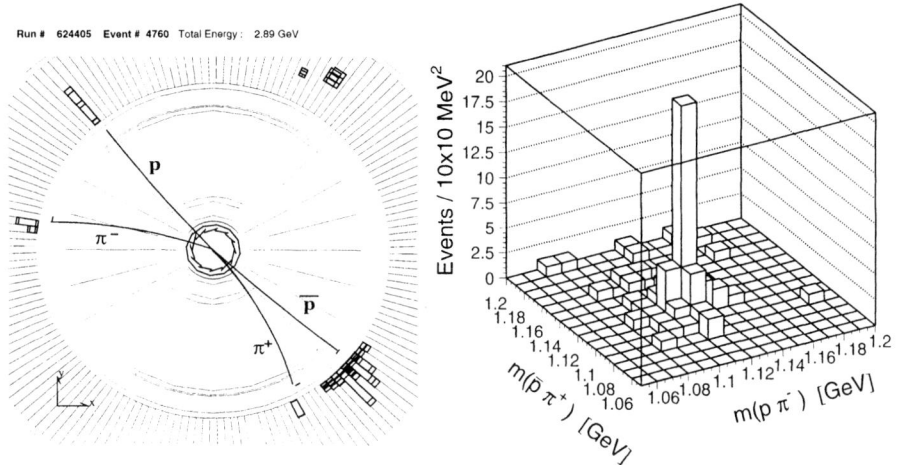

Run # 624405 Event # 4760 Total Energy : 2.89 GeV

FIGURE 3. (a) A typical event $e^+e^- \rightarrow e^+e^-\Lambda\bar{\Lambda}X$. (b) The signal of $\Lambda\bar{\Lambda}$ production.

signal region is chosen to be $\pm 2\sigma$ around the $\xi(2230)$ mass. In order to evaluate the background two sidebands of 2σ are considered. Using the standard method [14] for extracting an upper limit for a Poisson distribution with background, we determine the upper limit $\Gamma_{\gamma\gamma}(\xi(2230)) \times \mathrm{Br}(\xi(2230) \rightarrow K_S^0 K_S^0) < 1.4$ eV at 95% C.L.

Since gluons do not couple directly to photons, the two-photon width is expected to be small for a glueball. To make this statement more quantitative, a parameter called stickiness [20] is introduced. It is expected to be of the order of one for quarkonia and much larger for glueballs. Combining the results reported by BES and Mark III for the $\Gamma(J/\psi \rightarrow \gamma\xi(2230)) \times \mathrm{Br}(\xi(2230) \rightarrow K_S^0 K_S^0) = 1.9 \pm 0.5$ eV and our upper limit on $\Gamma(\xi(2230) \rightarrow \gamma\gamma) \times \mathrm{Br}(\xi(2230) \rightarrow K_S^0 K_S^0)$, we obtain a lower limit on the stickiness $S_{\xi(2230)} > 73$ at 95% C.L. This value is in agreement with the measurements by CLEO [21] and is much larger than the values measured for all the well established $q\bar{q}$ states and supports the interpretation of the $\xi(2230)$ as the tensor glueball. A further confirmation of its existence in gluon rich environments becomes now very important.

$\Lambda\bar{\Lambda}$ PRODUCTION

The production of $\Lambda\bar{\Lambda}$ pairs in two-photon collisions is studied using the decays $\Lambda \rightarrow p\pi^-$ and $\bar{\Lambda} \rightarrow \bar{p}\pi^+$. Events with four charged tracks, a net charge of zero and two secondary vertices are selected. For each secondary vertex the mass of the proton is assigned to the track with the largest transverse momentum. The ionization loss measurements in the tracking chamber must be consistent with the $p\pi^-\bar{p}\pi^+$ hypothesis. The track of the anti-proton candidate must be in correspondence

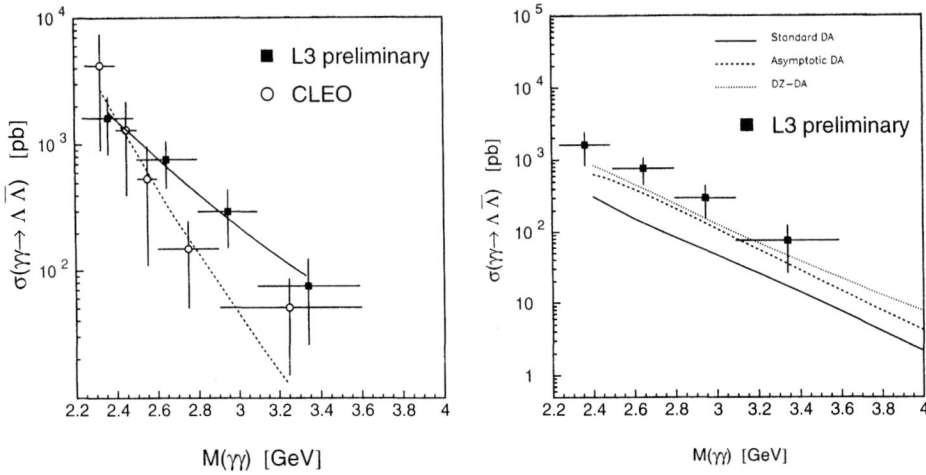

FIGURE 4. The measured cross section $\sigma(\gamma\gamma \to \Lambda\bar{\Lambda}X)$ is compared to CLEO data (a) and to quark di-quark model theoretical predictions (b).

with a large deposit of energy ($E_{em} > 0.35$ GeV) in the electromagnetic calorimeter produced by its annihilation. A typical event is shown in fig. 3(a) where the $\Lambda\bar{\Lambda}$ topology and in particular the annihilation of the anti-proton is clearly visible. In fig. 3(b) the mass distribution of the Λ candidate is shown versus the mass of the $\bar{\Lambda}$ candidate: the signal due to $\Lambda\bar{\Lambda}$ production is very clear and 44 events are selected inside a circle of 40 MeV radius around the peak of the $\Lambda\bar{\Lambda}$ signal.

Since no cut is applied to $|\vec{p_T}(p\pi^-\bar{p}\pi^+)|^2$ and events with photons are not removed, the inclusive cross section $\sigma(\gamma\gamma \to \Lambda\bar{\Lambda}X)$ is measured by deconvoluting the luminosity function. The result is shown in fig. 4(a) together with CLEO [10] data. The two measurements are in agreement within the large errors. Fig.4(a) shows also a fit of the form $\sigma \propto M^{-n}$ that gives $n = 8.7 \pm 3.8$ for our data and $n = 16.1 \pm 5.8$ for CLEO. According to the dimensional counting rule [22], the value of n is 12 for a baryon composed by three quarks and 8 for a baryon composed by a quark and a di-quark. Thus our data favor the quark di-quark model.

In fig.4(b) our measurement is compared to recent calculations of $\sigma(\gamma\gamma \to \Lambda\bar{\Lambda})$ in the framework of the quark di-quark model [23] for three different distribution amplitudes for the octet baryons. The mass dependence is in good agreement with our data that lie above the predictions. The excess in the data may be due to the fact that the contribution of Σ^0, $\bar{\Sigma}^0$ and other baryons is not removed.

ACKNOWLEDGMENTS

I would like to acknowledge all the members of the two-photon physics analysis group of the L3 Collaboration, in particular B. Echenard, J.H. Field, M.N. Focacci-Kienzle and M. Wadhwa.

REFERENCES

1. H. J. Lipkin, *Nucl. Phys.* **B 7** (1968) 321.
2. C. Michael, *Nucl. Phys.* **A 655** (1999) 12 and references therein.
3. C. Amsler and F. E. Close, *Phys. Rev.* **D 53** (1996) 295 and references therein.
4. P. Minkowski and W. Ochs, *Eur. Phys. J.* **C 9** (2000) 283.
5. U. Gastaldi et al., LNL-INFN(rep) 148/99 and QCD99 proceedings, Montpellier, France.
6. S. J. Brodsky and J. P. Lepage, *Phys. Rev.* **D 22** (1980) 2157.
7. CLEO Collab., M. Artuso et al., *Phys. Rev.* **D 50** (1994) 5484.
8. G. Farrar et al., *Nucl. Phys.* **B 259** (1985) 702; *Nucl. Phys.* **B 263** (1986) 746.
9. M. Anselmino, F. Caruso, P. Kroll and W. Schweiger, *Int. J. Mod. Phys.* **A 4** (1989) 5213.
10. CLEO Collab., S. Anderson et al., *Phys. Rev.* **D 56** (1997) 2485.
11. L3 Collab., B. Adeva et al., *Nucl. Inst. Meth.* **A 289** (1990) 35; L3 Collab., O. Adriani et al., *Phys. Rep.* **236** (1993) 1.
12. P. Béné et al., *Nucl. Inst. Meth.* **A 306** (1991) 150; D. Haas et al.,*Nucl. Inst. Meth.* **A 420** (1999)101.
13. F. L. Linde, *"Charm Production in Two-Photon Collisions"*, Ph. D. Thesis, Rijksuniversiteit Leiden, (1988).
14. Particle Data Group, D. E. Groom et al., *Eur. Phys. J.* **C 15** (2000) 1.
15. B. Schrempp-Otto et al., *Phys. Lett.* **36 B** (1971) 463; G. Köpp et al., *Nucl. Phys.* **B 70** (1974) 461; P. Grassberger and R. Kögerler, *Nucl. Phys.* **B 106** (1976) 451.
16. L3 Collab., M. Acciarri et al., *Phys. Lett.* **363 B** (1995) 118.
17. BES Collab., J. Z. Bai et al., *Phys. Rev. Lett.* **77** (1996) 3959.
18. C. R. Münz, *Nucl. Phys.* **A 609** (1996) 364.
19. BES Collab., J. Z. Bai et al., *Phys. Rev. Lett.* **76** (1996) 3502; Mark III Collab., R. M. Baltrusaitis et al., *Phys. Rev. Lett.* **56** (1986) 107.
20. M. Chanowitz, *"Resonances in Photon-Photon Scattering"*, Proceedings of the VIth International Workshop on Photon-Photon Collisions, World Scientific, 1984.
21. CLEO Collab., R. Godang et al., *Phys. Rev. Lett.* **79** (1997) 3829; CLEO Collab., M. S. Alam et al., *Phys. Rev. Lett.* **81** (1998) 3328.
22. S. J. Brodsky and J. P. Ferrar, *Phys. Rev. Lett.* **31** (1973) 1153.
23. C. Berger, B. Lechner and W. Schweiger, *Fizika* **B 8** (1999) 371 and hep-ph/9901338.

Search for the glueball candidates $f_0(1500)$ and $f_J(1710)$ in $\gamma\gamma$ collisions in ALEPH

Roger W. L. Jones[†], for the ALEPH Collaboration

[†]Dept. of Physics, University of Lancaster, Lancaster, UK

Abstract. The glueball candidates $f_0(1500)$ and $f_J(1710)$ have been sought in $\gamma\gamma$ interations in the ALEPH data *via* their decay to $\pi^+\pi^-$. No signal is observed; the products of $\gamma\gamma$ width and $\pi^+\pi^-$ branching ratio in the case of the $f_0(1500)$ and the $f_J(1710)$ are determined have upper limits of $\Gamma(\gamma\gamma \rightarrow f_0(1500)) \cdot \mathcal{BR}(f_0(1500) \rightarrow \pi^+\pi^-) < 0.31$ keV and $\Gamma(\gamma\gamma \rightarrow f_J(1710)) \cdot \mathcal{BR}(f_J(1710) \rightarrow \pi^+\pi^-) < 0.55$ keV , respectively, at 95% confidence level.

INTRODUCTION

Lattice calculations have predicted the lightest scalar glueball to be a scalar resonance with mass 1600 ± 150 MeV$/c^2$ [1,2]. Experimentally, two principal candidates for the scalar glueball, the $f_0(1500)$ and the $f_J(1710)$, have been observed in gluon-rich interactions such as J$/\psi$ decay. However, the $f_0(1500)$ would be interpreted not as a pure glueball but as a mixture with $q\bar{q}$ states, and the spin of the $f_J(1710)$ is variously reported as spin 2 and spin 0. In $\gamma\gamma$ interactions, the pure glueball state should be supressed, and hence determining the two-photon width $\Gamma_{\gamma\gamma}$ of a glueball candidate helps to indicate its quark content.

THE ALEPH DETECTOR, TRIGGER AND MONTE CARLOS

The ALEPH detector and trigger are described in detail elsewhere [4,5]. Tracks were measured using a silicon microvertex detector, a multiwire drift chamber and a large time projection chamber. The tracking system was immersed in a 1.5 T axial magnetic field. An an electromagnetic calorimeter lay between the time projection chamber and the coils, while luminosity monitors covered the small polar angle region. Outside of the coil lay a hadron calorimeter that provided a measurement of the energy of charged and neutral hadrons, and which with external muon chambers provided muon identification.

CP571, *PHOTON 2000*, edited by A. J. Finch

For the subsample used in this analysis, the main trigger was based on the dentification of two track candidates in the multiwire tracker with at least one track pointing to an energy deposit in excess of 200 MeV in the electromagnetic calorimeter. Earlier data relied on two track candidates being back-to-back within 11°. While this was a reasonably efficient trigger for the events in the mass region of interest, its efficiency fell steeply for lower mass candidates.

Fully simulated and reconstructed Monte Carlo event samples were used for the design of the event selection, background estimation and extraction of a limit on two-photon widths $\Gamma_{\gamma\gamma}^{f_0(1500)}$ and $\Gamma_{\gamma\gamma}^{f_J(1710)}$. Samples of the signal processes were generated using PHOT02 [6]. Both of the meson states were generated as a Breit-Wigner resonance of appropriate mass, width and spin; the $f_J(1710)$ was assumed to be spin zero. The experimental resolution of the masses of both resonances was much smaller than the natural width. Expected background processes $\gamma\gamma \to e^+e^-$, $\gamma\gamma \to \mu^+\mu^-$, and $\gamma\gamma \to \tau^+\tau^-$ were also generated using PHOT02, while background from $e^+e^- \to Z \to \tau^+\tau^-$ was estimated using KORALZ [7].

EVENT SELECTION

The analysis used 160.9 pb^{-1} of data taken near the Z resonance. Candidate $\gamma\gamma \to \pi^+\pi^-$ events were selected by requiring the following criteria, designed to ensure (quasi-)real photon ($Q^2 = 0$) collisions, and to suppress three- or more body decays:

- that the event contained only two good tracks, of equal and opposite charge and contained within the tracking system and passing close to the nominal interaction point;

- that the total energy summed over all reconstructed objects (after accounting for possible double-counting) equalled to the total energy of the two charged tracks;

- that the total energy observed in the event was less than 30 GeV and visible mass less than 10 GeV, thus suppressing Z events;

- that the final state showed a transverse momentum balance to within 0.1 GeV/c;

- that the absolute value of the cosine of the angle θ^* between the two tracks in their centre-of-mass system and the boost direction was less than 0.9.

At this stage the dominant background was the high cross-section $\gamma\gamma \to \mu^+\mu^-$ process. Events containing identified muons were rejected, giving a rejection efficiency of nearly 100% for events with invariant mass $W > 3$ GeV/c^2. Below $W = 3$ GeV/c^2, additional cuts cuts were imposed on the fractional energy deposition in the calorimeters, but the rejection efficiency still fell to $\sim$20% for $W = 1$ GeV/c^2. Specific ionisation ($\mathrm{d}E/\mathrm{d}x$) information for each track was used to

375

reject electrons, and some of the low rate of charged kaon pairs. Backgrounds from beam-gas interactions were made neglibible by the requirement that the event be within 2 cm of the nominal interaction point. The remaining background contribution from τ-pair production was also negligible.

Some 294141 $\pi^+\pi^-$ candidate events remained after selections, with an invariant mass spectrum as shown in Fig. 1. The steep rise of the spectrum around 0.5 GeV/c^2 was an artefact of the trigger efficiency. The clear peak in the spectrum above 1 GeV/c^2 was identified with the known tensor resonance $f_2(1270)$. No other structure was observed above the $f_2(1270)$ peak.

A fit to the invariant mass spectrum was performed. For the resonances $f_2(1270)$, $f_0(1500)$ and $f_J(1710)$, a Breit-Wigner function of the form

$$\frac{mm_0\Gamma(m)}{(m_0^2 - m^2)^2 + m^2\Gamma^2(m)} \tag{1}$$

was used, where the mass-dependent width $\Gamma(m)$, away from the two-pion threshold, had the form

$$\Gamma(m) = \Gamma_0 \left(\frac{m}{m_0}\right)^{(l+1/2)} \exp\left[\frac{-(m^2 - m_0^2)}{48\beta^2}\right] \tag{2}$$

and $\beta = 0.4$ GeV/c^2 [8]. For the $f_2(1270)$ the mass m_0, total width Γ_0 and overall normalisation were allowed to be free; for the $f_0(1500)$ and the $f_J(1710)$

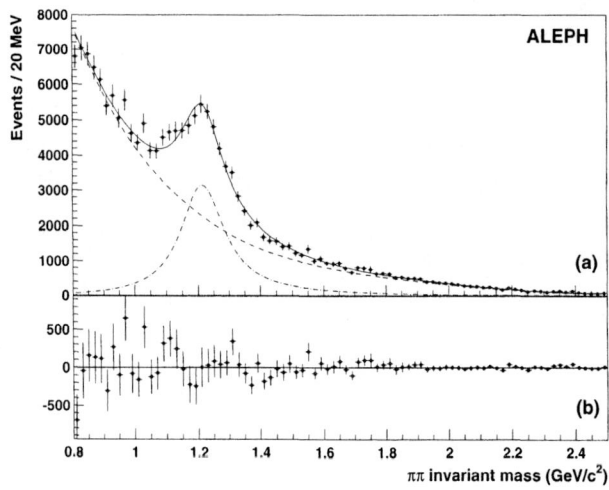

FIGURE 1. The invariant mass distribution for two-pion final states with a fit to data with a Breit-Wigner function for the $f_2(1270)$ (dot-dash line), a polynomial for the background (dashed line) and the combination of these functions (solid line) - the error bars indicate statistical errors only; and (b) the data after subtraction of the fitted curve. Error bars indicate statistical errors only.

TABLE 1. Table of parameters for fits using the Breit-Wigner form of Eqn.1 for resonances.

	(i) Fit for $f_2(1270)$ only
χ^2	74.96
Degrees of freedom	76
Fitted mass of $f_2(1270)$ (MeV/c^2)	1213.5 ± 3.7
Fitted width of $f_2(1270)$ (MeV/c^2)	178.3 ± 12.8
	(ii) Fit for $f_2(1270) + f_0(1500)$
χ^2	73.33
Degrees of freedom	75
Fitted mass of $f_2(1270)$ (MeV/c^2)	1214.1 ± 3.8
Fitted width of $f_2(1270)$ (MeV/c^2)	173.9 ± 13.8
No. of $f_0(1500)$ signal events	-808.3 ± 602.6
	(iii) Fit for $f_2(1270) + f_J(1710)$
χ^2	74.02
Degrees of freedom	75
Fitted mass of $f_2(1270)$ (MeV/c^2)	1213.9 ± 3.9
Fitted width of $f_2(1270)$ (MeV/c^2)	180.2 ± 15.9
No. of $f_J(1710)$ signal events	468.3 ± 476.6
	(iv) Fit for $f_2(1270) + f_0(1500) + f_J(1710)$
χ^2	73.21
Degrees of freedom	74
Fitted mass of $f_2(1270)$ (MeV/c^2)	1214.2 ± 3.8
Fitted width of $f_2(1270)$ (MeV/c^2)	175.5 ± 14.2
No. of $f_0(1500)$ signal events	-671.0 ± 690.0
No. of $f_J(1710)$ signal events	198.6 ± 541.9

only the normalisation was allowed to be free, with the mass and width fixed to 1500 MeV/c^2 and 112 MeV/c^2, and 1712 MeV/c^2 and 133 MeV/c^2 for the $f_J(1710)$, respectively [3]. The combination of the $\mu\mu$ and $\pi\pi$ continuum background was fitted with a fifth-order Chebyshev polynomial. Only the mass region with a high trigger efficiency from 0.8 to 2.5 GeV/c^2 was used in the fit. Four fit variations were attempted, each with the $f_2(1270)$ and the polynomial background and optionally the glueball candidates. The fit results are summarised in Table 1, including the number of events fitted for the glueball candidate signals. The fit including the $f_2(1270)$ alone already gave a very good description of the data. The χ^2 per degree of freedom hardly changes with the addition of glueball signals. The fits including the $f_J(1710)$ were also performed for $J = 2$. The results of these fits were essentially the same as for the $J = 0$ case.

The fitted width of the $f_2(1270)$ was in all cases in agreement with the world average value, but its mass was lower. A similar shift was observed by the MARKII and CELLO collaborations [9] and ascribed to an interference of the spin 2 resonant amplitude with other components in the background. The number of events fitted for the $f_0(1500)$ signal was negative, but consistent with zero. While destructive interference is possible, limits were set assuming no interference; the limited knowledge of the trigger response prevented a phase-shift analysis of the region.

The first fit was repeated excluding a region around 1.1 GeV/c^2, and a clear excess of data over the fitted curve extrapolated through that region was observed, consistent with the CELLO observation of a possible structure in the $\pi^+\pi^-$ invariant mass spectrum around 1.1 GeV/c^2 [10]. Accounting for this with a spin-0 resonance term made little difference to the conclusions of this study.

The fitted numbers of signal events from the processes $\gamma\gamma \rightarrow f_0(1500) \rightarrow \pi^+\pi^-$ and $\gamma\gamma \rightarrow f_J(1710) \rightarrow \pi^+\pi^-$ can be used to calculate upper limits on $\Gamma_{\gamma\gamma} \cdot \mathcal{BR}(\pi^+\pi^-)$ for both resonances. The acceptance and selection efficiency for $\gamma\gamma \rightarrow f_0(1500) \rightarrow \pi^+\pi^-$ and for $\gamma\gamma \rightarrow f_J(1710) \rightarrow \pi^+\pi^-$ were determined from Monte Carlo to be $(17.5\pm0.4)\%$ and $(16.3\pm0.4)\%$, where the quoted errors include systematics due to errors in the Monte Carlo simulation of the detector resolutions and those due to limited Monte Carlo statistics.

The fitted value of $\Gamma_{\gamma\gamma} \cdot \mathcal{BR}(\pi^+\pi^-)$ (which may be negative) and its error were used to define the mean and width of a Gaussian distribution: the area under the Gaussian was integrated above zero to obtain the 95% C.L. limit on $\Gamma_{\gamma\gamma}\cdot\mathcal{BR}(\pi^+\pi^-)$.

The upper limits on the product branching ratios at 95% C.L. are, for the individual fits, $\Gamma_{\gamma\gamma}\cdot\mathcal{BR}(\pi^+\pi^-) < 0.25$ keV and $\Gamma_{\gamma\gamma}\cdot\mathcal{BR}(\pi^+\pi^-) < 0.59$ keV for the $f_0(1500)$ and $f_J(1710)$ respectively, and for the combined fit are $\Gamma_{\gamma\gamma}\cdot\mathcal{BR}(\pi^+\pi^-) < 0.31$ keV and $\Gamma_{\gamma\gamma}\cdot\mathcal{BR}(\pi^+\pi^-) < 0.55$ keV for the $f_0(1500)$ and $f_J(1710)$ respectively. If an additional spin 0 resonance with a fixed mass of 1.1 GeV/c^2 was included in the combined fit, the limit on the product branching ratio for the $f_0(1500)$ worsened by ~30% and for the $f_J(1710)$ tightens by ~10%. The use of other Breit-Wigner forms gave considerably worse fits, but all of the qualitative conclusions remained unchanged, and the quantitative results agreed within the quoted errors.

From limits on $\Gamma_{\gamma\gamma}$ the *stickiness* [11] of each resonance can be calculated, by comparison of the two photon width with the particle's width for production in the glue-rich environment of radiative J/ψ decay. For a glueball a high value of stickiness is expected. Stickiness for a 0^{++} resonance X is defined as

$$ S_X(0^{++}) = \mathcal{N}\,\frac{m_X}{k_{J/\psi \rightarrow \gamma X}}\,\frac{\Gamma(J/\psi \rightarrow \gamma X)}{\Gamma(X \rightarrow \gamma\gamma)} $$

where m_X is the mass of the resonance; $k_{J/\psi \rightarrow \gamma X} = (m_{J/\psi}^2 - m_X^2)/2m_{J/\psi}$ is the energy of the photon from the radiative J/ψ decay, measured in the J/ψ rest frame; $\Gamma(J/\psi \rightarrow \gamma X)$ is the width for production of the resonance in radiative J/ψ decay; and $\mathcal{N}$ is the normalisation factor chosen such that the $q\bar{q}$ resonance $f_2(1270)$ resonance has unit stickiness. The branching ratio for $f_0(1500) \rightarrow \pi^+\pi^-$ was taken as 0.30 ± 0.07, while the branching ratio for $f_J(1710) \rightarrow \pi^+\pi^-$ is $0.026^{+0.001}_{-0.016}$ [3], giving, from the fit for both resonances, $\Gamma(f_0(1500) \rightarrow \gamma\gamma) = 1.08$ keV and $\Gamma(f_J(1710) \rightarrow \gamma\gamma) = 21.25$ keV. The lower limits on stickiness are then 1.4 and 0.3 for the $f_0(1500)$ and the $f_J(1710)$ respectively.

CONCLUSIONS

Production of the glueball candidates $f_0(1500)$ and $f_J(1710)$ in $\gamma\gamma$ collisions at LEP1 was studied via decay to $\pi^+\pi^-$. No signal from either resonance was seen, and the upper limits on the product of two-photon width and $\pi^+\pi^-$ branching ratio were calculated as $\Gamma_{\gamma\gamma} \cdot \mathcal{BR}(\pi^+\pi^-) < 0.31$ keV for the $f_0(1500)$ and $\Gamma_{\gamma\gamma} \cdot \mathcal{BR}(\pi^+\pi^-) < 0.55$ keV for the $f_J(1710)$, both at 95% C.L., from a simultaneous fit for both resonances.

ACKNOWLEDGEMENTS

I wish to thank my ALEPH collegues for the smooth and efficient running of ALEPH and our colleagues from the accelerator divisions for the successful operation of LEP.

REFERENCES

1. UKQCD Collaboration, *A Comprehensive Lattice Study of SU(3) Glueballs*, Phys. Lett. **B309** (1993) 378.
2. J. Sexton, A. Vaccarino and D. Weingarten, Phys. Rev. Lett **75** (1995) 4563.
3. Particle Data Group, Eur. Phys. J. **C3** (1998) 1.
4. ALEPH Collaboration, *ALEPH: A Detector for Electron-Positron Annihilations at LEP*, Nucl. Instrum. Meth **A294** (1990) 121.
5. ALEPH Collaboration, *Performance of the ALEPH Detector at LEP*, Nucl. Instrum. Meth **A360** (1995) 481.
6. ALEPH Collaboration, *An Experimental study of $\gamma\gamma \to$ hadrons at LEP*, Phys. Lett. **B313** (1993) 509.
7. S. Jadach, B.F.L. Ward and Z. Wąs, Comp. Phys. Comm. **79** (1994) 503.
8. A. Donnachie, private communication.
9. J. Boyer *et al.* (MARK II Collaboration), *Two-Photon Production of Pion Pairs*, Phys. Rev. **D42** (1990) 1350; J.E. Olsson, in Proc. of 5th Intern. Colloquium on Photon Photon Collisions (1983) Aachen.
10. CELLO Collaboration, *An Experimental Study of the Process $\gamma\gamma \to \pi^+\pi^-$*, Z. Phys. **C56** (1992) 381.
11. M. Chanowitz, in Proc. of the VI Intern. Workshop on Photon-Photon Collisions (1984) Lake Tahoe.

Resonance Production in Two-Photon Collisions at LEP with the L3 Detector

Valery P. Andreev

Department of Physics and Astronomy
202 Nicholson Hall
Louisiana State University
Baton Rouge, LA 70803, USA

Abstract. The $ee \to eeK_S^0 K^\pm \pi^\mp$, $ee \to ee\eta\pi^+\pi^-$ and $e^+e^- \to e^+e^-\pi^+\pi^-\pi^+\pi^-$ final states are studied with the L3 detector at LEP using data collected at centre–of–mass energies from 160 GeV up to 202 GeV. The mass spectrum of the $K_S^0 K^\pm \pi^\mp$ final state shows an enhancement around 1470 MeV, which is identified with the pseudoscalar meson $\eta(1440)$. This state is observed in $\gamma\gamma$ collisions for the first time and its two–photon width is measured. Clear evidence is also obtained for the formation of the axial vector mesons $f_1(1420)$ and $f_1(1285)$. In the $\eta\pi^+\pi^-$ channel the $f_1(1285)$ is observed, and upper limits for the formation of $\eta(1440)$ and $\eta(1295)$ are obtained. The process $e^+e^- \to e^+e^-\pi^+\pi^-\pi^+\pi^-$ is dominated by $\rho^0\rho^0$ production. A spin-parity analysis of the $\rho^0\rho^0$ system for $W_{\gamma\gamma} < 3GeV$ shows a dominance of $J^P = 2^+$ and helicity 2. A contribution of $J^P = 0^+$ is also observed whereas contributions of negative parity states and $(J^P, J_z) = (2^+, 0)$ are found to be negligible. For the $\gamma\gamma$ mass region $W_{\gamma\gamma} > 3GeV$ the $f_2(1270)$ meson is observed.

INTRODUCTION

Resonance formation in two–photon interactions offers a clean environment to study the spectrum of mesonic states. The L3 collaboration has been studied the reactions $\gamma\gamma \to K_S^0 K^\pm \pi^\mp$ and $\gamma\gamma \to \eta\pi^+\pi^-$ in Ref. [1].

The mass region between 1200 MeV and 1500 MeV is expected to contain several states [2]. For the pseudoscalar sector ($J^{PC} = 0^{-+}$), the $\eta(1440)$ meson, formerly known as E or ι, is expected to be seen. The $\eta(1440)$ was observed in hadron collisions and in radiative J/ψ decay, but not in two–photon collisions, and only upper limits of its two–photon width, $\Gamma_{\gamma\gamma}$, of the order of 1 keV exist [3,4]. There are, possibly, two pseudoscalars [5] in the 1440 MeV mass region: one at lower mass, $\eta_L(1418 \pm 1)$ MeV, which decays into $a_0\pi$ or directly into $\eta\pi\pi$, and another at higher mass, $\eta_H(1475 \pm 5)$ MeV, decaying to K^*K. Taking into account also the $\eta(1295)$, there are therefore three candidates for the first radial excitations of the pseudoscalar SU(3) nonet. If one of these states is a gluonium, its two–photon

CP571, *PHOTON 2000*, edited by A. J. Finch
© 2001 American Institute of Physics 0-7354-0010-5/01/$18.00

width is expected to be small with respect to a $q\bar{q}$ state. However, the gluonium interpretation is disfavoured by lattice gauge theories [6] which predict the lowest lying 0^{-+} gluonium state to be above 2 GeV.

Axial vector mesons ($J^{PC} = 1^{++}$) are also present in these final states in the 1440 MeV mass region. The $f_1(1420)$ was observed in two–photon collisions, in the $K\bar{K}\pi$ decay channel, by the CELLO [4], TPC-2γ [16], JADE [8] and Mark II [9] Collaborations, the $f_1(1285)$ was seen [9,16] in the $\eta\pi^+\pi^-$ decay channel.

The data used here were collected by the L3 detector [10] at LEP from 1997 to 1999 at centre–of–mass energies between 183 GeV and 202 GeV, corresponding to a total integrated luminosity of 449 pb^{-1}. The analysis makes use of the dependence of the signal yield on the total transverse momentum $P_T^2 = (\sum \vec{p_T})^2$, where the sum runs over all the observed particles. To a good approximation $P_T^2 = Q^2$, where Q^2 is the maximum virtuality of the two photons. According to the Landau–Yang theorem [20], for real photons ($Q^2 \simeq 0$), the production of a spin-0 state is allowed while that of a spin-1 state is suppressed.

A large cross section for the two-photon reaction $\gamma\gamma \to \rho^0\rho^0$ near threshold has been observed in several experiments [11,17]. The cross section exceeds the estimations from the vector dominance model (VDM) by almost one order of magnitude. Similar enhancements have been also found in the decay $J/\psi \to \gamma\rho\rho$ [18] and in inelastic anti-proton scattering on deuterium $\bar{p}d \to \pi^-(2\pi^+2\pi^-)$ [19].

Two Monte Carlo generators are used to describe two–photon resonance formation: EGPC [21] and GaGaRes [22].

RESULTS

The $e^+e^- \to e^+e^-K_S^0 K^\pm\pi^\mp$ channel

Events are selected by requiring four charged particles in the central tracker, two tracks ($K^\pm\pi^\mp$), coming from the interaction point, and a K_S^0 decaying into $\pi^+\pi^-$ at a secondary vertex. For charged particle identification, consistency of the corresponding dE/dx measurement, with a confidence level CL>1% is evaluated for each particle identity.

The selection results in the $K_S^0 K^\pm\pi^\mp$ effective mass spectra shown in Figure 1. A clear peak is seen in the 1440 MeV mass region. The data are subdivided into four P_T^2 intervals of similar statistics in the peak region, in order to study the $\eta(1440)$ and $f_1(1420)$ contributions. For the peak in the lowest P_T^2 interval, Figure 1a, the mass and width, $M = 1481 \pm 12 MeV$ and $\sigma = 48 \pm 9 MeV$, are compatible with those of the η_H. In the highest P_T^2 interval, $1 GeV^2 < P_T^2 < 7 GeV^2$, an $f_1(1285)$ signal is present. The differential cross section $d\sigma/dP_T^2$ is shown in Figure 2. It is fitted, using the predictions for the Q^2 dependence of $\eta(1440)$ and $f_1(1420)$ production given by the GaGaRes Monte Carlo. Pure pseudoscalar $\eta(1440)$ or pure axial vector meson $f_1(1420)$ hypotheses are excluded with a CL $\sim 10^{-3}$ and CL $\sim 10^{-5}$ respectively, while simultaneous presence of the $\eta(1440)$ and $f_1(1420)$ hypothesis

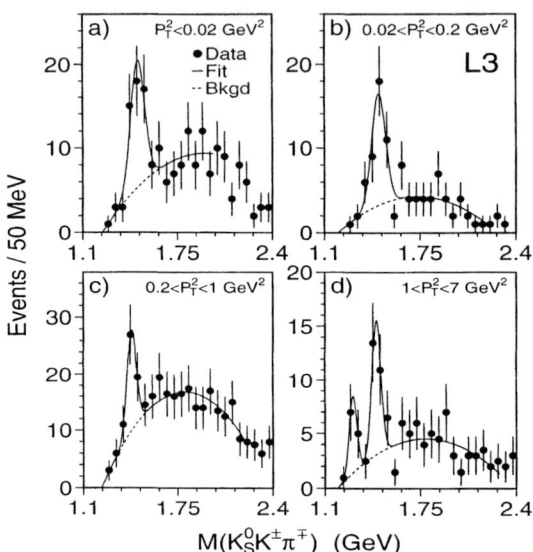

FIGURE 1. $K_S^0 K^\pm \pi^\mp$ spectra for different P_T^2 bins. The fits of a Gaussian plus polynomial background are superimposed on the data. In the highest P_T^2 bin, also the peak of the $f_1(1285)$ is present and the fit includes two Gaussians.

leads to a CL $\simeq 9\%$, with 74 ± 10 events for the spin–0 particle and 42 ± 10 events for the spin–1.

Taking into account the branching ratio values [2], BR($K_S^0 \rightarrow \pi^+\pi^-$) and BR($K^0 \rightarrow K_S^0$), and the isospin factor $K\bar{K}\pi / K^0 K^\pm \pi^\mp$ the two–photon width for the $K\bar{K}\pi$ decay channel is: $\Gamma_{\gamma\gamma}\left((\eta(1440))\right) \times \text{BR}\left((\eta(1440) \rightarrow K\bar{K}\pi)\right) = 212 \pm 50\text{stat} \pm 23\text{sys} \text{eV}$. This value is consistent with the upper limit of 1.2 keV reported by the CELLO Collaboration [4].

To search for signals due to K*(892) and $a_0(980)$, the $K\pi$ and $K_S^0 K^\pm$ mass spectra in the $\eta(1440)$ region, $1370 MeV < \text{M}(K_S^0 K^\pm \pi^\mp) < 1560 MeV$, are investigated. A clear K*(892) signal is seen in the $K_S^0 \pi^\pm$ and $K^\pm \pi^\mp$ spectra. The $K_S^0 K^\pm$ spectrum shows no clear evidence for the presence of the $a_0(980)$. With the present limited statistics no firm conclusion can be drawn concerning the possible presence of the $\eta_L \rightarrow a_0\pi$ state in the data. The mass and width of $\eta(1440)$ meson as well as the observation of a dominant K*(892)K decay are compatible with the characteristics of the η_H.

The $e^+e^- \rightarrow e^+e^-\eta\pi^+\pi^-$ channel

The η is detected via its decay $\eta \rightarrow \gamma\gamma$. After all cuts, 6444 events are selected with $\eta\pi^+\pi^-$ masses below 1750 MeV. The $\eta\pi^+\pi^-$ mass spectrum, shown

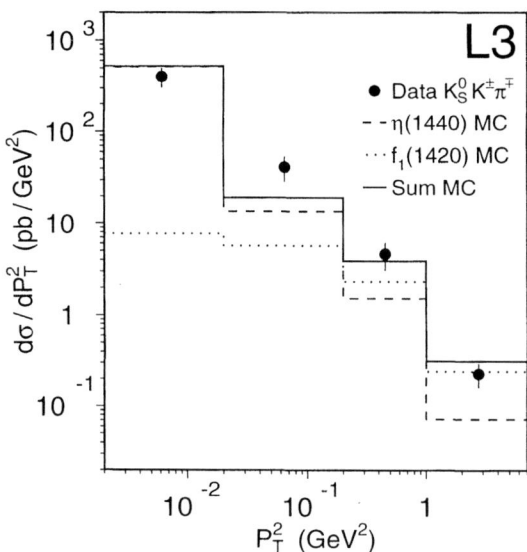

FIGURE 2. Differential cross section for the 1440 MeV mass region, as estimated from the Gaussian fit, as a function of P_T^2 in the $K_S^0 K^\pm \pi^\mp$ channel. The solid line is the sum of the $\eta(1440)$ and $f_1(1420)$ simulations fitted to the data. The partial contributions of $\eta(1440)$ (dashed line) and $f_1(1420)$ (dotted line) are also shown.

in Figure 3a, is dominated by the $\eta'(958)$ resonance. The $\eta\pi^+\pi^-$ mass spectrum for $P_T^2 < 0.02 GeV^2$ is shown in Figure 3b. No peak is seen in the region $1200-1480$ MeV.

The absence of signals of $\eta(1440)$ and $\eta(1295)$ allows to calculate upper limits, at 95% CL on the $\eta\pi\pi$ decay. Taking into account the branching ratio $BR(\eta \to \gamma\gamma)$ [2] and the isospin factor $\eta\pi\pi / \eta\pi^+\pi^-$, they are: $\Gamma_{\gamma\gamma}\left((\eta(1440))\right) \times BR\left((\eta(1440) \to \eta\pi\pi)\right) < 95eV$ and $\Gamma_{\gamma\gamma}\left((\eta(1295))\right) \times BR\left((\eta(1295) \to \eta\pi\pi)\right) < 66eV$. They improve the limits of the Crystal Ball Collaboration [3].

The peak around 1285 MeV, seen in Figure 3a and absent in Figure 3b at $Q^2 \simeq 0$, is identified with the spin–1 state $f_1(1285)$. A Gaussian fit plus polynomial background gives $M = 1280 \pm 4$ MeV and $\sigma = 21 \pm 4$ MeV.

Test on glueball

The two–photon width of the $\eta(1440)$, calculated under the assumption that it is a member of the first radial excitation of the pseudoscalar nonet [23], is of the order of 0.1 keV. The $s\bar{s}$ content gives a negligible contribution. This is in agreement with our measurement, if we assume that $BR\left((\eta(1440) \to K\bar{K}\pi)\right) \sim 100\%$.

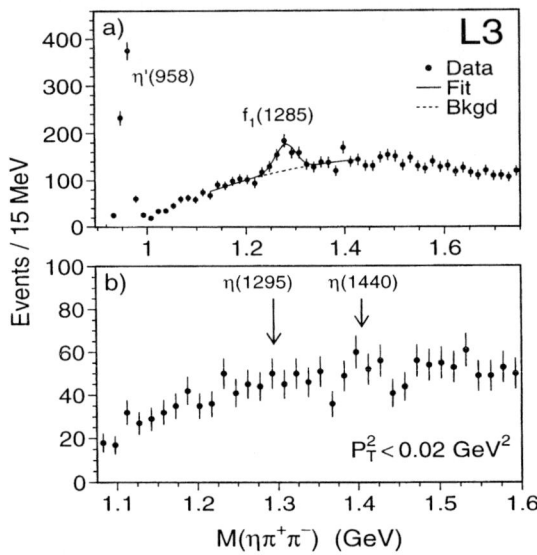

FIGURE 3. The $\eta\pi^+\pi^-$ mass spectra: a) total spectrum, the fit of a Gaussian plus polynomial background for the $f_1(1285)$ region is superimposed on the data; b) for $P_T^2 < 0.02 GeV^2$, arrows show the location of $\eta(1295)$ and $\eta(1440)$.

If the $\eta(1440)$ is a gluon rich state, its stickiness parameter [24] is expected to be large. Using the present measurement of $\Gamma_{\gamma\gamma}\left((\eta(1440))\right) \times BR\left((\eta(1440) \to K\bar{K}\pi)\right)$ and $\Gamma(J/\psi \to \gamma\eta(1440)) = 79 \pm 16eV$ [2], a value of the stickiness $S_{\eta(1440)} = 79\pm26$ is obtained. Another parameter, the gluiness (G), was introduced [25,26] to quantify the ratio of the two–gluon and two–photon coupling of a particle. Whereas stickiness is a relative measure, the gluiness is a normalised quantity and is expected to be near unity for a q$\bar{\text{q}}$ meson. Using the present measurement of $\Gamma_{\gamma\gamma}\left((\eta(1440))\right) \times BR\left((\eta(1440) \to K\bar{K}\pi)\right)$ and the average value of $\alpha_s(1440MeV) = 0.369 \pm 0.022$ [2], a value $G_{\eta(1440)} = 41 \pm 14$ is obtained.

From the upper limit in the $\eta\pi^+\pi^-$ channel and the value $\Gamma(J/\psi \to \gamma\eta(1440)) = 29 \pm 6eV$ [2] the limits $S_{\eta(1440)} > 87$ and $G_{\eta(1440)} > 45$ at 95% confidence level are obtained, compatible with the values obtained for the $K_S^0 K^\pm \pi^\mp$ decay channel. Both the stickiness and gluiness of the $\eta(1440)$ point to a large gluonium content in this resonance. For comparison, the η' pseudoscalar meson has $S_{\eta'} = 3.6 \pm 0.3$ and $G_{\eta'} = 5.2 \pm 0.8$ for $\alpha_s(958MeV) = 0.56 \pm 0.07$ [2].

The $e^+e^- \rightarrow e^+e^-\pi^+\pi^-\pi^+\pi^-$ channel

In order to select $\gamma\gamma \rightarrow \pi^+\pi^-\pi^+\pi^-$ events exactly four charged tracks in the tracking chamber with a net charge of zero are required. Events with photons are rejected. Pion identification is based on the dE/dx information. After applying these cuts a sample of 56359 events is obtained, of which 55120 events lie in the mass region between 1.0 and 3.0GeV. Data sample is divided in two parts, low mass data ($W_{\gamma\gamma} < 3.0GeV$) and high mass data ($W_{\gamma\gamma} > 3.0GeV$). The spin-parity and helicity analysis is performed only on the low mass data. The $\pi^+\pi^-$ mass

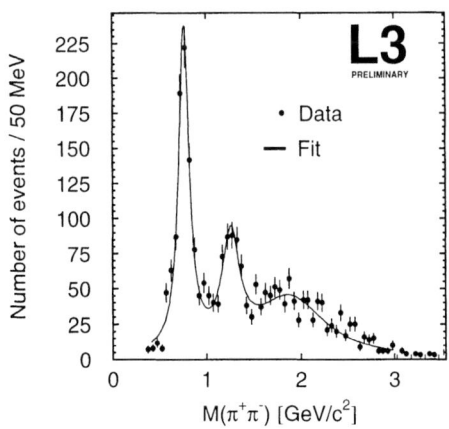

FIGURE 4. The two-pion mass after combinatorial background subtraction for the high mass data $W_{\gamma\gamma} > 3GeV$.

spectrum after subtraction of the $\pi^\pm\pi^\pm$ combinatorial background is presented in Fig. 4 for $W_{\gamma\gamma} > 3.0GeV$. There is still a strong production of the ρ^0 meson but the $f_2(1270)$ meson is also seen. Since the f_2 is a tensor meson $I^G(J^{PC}) = 0^+(2^{++})$ it cannot be produced diffractively because its quantum numbers differ from the ones of the photon. One possibility to produce the f_2 meson is an intermediate resonance decaying into a pair of f_2 mesons : $\gamma\gamma \rightarrow R \rightarrow f_2 f_2$. Additional studies are needed to clarify this hypotheses.

According to the model proposed by TASSO [13] in the analysis of the reaction $\gamma\gamma \rightarrow \pi^+\pi^-\pi^+\pi^-$ we consider the $\rho^0\rho^0$ production in different spin-parity and helicity states (J^P, J_z) and isotropic production of $\rho^0\pi^+\pi^-$ and $\pi^+\pi^-\pi^+\pi^-$. All states are assumed to be produced incoherently and therefore no interference effects between these three final states are taken into account. In each $W_{\gamma\gamma}$ bin a maximum

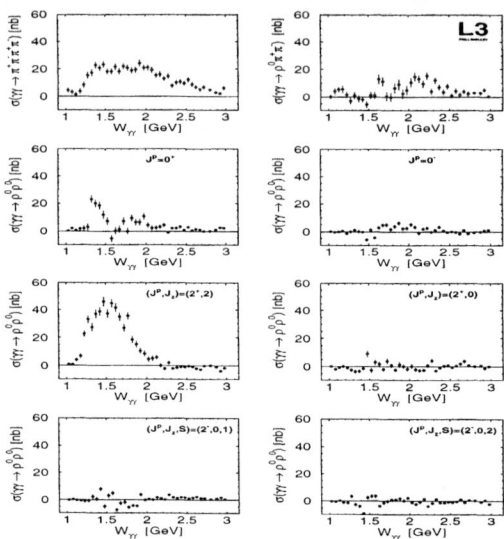

FIGURE 5. Cross sections for different spin-parity and helicity states (J^P, J_z) of $\rho^0\rho^0$ and for the isotropic $\rho^0\pi^+\pi^-$ and $\pi^+\pi^-\pi^+\pi^-$ states as determined by the 8 parameter fit.

likelihood fit is used to determine the contribution λ_j of the eight amplitudes to the data sample. The fit maximizes the logarithm of the likelihood function:

$$\log \Lambda = \sum_{i=1}^{N} \left[\log \left(\sum_j \lambda_j \frac{|g_j(\xi_i)|^2}{|g_j|^2} \right) - \sum_j \lambda_j \right]$$

with the constraint $\sum_j \lambda_j = 1$, where N is the number of events in the bin and $g_{\rho^0\rho^0} = BW(m_{\rho^0_1})BW(m_{\rho^0_2})\Psi_{J^P J_z}(1,2)$ + permutations, $g_{\rho^0\pi^+\pi^-} = BW(m_{\rho^0_1})$ + permutations and $g_{4\pi} = 1$, where $BW(m_{\rho^0})$ is the relativistic Breit-Wigner amplitude for the ρ^0 meson [27] and Ψ_{J^P, J_z} is the angular part of amplitude which describe the rotational properties of the $\rho^0\rho^0$ state with definite spin-parity J^P and helicity J_z.

The cross sections derived from the eight parameter fit are shown in Fig. 5. The $\rho^0\rho^0$ state with spin-parity 2^+ and helicity 2 dominates. The $J^P = 0^+$ state is also significant in the $\gamma\gamma$ mass region from $1.3 GeV$ to $2. GeV$. The contributions of the negative parity states and of the state with $(2^+, 0)$ are negligible. These results agree well with ARGUS [17] analysis. To check the stability of the fit and the possible migration of the events between different states the fit is repeated with only the four significant states: $\pi^+\pi^-\pi^+\pi^-$, $\rho^0 2\pi$ and $\rho^0\rho^0$ with $(J^P, J_z) = 0^+, (2^+, 2)$. Similar cross-sections are obtained.

ACKNOWLEDGEMENTS

I would like to thank Alex Finch and his colleagues for the perfect organisation. I wish to thank my L3 colleagues of the two-photon physics group Igor Vodopianov and Oleg Fedin performed the studies presented in this proceedings.

REFERENCES

1. L3 Collaboration, M. Acciarri et al., CERN-EP-2000-129, submit. to Phys. Lett. B .
2. D. E. Groom et al., (Particle Data Group), Eur. Phys. J. C 15, 1 (2000).
3. Crystal Ball Collaboration, D. Antreasyan et al., Phys. Rev. D 36, 2633 (1987).
4. CELLO Collaboration, H. J. Behrend et al., Z. Phys. C 42, 367 (1989).
5. OBELIX Collaboration, C. Cicalo et al., Phys. Lett. B 462, 453 (1999);
   M. G. Rath et al., Phys. Rev. D 40, 693 (1989);
   DM2 Collaboration, J. E. Augustin et al., Phys. Rev. D 46, 1951 (1992);
   Mark III Collaboration, Z Bai et al., Phys. Rev. Lett. 65, 2507 (1990).
6. C. J. Morningstar and M. Pearon, Phys. Rev. D 60, 034509 (1999);
   UKQCD Collaboration, G. Bali et al., Phys. Lett. B 309, 378 (1993).
7. TPC-2γ Collaboration, H. Aihara et al., Phys. Lett. B 209, 107 (1988); Phys. Rev. D 38, 1 (1988).
8. JADE Collaboration, P. Hill et al., Z. Phys. C 42, 355 (1989).
9. Mark II Collaboration, G. Gidal et al., Phys. Rev. Lett. 59, 2012 (1987); Phys. Rev. Lett. 59, 2016 (1987).
10. L3 Collaboration., B. Adeva et al., Nucl. Inst. Meth. A 289, 35 (1990).
11. TASSO Coll., R. Brandelik et al., Phys. Lett. 97 B (1980) 448.
12. MARK II Coll., D.L. Burke et al., Phys. Lett. 103 B (1981) 153.
13. TASSO Coll., M. Althoff et al., Z. Phys. C 16 (1982) 13.
14. CELLO Coll., H.-J. Behrend et al., Z. Phys. C 21 (1984) 205.
15. PLUTO Coll., Ch. Berger et al., Z. Phys. C 38 (1988) 521.
16. TPC/Two-Gamma Coll., H. Aihara et al., Phys. Rev. 37 D (1988) 28.
17. ARGUS Coll., H. Albrecht et al., Z. Phys. C 50 (1991) 1.
18. MARK III Coll., R.M.Baltrusaitis et al., Phys. Rev. D 33 (1986) 1222.
19. D. Bridges et al., Phys. Rev. Lett. 56 (1986) 211, Phys. Rev. Lett. 56 (1986) 215.
20. L. D. Landau, Dokl. Akad. Nauk. USSR 60, 207 (1948);
    C. N. Yang, Phys. Rev. 77, 242 (1950).
21. F. L. Linde, "Charm Production in γγ Collisions", PhD Thesis, Leiden, (1988).
22. R. van Gulik, Nucl. Phys. B 82 (Proc. Suppl.), 311 (2000).
23. A. V. Anisovich et al., Eur. Phys. J. A 6 (1999) 247.
24. M. Chanowitz, Proc. of the VI Int. Workshop on γγ Collisions, Lake Tahoe, California, 1984, edited by R. L. Lander (World Scientific, 1985).
25. F. E. Close, G. R. Farrar, Z. Li, Phys. Rev. D 55 (1997) 5749.
26. H. P. Paar, Nucl. Phys. B 82 (Proc. Suppl.) (2000) 337.
27. J.D. Jackson, Nuovo Cimento 34, (1964) 1644.

Is the $\sigma(600)$ a Glueball?
Two photon reactions can tell us

M.R. Pennington

Centre for Particle Theory, University of Durham, Durham DH1 3LE, U.K.

Abstract: Minkowski and Ochs have recently argued that the small two photon coupling of a conjectured $\sigma(600)$ is so small that it is likely to be a glueball. We ask whether this can be so or whether it is simply gauge invariance that produces the observed low mass suppression?

THE SCALAR GLUEBALL

QCD predicts that there should exist bound states of glue: states we know as *glueballs* [1]. All modellings of non-perturbative QCD, including lattice computations [2,3], indicate that the lightest glueball should be spinless. It is then not surprising that if we look at the Particle Data tables [4], there are more light isosinglet scalars than can fit into one $q\bar{q}$ nonet: $f_0(400 - 1200)$ (or σ), $f_0(980)$, $f_0(1370)$, $f_0(1500)$ and $f_0(1710)$. But which one is the glueball?

Each state has its own protagonists. Lattice calculations expect the scalar glueball has a mass (in the quenched approximation) between 1600 and 1700 MeV [2,3]. Amsler and Close [5] claim the $f_0(1500)$, studied extensively (but not exclusively) in $\bar{p}p$ annihilation with Crystal Barrel at LEAR, is the glueball, while others [3], even more vehemently, insist the scalar component of the $f_J(1710)$ [4] observed in J/ψ radiative decays is gluish. Here we concentrate on the claim that it is the $\sigma(600)$ that is predominantly a glueball [6,7]. Though Minkowski and Ochs [6] are not the first to make such a proposal, it is the appearance of the $\sigma(600)$ in $\gamma\gamma$ reactions that is central to their claim, and so most appropriate for this meeting.

TWO PHOTON v. DI-PION PRODUCTION OF PIONS

In Fig. 1, we compare the cross-section for $\pi^+\pi^- \to \pi^0\pi^0$ from the very recently published results from the E852 experiment at Brookhaven [8] with the Crystal Ball data [9] on $\gamma\gamma \to \pi^0\pi^0$. Each is dominated by the spin two $q\bar{q}$ state, the $f_2(1270)$. Consequently, it is at this peak that these cross-sections (or rather the squares of the moduli of the corresponding amplitudes) are normalised. Below 1 GeV we see the well-known enhancement in the $\pi\pi$ elastic cross-section, identified as the $\sigma(600)$, while in the $\gamma\gamma$ reaction little is seen. It is this smallness around 600 MeV that Minkowski and Ochs [6] claim points to the σ being predominantly glue.

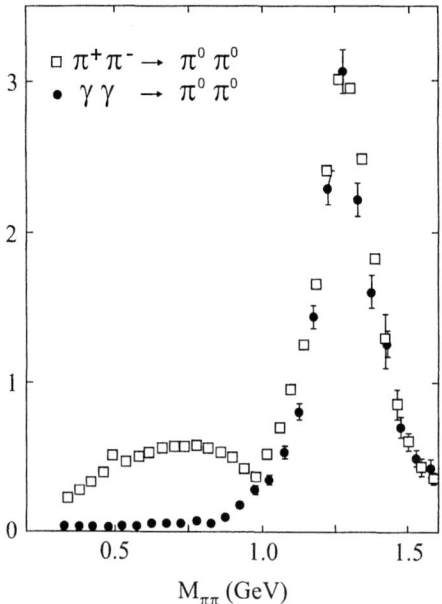

Figure 1: The modulus squared of the amplitudes for $\pi^+\pi^- \to \pi^0\pi^0$ and $\gamma\gamma \to \pi^0\pi^0$, deduced from BNL-E852 [8] and Crystal Ball [9] data, respectively.

In the quenched approximation this is natural. Photons, of course, couple to electric charge. Consequently, a neutral $q\bar{q}$ meson can readily decay into photons. In contrast, a glueball can only decay radiatively through quark loops, Fig. 2. This means that, in the quenched approximation, a glueball doesn't have photonic decays. In perturbative QCD, quark loops (Fig. 2) are suppressed by factors of α_s. But are these arguments relevant at 600 MeV? Indeed, if the $\sigma(600)$ were to be a glueball, it would have to have large couplings to quarks, if its decay width is to be 300-500 MeV as experiment indicates (Fig. 1).

The idea that the $\gamma\gamma \to \pi^0\pi^0$ cross-section directly determines the σ couplings to two photons requires a modelling of the $\gamma\gamma$ reaction. The simplest would be to imagine that the two photon amplitude is just described by the pion Born term plus the σ−resonance component, as in Fig. 3. Since the Born term does not contribute to the $\pi^0\pi^0$ channel, the cross-section in Fig. 1 then fixes the magnitude

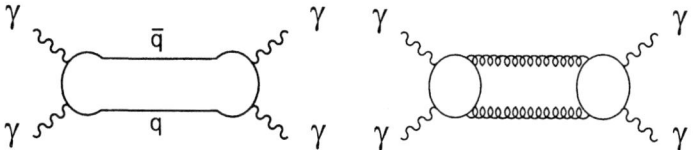

Figure 2: Modulus squared of the amplitude for a $\bar{q}q$ state and a glueball, respectively, to decay to two photons — at lowest order in perturbative QCD.

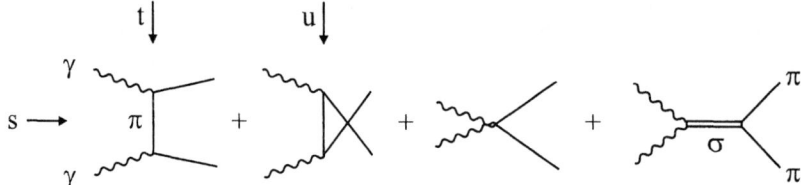

Figure 3: Feynman diagram modelling of $\gamma\gamma \to \pi\pi$: the Born term plus direct channel σ formation.

of this σ−component. But if this model makes any sense it must also explain the $\pi^+\pi^-$ cross-section [10] too. Its S−wave component [11], then determines the orientation of this σ−component. Moreover, unitarity requires that the phase of each $\gamma\gamma \to \pi\pi$ partial wave amplitude with definite isospin must equal the phase of the corresponding $\pi\pi$ elastic partial wave. As detailed in [12], this also fixes the orientation of the σ−component in the model of Fig. 3. These determinations don't agree. This teaches us that the modelling shown in Fig. 3 is far too simplistic. There are in fact many other contributions that must be included, like those in Fig. 4, before we can be sure that unitarity is fulfilled and so extract resonance couplings in a meaningful way.

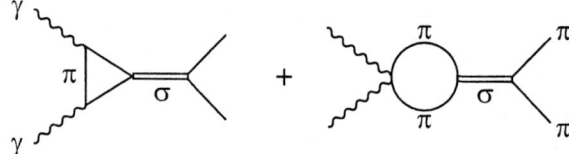

Figure 4: Two of the additional contributions to the model amplitudes for $\gamma\gamma \to \pi\pi$ of Fig. 3 essential for ensuring the final state interaction theorem is satisfied.

Over ten years ago, David Morgan and I [13] worked out a procedure for ensuring that $\gamma\gamma \to \pi\pi$ amplitudes satisfy the constraints of analyticity and unitarity, as well obeying the Thomson limit of QED at low energies [14]. Applying this to all recent experimental data, Elena Boglione and I [11] have found that the $I = 0$ $\gamma\gamma \to \pi\pi$ S−wave cross-section has a quite different shape than that for $\pi\pi$ scattering, as shown in Fig. 4. Indeed, while the Born component dominates the $\gamma\gamma$ channel close to threshold, we see that any resonance contribution is pushed to higher mass away from 600 MeV. There must be some reason for this.

GAUGE INVARIANCE

The simplest Lagrangian describing a scalar field ϕ coupling to two photons is

$$\mathcal{L}_I = g\,\phi\,\mathcal{F}^{\mu\nu}\,\mathcal{F}_{\mu\nu} \quad,$$

where $\mathcal{F}^{\mu\nu}$ is the electromagnetic tensor. The scalar's radiative decay rate is proportional to g^2. Dimensional analysis requires g to have dimension of $1/M$. So g^2

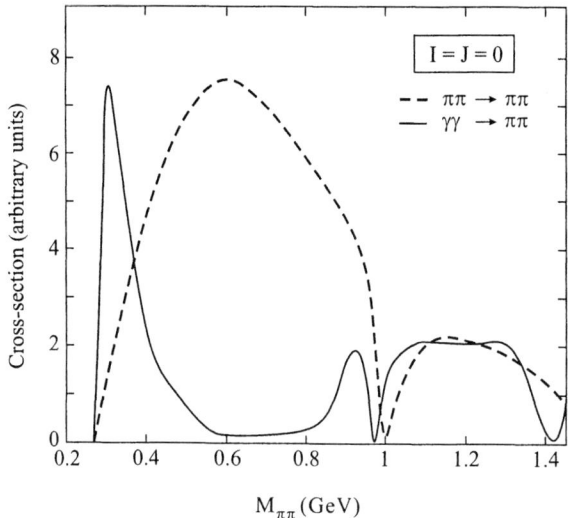

Figure 5: The $I = J = 0$ components of the cross-sections for $\pi\pi \to \pi\pi$ (dashed line) based on data analysed by Ochs [15] and by Bugg *et al.* [16], and for $\gamma\gamma \to \pi\pi$ (solid line) from the Amplitude Analysis by Boglione *et al.* [11] — *dip* solution.

must be multiplied by a mass cubed to give the radiative width. Only a detailed non-perturbative calculation would tell us exactly what this mass scale is, but simple phenomenology indicates it must the mass of the scalar [1]. This intuition is confirmed in the pseudoscalar sector. The dimensional analysis is the same [17]. In the quark model, the $\gamma\gamma$ couplings are related to the fourth power of the charges of the constituents (Fig. 2), but for the π, η and η' this has to be corrected by a factor of their mass cubed to bring agreement with experiment. Without this factor, the quark model predictions would be orders of magnitude wrong [2].

It is then natural to compare the cross-sections of Fig. 1 with the $\gamma\gamma \to \pi^0\pi^0$ one divided by a factor of $M_{\pi\pi}^3$, normalised to unity at the $f_2(1270)$. The result is displayed in Fig. 6. We see that the $\pi\pi$ and $\gamma\gamma$ cross-sections are now much more equal across the whole mass region below 1.5 GeV with no further suppression of the σ at 600 MeV. Clearly the factor of mass cubed cannot be exact as $M_{\pi\pi}$ increases indefinitely. However, the phenomenology in the pseudoscalar sector indicates that the factor is indeed mass cubed, at least up to the η' around 1 GeV, which is sufficient for this illustration.

We see that it is wholly a consequence of gauge invariance that the σ contribution is suppressed at 600 MeV, just as the radiative width of the pion is reduced

---

[1] Just as for the Higgs boson.

[2] Arguments from non-relativistic quark models would give other mass dependent factors too. While such models are applicable to heavy quark systems, it is not clear that they have much relevance to light quark systems, where, as stressed in [18], for instance, genuine bound state calculations, which are inevitably non-perturbative, are essential for meaningful results.

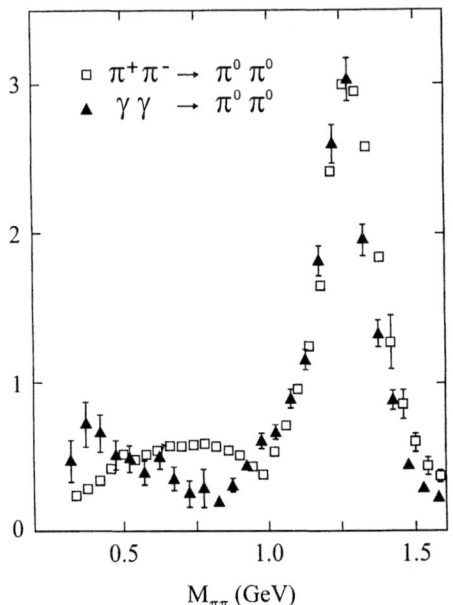

Figure 6: The modulus squared of the amplitudes for $\pi^+\pi^- \to \pi^0\pi^0$ and $\gamma\gamma \to \pi^0\pi^0$, (data as in Fig. 1). Here the $\gamma\gamma$ results are scaled by a factor of $M_{\pi\pi}^{-3}$ normalized to 1 at 1270 MeV to be compared with Fig. 1.

compared to the η and η'. The $I = 0$ S–wave $\gamma\gamma \to \pi\pi$ cross-section is skewed to higher masses. The $f_0(400 - 1200)$ seen in Fig. 5 has a radiative width of about 4 keV, as discussed in [11]. This is just as expected for a scalar composed of non-strange quarks, i.e. a $(u\bar{u} + d\bar{d})$ state [19]. Thus the $\sigma(600)$ is not, as claimed by Minkowski and Ochs [6], the glueball. For this, we have to look above 1.4 GeV.

It is an important feature of the scalar sector and its relation to the vacuum that the glueball lies amongst the lightest $\bar{q}q$ isoscalars and inevitably mixes with these. In contrast, lattice calculators [1] compute the tensor glueball mass is above 2 GeV and so far above the lightest quark nonet, containing the $f_2(1270)$. Consequently, any mixing is likely to be less (ignoring the effect of radial excitations) and the glueball nearly naked [1]. While the $\xi(2230)$ [4,20] may be a candidate for such a state, it is essential to confirm its existence and then to determine it does have spin 2, before speculating further about its composition [21].

QCD predicts that bound states of glue should exist. The scalar is the lightest. Surely soon we will know which of the many f_0's [4] it is. Two photon reactions tell us it is not the $\sigma(600)$.

ACKNOWLEDGEMENTS

It is a pleasure to thank the organisers, particularly Alex Finch, for this de-lightful meeting in such a wonderful setting. I acknowledge support from the EU-TMR Programme, Contract No. CT98-0169, EuroDAΦNE.

REFERENCES

1. see, for instance, Pennington, M.R., *Proc. Workshop on Photon Interactions and the Photon Structure* (Lund, September 1998), eds G. Jarlskog, T. Sjöstrand, pub. Lund Univ, pp. 312-328.

2. Teper, M., *Proc. Int. Europhysics Conf. on High-Energy Physics (HEP 97)* (Jerusalem, Israel, August 1997) pp. 384-387; "Glueball masses and other physical properties of SU(N) gauge theories in D=(3+1): A review of lattice results for theorists", hep-th/981217.

3. Sexton, J., Vaccarino, A., and Weingarten, D., *Phys. Rev. Lett.* 75, 4563-4566 (1995);

Lee, W., and Weingarten, D., *Nucl. Phys. B (Proc. Suppl.)* 53, 236-238 (1997), 63, 194-196 (1998);

Morningstar, C., and Peardon, M., *Nucl. Phys. B (Proc. Suppl.)* 53, 917-920 (1997).

4. Groom, D.E., et al. (PDG) *Eur. Phys. J.* C15, 1 (2000).

5. Amsler, C., and Close, F.E., *Phys. Lett.* B353, 385-390 (1995), *Phys. Rev.* D53, 295-311 (1996).

6. Minkowski, P., and Ochs, W., hep-ph/9811518, *Eur. Phys. J.* C9, 283-312 (1999).

7. Kisslinger, L.S., Gardner, J., and Vanderstraeten, C., *Phys. Lett.* B410, 1-5 (1997);

Kisslinger, L.S., and Li, Z., *Phys. Lett.* B445, 271-273 (1999);

see also, for instance, Narison, S., *Nucl. Phys.* B509, 312-356 (1998).

8. Gunter, J., et al., (E852 Collab.) hep-ex/0001038.

9. Marsiske, H., et al., *Phys. Rev.* D41, 3324-3335 (1990);

Bienlein, J.K., *Proc. IXth Int. Workshop on Photon-Photon Collisions* (San Diego, 1992) eds. D. Caldwell and H.P. Paar, pub. World Scientific, pp. 241-257.

10. Boyer, J., et al. (MarkII), *Phys. Rev.* D42, 1350-1367 (1990);

Behrend, H.J., et al. (CELLO), *Z. Phys.* C56, 381-390 (1992).

11. Boglione, M., and Pennington, M.R., *Eur. Phys. J.* C9, 11-29 (1999).

12. Pennington, M.R., *Proc. Workshop on Hadron Spectroscopy* (Frascati, March 1999), eds T. Bressani, A. Feliciello and A. Filippi, pub. INFN, 1999, pp. 95-114.

13. Morgan, D., and Pennington, M.R., *Z. Phys.* C37, 431-447 (1988); C39, 590 (1988); C48, 623-632 (1990).

14. Pennington, M.R., *DAΦNE Physics Handbook*, eds. L. Maiani, G. Pancheri and N. Paver, pub. INFN, Frascati, 1992, pp. 379-418; *Second DAΦNE Physics Handbook*, eds. L. Maiani, G. Pancheri and N. Paver, pub. INFN, Frascati, 1995, pp. 531-558.

15. Ochs, W., thesis submitted to the University of Munich (1974).

16. Bugg, D.V., Zou, B.S., and Sarantsev, A.V., *Nucl. Phys.* B471, 59-89 (1996).

17. Hayne, C., and Isgur, N., *Phys. Rev.* D25, 1944-1950 (1982);

Schuler, G.A., Berends, F.A., and van Gulik, R.,, *Nucl. Phys.* B523, 423-438 (1998).

18. Pennington, M.R., *Proc. Int. Conf. on the Structure and Interactions of the Photon* (Freiburg, May 1999), ed. S. Söldner-Rembold, *Nucl. Phys. B (Proc. Suppl.)* 82, 291-299 (2000).

19. Li, Z.P., Close, F.E., and Barnes, T., *Phys. Rev.* D43, 2161-2170 (1991).

20. Baltrusaitis, R.M. et al. (Mark III), *Phys Rev. Lett.* 56, 107-110 (1986);

Bai, J.Z., et al. (BES), *Phys. Rev. Lett.* 76, 3502-3505 (1996).

21. Chao, K.T., *Commun. Theor. Phys.* 24, 373-376 (1995);

Huang, T., et al., *Phys. Lett.* B380, 189-192 (1996); Paar, H.P., *Nucl. Phys. B (Proc. Suppl.)* 82, 337-343 (2000).

Exclusive electroproduction of ρ^0 mesons with the ZEUS detector at HERA

S. KANANOV [1]

For the ZEUS Collaboration

School of Physics and Astronomy,
Raymond and Beverly Sackler Faculty of Exact Sciences
Tel Aviv University, Tel Aviv, Israel.

Abstract. Exclusive production of ρ^0 mesons has been studied in ep collisions with the ZEUS detector at HERA using an integrated luminosity of 38 pb^{-1}. The cross section for $\gamma^* p \to \rho^0 p$ has been measured for center-of-mass energies $32 < W < 160$ GeV and for photon virtualities $5 < Q^2 < 80$ GeV^2. The ratio of the production cross sections, $R = \sigma_L/\sigma_T$, for longitudinally (σ_L) and transversely (σ_T) polarized virtual photons is determined from the angular distribution of the decay products of the ρ^0. The Q^2 and W dependence of the total ρ^0 production cross section and of R are presented.

INTRODUCTION

In the past few years the description of ep interactions at low values of Bjorken x, $x \approx Q^2/W^2$, has shifted to the frame in which the proton is at rest (see e.g. [1] and references therein). In this frame, the emitted virtual photon fluctuates into a quark-antiquark state, which lives long enough such that the $q\bar{q}$ color dipole interacts with the proton. In this picture, two components are necessary to calculate the $\gamma^* p$ cross section, the $q\bar{q}$ wave-functions of the transversely (T) and longitudinally (L) polarized virtual photons, $\Psi_{T,L}$ (known from QED), and the dipole cross section, $\sigma_{q\bar{q}p}$, characterizing the interaction of $q\bar{q}$ pair with the proton [2]. This cross section depends on the transverse size of the dipole and could be considered as a universal cross section over a wide range of small-x inclusive and exclusive processes. There are several models available for this interaction cross section (see [3] and references therein). For the small transverse size dipoles the cross section, $\sigma_{q\bar{q}p}$, has been calculated in perturbative QCD (pQCD) [4] and was found to be proportional to the gluon distribution of the proton at appropriate Q^2 and Bjorken x. For large transverse size dipoles, the $\sigma_{q\bar{q}p}$ behavior is thought to be determined by soft physics, as known, for example, from πp scattering in particular.

---

[1] Supported by the German-Israeli Foudation, the U.S.-Israel Binational Science Foundation, the Israel Science Foundation and by the Israeli Ministry of Science

A small $q\bar{q}$ dipole is most likely to be produced if the virtual photon is longitudinally polarized or if the dipole consists of heavy quarks such as $c\bar{c}$ or $b\bar{b}$. Its properties may therefore be studied in reactions such as exclusive light vector-meson (V) production at high Q^2 [5] or exclusive J/ψ [6] and Υ photoproduction [7,8] [2].

In the region where pQCD is expected to dominate V production, the following features are expected [9,10]: the dependence of the cross section $\sigma(\gamma^*p \to Vp)$ on the γ^*p center of mass energy W should become steeper as Q^2 or the mass m_V of the V become larger, due to the steep growth of the gluon distribution with increasing Q^2 and decreasing x; the steep increase of the gluon distribution with increasing Q^2 will tame the Q^2 dependence ($\sim Q^{-6}$) of $\sigma(\gamma^*p \to Vp)$; at fixed W, the slope of the t distribution, where t is the squared four-momentum transfer at the proton vertex, should decrease, as Q^2 or m_V increases.

All such features have been observed at HERA [11–17].

To disentangle the various components of the pQCD calculation of the $\sigma(\gamma^*p \to Vp)$ cross section, a large statistics data sample is required. In this paper, preliminary results from a study of exclusive ρ^0 production, $e^+p \to e^+\rho^0p$, are presented with data collected in 1996/97 in the ZEUS detector. The integrated luminosity of the sample corresponds to $38\,\mathrm{pb}^{-1}$, a six-fold increase in statistics compared to the previous ZEUS publication [11]. The data were collected with a proton energy of 820 GeV and a positron energy of 27.5 GeV.

EVENT SELECTION

The ZEUS detector has been described in detail elsewhere [18]. The components of particular importance in this analysis were the central tracking detector (CTD) and the high-precision uranium-scintillator calorimeter (CAL).

Candidate events for the reaction $e^+p \to e^+\rho^0p$, where the ρ^0 decays into $\pi^+\pi^-$, were selected with the following requirements: the presence of a scattered positron with energy greater than 10 GeV in the CAL; two oppositely-charged tracks in the CTD with the invariant mass of the two tracks, assumed to be pions, $m_{\pi\pi}$, within the range $0.6 < m_{\pi\pi} < 1.2$ GeV; no additional energy deposits in the CAL with energy of the cluster greater than 200 MeV.

About 7500 events were selected in the kinematic range $5 < Q^2 < 80$ GeV^2, $32 < W < 160$ GeV and $|t| < 0.6$ GeV^2.

The acceptance corrections and resolution effects were evaluated using a dedicated Monte Carlo generator ZEUSVM [19] interfaced to HERACLES [20].

The selected sample is contaminated by events in which the final state proton dissociates into a low-mass diffractive state (typically below 5 GeV) and deposits no energy in the calorimeter. In order to account for this contribution, all measured cross sections have been scaled down by $24^{+9}_{-5}\%$, the level of proton dissociation

---

[2] These processes are also sensitive to the model dependence associated with the wave-function of the produced vector meson

contamination determined in the previous analysis [11] and found to be independent of Q^2 and W.

RESULTS

The W dependence of $\sigma(\gamma^*p \to \rho^0 p)$ is shown in Fig. 1 for various Q^2 values along with previous ZEUS measurements at lower Q^2 [21,11]. For each Q^2 value, the W dependence of the cross section is assumed to be of the form W^δ (solid line), and the value of δ is extracted from the fit. A marked increase of δ with Q^2 is observed.

The Q^2 dependence of $\sigma(\gamma^*p \to \rho^0 p)$ is shown in Fig. 2 for $W = 75$ GeV. A fit of the propagator form $(Q^2 + m_\rho^2)^{-n}$ over the full Q^2 range (solid line) including photoproduction [21] and low Q^2 [11] data leads to $n = 2.000 \pm 0.005$ with very poor $\chi^2/\mathrm{ndf} = 142/12$. Restricting the fit to the range $Q^2 > 5$ GeV^2 (dashed line) leads to $n = 2.37 \pm 0.02$ with an improved $\chi^2/\mathrm{ndf} = 13.2/5$. This Q^2 behaviour was found to be fairly independent of W. The dependence of n on Q^2 could be the first indication that the propagator term $(Q^2 + m_\rho^2)^{-n}$ is substantially modified by other effects, such as the increase of the gluon density with Q^2 or Q^2-dependent V wave-function effects.

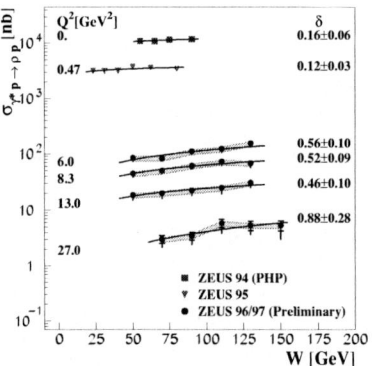

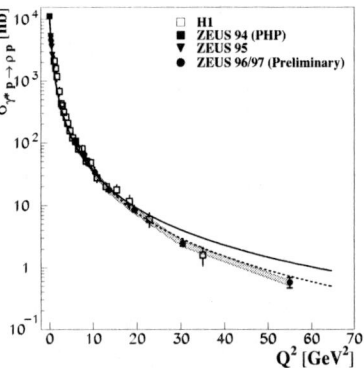

FIGURE 1. W dependence of the cross section $\sigma(\gamma^*p \to \rho^0 p)$ for various Q^2 values as denoted in the figure. The shaded area indicates additional normalization uncertainties due to proton dissociation background.

FIGURE 2. The Q^2 dependence of $\sigma(\gamma^*p \to \rho^0 p)$ at $W = 75$ GeV for the present measurements, compared to the previous ZEUS measurements at low Q^2 and the H1 measurements. The shaded area indicates additional normalization uncertainties due to proton dissociation background.

In the dipole picture, the transverse size of the dipoles is expected to be smaller for longitudinally polarized virtual photons, γ_L^*, than for the transversely polarized ones, γ_T^*. The separation of the γ^*p cross section into the individual contributions from γ_L^* and γ_T^* can be performed by measuring the ρ^0 spin- density matrix elements. Assuming s-channel helicity conservation (SCHC) the ratio of these cross sections, $R = \sigma_L/\sigma_T$, is related to the spin-density matrix elements of the ρ^0 through the expression [22]:

$$R = \frac{r_{00}^{04}}{\varepsilon(1 - r_{00}^{04})} \tag{1}$$

where r_{00}^{04} is a linear combination of the ρ^0 spin-density matrix elements and in the kinematic range of this analysis, $\varepsilon \simeq 0.99$. The value of r_{00}^{04} was obtained directly from the distribution of $\cos\theta_h$, where θ_h is the polar angle of the π^+ in the helicity frame [22]. A small breaking of SCHC has been predicted [23], measured [12,24] and was found to have a negligible effect on R.

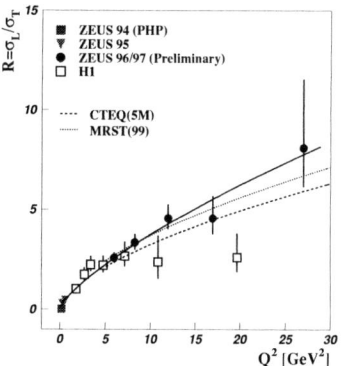

FIGURE 3. The value of R as a function of Q^2, compared to previous measurements from ZEUS and H1 as denoted in the figure (see text for description of curves).

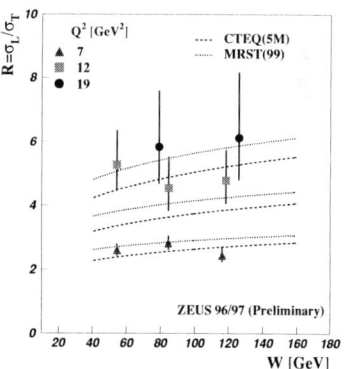

FIGURE 4. The value of R as a function of W for different Q^2 values (see text for description of curves).

The values of R, derived from Eq. (1) as a function of Q^2 and W are presented in Fig. 3 and Fig. 4, respectively. The value of R is seen to increase with Q^2, with no marked W dependence. The Q^2 dependence has been fitted (solid line) assuming $R = \xi(Q^2/m_\rho^2)^\kappa$ with ξ and κ treated as free parameters. The fit yielded $\xi = 0.46 \pm 0.04$ and $\kappa = 0.75 \pm 0.03$ with $\chi^2/\mathrm{ndf} = 3.1/6$. The data presented in Fig. 3 are compared to other HERA measurements [21,11,12]. While the present data are compatible within errors with previous measurements, they indicate a continuous rise of R with Q^2. The dashed and dotted lines correspond to recent pQCD [25] calculations using different parameterizations of the gluon density [3]. The increase of R with Q^2 can also be seen when σ_L and σ_T are plotted as functions of Q^2, as shown in Fig. 5.

---

[3] We thank T. Teubner for providing the files used to produce the curves in Fig. 3 and Fig. 4.

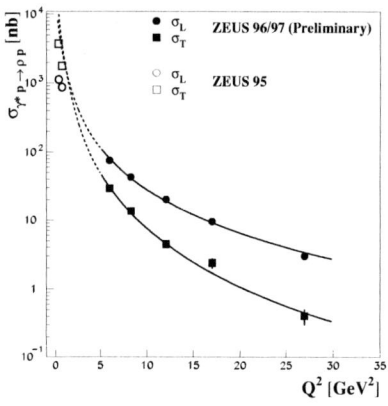

FIGURE 5. The cross section for elastic production of ρ^0 mesons by longitudinally polarized (σ_L) and transversely polarized (σ_T) virtual photons, as a function of Q^2. Also included are the low-Q^2 ZEUS data (see text for description of lines).

The decrease of σ_L with increasing Q^2 is evidently slower than that of σ_T. The Q^2 dependence of $\sigma_{L(T)}$ for $Q^2 > 5$ GeV^2 was fitted with the form $(Q^2 + m_\rho^2)^{-n}$. Values of $n_L = 2.21 \pm 0.03$ and $n_T = 2.94 \pm 0.06$ were found for the longitudinal and transverse cross sections, respectively. The fitted curves and the extrapolation of the Q^2 dependence towards low Q^2 measurements are also shown in Fig. 5. Here again, a single value of n_L or n_T does not reproduce the full Q^2 range. The lack of W dependence of R (Fig. 4) implies that the W dependence of σ_L and σ_T remains the same throughout the measured Q^2 range. This suggests that, despite the large difference expected in the dipole sizes for longitudinally and transversely polarized virtual photons, the typical dipole sizes which in fact contribute to the production of the ρ^0 are of similar order.

SUMMARY

Exclusive ρ^0 electroproduction has been studied in the kinematic region $5 < Q^2 < 80$ GeV^2, $32 < W < 160$ GeV and $|t| < 0.6$ GeV^2, using the ZEUS detector at HERA.

The measurements confirm the previously observed increase of the cross section with W, incompatible with the dominance of soft physics, with a marked tendency for the power of the W dependence to increase with Q^2.

The Q^2 dependence, over the full Q^2 range covered by the present data and the previously published low-Q^2 data, cannot be described by a simple $(Q^2 + m_\rho^2)^{-n}$ functional dependence. The value of n increases in the higher Q^2 range. This behavior is fairly W independent. This could be the first indication for the importance of the Q^2 dependence of the gluon distribution and/or of ρ^0 wave-function effects on the cross section behavior.

The measurement of the spin density matrix element, r_{00}^{04}, as a function of Q^2 and W shows that the contribution of σ_L increases with Q^2; however, the W dependence of σ_L and σ_T remains the same throughout the measured Q^2 range. Therefore, despite the large difference in the dipole size expected for longitudinally and transversely polarized photons, the typical dipole size contributing to ρ^0 production appears to be independent of the photon polarization.

REFERENCES

1. M. F. McDermott, DESY-00-126, (2000).
2. A. H. Mueller, Nucl. Phys. B 335 (1990) 115;
   N. N. Nikolaev and B. G. Zakharov, Z. Phys. C 49 (1991) 607.
3. K. Golec-Biernat, M. Wüsthoff, Phys. Rev. D 59 (1999) 014017;
   G. R. Forshaw et al., Phys. Rev. D 60 (1999) 074012;
   E. Gotsman et al., Eur. Phys. J. C 10 (1999) 689;
   M. McDermott et al., hep-ph/9912547.
4. B. Blaettel et al., Phys. Rev. Lett. 70 (1993) 896.
5. S. J. Brodsky et al., Phys. Rev. D 50 (1994) 3134 .
6. M. G. Ryskin, Z. Phys. C 57 (1993) 89.
7. L. Frankfurt, M.McDermott and M. Strikman, JHEP 02 (1999) 002.
8. A. D. Martin, M. G. Ryskin and T. Teubner, Phys. Lett. B 454 (1999) 399.
9. H. Abramowicz, L. Frankfurt and M. Strikman, Surveys High Energy Phys. 11 (1997) 51.
10. L. Frankfurt, W. Koepf and M. Strikman, Phys. Rev. D 54 (1996) 3194, Phys. Rev. D 57 (1998) 512.
11. ZEUS Collaboration, J. Breitweg et al., Eur. Phys. J. C 6 (1993) 603.
12. H1 Collaboration, C. Adloff et al., Eur. Phys. J. C 13 (2000) 371.
13. ZEUS Collaboration, paper submitted to the 29th International Conference on High Energy Physics, abstract 793, Vancouver, 1998.
14. H1 Collaboration, C. Adloff et al., hep-ex/0005010, 2000.
15. ZEUS Collaboration, J. Breitweg et al., Z. Phys. C 75 (1997) 215.
16. H1 Collaboration, C. Adloff et al., Eur. Phys. J. C 10 (1999) 373.
17. ZEUS Collaboration, J. Breitweg et al., hep-ex/0006013, 2000 (accepted by Phys.Lett. B).
18. ZEUS Collaboration, M. Derrick et al., The ZEUS Detector Status Report 1993, DESY.
19. K. Muchorowski, Ph.D. Thesis, Warsaw University, 1998.
20. A. Kwiatkowski, H. Spiesberger and H.J. Moehring, Comp.Phys.Commun. 69 (1992) 155.
21. ZEUS Collaboration, J. Breitweg et al., Eur. Phys. J. C 2 (1998) 247.
22. K. Schilling and G. Wolf, Nucl. Phys. B 61 (1973) 381.
23. D. Y. Ivanov and R. Kirschner, Phys. Rev. D 58 (1998) 114026.
24. ZEUS Collaboration, J. Breitweg et al., Eur. Phys. J. C 12 (2000) 393.
25. A. D. Martin, M. G. Ryskin and T. Teubner, Phys. Rev. D 62 (2000) 014022.

FUTURE LINEAR COLLIDER

Introduction and recent developments in $\gamma\gamma,\gamma$e colliders

Valery Telnov

Institute of Nuclear Physics, 630090, Novosibirsk, Russia
and DESY, Germany

Abstract.
High energy photon colliders ($\gamma\gamma$, γe) based on backward Compton scattering of laser light is a very natural addition to e^+e^- linear colliders. In this report we consider mainly this option for the TESLA project. Recent study has shown that the horizontal emittance in the TESLA damping ring can be further decreased by a factor of four. In this case the $\gamma\gamma$ luminosity luminosity in the high energy part of spectrum can reach 0.3–0.5 $L_{e^+e^-}$. Typical cross sections of interesting processes in $\gamma\gamma$ collisions are higher than those in e^+e^-collisions by about one order of magnitude, so the number of events in $\gamma\gamma$ collisions will be more that in e^+e^- collisions. The key new element in photon colliders is a very powerful laser system. The most straightforward solution is "an optical storage ring (optical trap)" with diode pumped laser injector which is today technically feasible. This paper briefly review the status of a photon collider based at TESLA, its possible parameters.

I INTRODUCTION

Over the last decade, several laboratories in the world have been working on linear e^+e^- collider projects with an energy from several hundreds GeV up to several TeV. Beside e^+e^- collisions, linear colliders can "convert" electrons to high energy photons using the Compton backscattering of laser light, thus obtaining $\gamma\gamma$ and γe collisions with energies and luminosities close to those in e^+e^- collisions [1,2]. Recently the International Workshop on High Energy Photon Colliders has been held at DESY. There are very good news, which I will discuss below.

The basic scheme of a photon collider is shown in Fig. 1.

The maximum energy of the scattered photons (in direction of electrons) is [1]

$$\omega_m = \frac{x}{x+1}E_0; \quad x = \frac{4E_0\omega_0 \cos^2 \alpha/2}{m^2c^4} \simeq 15.3 \left[\frac{E_0}{\text{TeV}}\right]\left[\frac{\omega_0}{eV}\right], \tag{1}$$

where E_0 is the electron energy, ω_0 the energy of the laser photon, α is the angle between electron and laser beam (see, fig.1a). For example: E_0 =250 GeV, $\omega_0 =$

CP571, *PHOTON 2000,* edited by A. J. Finch
© 2001 American Institute of Physics 0-7354-0010-5/01/$18.00

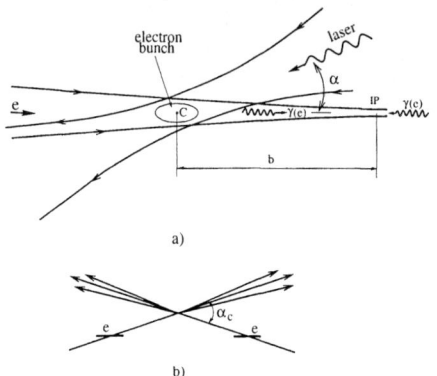

FIGURE 1. Scheme of $\gamma\gamma$, γe collider.

1.17 eV ($\lambda = 1.06~\mu$m) (Nd:Glass and other powerful solid state lasers) $\Rightarrow x = 4.5$ and $\omega/E_0 = 0.82$. For an introduction to photon colliders see refs [1,2].

Below I will discuss most important issues connected with a photon collider based on the TESLA [3]. The following essential topics are discussed: physics motivation, parameters of the photon collider at TESLA, lasers, optics.

II PHYSICS

In general, physics in e^+e^- and $\gamma\gamma$, γe collisions is quite similar because the same particles can be produced. However, it is always better to study new phenomena in various reactions because they give complementary information. Some phenomena can best be studied at photon colliders due to better accuracy or larger accessible masses.

The second aspect important for physics motivation is the luminosity. Typical luminosity distribution in $\gamma\gamma$ and γe collisions has a high energy peak and some low energy part (see the next section). This peak has a width at half maximum of about 15% (in $\gamma\gamma$ collision) and photons here can have a high degree of polarization. In the next section we will see that in the current TESLA designs the $\gamma\gamma$ luminosity in the high energy peak can be up to 30-40% of the e^+e^- luminosity at the same beam energy.

Higgs boson

The present Standard Model (SM) assumes existence of a very unique particle, the Higgs boson. It has not been found yet, but from existing experimental information it follows that, if it exists, its mass is higher then 112 GeV (LEP200) and is below 200 GeV, i.e. lays in the region of the next linear colliders. In the simplest extensions to the SM the Higgs sector consists of five physics states: h^0, H^0, A^0 and $H^\pm$. All these particles can be studied at photon colliders and some characteristics can be measured better than in e^+e^- collisions.

The process $\gamma\gamma \to H$ goes via the loop with heavy virtual charged particles and its cross section is very sensitive to the contribution of particles with masses far beyond the energies covered by present and planned accelerators. For the integrated luminosity 50 fb^{-1} (in the peak) the number of produced Higgs will be 50–150 thousands (depending on the mass). As a result, one can measure the $\Gamma_{\gamma\gamma}(H)$ width at photon colliders with an accuracy better than 2-3% [5]. This is sufficient for distinguishing between Higgs models [6].

Moreover, in the models with several neutral Higgs bosons, heavy H^0 and A^0 bosons have almost equal masses and at certain parameters in the theory are produced in e$^+$e$^-$ collisions only in associated production e$^+$e$^-$ $\rightarrow HA$, while in $\gamma\gamma$ collision they can be produced singly with sufficiently high cross section. Correspondingly, in $\gamma\gamma$ collisions one can produce Higgs bosons with about 1.5 times higher masses.

Charge pair production

The second example is the charged pair production. Cross sections for the production of charged scalar, lepton, WW pairs in $\gamma\gamma$ collisions are larger than those in e$^+$e$^-$ collisions by a factor of approximately 5–20. The corresponding graphs can be found elsewhere [2,4].

Note, that in e$^+$e$^-$ collisions two charged pairs are produced both via annihilation diagram with virtual γ, Z and also via exchange diagrams where some new particles can give contributions, while in $\gamma\gamma$ collisions it is pure a QED process which allows the charge of produced particles to be measured unambiguously.

Accessible masses

In γe collisions, charged particle with a mass higher than that in e$^+$e$^-$ collisions can be produced (a heavy charged particle plus a light neutral); for example, supersymmetric charged particle plus neutralino or new W boson and neutrino. Also, $\gamma\gamma$ collisions provide higher accessible masses for particles which are produced as a single resonance in $\gamma\gamma$ collisions (such as neutral Higgs bosons).

Search for anomalous interactions

Precise measurement of cross sections allow the observation of effects of anomalous interactions. The process $\gamma\gamma$ $\rightarrow$ WW has large cross section (about 80 pb) and it is one of most sensitive processes for a search for a new physics. The vertex γWW can be studied much better than in e$^+$e$^-$ collisions because in the latter case the cross section is much smaller and this vertex gives only 10% contribution to the total cross section. The two factors together give about 40 times difference in the cross sections. Besides that, in $\gamma\gamma$ collisions the $\gamma\gamma WW$ vertex can be studied.

Quantum gravity effects with Extra Dimensions

This new theory suggests a possible explanation of why gravitation forces are so weak in comparison with electroweak forces. It is suggested that the gravitation constant is equal to the electroweak but in a space with extra dimensions. It can be tested at photon colliders and a two times higher mass scale than in e$^+$e$^-$ collisions can be reached [8].

Many other examples can be found in proceedings of GG2000 Workshop [7].

III POSSIBLE LUMINOSITIES OF $\gamma\gamma,\gamma$e COLLISIONS AT TESLA

As it is well known in e^+e^- collisions the luminosity is restricted by beam-strahlung and beam instabilities. Due to the first effect the beams should be very flat. In $\gamma\gamma$ collisions these effects are absent, therefore one can use beams with much smaller cross section. At present TESLA beam parameters the $\gamma\gamma$ luminosity is determined only by the attainable geometric L_{ee} luminosity. So, the $\gamma\gamma$ luminosity depends on emittances of electron beams.

Specially for photon collider the TESLA group has studied the possibility of decreasing emittances at the TESLA damping ring. The conclusion is that the horizontal emittance can be reduced by a factor of 4 in comparison with the previous design.

The resulting parameters of the photon collider at TESLA for 2E=500 GeV and H(130) are presented in Table 1 [9]. It is assumed that electron beams have 85% longitudinal polarization and laser photons have 100% circular polarization and that $L \propto E$. The thickness of laser target is one collision length (so that $k^2 \approx 0.4$).

TABLE 1. Parameters of the $\gamma\gamma$ collider based on TESLA. Left column for 2E=500 GeV, next two columns for Higgs with M=130 GeV, two options.

	2E=500 $x = 4.6$	2E=200 $x = 1.8$	2E=158 $x = 4.6$
$N/10^{10}$	2	2	2
σ_z, mm	0.3	0.3	0.3
$f_{rep} \times n_b$, kHz	14.1	14.1	14.1
$\gamma\epsilon_{x/y}/10^{-6}$,m·rad	2.5/0.03	2.5/0.03	2.5/0.03
$\beta_{x/y}$,mm at IP	1.5/0.3	1.5/0.3	1.5/0.3
$\sigma_{x/y}$,nm	88/4.3	140/6.8	160/7.6
L_{ee}(geom), 10^{33}	120	48	38
$L_{\gamma\gamma}(z > 0.8z_{m,\gamma\gamma})$, 10^{33}	11.5	3.5	3.6
$L_{\gamma e}(z > 0.8z_{m,\gamma e})$, 10^{33}	9.7	3.1	2.7

For these luminosities the rate of production of the SM Higgs boson with M_H=130(160) GeV in $\gamma\gamma$ collisions is 0.9(3) of that in e^+e^- collisions at 2E = 500 GeV (both reactions, ZH and H$\nu\nu$).

Comparing the $\gamma\gamma$ luminosity with the e^+e^- luminosity ($L_{e^+e^-} = 3 \times 10^{34}$ cm^{-2}s^{-1} for $2E = 500$ GeV) we see that for the same energy

$$L_{\gamma\gamma}(z > 0.8z_m) \sim 0.4 L_{e^+e^-}. \tag{2}$$

For example, the cross section for production of H^+H^- pairs in collisions of polarized photons is higher than that in e^+e^- collisions by a factor of 20 (not far from the threshold); this means 8 times higher production rate for the luminosities given above. The relation (2) is valid only for the considered beam parameters. A more universal relation is (for $k^2 = 0.4$)

$$L_{\gamma\gamma}(z > 0.8z_m) \sim 0.1L_{ee}(geom).\qquad(3)$$

Figures for the luminosity distribution in $\gamma\gamma$ and γe collisions can be found elsewhere [9].

IV   LASERS, OPTICS

A key element of photon colliders is a powerful laser system which is used for $e \to \gamma$ conversion. Lasers with the required flash energies (several J) and pulse duration ~ 1 ps already exist and are used in several laboratories. The main problem is the high repetition rate, about 10–15 kHz, with a pulse structure repeating the time structure of the electron bunches.

The most attractive and reliable solution at this moment is an "optical storage ring" with a diode pump laser injector. This new approach can be considered as a base-line solution for the TESLA photon collider [9]. This scheme with multiple use of each laser bunch is shown in fig. 2.

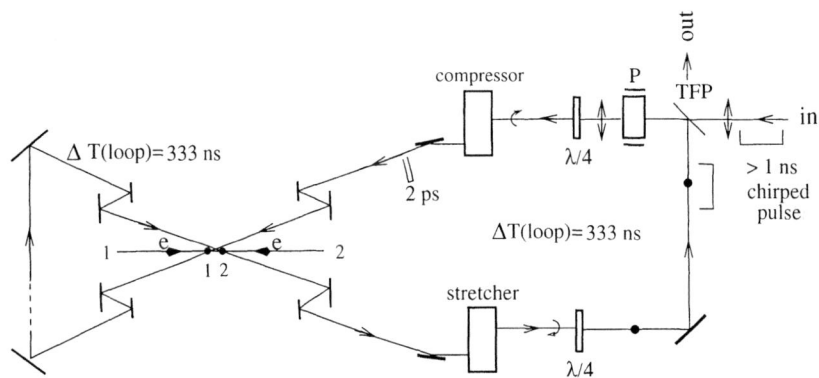

FIGURE 2. Optical storage ring for $e \to \gamma$ conversions. P is a Pockels cell, TFP is a thin film polarizer, thick dots and double arrows show the direction of polarization.

The laser pulses are send to the interaction region where they are trapped in an optical storage ring. This can be done using Pockels cells (P), thin film polarizers (TFP) and 1/4-wavelength plates ($\lambda/4$). Each bunch makes several (n) round trips (I assume n=10 for further example) and then is deleted from the ring. All these tricks can be done by switching one Pockels cell.

During one total loop each bunch is used for conversion twice. To avoid problems of non-linear effects (self-focusing) in optical elements, laser pulses which are compressed before collisions down to about 2 ps using grating pairs are then stretched again up to previous length before passing through optical elements. Such system can be done in the best way only at the TESLA due to a large spacing between bunches and a long train.

A laser system required for a such optical storage ring with 10 round trips can consist of about 8 lasers of 0.9 kW average powereach. Due to the high average power the lasers should be based on diode pumping. Diodes have much higher efficiency than flash lamps and much more reliable. This technology is developed very actively for other application, such as inertial fusion. Present cost of diodes for such laser system is about 15 M$ (for 10-fold use of one laser bunch as described above) and it is expected that their cost will be further decreased several times. Such system can be done now: all technologies exist.

V CONCLUSION

The luminosity in $\gamma\gamma$ collisions (in the high energy peak) can reach about 40% of e^+e^- luminosity. Since cross sections in $\gamma\gamma$ collisions are typically higher by one order of magnitude than those in e^+e^- collisions and because of access to higher masses for some particles, the photon collider now has very serious physics motivation.

There is good suggestion for the laser system, which, it seems, can be build now.

REFERENCES

1. I.Ginzburg, G.Kotkin, V.Serbo, V.Telnov,*Nucl.Instr. & Meth.* **205** (1983) 47 (Prepr. INP 81-102, Novosibirsk, 1991).
2. V.Telnov, *Nucl.Instr.& Meth.* **A 294** (1990) 72;*Nucl.Instr.&Meth.* **A 355** (1995) 3.
3. R.Brinkmann et al., *Nucl. Instr. &Meth.* **A 406** (1998) 13.
4. V. Telnov, Proc. of the International Conference on the Structure and Interactions of the Photon (Photon 99), Freiburg, Germany, 23-27 May 1999, to be published in Nucl. Phys. Proc. Suppl. B, e-print: hep-ex/9908005.
5. G.Jikia, S.Soldner-Rembold, Proceedings of 4th International Workshop on Linear Colliders (LCWS 99), Sitges, Barcelona, Spain, 28 Apr - 5 May 1999. e-print: hep-ph/9910366.
6. I.Ginzburg, M.Krawczyk, Proceedings of 4th International Workshop on Linear Colliders (LCWS 99), Sitges, Barcelona, Spain, 28 Apr - 5 May 1999, hep-ph/9909455 and these proceedings.
7. Proc. of Intern. Workshop on High Energy Photon Colliders, 14-17 June, 2000, DESY, Hamburg, Germany, to be published in Nucl. Inst. and Methods A.
8. Thomas Rizzo, *Proceedings of 4th International Workshop on Linear Colliders (LCWS 99)*, Sitges, Barcelona, Spain, 28 Apr - 5 May 1999, SLAC-PUB-8204, e-Print Archive: hep-ph/9907401, also these proceedings.
9. V.Telnov, Proc. of Intern. Workshop on High Energy Photon Colliders, 14-17 June, 2000, DESY, Hamburg, Germany, to be published in Nucl. Inst. and Methods A.

Why Photon Colliders are necessary in future collider program

Ilya F. Ginzburg

Sobolev Institute of Mathematics SB RAS, 630090, Novosibirsk, Russia

Abstract.

Subjects of discussion

1. Different scenarios
2. Higgs window to a New Physics
3. Anomalies in the interactions of gauge and Higgs bosons
4. Some problems in QCD and hadron physics.
5. By-product: Production of axions, etc. from region of conversion $e \to \gamma$.

I DIFFERENT SCENARIOS

Photon Colliders in the widespread scenario.

When discuss the program for future high energy colliders, the basic point is usually that

the Nature is so favorable to us that to dispose the essential fraction of new particles and the thresholds of new interactions (e.g., effects of new higher dimensions, compositeness, etc.) *within the LHC operation domain.*

In this case main discoveries will be made at the Tevatron and the LHC. The e^+e^- Linear Colliders (LC) in their first stages will be machines for measuring precise values of coupling constants and exploring in detail supersymmetry. The high luminosity expected for e^+e^- Linear Colliders provides opportunity to obtain parameters of models realized with very high accuracy, see, e.g., [2].

With the Photon Collider mode of LC [1] having roughly the same luminosity as that for e^+e^- mode [3] these results will be improved. Indeed,

• The cross sections of production of pairs of charged particles in the $\gamma\gamma$ collision are $5 \div 8$ times higher than that in the e^+e^- collision in its maxima. This ratio increases with growth of energy.

• The set of final states at Photon Collider is reacher than that at e^+e^- mode.

• One can vary polarizations of photon beams relatively easily.

Unfortunately, this Photon Collider opportunity was not explored by community in necessary details. It can be assured that the forthcoming studies of many physi-

CP571, *PHOTON 2000*, edited by A. J. Finch

cists with detail simulation will show an exceptional potential of Photon Collider mode in both new problems and all problems considered till now.

Possible $\mathcal{SM}$ – like scenario. It can also happen that the opposite scenario will be realized: **No new particles and interactions will be discovered at the Tevatron, LHC and e^+e^- LC except Higgs boson.** If additionally its coupling constants to quarks and gauge bosons will be close to their values in the $\mathcal{SM}$, we meet $\mathcal{SM}$ – like picture of the World. It can be realized both if our World is really described by simple $\mathcal{SM}$ till to very small distances and if some other model is realized and New Physics is *round the corner*. In this case the main goal of studies at new colliders will be the hunting for indirect signals of New Physics – deviations of observed quantities from $\mathcal{SM}$ predictions.

Photon Colliders are the best machines for the hunting for signals of New Physics if the $\mathcal{SM}$ – like scenario is realized.

II   HIGGS WINDOW TO A NEW PHYSICS

The study of Higgs-boson couplings with photons ($h\gamma\gamma$ and $hZ\gamma$) looks like a very promising tool for resolving of models of New Physics.

• These couplings are absent in the $\mathcal{SM}$ at tree level, appearing only at the loop level. Therefore, the background for signals of New Physics will be relatively lower here than in other processes which are allowed at tree level of the SM.

• All fundamental charged particles contribute to these effective couplings. The whole structure of the theory influences the corresponding Higgs-boson decays.

• The expected accuracy in the two-photon width is $\sim 2\%$ at $M_h \leq 150$ GeV and even at the luminosity integral ~ 30 fb^{-1} in the high energy luminosity peak [4].

In the $2\mathcal{HDM}$ and $\mathcal{MSSM}$, the observed Higgs boson will be either the lightest Higgs boson h or heavier one H. Assuming the coupling constants of observed Higgs boson to quarks and gauge bosons are close to their values in the $\mathcal{SM}$, other Higgs bosons can easily avoid observation at LHC and e^+e^- LC due to small couplings to standard matter. These neutral Higgses can be seen in the process $\gamma\gamma \to h$ [5] and charged Higgses in the process $\gamma\gamma \to H^+H^-$.

Even if these additional Higgs bosons are so heavy that they cannot be produced at the first stage of Photon Collider, the models can be distinguished well via precise measurement of two–photon width of observed $\mathcal{SM}$ –like Higgs boson [6]. In this paper we assumed that the discussed $\mathcal{SM}$ –like scenario is realized with accuracy estimated for $\mathcal{SM}$ Higgs boson at e^+e^- Linear Collider [2]. The two photon width is calculated via the Higgs couplings with the matter measured at e^+e^- LC. In the $2\mathcal{HDM}$ deviation from $\mathcal{SM}$ is obliged mainly by contribution of charged Higgs. In the general case the two-photon width is about 10% less than that in the $\mathcal{SM}$ (see Figs.1). This difference is considerably larger than the expected experimental inaccuracy. Therefore, measurement of two-photon width at Photon collider can resolve these models reliably.

In the $\mathcal{MSSM}$ the lightest Higgs boson is decoupled with superpartners and

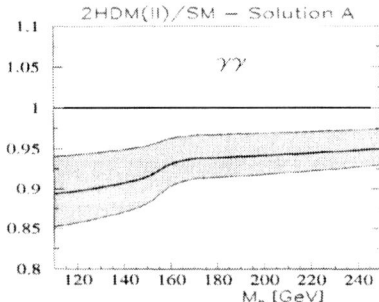
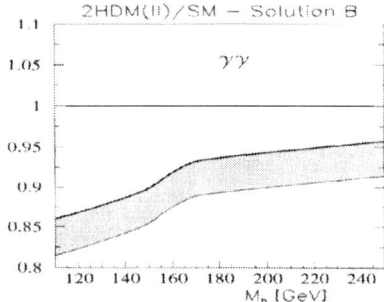

FIGURE 1. *The ratio of two–photon width of Higgs boson in $2\mathcal{H}\mathcal{D}\mathcal{M}$ to its $\mathcal{S}\mathcal{M}$ value at $\lambda_5 = 0$ and $M_h \geq 800$ GeV. Shaded zones correspond experimental uncertainties expected in the experiments at e^+e^- LC. Deviation from $\mathcal{S}\mathcal{M}$ depends on λ_5 as $\propto (1 - \lambda_5 v^2/M^2_{H^\pm}$ with $v = 246$ GeV. Solution A is really close to $\mathcal{S}\mathcal{M}$. For the solution B some of couplings of Higgs field with matter at close to the $\mathcal{S}\mathcal{M}$ values, the other have the opposite sign.*

other members of Higgs multiplet if they are very heavy. More detail calculations will give us the upper bounds for masses of superparticles which can influence for photon widths so that the difference will be seen in the experiment. Preliminary estimates show that these values are higher than the discovery limits at LHC.

In many variants of $2\mathcal{H}\mathcal{D}\mathcal{M}$ or $\mathcal{M}\mathcal{S}\mathcal{S}\mathcal{M}$ the masses of heavy scalar Higgs H and Higgs pseudoscalar A are close each other. In some other variants they are mixed ($\mathcal{C}\mathcal{P}$ violated scenarios). The observations at LHC and e^+e^- Linear Collider cannot often resolve these opportunities due to low resolution for these bosons. The polarization asymmetries in Higgs boson production at Photon Collider can resolve these variant, i.e. establish, whether $\mathcal{C}\mathcal{P}$ parity at Higgs level is violated or not.

III ANOMALIES IN THE INTERACTIONS OF GAUGE AND HIGGS BOSONS

Before discovery new heavy particles inherent New Physics, it reveals itself at lower energies as some anomalies in the interactions of known particles. Our goal is to find these anomalies and discriminate as better as possible. *The correlation between coefficients of different anomalies will be the key for understanding what is nature of New Physics.*

Interactions of gauge bosons. The practically unique process under interest in the e^+e^- mode is $e^+e^- \to W^+W^-$. At suitable electron polarization the neutrino exchange contribution (having small interest) disappear, and residual cross section (obliged by photon and Z boson exchange) has maximum about 2 pb within LEP operation interval and decreases with energy after that. The cross sections of other processes with W production ($e^+e^- \to e^+e^- WW$, $e^+e^- \to e\nu W$, ...) are small at $\sqrt{s} < 1$ TeV.

At Photon collider main processes are – $\gamma\gamma \to W^+W^-$, $e\gamma \to \nu W$. Their cross sections are about 80 pb and energy independent at $\sqrt{s} > 200$ GeV. That is at least 40 times higher than for e^+e^- collisions, it gives about $(1 \div 5) \cdot 10^7$ W's per year, that is comparable with Z production at LEP. Due to high value of these basic cross sections, many processes of 3-rd and 4-th order have large enough cross sections: $\gamma e \to eWW$, $\gamma\gamma \to ZWW$, $\gamma e \to \nu WZ$, $\gamma\gamma \to WWWW$, $\gamma\gamma \to WWZZ$,...

Large variety of these processes permit to discover and separate well anomalies in specific processes and (or) distributions. This subject needs for studies of all processes enumerated with details of behavior in different regions of phase space and polarization dependence there.

• For example, the following line of procedures is almost evident:

(a) To extract γWW anomalies from $\gamma e \to \nu W$.

(b) To extract ZWW anomalies from $e^+e^- \to W^+W^-$.

(c) To extract $\gamma\gamma WW$ anomaly from $\gamma\gamma \to W^+W^-$.

• $e\gamma \to \nu W$. The cross section of process is $\propto (1 - 2\lambda_e)$, it is switched on or off with variation of electron helicity λ_e. It gives precise test of absence of right handed currents in the interaction of W with the matter.

• $\gamma e \to eWW$. The cross section of process at $\sqrt{s} = 500$ GeV is about 10 pb, corresponding counting rate is few millions events per year.

⋆ In events with transverse momentum of scattered electron $p_\perp \geq 30$ GeV one can study here anomaly γZWW.

⋆ The charge asymmetry of produced W's looks most sensitive key for the study of strong interaction in Higgs sector.

• $\gamma\gamma \to W^+W^-$ and $e\gamma \to \nu W$. The two–loop radiative corrections to these processes are measurable and sensitive to the problems:

⋆ *construction of S–matrix of theory with unstable particles;*

⋆ *gluon corrections like Pomeron exchange between quark components of W's.*

Interactions of Higgs boson with light ($\gamma\gamma \to h$, $\gamma e \to eh$).

All observable anomalous $\mathcal{CP}$ – even and $\mathcal{CP}$ – odd interactions of Higgs boson with light are summarized in the form of an effective Lagrangian:

$$\Delta\mathcal{L} = 2Hv \left(\theta_\gamma \frac{F_{\mu\nu}F^{\mu\nu}}{2\Lambda_\gamma^2} + \theta_Z \frac{Z_{\mu\nu}F^{\mu\nu}}{\Lambda_Z^2} + i\theta_{P\gamma} \frac{F_{\mu\nu}\tilde{F}^{\mu\nu}}{2\Lambda_{P\gamma}^2} + i\theta_{PZ} \frac{Z_{\mu\nu}\tilde{F}^{\mu\nu}}{\Lambda_{PZ}^2} \right), \quad \left(\theta_i = e^{i\xi_i} \right).$$

Here $F^{\mu\nu}$ and $Z^{\mu\nu}$ are the standard field strengths for the electromagnetic and Z field, $\tilde{F}^{\mu\nu} = \varepsilon^{\mu\nu\alpha\beta}F_{\alpha\beta}/2$ and ξ_i *are the phases of couplings*, generally $\neq 0$ or π even in the C–even case (due to possible contributions of some light particles) [7].

While $\mathcal{CP}$ even anomalies can be seen via the values of measured cross sections, the $\mathcal{CP}$ anomalies will be seen via the polarization asymmetries ("longitudinal" – variation of cross sections with change of sign of collided photon helicities, or "transverse" – variation of cross section in dependence on the angle between directions of linear polarization of collided photons). The effects are large enough to see them at reasonable values of anomaly scales Λ_i and phases ξ_i (see Fig. 2).

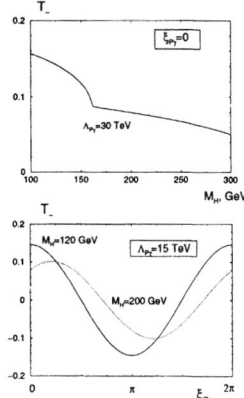

FIGURE 2. *"Longitudinal" asymmetry for $\gamma\gamma \to H$ process*

The specific example of anomaly presents $\mathcal{CP}$ violating mixing in Higgs sector of $2\mathcal{HDM}$ (scalar – pseudoscalar) with mixing angle α_2. We present results in Fig. 3 fixing other model parameters to be close to the $\mathcal{SM}$ case.

- It can happen that the masses of heavy scalar Higgs H and pseudoscalar A are so close each other that they cannot be resolved in the mass spectrum. The study of decay products cannot distinguish this overlapping from the case of their mixing with $\mathcal{CP}$ violation. In the case when these Higgses are overlap in the true mass spectrum, the correlations in the decay products will be identical if $\mathcal{CP}$ violated mixing exists or not. The study of polarization asymmetries in the Higgs boson production at Photon Collider distinguish these two opportunity well.

IV   QCD AND HADRON PHYSICS

All problems studied at HERA and LEP will be studied there but in much more wide interval of parameters and with much better accuracy. Among them I underline those which look most interesting now.

- Study of photon structure function(s) at small x.
- Nature of growth of **total cross sections**. The widespread concepts assume standard Regge type factorization and universal energy behavior for different processes. With Photon colliders, *for the first time in particle physics*, one can have the set of mass shell cross sections of very high energy processes, appropriate for the testing of factorization or level of its violation. That are σ_{pp}, measured at Tevatron and LHC, $\sigma_{\gamma p}$, measured at HERA, $\sigma_{\gamma\gamma}$, measurable at Photon collider. For this goal *the preliminary stage of operations with low luminosity can be used* to observe large enough cross sections at small scattering angles.
- In this very low luminosity stage the study of events with particle production in the center of rapidity scale and with rapidity gaps for two vector mesons or dijets originated from initial photons will be crucial for the understanding of **double**

413

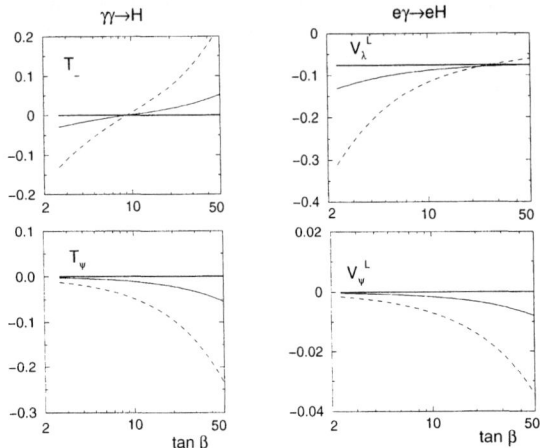

FIGURE 3. *Polarization asymmetries in* $\gamma\gamma \rightarrow h$ *and* $\gamma e \rightarrow eh$ *processes due to scalar-pseudoscalar mixing in* $2\mathcal{H}\mathcal{D}\mathcal{M}$ *(II),* $M_h = 110$ *GeV; for* $e\gamma \rightarrow eH$ *process* $E_{tot} = 1.5$ *TeV. The thick solid lines show the* $\mathcal{S}\mathcal{M}$ *values; thin solid and dashed lines refer to* $\sin\alpha_2 = 0.1$ *and* 0.5 *respectively (* α_2 *is light scalar-pseudoscalar mixing angle).*

Pomeron effects having no reliable explanation till now.

V CONVERSION REGION – γe COLLIDER

The conversion region is γe collider with c.m.s. energy about 1.2 MeV but with luminosity about 0.1 fb^{-1}/sec(!). It will be unique source of light axions, majorons, etc. a, expected in numerous schemes [9].

The production processes are

$$e\gamma_0 \rightarrow ea, \quad \gamma\gamma_0 \rightarrow a. \tag{1}$$

The energy of a is limited from above as

$$E_a \leq \frac{x + a^2 + \sqrt{(x - a^2)^2 - 4a^2}}{2(x + 1)} E, \quad a = \frac{m_a}{m_e}. \tag{2}$$

The angular spread of these a is even narrower than that of high energy photons. As compare with neutrinos produced by photons in the final wall, the main part of these "axions" is concentrated in $\mathbf{10^6 \div 10^8}$ times more narrow solid angle.

These a interact with the matter very weakly. So, the registration scheme can be of the type shown in Fig. 4: The produced a after concrete wall moves to detector: Disposed in vacuum 300-500 m length lead wire of diameter 3-5cm. Here it can interact with hadrons in the nucleus of lead and produce hadrons with typical mean transverse momentum about 300-500 MeV. They can be recorded in somewhat like scintillator of diameter 3-5 m in the end of wire.

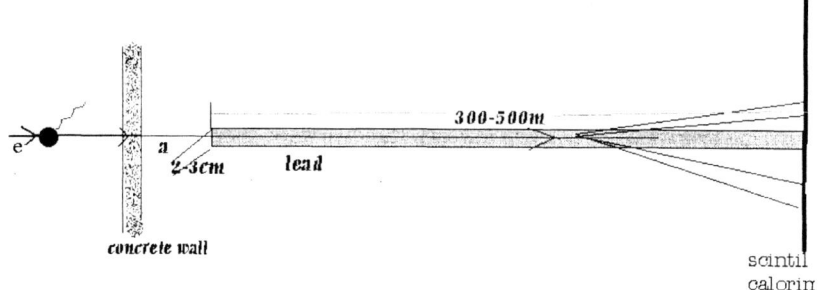

FIGURE 4. Possible recording scheme for "axion"

VI CONCLUSION

This discussion shows that *R&D for Photon collider mode should be performed simultaneously with that of e^+e^- mode of TESLA, etc. Final decision about the turn of different stages of Linear collider should be made only* **AFTER** *first operations of LHC.* It can imagine even the opportunity that the Photon collider mode will be switched on before e^+e^- mode. The advantages of this way are:

• The basic electron energy is lower. • Positron beam is unnecessary.

I am thankful V.Ilyin, I. Ivanov, M.Krawczyk, P.Olsen for collaboration related different parts of paper and A.Djouadi, V. Serbo, M. Spira, V. Telnov, P.Zerwas for useful discussions. I am also grateful to RFBR (grants 99-02-17211 and 00-15-96691) for support.

REFERENCES

1. I.F. Ginzburg, G.L. Kotkin, V.G. Serbo, V.I. Telnov, *Nucl. Instrum. Methods* **205** (1983) 47; I.F. Ginzburg, G.L. Kotkin, S.L. Panfil, V.G. Serbo, V.I. Telnov, *Nucl. Instrum. Methods* **A 219** (1984) 5; *Zeroth-order Design Report for the NLC*, SLAC Report 474 (1996); R. Brinkmann et al., *Nucl. Instrum. Methods* **406** (1998) 13.

2. M. Battaglia, *Proc. 4th Int. Workshopon Linear Colliders, April 28-May 5, 1999; Sitges (Spain)*, hep-ph/9910271; A. Djouadi, *ibid.*, hep-ph/9910449; P. P.Zerwas, DESY 99-178, hep-ph/003221

3. V.Telnov et al. Report at the GG2000.

4. G. Jikia, S. Söldner-Rembold, *Nucl. Phys. B (Proc. Suppl.)* **82** (2000) 373; M. Melles, W.J. Stirling, V.A. Khoze,*Phys. Rev.* **D61**(2000) 054015.

5. M. Muhlleitner, report at the GG2000.

6. I.F. Ginzburg, M. Krawczyk, P.Osland, LC note LC-TH-2000-039 (hep-ph/9909455).

7. I.F. Ginzburg, I.P. Ivanov, Phys. Rev. **B59** (1999) 115001; hep-ph/0004069.

8. I.F. Ginzburg, V.A. Ilyin, in preparation.

9. S.I. Polityko, *Sov. Yad. Fiz.* **43** (1986) 146, **56** (1993) 144.

Two Photon Physics at Future Linear Colliders

A. De Roeck*

* CERN 1211 Geneva 23 Switzerland
e-mail: deroeck@mail.cern.ch

Abstract. Prospects for QCD studies in two-photon interactions at a future linear e^+e^- and $\gamma\gamma$ collider are discussed.

I INTRODUCTION

Traditionally e^+e^- colliders provide a wealth of two-photon data. The photons are produced via bremsstrahlung [1] from the electron and/or positron beam, which leads to a soft energy spectrum for the photons. Such processes will also occur at future high energy (0.5-1 TeV) e^+e^- colliders, but due to the "one time" use of the colliding beams these will allow other operation modes, such as a photon collider mode. A photon collider [2,3], where the electron beams of a linear e^+e^- collider are converted into photon beams via Compton laser backscattering, offers an exciting possibility to study two-photon interactions at the highest possible energies, with high luminosity. A plethora of QCD physics topics in two-photon interactions can be addressed with a linear e^+e^- collider or photon collider. In this mini review we discuss perspectives for measurements of the total cross-section, the photon structure function and the onset of large logarithms in the QCD evolution.

II TOTAL CROSS SECTION

At a linear e^+e^- collider and $\gamma\gamma$ collider detailed properties of two-photon collisions can be studied. A key example is the total $\gamma\gamma$ cross-section, which is not understood from first principles. Fig. 1 shows the present photon-photon cross-sections data in comparison with recent phenomenological models [4]. All models predict a rise of the cross-section with the collision energy, $\sqrt{s_{\gamma\gamma}}$, but the amount of the rise differs and predictions for high photon-photon energies show dramatic differences. In *proton-like-models* (dash-dotted [5,6], dashed [8], dotted [9] and solid [10] curves), the curvature follows closely that of proton-proton cross-section, while in *QCD based* models (upper [7] and lower [4,6] bands), the rise is obtained using the eikonalized PQCD jet cross-section.

The figure demonstrates that large differences between the models become apparent in the energy of a future 0.5-1 TeV e⁺e⁻ collider. A detailed comparison of the predictions [4] reveals that in order to distinguish between all the models the cross-sections need to be determined to a precision of better than 10%. This is difficult to achieve in the e⁺e⁻ collider mode, since the variable $\sqrt{s}_{\gamma\gamma}$ needs to be reconstructed from the visible hadronic final state in the detector. At the highest energies, the hadronic final state extends in pseudorapidity $\eta = \ln\tan\theta/2$ in the region $-8 < \eta < 8$, while the detector covers roughly the region $-3 < \eta < 3$. However, for a photon collider the photon beam energy can be tuned with a spread of less than 10%, such that measurements of $\sigma^{tot}_{\gamma\gamma}$ can be made at a number of different energy values in the range $50 < \sqrt{s}_{\gamma\gamma} < 400$ GeV. The absolute precision with which these cross-sections can be measured ranges from 5% to 10%, where the largest contributions to the errors are due to the control of the diffractive component of the cross-section, Monte Carlo models used to correct for the event selections, the absolute luminosity and knowledge on the shape of the luminosity spectrum. It will be necessary to constrain the diffractive component in high energy two-photon data. A technique to measure diffractive contributions separately, mirrored to the rapidity gap methods used at HERA, has been proposed in [11].

III PHOTON STRUCTURE

The nature of the photon is complex. A high energy photon can fluctuate into a fermion pair or even into a bound state, i.e. a vector meson with the same quantum numbers as the photon $J^{PC} = 1^{--}$. These quantum fluctuations lead to the so-called hadronic structure of the photon. In contrast to the structure function of the proton the structure function of the photon is predicted to rise linearly with the logarithm of the momentum transfer Q^2, and to increase with increasing Bjorken-x [12]. The absolute magnitude of the photon structure function is asymptotically determined by the strong coupling constant [13].

The classical way to study the structure of the photon is via deep inelastic electron-photon scattering, i.e. two-photon interactions with one quasi-real (virtuality $Q^2 \sim 0$) and one virtual ($Q^2 >$ few GeV²) photon. The unpolarised $e\gamma$ DIS cross-section is

$$\frac{d\sigma(e\gamma \to eX)}{dQ^2 dx} = \frac{2\pi\alpha^2}{Q^4 x} \cdot \left[\{1 + (1-y)^2\}F_2^\gamma(x, Q^2) - y^2 F_L^\gamma(x, Q^2)\right], \quad (1)$$

where $F_{2,L}^\gamma(x, Q^2)$ denote the structure functions of the real photon.

To measure F_2^γ it is important to detect (tag) the scattered electron which has emitted the virtual photon. Background studies suggest that these electrons can be detected down to 25 mrad and down to 50 GeV. $e\gamma$ at a photon collider resembles experimentally ep at HERA, i.e. the energy of the probed quasi-real photon is

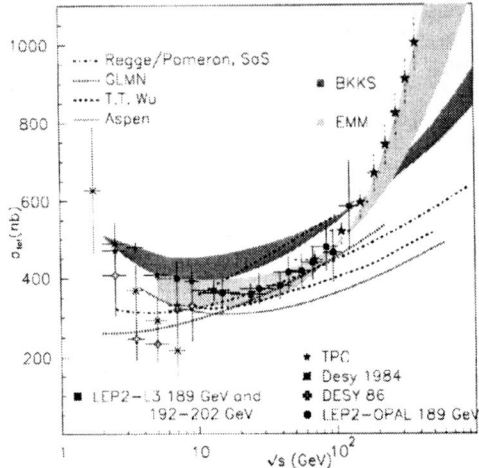

FIGURE 1. The total $\gamma\gamma$ cross-section as function of the collision energy, compared with model calculations: BKKS band (upper and lower limit correspond to different photon densities); SAS lines (Regge Pomeron exchange, upper and lower limits as given by SAS); ASPEN (QCD inspired model, satisfying factorization); EMM band (Eikonal Minijet Model for total and inelastic cross-section, with different photon densities and different minimum jet transverse momentum)

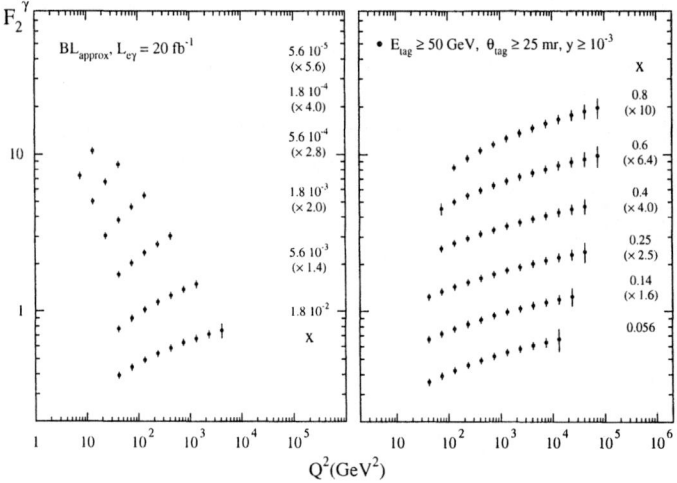

FIGURE 2. The kinematic coverage of the measurement of F_2^γ for the backscattered $e\gamma$ mode at a 500 GeV linear collider.

known (within the beam spread of 10%) and the systematic error can be controlled to about 5%. Fig. 2 shows the measurement potential for a photon collider [14]. The measurements are shown with statistical and (5%) systematical error, for 20 fb^{-1} photon collider luminosity, i.e. about a year of data taking. Measurements can be made in the region $5.6 \cdot 10^{-5} < x < 0.56$, i.e. in a region similar to the HERA proton structure function measurements, and $10 < Q^2 < 8 \cdot 10^4$ GeV^2. For the e$^+$e$^-$ collider mode the hadronic final state needs to be measured accurately in order to reconstruct x. This will limit the lowest reachable x value will be around 10^{-3}.

The Q^2 evolution of the structure function at large x and Q^2 has also been often advocated as a clean measurement of α_s. A 5% change on α_s results however in a 3% change in F_2^γ only, hence such a α_s determination will require very precise F_2^γ measurements.

At high Q^2 values, apart from γ exchange, also Z and W exchange will become important, the latter leading to charged current events [15] which leads to spectacular signals due to the escaping neutrino with high transverse momentum. By measuring the electroweak neutral and charged current structure functions, the up and down type quark content of the photon can be determined separately.

While $e\gamma$ scattering allows to measure the quark distributions it only constrains the gluon distribution via the QCD evolution of the structure functions. Direct information on the gluon in the photon can however be obtained from measurements of jet and charm production [16] in $\gamma\gamma$ collisions. The x_γ distribution is shown for two different assumptions of the parton distributions in dijet production in Fig. 3. x_γ values down to a few times 10^{-3} can be reached with charm and di-jet measurements [17].

A linear collider also provides circularly polarised photon beams This offers a unique opportunity to study the polarised parton distributions of the photon, for which no experimental data are available to date.

Information on the spin structure of the photon can be obtained from inclusive polarised deep inelastic $e\gamma$ measurements and from jet and charm measurements [18,19] in polarised $\gamma\gamma$ scattering. An example of a jet measurement is presented in Fig. 4 which shows the asymmetry measured for dijet events, for the e$^+$e$^-$ and photon collider modes separately. Two extreme models are assumed for the polarised parton distributions in the photon. Already with very modest luminosity significant measurements of the polarised parton distributions become accessible at a linear collider. The extraction of the polarised structure function $g_1(x, Q^2) = \Sigma_q e_q^2 (\Delta q^\gamma(x, Q^2) + \Delta \bar{q}^\gamma(x, Q^2))$, with Δq the polarised parton densities, can however be best done at a $e\gamma$ collider. Measurements of g_1, particularly at low x, are extremely important for studies of the high energy QCD limit, or BFKL regime [20]. Indeed, the most singular terms of the effects of small x resummation on $g_1(x, Q^2)$ behave like $\alpha_s^n \ln^{2n} 1/x$, compared to $\alpha_s^n \ln^n 1/x$ in the unpolarised case of F_2^γ. Thus large $\ln 1/x$ effects are expected to set in much more rapidly for polarised than for unpolarised structure measurements. Fig. 4 shows that for leading order calculations, including kinematic constraints, the differences in predictions of

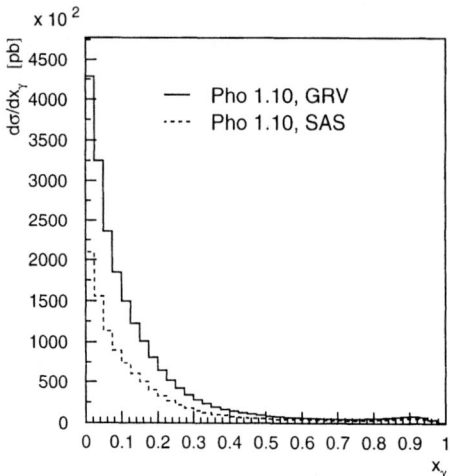

FIGURE 3. Jet cross-sections versus x_γ for the backscattered $\gamma\gamma$ mode at a 500 GeV linear collider, for two assumptions of parton distributions of the photon.

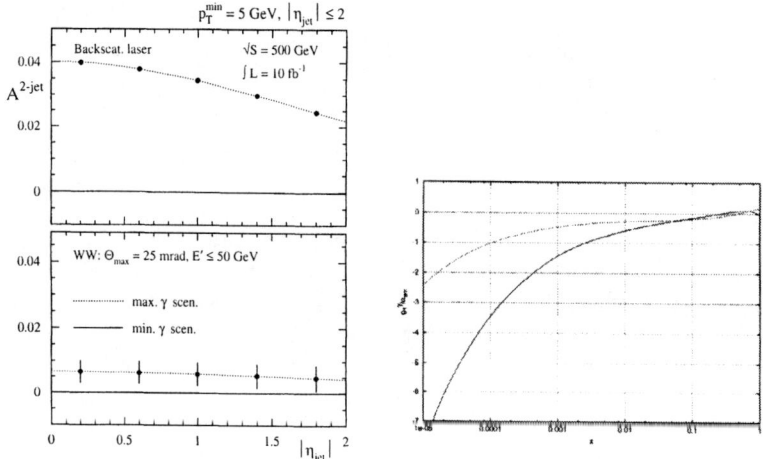

FIGURE 4. (left) di-jet spin asymmetry for events with $x_\gamma^\pm < 0.8$, $p_T^{jet} = 5$ GeV and $|\eta_{jet}| < 2$ for $e\gamma$ (top) and $\gamma\gamma$ (bottom) collisions. Predictions are shown for two different assumptions for the polarized parton distributions of the photon. (right) Predictions for g_1 calculated without (dashed line) and with (full line) $\alpha_s^n \ln^{2n} 1/x$ effects.

g_1 with and without these large logarithms can be as large as a factor 3 to 4 for $x = 10^{-4}$ and could thus be easily measured with a few years of data taking at a photon collider.

IV   TESTING OF BFKL DYNAMICS

Apart from the inclusive polarised structure function measurements, discussed in the previous section, several dedicated measurements exist for detecting and studying the large $\ln 1/x$ logarithm resumation effects in QCD, also called BFKL dynamics.

The most promising measurement for observing the effect of the large logarithms is the total $\gamma^*\gamma^*$ cross-section, i.e. two-photon scattering of virtual photons with approximately equal virtualities for the two photons. Recent calculations, taking into account higher order effects, confirm that this remains a gold-plated measurement, which can be calculated essentially entirely perturbatively and has a sufficiently large cross-section. The events are measured by tagging both scattered electrons. At a 500 GeV e^+e^- collider about 3000 events are expected per year (200 fb^{-1}) and a factor of 3 less in the absence of BFKL effects in the data [21]. Tagging electrons down to as low angles as possible (e.g. 25 mrad) is however a crucial requirement for the experiment.

Closely related to the $\gamma^*\gamma^*$ measurement is vector meson production, e.g $\gamma\gamma \rightarrow J/\psi J/\psi$ or (at large t) $\gamma\gamma \rightarrow \rho\rho$, where the hard scale in the process is given by the J/ψ mass or the momentum transfer t. J/ψ's can be detected via their decay into leptons, and separated from the background through a peak in the invariant mass. Approximately 100 fully reconstructed 4-muon events are expected for 200 fb^{-1} of luminosity for a 500 GeV e^+e^- collider [22]. For this channel it is crucial that the decay muons and/or electrons can be measured to angles below 10 degrees in the experiment.

A process similar to the 'forward jets' at HERA can be studied at a linear collider in $e\gamma$ scattering, with a forward jet produced in the direction of the real photon. The measurements can reach out to smaller x values than presently reachable at HERA, due to the more favourable kinematics of the final state [23].

Finally the processes $e^+e^- \rightarrow e^+e^-\gamma X$ and $\gamma\gamma \rightarrow \gamma X$ have been studied [24], and found to be very sensitive to BFKL dynamics. Event rates for events with photons with energy larger than 5 GeV and p_T larger than 1 GeV are large. At an e^+e^- collider several thousand events will be collected per year, while at a photon collider the event rate is about a factor ten larger.

In all, the study of these processes will provide new fundamental insight in small x QCD physics.

V CONCLUSION

Future linear e^+e^- and $\gamma\gamma$ colliders offer a great opportunity to study photon interactions and QCD processes in detail.

REFERENCES

1. C. F. Weizsäcker, Z. Phys. **88** (1934) 612; E.J. Williams, Phys. Rev. **45** (1934)729.
2. I.F.Ginzburg, G.L.Kotkin, V.G.Serbo and V.I.Telnov, Nucl. Instr. and Meth. **205** (1983) 47; I.F.Ginzburg, G.L.Kotkin, S.L.Panfil, V.G.Serbo and V.I.Telnov, Nucl. Instr. and Meth. **219** (1984) 5.
3. V.I.Telnov, Nucl. Instr. and Meth. **A294** (1990) 72.
4. R.M. Godbole and G. Pancheri, LC-TH note in preparation.
5. G. Schuler and T. Sjostrand, Zeit. Phys. **C68** (1995) 607
6. A. Corsetti, R.M. Godbole and G. Pancheri, Phys. Lett. **B435** (1998) 441, hep-ph/9807236
7. B. Badelek, M. Krawczyk, J. Kwiecinski and M. Stasto, hep-ph/0001161.
8. T.T. Wu, Mod. Phys. Lett. **A15** (2000) 9.
9. E. Gotsman, E. Levin, U. Maor, E. Naftali, Eur.Phys. J. **C14** (2000) 5, hep-ph/0001080.
10. M. Block, E. Gregores, F. Halzen and G. Pancheri, Phys. Rev. **D60** (1999) 54024.
11. A. De Roeck, R. Engel and A. Rostovtsev, hep-ph/9710366.
12. T.F. Walsh and P.M. Zerwas, Phys. Lett., **B44** (1973) 196.
13. E. Witten, Nucl. Phys. **B120** (1977) 189.
14. A. Vogt, Nucl. Phys. Proc. Suppl. **82** (2000) 394;
    A. De Roeck, Proc. of the International Workshop on Linear Colliders (LCWS99) Sitges, May 1999; LC-TH note in preparation
15. A. Gehrmann-De Ridder, H. Spiesberger, P.M. Zerwas, Phys. Lett. **B469** (1999) 259.
16. P. Jankovski, M. Krawczyk, and A. De Roeck, LC-TH-2000-034, hep-ph/0002169.
17. T. Wengler, A. De Roeck, Proc. of the Workshop on Photon Colliders, DESY, Hamburg, June 2000.
18. M. Stratmann, Nucl. Phys. Proc. Suppl. **82** (2000) 400.
19. J. Kwiecinski and B. Ziaja, hep-ph/0006292.
20. E.A. Kuraev, L.N. Lipatov, V.S. Fadin, Sov. Phys. JETP **45** (1972) 199;
    Y.Y. Balitsky, L.N. Lipatov, Sov. J. Nucl. Phys. **28** (1978) 822.
21. J.Kwiecinski, L.Motyka, Phys. Lett. **B462** (1999) 203.
22. J.Kwiencinski, L.Motyka, A. De Roeck, LC-TH-2000-012, hep-ph/0001180.
23. G. Contreras and A. De Roeck, LC-TH note in preparation.
24. N.Evanson and J.Forshaw, LC-TH-2000-010, hep-ph/9912487.

Charm Quark Production
at Linear e^+e^- and Photon Colliders

P.Jankowski

Institute of Theoretical Physics, Warsaw University
ul. Hoża 69, 00-681 Warsaw, Poland

Abstract. LO analysis of the charm quark production in unpolarized $\gamma\gamma$ collisions at a e^+e^- Linear Collider (LC) and a Photon Collider (PC) is presented. The rapidity and x_γ distributions of the charm quark production are shown. The sensitivity to the gluonic content of the real photon, very weekly constrained by the present experiments, is studied in detail. A higher sensitivity of the PC in comparison to the LC to the partonic content of the real photon was found.

INTRODUCTION

Experimental results show high importance of the resolved photon reactions. As a good example, such contributions are required to describe the charm production cross-section in two photon collisions measured by the L3 collaboration [1]. The dominant subprocess for this process is $\gamma g \to c\bar{c}$. On the other hand the gluon density is the most weakly constrained part of the real photon structure. The recent H1 analysis of the di-jet cross-section in photoproduction [2] indicates that the GRV photon parametrization [3] reproduces the experimental data better than the others. Still, we can not say that other parametrizations are excluded by those results. Therefore it is highly important to find and analyze possible future constraints of the gluonic content of the real photon. In this talk I try to answer the question: 'What we can learn about the gluonic content of the photon at future linear colliders?'. The talk partly includes results presented in the article [4]. More detailed work is in preparation [5].

A future linear e^+e^- collider (LC) running at energy of the few hundred GeV will extend kinematic range of photon structure measurements as compared to the LEP experiments. It also offers the opportunity to build a high energy $\gamma\gamma$ collider (Photon Collider - PC) based on Compton-backscattering high energy laser beams on the electron and positron beams, [6].

Our main goal was to test whether the LC and PC colliders are sensitive to partonic, especially gluonic, content of real photon. If they are then the crucial question is: "Which collider proves more sensitive?". To perform our test we

CP571, *PHOTON 2000*, edited by A. J. Finch
© 2001 American Institute of Physics 0-7354-0010-5/01/$18.00

chose the heavy quark production process which is a promising tool for a study of gluonic content of the photon, see e.g. [7], [8] and [9]. In this talk results of LO QCD calculation of charm quark inclusive production in two unpolarized photon interaction $e^+e^- \to \gamma\gamma \to e^+e^- c\bar{c} + X$ will be presented for e^+e^- LC and PC colliders.

THEORETICAL FRAMEWORK

One of main differences between the LC and PC colliders is in its photon spectra, $f_\gamma(y)$ with $y = E_\gamma/E_e$. First provides us with quasi-real ($P^2 < 1$) photon events. Very soft photon spectrum may be described then by the effective Williams-Weizsäcker (WW) formula. On contrary, photons produced in the PC collider are real with the resulting spectrum much harder than the WW one. We chose the laser conditions so that the unpolarized spectrum discussed by Ginzburg *et al* in ref. [6] shows a narrow pick close to the y value of 0.8. Photons with energies greater then 83% ($y = 0.83$) of the original electron/positron beams can not be produced. In our recent calculations also an additional cut on the spectrum was made ($E_\gamma/E_e > 0.6$) to retain only the high energy part, [10].

Heavy quarks can be produced in $\gamma\gamma$ collisions through three mechanisms shown in Fig.1. Direct (DD) production occurs when both photons couple directly to the $c\bar{c}$ pair. In single resolved photoproduction processes (DR) one of photons interacts via its partonic structure with the second photon. When both photons split into a flux of quarks and gluons, the process is labelled a double resolved photon (RR) process.

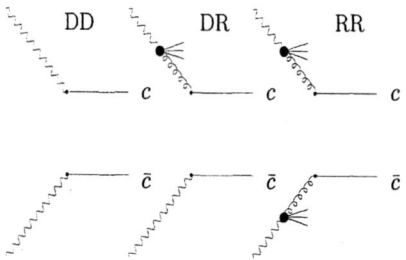

FIGURE 1. Three types of contributions to heavy quark (c quark) production in $\gamma\gamma$ collision.

Our LO QCD calculations are performed in two schemes of heavy quark treatment: Fixed Flavour Number Scheme (FFNS) and Variable Flavour Number Scheme (VFNS). They can be also called massive and massless schemes respectively [11]. They differ by the number of active quark flavours (N_f) which can take part in the process as photon partons. In massive FFNS photon consists of gluons and only light quarks, therefore here $N_f = 3$. The produced heavy quark, here being the charm quark, is massive (we used value $m_c = 1.6$ GeV) in this approach. The massless VFNS considers apart from the gluons and u, d and s quarks also heavy quarks as potentially active flavours. Charm quark becomes active ($N_f = 4$) for $\mu \gg m_c$, with μ being the hard (factorisation, renormalization) scale of the process. Mass of c in VFNS is kept equal to zero througout the calculations.

The hard scale is chosen to be equal to the transverse mass of the produced charm quark $m_T = \sqrt{m_c^2 + p_T^2}$ (giving $\mu = p_T$ in VFNS). Hence the FFNS scheme is expected to be valid for $p_T \leq m_c$ while the VFNS for $p_T \gg m_c$.

We calculate in the LO QCD the production rates for c produced with large p_T for the e^+e^- colliders LEP and LC at energies of 180 GeV and of 300, 500 and 800 GeV, respectively, and for the $\gamma\gamma$ PC based on the corresponding (e^+e^-) LC collider. We test the sensitivity of the considered processes to the gluonic content of the photon by using two different parton density parametrizations for the real photon: GRV [3] and SaS1d [12]. The QCD energy scale Λ_{QCD}, which appears in the one loop formula for the strong coupling constant α_s, is affected by change of the number of active flavours. Therefore we denote it as $\Lambda_{QCD}^{N_f}$. We take this scale to be: $\Lambda_{QCD}^3 = 232$ MeV and $\Lambda_{QCD}^4 = 200$ MeV as in [3].

RESULTS

If not stated otherwise the results shown in this talk are for the differential LO cross-sections $\frac{d^2\sigma}{dp_T^2 d\eta}$ calculated in the VFNS scheme, with η being the rapidity of the produced charm quark. Values of the central-mass energy of the e^+e^- system and the transverse momentum of c where chosen to be $\sqrt{s} = 500$ GeV and $p_T = 10$ GeV.

First we observe domination of the resolved photon processes in case of the PC in contrary to the LC, presented in Fig.2 for the VFNS scheme. Similar result holds in the FFNS case though the double resolved (RR) contributions become then much smaller than the single resolved (DR) ones.

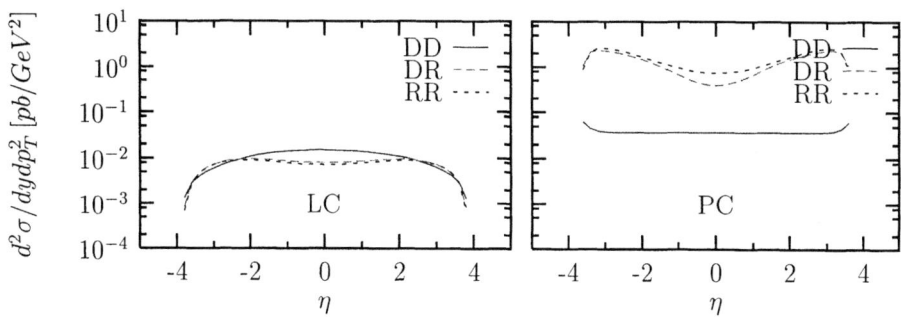

FIGURE 2. $d^2\sigma/dp_T^2 d\eta$ distribution for $c/\bar{c}$ production in the VFNS at $\sqrt{s} = 500$ GeV, $p_T = 10$ GeV, with the GRV photon parametrization in the LC and PC photon spectra.

Another crucial characteristic is domination of the gluon induced processes over other resolved reactions, Fig.3. It is important for both the LC and PC but more striking in the latter case.

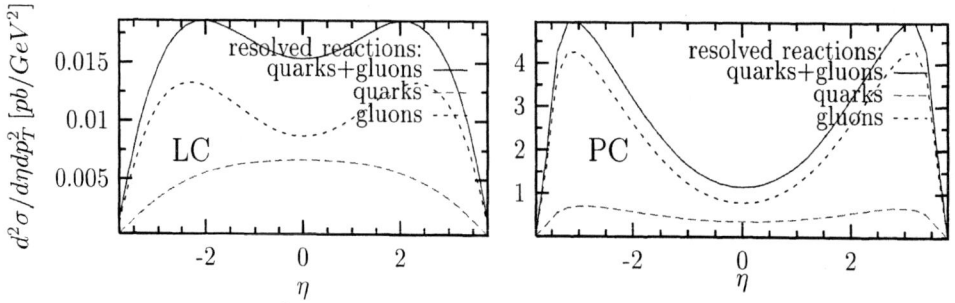

FIGURE 3. Comparison of contributions initiated by gluons with other partonic contributions to the resolved (DR+RR) part of the cross-section $\frac{d^2\sigma}{dp_T^2 \cdot d\eta}$ $(e^+e^- \to e^+e^- c/\bar{c}X)$ calculated in the VFNS scheme with the GRV photon parametrization at $\sqrt{s} = 500$ GeV, $p_T = 10$ GeV.

In linear colliders running at high energy small x_γ can be reached in the photoproduction, what is ilustrated in Fig.4(left), where $x_{\gamma,min}$ at various energies are shown. This means that in the LC and PC one can explore regions of gluon dominance being at the same time regions of greatest difference among the gluon distributions in the photon. This is ilustrated in Fig.4(right), where u quark and g densities are shown. As an effect of this feature we observed rise of the resolved photon process contribution with increasing energy [4].

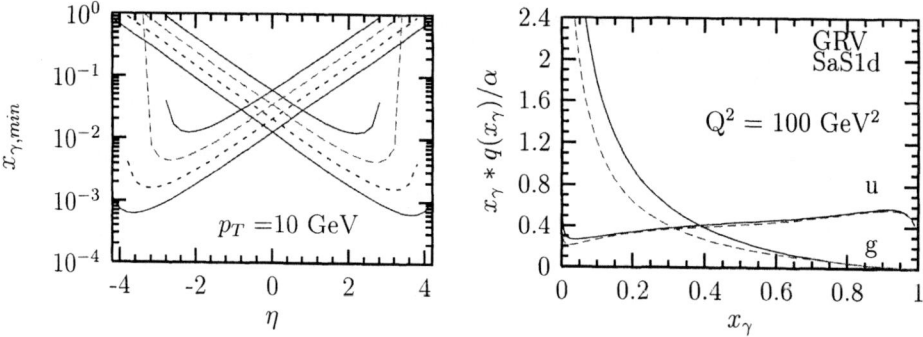

FIGURE 4. Left plot presents minimal x_γ values accesible in linear e^+e^- collider as a function of rapidity η: $x_{\gamma,min}(\eta) = \frac{p_T \exp(\eta)}{2E_e - p_T \exp(-\eta)}$ at $p_T =10$ GeV. Upper line corresponds to the LEP energy $\sqrt{s} = 180$ GeV, following lines to the LC energies of 300, 500 and 800 GeV. Right plot shows u quark and gluon distributions in the photon at $Q^2 =100$ GeV^2. Solid lines correspond to the GRV and dashed ones to the SaS1d photon parametrizations.

The sensitivity of the considered process to the gluon distribution in the photon is studied further by comparing the predictions obtained using two different parton parametrizations. In Fig.5 the ratio of the relative difference of cross-sections $\frac{d^2\sigma}{dp_T^2 d\eta}$ is presented, obtained using the GRV and SaS1d parton distribution parametrizations in the VFNS scheme. As it can be expected after the above analysis the PC collider option leads to a larger sensitivity to the gluonic density in γ than the LC one for a given energy of the e^+e^- collider: the difference between results obtained using the two structure function parametrizations is 5-20% for the LC and 25-40% for the PC photon spectrum. The sensitivity increases with the energy of the collider. Plots for the case of the FFNS scheme are very similar, see [4].

We also performed the analysis with the additional $y > 0.6$ cut on the PC photon spectrum. Sensitivity to the parametrization choice grows then by about 5% for all contributions to the process. Fig.6 shows the $\frac{d^2\sigma}{dp_T^2 d\eta}$ cross-sections obtained with $y > 0.6$ cut.

It is also worth mention that the same predictions for the beauty quark production give an increase of the tested ratio by about 10% for both LC and PC colliders. However the b production cross-sections are much smaller than the corresponding c ones resulting in worse experimental precision, [4].

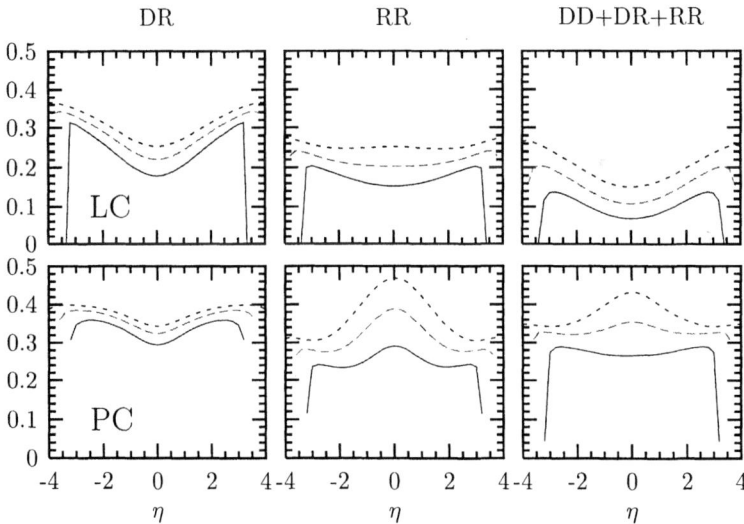

FIGURE 5. The ratio $\frac{GRV-SAS1d}{GRV}$ of the cross-section $\frac{d^2\sigma}{dp_T^2 dy}$ $(e^+e^- \to e^+e^- c/\bar{c}X)$ for $p_T = 10$ GeV, in the VFNS scheme. The lower, middle and upper lines correspond to $\sqrt{s} = 300, 500, 800$ GeV respectively.

At the very end, Fig.7 shows the importance of the resolved photon contributions

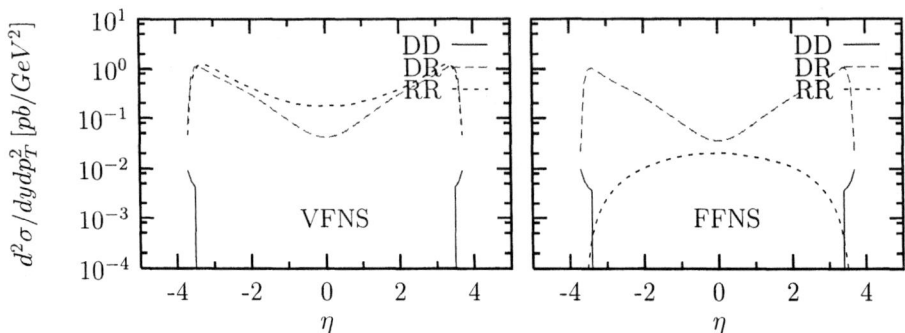

FIGURE 6. $d^2\sigma/dp_T^2 d\eta$ distribution for $c/\bar{c}$ production in the PC with an extra $y = E_\gamma/E_e > 0.6$ cut at $\sqrt{s} = 500$ GeV, $p_T = 10$ GeV with the GRV photon parametrization.

to the charm production at PC in small x_γ region. The x_γ distribution was calculated with the additional cut on the photon spectrum ($E_\gamma/E_e > 0.6$). The double resolved contribution (where each event has two entries, with an x_γ value from each photon) dominates except at smallest x_γ values, where the single resolved one becomes important too.

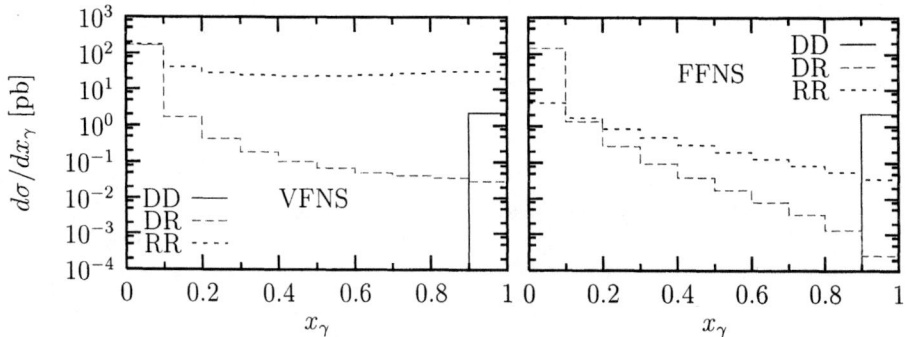

FIGURE 7. $d\sigma/dx_\gamma$ distribution for $c/\bar{c}$ production in the PC with an extra $y = E_\gamma/E_e > 0.6$ cut at $\sqrt{s} - 500$ GeV, $p_T > 10$ GeV with the GRV photon parametrization.

CONCLUSIONS

Charm quark production in two real photon collisions at future linear colliders will give an access to small x_γ which is the gluon domination region of the photon structure being also the region of greatest gluon distribution uncertainty. We performed the LO QCD calculations of the $e^+e^- \to \gamma\gamma \to e^+e^- c\bar{c} + X$ cross-sections for the cases of LC and PC photon spectra. We used two schemes of calculation of heavy quark production: the FFNS and VFNS. The calculations showed the following. In both, the e^+e^- LC and PC ($\gamma\gamma$) resolved photon contributions will play an important role. Those contributions will be dominated by the gluon initiated subprocesses. The above features *do not* depend on the particular scheme used to calculate the heavy quark (c) cross-sections. They *do* depend on the choice of the collider mode. The PC proves to be much more sensitive to the resolved photon processes than the LC with WW spectrum of the quasi-real photons. Also in the PC resolved photon contributions are more strongly dominated by the gluon initiated subprocesses. Therefore we conclude that the charm quark production provides a sensitive probe of the gluon content of the photon, favouring the PC option for future photon structure research.

I thank M.Krawczyk and A.De Roeck for many discussions and M.Krawczyk for reading of this manuscript and critical comments.

I would like to thank organizers of this conference for the financial support.

This work has been partly supported by the Polish Committee for Research, KBN grant No 2P03B05119.

REFERENCES

1. L3 Coll., Acciarri, M., et al., *Phys. Lett.* **B453** 83-93 (1999)
2. H1 Coll., Adloff, C., et al., *Phys. Lett.* **483** 36-48 (2000)
3. Glück, M., Reya, E., and Vogt, A., *Phys. Rev.* **D46**, 1973-1979 (1992).
4. Jankowski, P., Krawczyk, M., and De Roeck, A., hep-ph/0002169 (2000); LC Note LC-TH-2000-034
5. Jankowski, P., Krawczyk, M., and De Roeck, A., paper in preparation
6. Ginzburg, I.F., Kotkin, G.L., Serbo, V.G., and Telnov, V.I., *Nucl. Instr. and Meth.* **205** 47 (1983); Ginzburg, I.F., Kotkin, G.L., Panfil, S.L., Serbo, V.G., and Telnov, V.I, *Nucl. Instr. and Meth.* **219** 5-24 (1984); Telnov, V.I., *Nucl. Instr. and Meth.* **A294** 72-92 (1990); Brinkmann, R., *et al*, *Nucl. Instrum. Meth.* **A406** 13-49 (1998)
7. Drees, M., Krämer, M., Zunft, J., and Zerwas, P.M., *Phys. Lett.* **B306** 371-378 (1993).
8. Cacciari, M., Greco, M., Kniehl, B.A., Krämer, M., Kramer, G., and Spira, M., *Nucl. Phys.* **B466**, 173-188 (1996).
9. Doncheski, M., Godfrey, S., and Peterson, K.A., *Phys. Rev.* **D55**, 183-189 (1997).
10. Ginzburg, I.F., Kotkin, G.L., *Eur. Phys. J.* **C13** 295-300 (2000).
11. Kniehl, B.A., Krämer , M., Kramer, G., and Spira, M., *Phys. Lett.* **B356**, 539-545 (1995)
12. Schuler, G.A., and Sjöstrand, T., *Z. Phys.* **C68**, 607-624 (1995); *Phys. Lett.* **B376**, 193-200 (1996).

Progress towards a
$\gamma\gamma \to 4$ leptons Monte Carlo

C. Carimalo, W. Da Silva, F. Kapusta

Laboratoire de Physique Nucléaire et de Hautes Energies,
IN2P3-CNRS Universités Paris VI et VII,
4, PlaceJussieu - BP200, Tour 33 - Rez de Chaussée,
75252 PARIS Cedex 05, FRANCE

Abstract. A Monte Carlo based on an exact computation of peripheral diagrams is presented. Total cross sections and angular distributions with experimental cuts are shown. This is an updated version of a previous talk at PHOTON'99 [1].

INTRODUCTION

An analytic computation of the $\gamma\gamma \to e^+e^-\mu^+\mu^-$ cross section which has a large value at high energy is performed and the results are compared to existing theoretical predictions. The effect of detector acceptance has to be checked via a Monte Carlo generator, to get realistic numbers of events expected at TESLA, and check whether the $\gamma\gamma$ luminosity could be obtained through 4 lepton processes. The forward region instrumentation has to be studied accordingly. Moreover these processes could be a background to inclusive J/ψ production. GRACE has been used to cross-check the size of the neglected s-channel contribution, as shown in figure 1.

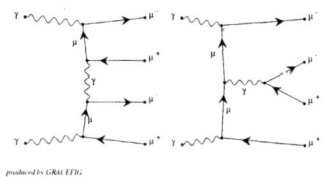

produced by GRACEFIG

FIGURE 1. Example of t-channel and s-channel contribution for 4 muon production.

COMPUTATION

The differential cross section corresponding to the first diagram of fig. 1 can be factorized as $\frac{1}{t^2} G^{\lambda\lambda'} D_{\lambda\lambda'}$, where $t = -q^2$, q being the 4-momentum of the space-like exchanged photon and $G^{\lambda\lambda'}$ and $D_{\lambda\lambda'}$ are vertex amplitudes associated with the upper vertex and lower vertex respectively. We define $T = \frac{K + K'}{2\omega}$, $Z = \frac{K - K'}{2\omega}$, where $K(\omega, \vec{\omega})$ and $K'(\omega, -\vec{\omega})$ are the 4-momenta of initial photons γ_1 (upper vertex) and γ_2 (lower vertex) respectively, and expand $g_{\alpha\beta}$ as $\frac{1}{2}(T + Z)_\alpha(T - Z)_\beta + \frac{1}{2}(T - Z)_\alpha(T + Z)_\beta + \epsilon_\alpha^+\epsilon_\beta^- + \epsilon_\alpha^-\epsilon_\beta^+$, ϵ^+ and ϵ^- being transverse polarisations of initial photons.

Following the well-known procedure of Impact Factor Method (cf. for instance [2]), we insert such an expansion into the above tensor product to obtain
$$G^{\lambda\lambda'} D_{\lambda\lambda'} = \tfrac{1}{4} G_{+\,+} D_{-\,-} + \tfrac{1}{4} G_{-\,-} D_{+\,+} + \tfrac{1}{2} G_{+\,-} D_{-\,+} + G_{+\,T} D_{-\,T} + G_{-\,T} D_{+\,T} + G_{T\,T'} D_{T\,T'}$$
where : $G_{+\,+} = G^{\alpha\beta} (T - Z)_\alpha(T - Z)_\beta$ and $D_{-\,-} = D^{\alpha\beta} (T + Z)_\alpha(T + Z)_\beta$.
Some extra notation : $l = e^-$, $\bar{l} = e^+$, $L = \mu^-$, $\bar{L} = \mu^+$
In this way, light cone variables : $p_+ = p.(T - Z)$, $p_- = p.(T + Z)$ and $\vec{p_T}$ appear in a very natural way. At high energy and low angles, the first term of the above expansion is the dominant one, so we concentrate on the computation of $G_{+\,+}(l , \bar{l})$ and $D_{-\,-}(L , \bar{L})$.

Without any approximation we obtain :
$$G_{+\,+}(l , \bar{l}) = 4\,\omega\,(1_+ P_{\bar{l}/\gamma}(\bar{x}) + \bar{l}_+ P_{l/\gamma}(x)) + 8\,\vec{l_T} . \vec{l_T}\, (x\,(1 - \bar{x}) + \bar{x}\,(1 - x)) - 8\,\frac{m^2}{l_- \bar{l}_-}$$
where $P_{l/\gamma}(x)$ is the lepton spectrum distribution :
$$P_{l/\gamma}(x) = \frac{\alpha}{2\pi}\left[\frac{2\,m^2\,x}{\omega^2\,l_-^2} + \frac{1}{\omega\,l_-} (x^2 + (1 - x)^2)\right]$$ where $x = 1 - \frac{l_+}{2\omega}$ is at low angle the fraction of the photon energy taken away by the virtual lepton at the upper vertex. The computation of $D_{-\,-}(L , \bar{L})$ is obtained with the exchange of photon γ_1 with photon γ_2.

TOTAL CROSS SECTION

The integration down to very small angles requires special treatments because of convergence problems.In this respect, using light-cone variables is very helpful, especially variables like tranverse momenta and rapidity of final particles. Keeping only the first term of the above expansion and adding the term coming from the exchange of lepton pairs $(l , \bar{l}) \iff (L , \bar{L})$, we obtain the total cross-sections $\sigma(W_{\gamma\gamma})$ plotted in figure 2. One notices that we obtain in this way the right asymptotic behavior and numerical values. Figure 3 clearly shows that the bulk of the cross section lies in the low p_T region . As already quoted some time ago by V. Serbo, the relevant p_T values are located below 2 m_μ.

Requiring 1 or 2 muons above 2° makes the cross section drop rather fast as shown in figure 4.

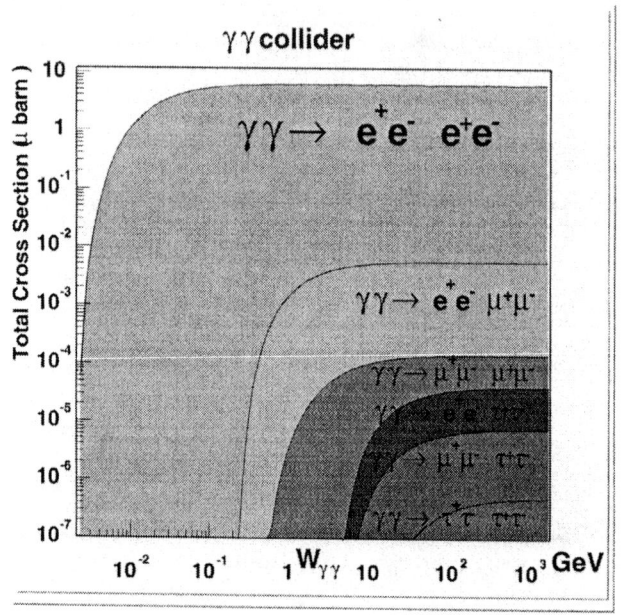

FIGURE 2. Total cross section.

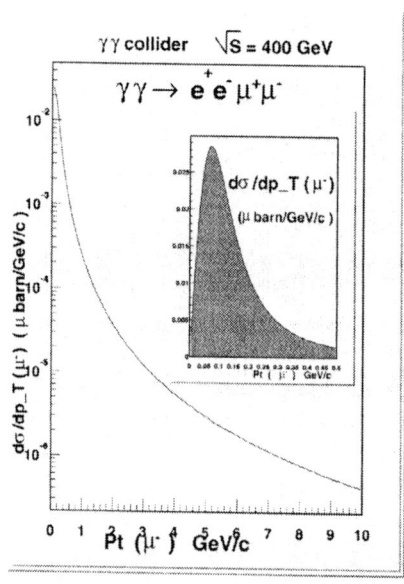

FIGURE 3. Single muon inclusive transverse momentum distribution

432

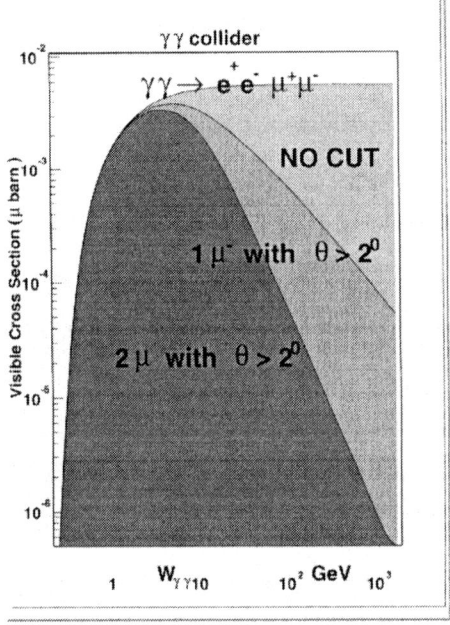

FIGURE 4. Acceptance cut effects on the total cross section.

VISIBLE CROSS SECTION

Choosing a centre of mass energy of 400 GeV a few comparisons with previous work [3] are made. First, from the analytical point of view, the low transfer approximation: $t = -q^2 \sim q_T^2$ gives

$$G_{++} \sim 4\, l_+\, \bar{l}_+ \left(\frac{q^2}{(m^2 + \vec{l}_T^{\,2})\,(m^2 + \vec{l}_T^{\,2})} - 2\,(1-x)\,(1-\bar{x})\,\vec{Q}_T^{\,2} \right) \quad \text{with}$$

$$\vec{Q}_T = \frac{\vec{l}_T}{m^2 + \vec{l}_T^{\,2}} + \frac{\vec{l}_T}{m^2 + \vec{l}_T^{\,2}} \quad \text{in agreement with [3].}$$

The agreement is still there in figure 5 after requiring 2 undetected electrons at low angle ($\theta_{e^+,e^-} \leq 2^\circ$) and 2 detected muons ($\theta_{\mu^+,\mu^-} \geq \theta_{MIN}$) with energy ≥ 1 GeV .

With one detected muon ($\theta_{\mu^-} \geq \theta_{MIN}$) with energy ≥ 1 GeV, the comparison shown in figure 6 is made with reference [4], using $\frac{d\sigma}{dz} = \frac{2\alpha^4}{3\pi M_{\mu^-}^2} \frac{1}{z^2} \int_{(z\,x_0)^2}^{z^2} \sqrt{\rho}\, F_0(\rho, \frac{\sqrt{\rho}}{z})\, d\rho$ where $\rho = \frac{l_{T\mu^-}^2}{M_{\mu^-}^2}, z = \frac{\omega\,\theta_{\mu^-}}{M_{\mu^-}}, x_0 = \frac{E_{\mu^-}^{MIN}}{\omega}$ and $\sigma = \int_{z_{MIN}}^{z_{max}} \frac{d\sigma}{dz}\, dz$

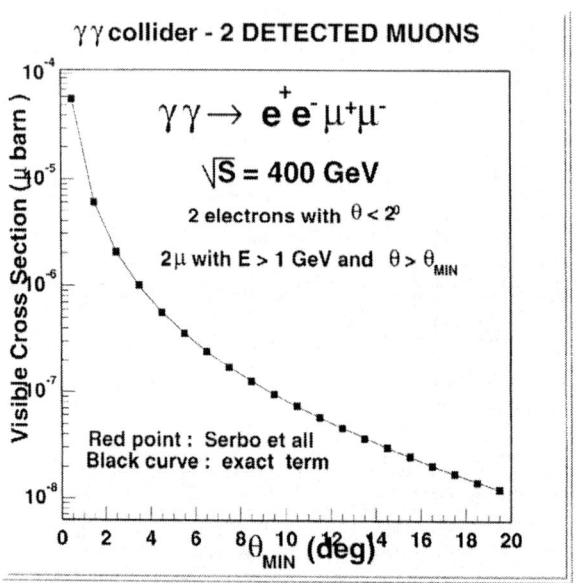

FIGURE 5. Visible cross section

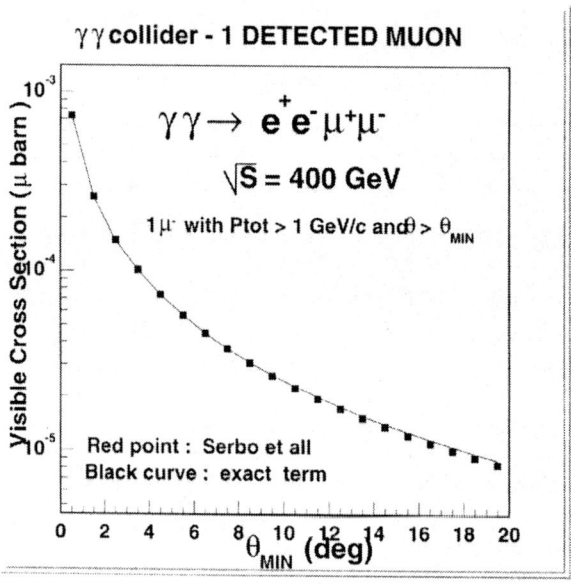

FIGURE 6. One visible muon

OUTLOOK

The integration at very low angle is under control. The numerical stability of the results have still to be fully tested. The next step is to improve the integration method with an analytic computation of helicity amplitudes, in order to try to estimate the numerical contribution of the time-like diagram and interferences at large angle. Such an estimate is shown in figure 7, when two muons with energy above 1 GeV are seen above some angle greater than 2°

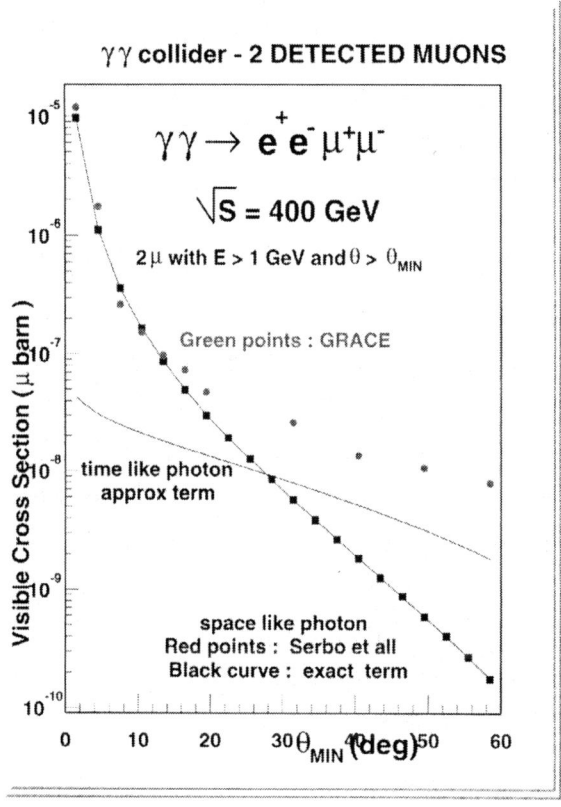

FIGURE 7. Two visible muons

REFERENCES

1. C. Carimalo, W. Da Silva, F. Kapusta *Nucl. Phys.* **B** (Proc. Suppl.) 82 (2000) 391-393
2. I.F. Ginzburg, S.L. Panfill and V.G. Serbo *Nucl. Phys.* **B284** (1987) 635
3. E.A Kuraev, A. Schiller, D.V Serebryakova and V.G. Serbo *Nucl. Phys.* **B326** (1985), *Eur. Phys. J.* **C4** (1998)
4. E.A Kuraev, A. Schiller and V.G. Serbo *Nucl. Phys.* **B2566** (1985) 189

RELATED PROCESSES

Results from the OPAL experiment using photonic final states

Anna Macchiolo*

*Université de Montréal, Montréal, Quebec H3C 3J7, Canada

Abstract. Event topologies characterised by photonic final states at LEP2 energies are analysed by the OPAL Collaboration in order to study gauge boson couplings. Results on neutral triple gauge boson couplings are presented, based on $e^+e^- \rightarrow Z\gamma$ events collected at 189 GeV.

$e^+e^- \rightarrow W^+W^-\gamma$ and $e^+e^- \rightarrow \nu\bar{\nu}\gamma$ events are studied to measure charged quartic gauge boson couplings, whereas neutral quartic boson vertices are investigated by using $e^+e^- \rightarrow q\bar{q}\gamma\gamma$ events.

Searches for Higgs bosons with an enhanced coupling to photon pairs are also based on event topologies characterised by two photons in the final state, at centre-of-mass energies between 91 and 108 GeV.

The prospects for these analyses at future colliders are discussed.

Introduction

Events with photonic final states play a key role in e^+e^- interactions. They allow a direct experimental test of radiative corrections in the Electroweak theory and the measurement of gauge boson couplings, both in the charged and in the neutral sector. In addition photonic final states are the signature of possible New Physics (NP) processes in different scenarios, from supersymmetric models to extra dimension theories. They also represent the test ground for models with enhanced Higgs boson decay couplings to photons pairs.

Among several OPAL analyses based on photonic final states we will concentrate on those where we can expect a deeper insight from studies at the Future Linear Collider (FLC) and the Photon Linear Collider (PLC), thanks to a higher luminosity and a wider accessible kinematic range: measurements of triple and quartic gauge boson couplings and the search for a Higgs boson decaying into two photons. For most of these analyses the dominant background is constituted by fermion pair production with single or double initial state radiation (ISR). The high cross-section for such events is explained by the effect called Z radiative return: at LEP2 the fast decrease of the cross-section as a function of energy favours the radiation of hard

CP571, *PHOTON 2000*, edited by A. J. Finch

photons that boost the effective two-fermion centre-of-mass energy back to the Z mass.

Measurement of neutral gauge boson couplings

The final states $q\bar{q}\gamma$ and $\nu\bar{\nu}\gamma$ are studied to search for $Z\gamma Z$ and $Z\gamma\gamma$ couplings. The corresponding triple vertices are not generated at tree level in the Standard Model (SM) and contributions from higher order corrections ($\mathcal{O}(10^{-4})$) are far too small for the present experimental sensitivity.

Neutral triple gauge boson couplings (NTGCs) can anyhow arise in the effective Lagrangian for the effects of NP, for example in compositeness models or if new particles enter in higher order corrections [1]. Assuming only Lorentz and $U(1)_{em}$ gauge invariance, the most general $Z\gamma Z$ and $Z\gamma\gamma$ vertices can be parametrised by means of eight couplings, four of them ($h_i^{Z,\gamma}$ ($i = 1,3$)) are associated to dimension six operators and the other four ($h_i^{Z,\gamma}$ ($i = 2,4$)) to dimension eight operators.

OPAL has performed an analysis of NTGCs using a data sample corresponding to an integrated luminosity of about 177 pb^{-1}, with an average centre-of-mass energy of 188.6 GeV [2].

The selection of $\nu\bar{\nu}\gamma$ and $q\bar{q}\gamma$ events

The selection of single-photon topologies is designed to identify events with one photon and significant missing transverse energy, indicating the presence of at least one neutrino-like invisible particle [3]. In the SM this kind of events are expected from the process $e^+e^- \rightarrow \nu\bar{\nu}\gamma(\gamma)$.

The kinematic acceptance of the selection is defined in terms of the photon energy E_γ, and the photon polar angle θ_γ: 50 GeV $< E_\gamma <$ 90 GeV ; 15° $< \theta_\gamma <$ 165°. The selection retains acceptance for events with additional photons in which the resulting photonic system is still consistent with the presence of significant missing energy. This reduces the sensitivity of the measurement to the modelling of higher order contributions. High selection efficiency is obtained also for photon conversion candidates, identified on the base of tracking chamber information. Additional cuts on cluster quality, forward energy, muon chambers and hadron calorimeter signals are used to remove cosmic rays backgrounds, beam related backgrounds and SM backgrounds that have no missing energy.

Events with an energetic photon in an hadronic environment are identified by looking at electromagnetic clusters without associated tracks and characterised by shape patterns typical of photons. The photon candidate is also required to be isolated with respect to the hadronic system, by applying cuts on the energy deposits and the momenta of tracks falling in a 15° cone built around the photon flight direction. Systems of one or two tracks associated with electromagnetic clusters are included if consistent with a photon conversion and if they satisfy the criteria listed above.

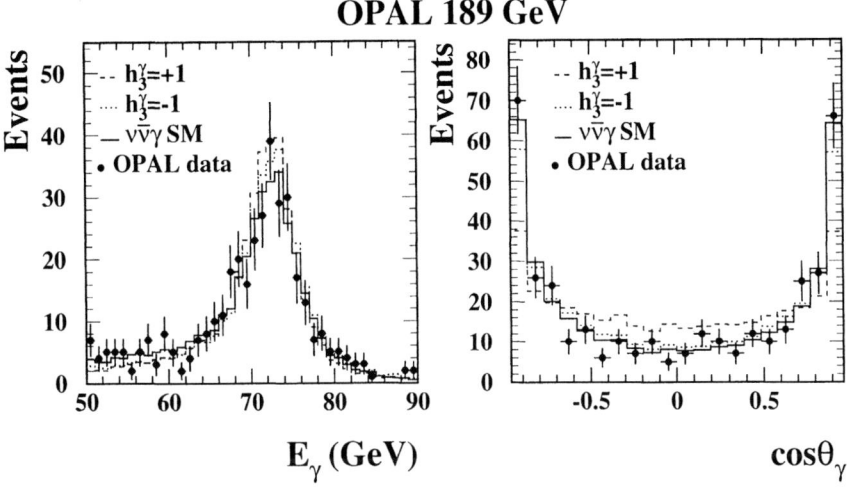

FIGURE 1. Distributions of E_γ and $\cos\theta_\gamma$ measured in $\nu\bar{\nu}\gamma$ events in the data, compared with the SM expectation (solid line) and two anomalous values for the h_3^γ parameter. The MC distributions are normalised to the number of events selected in the data.

After the photon identification the same kinematic cuts of the single photon analysis are applied to the most energetic photon in the event. A kinematic fit is also performed, imposing energy and momentum conservation, in order to improve the photon energy resolution.

NTGCs extraction

NTGCs are extracted with a maximum likelihood fit to the measured event rate and to differential distributions. As a general property, the Z and γ produced at the anomalous vertices are more isotropically distributed than the dominant SM background, which is characterised by the strongly forward peaked angular distribution of the ISR (see Fig.1). Therefore deviations from the Standard Model predictions due to NTGCs would be more pronounced for large angles between the beam direction and the photon. In addition the energy spectrum of the photon is used in both channels. In the $q\bar{q}\gamma$ channel the fit is performed also to the cosine of the angle between the photon and the closest jet since this variable depends on the polarisation of the Z boson, that can be affected by NTGCs.

The fit is performed in the two channels independently, under the assumption that only one coupling at the time is different from zero. The 95% C.L. for the eight

95% C.L. bounds on anomalous NTGCs		
	OPAL	LEP
h_1^γ	$[-0.115, \ +0.115]$	$[-0.10, \ +0.03]$
h_2^γ	$[-0.077, \ +0.077]$	$[-0.036, \ +0.061]$
h_3^γ	$[-0.164, \ -0.006]$	$[-0.075, \ -0.004]$
h_4^γ	$[+0.007, \ +0.134]$	$[+0.005, \ +0.056]$
h_1^Z	$[-0.190, \ +0.190]$	$[-0.13, \ +0.04]$
h_2^Z	$[-0.128, \ +0.128]$	$[-0.041, \ +0.080]$
h_3^Z	$[-0.269, \ +0.119]$	$[-0.16, \ +0.07]$
h_4^Z	$[-0.084, \ +0.175]$	$[-0.04, \ +0.10]$

TABLE 1. The 95% C.L. intervals on NTGCs obtained from OPAL (first column) and after the preliminary combination with the results from DELPHI and L3 (second column).

couplings, from OPAL and after combining with measurements from DELPHI and L3 [4], are listed in Tab.1.

Measurement of Quartic Gauge Boson Couplings

In the SM charged triple and quartic gauge boson couplings (TGCs and QGCs), W^+W^-V and W^+W^-VV (where V = Z, W or γ), arise as a consequence of the non-Abelian SU(2) gauge symmetry.

Events with photons in the final states can be sensitive to charged TGCs through the t-channel process of W fusion, $e^+e^- \to \nu\bar{\nu}\gamma$. Their contribution to TGCs determination is anyhow much less relevant with respect to W pair production events, mainly for statistical reasons. On the contrary events with photons in the final state are the only way to investigate quartic gauge couplings at LEP2, since final states involving three massive gauge bosons are beyond the kinematic reach.

Four quartic gauge boson vertices are predicted in the SM, with fixed couplings, $W^+W^-W^+W^-$, W^+W^-ZZ, $W^+W^-Z\gamma$, $W^+W^-\gamma\gamma$, but their contributions to processes studied at LEP2 is negligible. However it is possible to set the first direct limits on possible anomalous contributions to QGCs.

The $W^+W^-\gamma\gamma$ vertex is investigated through $e^+e^- \to W^+W^-\gamma$ production, and $e^+e^- \to \nu\bar{\nu}\gamma\gamma$ events occurring via the W^+W^- fusion process, the first and the second diagrams in Fig.2, respectively. The neutral quartic vertex $ZZ\gamma\gamma$, forbidden in the SM, is studied with $e^+e^- \to Z\gamma\gamma$ production, the third diagram in Fig.2. After neglecting operators leading also to TGCs, anomalous QGCs contribution to the $W^+W^-\gamma\gamma$ and $ZZ\gamma\gamma$ vertices are parametrised by mean of two additional terms of dimension six in the Lagrangian [5]:

$$\mathcal{L}_0 = -\frac{e^2}{16}\frac{a_0}{\Lambda^2}F^{\mu\nu}F_{\mu\nu}\vec{W}^\alpha\vec{W}_\alpha, \quad \mathcal{L}_c = -\frac{e^2}{16}\frac{a_0}{\Lambda^2}F^{\mu\alpha}F_{\mu\beta}\vec{W}^\beta\vec{W}_\alpha, \qquad (1)$$

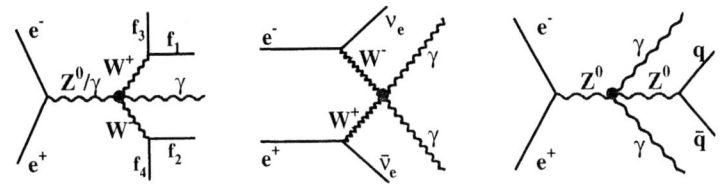

FIGURE 2. The processes involving quartic gauge boson couplings at LEP2.

where a_0/Λ^2 and a_c/Λ^2 are the anomalous QGCs and Λ represents the energy scale for NP. The two parameters a_0/Λ^2 and a_c/Λ^2, which are separately C and P conserving, are in general independent in the two vertices.

Analysis of QGCs in $e^+e^- \rightarrow Z\gamma\gamma$ events

OPAL has measured the cross-section for the process $e^+e^- \rightarrow Z\gamma\gamma$, arising in the SM mainly from doubly-radiative return to the Z, using 0.58 fb^{-1} of data recorded at centre-of-mass energies between 130 GeV and 208 GeV [6]. The signal definition is based on cuts on the photon energies and polar angles and on the di-jet invariant mass, which is required to be close to M_Z. In addition to the event rate, shown in the first plot of Fig.3, the shape of the distributions for the energy of the second most energetic photon, and for the polar angle of the most forward photon are used to extract QGCs. OPAL results and their combination with those from the other LEP experiments [4] are shown in Tab.2.

Analysis of QGCs in the $W^+W^-\gamma\gamma$ vertex

In the SM $W^+W^-\gamma$ final states arise from W pair production with ISR, FSR and radiation from the W boson itself. The signal of anomalous QGCs would be an enhancement of high energy photons (see second plot in Fig.3) and OPAL has set limits on these couplings using data at 189 GeV with a fit to the distributions of E_γ and $\cos\theta_\gamma$ [7].

In the SM the reaction $e^+e^- \rightarrow \nu\bar{\nu}\gamma\gamma$ proceeds predominantly through s-channel Z exchange and t-channel W exchange, whereas the SM contribution from W^+W^- fusion is negligible at LEP. Anomalous QGCs would enhance the $\nu\bar{\nu}\gamma\gamma$ production rate, especially for the hard tail of the photon energy distribution. Limits on the couplings are set in OPAL [8] by restricting the signal acceptance to values of the invariant mass, recoiling against the two-photon system, lower than 70–80 GeV where the SM predicts a negligible cross-section. Bounds on the a_0/Λ^2 and a_c/Λ^2 parameters relative to the $W^+W^-\gamma\gamma$ vertex are shown in Tab.2.

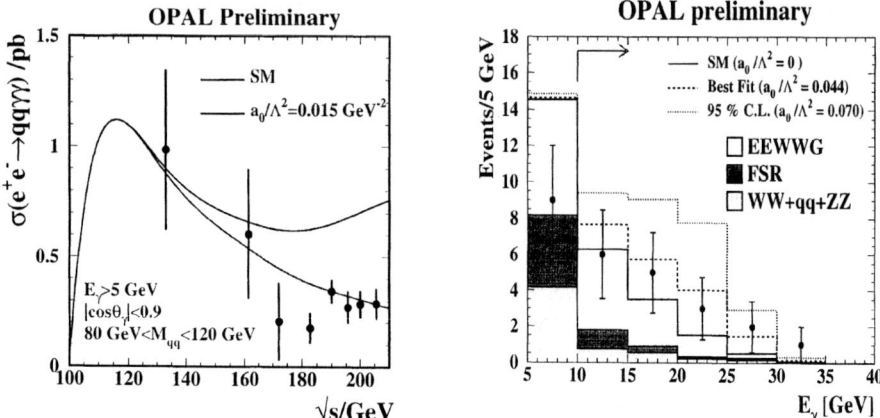

FIGURE 3. The cross section for $q\bar{q}\gamma\gamma$ production is shown in the first plot as a function of the centre-of-mass energy. The second plot illustrates the energy spectrum of photons in candidate $W^+W^-\gamma$ events. The data are compared with the expected distributions for different values of the anomalous couplings.

95% C.L. bounds on anomalous QGCs			
Charged vertex $W^+W^-\gamma\gamma$			
		a_0/Λ^2	a_c/Λ^2
$WW\gamma$	L	[-0.045, 0.045]	[-0.08,0.13]
	O	[-0.070, 0.070]	[-0.13,0.19]
$\gamma\gamma$	A	[-0.045,0.042]	[-0.115,0.115]
$+\not{E}_T$	L	[-0.041,0.040]	[-0.12,0.12]
	O	[-0.086,0.085]	[-0.23,0.23]
	All	[-0.037,0.036]	[-0.077,0.095]
Neutral vertex $Z^0Z^0\gamma\gamma$			
		$(a_0'/\Lambda^2)\cdot 10^2$	$(a_c'/\Lambda^2)\cdot 10^2$
$Z\gamma\gamma$	L	[-0.6,0.6]	[-0.6,1.0]
	O	[-0.6,0.8]	[-0.8,1.2]
	All	[-0.48,0.56]	[-0.52,0.99]

TABLE 2. Preliminary constraints on anomalous QGCs from Aleph (A), L3 (L) and Opal (O). The preliminary combined results (All) are supplied by the LEP TGC working group. Units are GeV^{-2}.

Measurement of gauge boson couplings at future colliders

At the FLC the statistical sensitivity to the NTGCs should be improved by more than two orders of magnitudes with respect to LEP2, thanks to the increase in luminosity and to the growing contributions of the NP amplitudes [1] as a function of the centre-of-mass energy.

We can also expect much improved constraints on charged TGCs. The sensitivity is enhanced by the use of beam polarisation. The estimated limits on possible deviations of these couplings are in the range 10^{-3}–10^{-4}, where we can expect contributions of NP interactions in the effective Lagrangian [9].

The reaction $\gamma\gamma \to W^+W^-$ would be the dominant source of W pairs at FLCs, provided that the PLC option will be realised. A PLC can be really considered as a W-factory and an ideal place to conduct precision tests on quartic couplings of the W boson. The one-loop SM contribution to $\gamma\gamma \to W^+W^-$ will be measurable and the limits on possible deviations of the a_0/Λ^2 and a_c/Λ^2 parameters from their SM values will be improved by at least two orders of magnitude with respect to LEP2 both for the charged vertex, $\gamma\gamma \to W^+W^-$, and for the neutral one, $\gamma\gamma \to ZZ$ [10].

Search for a fermiophobic Higgs boson

In the case of SM Higgs boson, $h \to \gamma\gamma$ proceeds via a top quark or a W boson loop and the relative branching ratio $(\mathcal{O}(10^{-3}))$ is too small to observe the process at LEP2. On the other side an enhanced photonic decay of Higgs bosons could be detected as a resonance in the di-photon mass spectrum. The results of this analysis have been interpreted in the context of the fermiophobic scenario, in which the coupling of the Higgs boson to fermions is strongly suppressed and the dominant decay of a light Higgs boson would be to two photons [11]. OPAL has performed at centre-of-mass energies up to 208 GeV a search for the process $e^+e^- \to XY$, with $X \to \gamma\gamma$ and $Y \to f\bar{f}$, where $f\bar{f}$ may be quarks, charged leptons, or a neutrino pair [12]. In the SM, Y is a Z and X is a Higgs boson decaying into two photons. The results of this analysis are combined with the ones from other LEP experiments to yield a limit of 106.4 GeV [13], for the mass of a fermiophobic Higgs boson, under the hypothesis of SM production cross-section and partial widths for the decay to fermions set to zero.

OPAL at 189 GeV has also performed a more general search by removing the restriction that Y is a Z, achieving sensitivity to the Higgs boson pair production process $e^+e^- \to hA$, where h decays into a pair of photons and A into a pair of fermions [14]. No excess over the SM background has been found and limits have been derived as a function of (M_X, M_Y).

At the future accelerators production and decays channels involving photons will have a fundamental role in the search and possibly in the study of the Higgs boson properties. At the LHC, for masses of the Higgs boson less than 140 GeV, the decay channel with the highest signal-to-background ratio is $h \to \gamma\gamma$. The signal

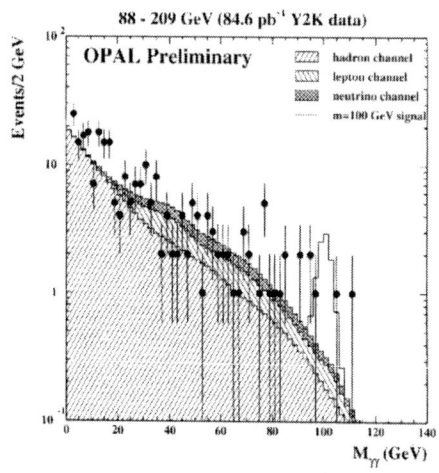
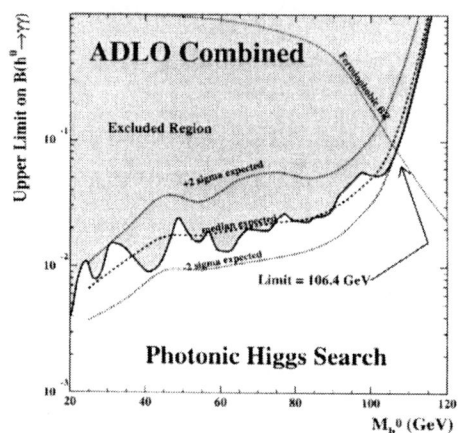

FIGURE 4. The first plot shows the distribution of the invariant mass of the two highest-energy photons in the hZ search. The white histogram represents the expected contribution of the signal for a fermiophobic Higgs boson with M_H=100 GeV. In the second plot the LEP combined 95% confidence level upper limit on the di-photon branching fraction is given. The shaded region is excluded. The intersection of the dotted line with the exclusion curve gives a lower limit of 106.4 GeV for the fermiophobic Higgs model.

significance in each of the two experiments, ATLAS and CMS, should be larger than five for $130 < M_H < 140$ GeV and an integrated luminosity of 100 fb^{-1} per experiment [15].

The PLC offers a unique possibility to produce the Higgs boson as an s-channel resonance: $h \rightarrow b\bar{b}$, WW^* or ZZ, through one-loop level diagrams involving all heavy charged particles which obtain their masses from Electroweak symmetry breaking. The measurement of $\Gamma(h \rightarrow \gamma\gamma)$ at the PLC, combined with the one of BR($h \rightarrow \gamma\gamma$) at the LHC and the FLC, will allow to calculate the total Higgs boson width with an accuracy of about 14% [16].

Conclusions

We have presented analyses of photonic final states at LEP2 aimed to study gauge boson couplings. All the measured values for the neutral triple gauge boson couplings and the quartic gauge boson couplings are in agreement with the SM predictions. We expect that the limits on possible deviations of these parameters will be improved by at least two orders of magnitude at the FLC and the PLC.

A search for a fermiophobic Higgs boson has been performed with events with two photons in the final state. No excess has been found over the expectation for the SM background and limits have been set on the mass of the fermiophobic Higgs boson.

Studies of production and decay channels of the Higgs boson involving photons are an important issue since they will be fundamental at the LHC and the PLC in the understanding of the mechanism of the Electroweak symmetry breaking.

REFERENCES

1. G.J.Gounaris, J. Layssac and F.M.Renard, *Phys. Rev.* **D61** (2000) 73013.
   G.J.Gounaris, J. Layssac and F.M.Renard, "New and Standard Physics contributions to anomalous Z and γ self-couplings", hep-ph/0003143, March 2000.
2. OPAL Collab., G. Abbiendi *et al.*, "Search for Trilinear Neutral Gauge Boson Couplings in Zγ production at $\sqrt{s} = 189$ GeV at LEP", CERN-EP-2000-0607; Submitted to *Eur. Phys. J.*.
3. OPAL Collab., G. Abbiendi *et al.*, "Photonic Events with Missing Energy in e^+e^- Collisions at $\sqrt{s} = 189$ GeV", CERN-EP-2000-060; Submitted to *Eur. Phys. J.*.
4. "Combined Preliminary Results on Electroweak Gauge Boson Couplings Measured by the LEP Experiments", the LEP Collaborations ALEPH, DELPHI, L3, OPAL and the LEP TGC Working Group.
5. W.J.Stirling, A. Werthenbach, *Eur. Phys. J.* **C14** (2000) 103 and references therein.
   G. Bélanger *et al.*, *Eur. Phys. J.* **C13** (2000) 283.
6. The OPAL Collaboration, "Measurement of the $e^+e^- \rightarrow q\bar{q}\gamma\gamma$ Cross-Section and Limits on Anomalous Electroweak Quartic Gauge Couplings", OPAL PN452
7. The OPAL Collaboration, *Phys. Lett.* **B 471** (1999) 293-307
8. The OPAL Collaboration, "Constraints on Anomalous Quartic Gauge Boson Couplings from Acoplanar Photon Pair Events", OPAL PN410.
9. K. Mönig "Beam polarisation for Electroweak Physics", 2nd ECFA/DESY study on Physics and Detectors for a Linear Electron-Positron Collider, Obernai, France, 16-19 Oct 1999.
10. S. Godfrey "Quartic Gauge Boson Couplings", proceedings of the "International Symposium on Vector Boson Self-Interactions", *American Inst. Phys.*, (1996) 410.
11. A.G. Akeroyd, *Phys. Lett.* **B368** (1996) 89.
    A. Stange, W. Marciano and S. Willenbrock, *Phys. Rev.* **D49** (1994) 1354.
    H. Haber, G. Kane and T. Sterling, *Nucl. Phys.* **B161** (1979) 493.
    J.F.Gounion, R. Vega and J. Wudka, *Phys. Rev.* **D42** (1990) 1673.
12. OPAL Collab., "Search for Higgs Bosons in e^+e^- Collisions at the Highest LEP energies " OPAL PN450 (2000)
13. "Searches for Higgs bosons: Preliminary combined results using LEP data collected at energies up to 209 GeV", the LEP Collaborations ALEPH, DELPHI, L3, OPAL and the LEP working group for Higgs boson searches
14. The OPAL Collaboration, *Phys. Lett.* **B 464** (1999) 311-322
15. CMS Collaboration, CMS Technical Proposal, CERN/LHCC/94-38 (1994)
16. "Light Higgs production at the Compton Collider", G. Jikia and S. Soldner-Rembold, *Nucl. Phys. Proc. Suppl.* **82** (2000) 873.

VECTOR PROPERTIES OF THE CENTRAL PRODUCTION POMERON

Alexander Singovski[a]

on behalf of the WA102 collaboration

[a]LAPP, Annecy-le-Vieux, France
(now in University of Minnesota)

Abstract. The data coming from the CERN WA102 experiment, update our understanding of the central production process. The "naïve" Double Pomeron Exchange (DPE) picture with the "vacuum Pomeron" do not fit to the experimental data. The non-uniform azimuthal angle distributions require the exchange particle spin grater that 0. The model of Close and Schuller with the exchange particle (Pomeron) transforming like a non-conserving vector current, fits well to the experimental data.

Introduction

The WA102 [1] experiment at CERN SPS was designed to study the meson states production in the central proton-proton collisions at 450 GeV. The mesons where produced in the reaction:

$$pp \rightarrow p_f X p_s,$$

where subscripts f and s refer to the fastest and slowest particle in the laboratory frame respectively and X represents the central system. The trigger required a certain rapidity gap between the fast and slow protons and any of the meson state decay particles (fig.1), which means that all reaction products except the fast and slow protons should be grouped at the *center* of the rapidity distribution (i.e. *central production* reaction)

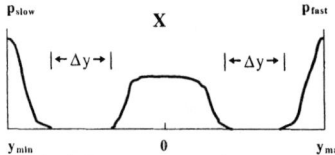

FIGURE 1. Sketch of the typical Central Production rapidity distribution

On the Regge model language such a rapidity gap selection favors the Double Reggeon exchange process [2]. Regge Model predicts that all exclusive Reggeon-Reggeon process cross-sections are diminish with the energy as $\sigma(s) \sim s^{-1}$, except for the Pomeron-Pomeron one, where $\sigma(s) \sim Const$. It was suggested that the Double

CP571, *PHOTON 2000*, edited by A. J. Finch
© 2001 American Institute of Physics 0-7354-0010-5/01/$18.00

Pomeron process is dominated the Central Production already at the CERN SPS energies with $\sqrt{s} \approx 30$ GeV. And once the Pomeron was thought to be essentially a multi-gluon exchange, the Central Production was considered as a process, enriched by the gluonic states, hence was the glueballs production process.

The WA102 experiment, last one in the CERN Central Production series, has supplied a reach sample of the high quality data [3] which can be used to verify a consistency of the initial suggestions.

WA102 experiment results

The total central production cross-sections for mesons investigated by the WA102 experiment , summarized by A.Kirk [4], are presented in table 1. It is clear that the

J^{PC}	Resonance	σ(nb) at $\sqrt{s}$=29.1 GeV	DPE compatible
0^{-+}	π_0	22 011±3 267	
	η	3 859± 368	
	η'	1 717±184	Yes
0^{++}	$a_0(980)$	638 ± 60	
	$f_0(980)$	5 711 ± 450	Yes
	$f_0(1370)$	1 753 ± 580	
	$f_0(1500)$	2 914 ± 301	Yes
	$f_0(1710)$	245 ± 65	
	$f_0(2000)$	3 139 ± 480	
1^{++}	$a_1(1260)$	10 011 ± 900	
	$f_1(1285)$	6 857 ± 1 306	Yes
	$f_1(1420)$	1 080 ± 385	Yes
1^{--}	$\rho(770)$	3 102 ± 250	No
	$\omega(782)$	7 440 ± 553	
	$\phi(1020)$	60 ± 21	No
2^{-+}	$\pi_2(1670)$	1 505 ± 145	
	$\eta_2(1645)$	1 907 ± 152	
	$\eta_2(1870)$	1 940 ± 185	
2^{++}	$a_2(1320)$	1684 ± 134	
	$f_2(1270)$	3 275 ± 422	Yes
	$f_2'(1520)$	68 ± 9	
	$f_2(1910)$	528 ± 40	
	$f_2(1950)$	2 788 ± 175	Yes
	$f_2(2150)$	121 ± 12	

Table 1. Summary of resonance central production. The "DPE compatible" value is NO for the resonances with the central production cross-section decreasing with energy and YES ones with rising up or constant.

DPE is not the dominant process at 29.1GeV. For example, the production cross-section for the $\omega(782)$ meson, which cannot be produced via DPE, is more than twice larger than the cross-section for the $f_0(1500)$, the most solid glueball candidate.

One of the consequences of the Double Pomeron description of the Central meson production is the intrinsic symmetry of this process. Fusion of two identical objects with the vacuum quantum numbers can create only the states with $J^{PC}=(Even)^{++}$ and should not have any production angle asymmetry.

This symmetric picture of the Central Production was broken first by the WA102 observation of the resonance production dependence on the fast and slow protons transverse momenta [3]. The best variable to describe this effect was found to be dp_T, difference in the Pomerons transverse momenta (fig. 2). The mass spectra for the low and high dp$_T$ regions where found significantly different, some states well present at low dp$_T$ disappeared at high ones and vice-versa.

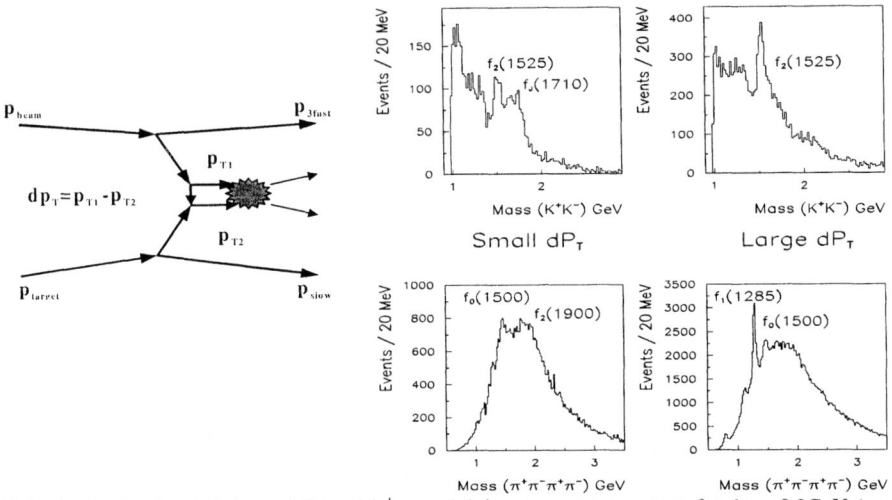

FIGURE 2. The dp$_T$ definition (left) and K$^+$K- and 4pi systems mass spectra for dp$_T$ <0.2GeV (small dp$_T$) and dp$_T$>0.5GeV (large dp$_T$)

To explain this effect, Close and Kirk have suggested that the dp$_T$ parameter is sensitive to the Pomeron-Pomeron fusion mechanism, which can be mediated by quark or gluon. The first case is enhanced at large dp$_T$, the second one – at small ones.

Even more dramatic deviation from the naïve "vacuum Pomeron"picture was observed in the azimuthal angle ϕ-distributions, where ϕ is defined as an angle between the momentum vectors of the outgoing protons. The ϕ-distribution should be flat for the vacuum quantum numbers exchange, which is not the case for the experimental data (see fig. 3-6).

Several theoretical papers have been published on these effects [5,6]. All agree that the exchanged particle must have J>0 and that J=1 is the simplest explanation for the observed ϕ distributions. We will discuss in more details the calculations of Close and Schuller [6]. Their model not simply described experimental distributions rather well, but gave a certain predictions, which were checked with the data [6].

Model of the Vector Pomeron

The model of Close and Schuller consider the Central Production of meson M in the scattering of two fermions proceeding through the fusion of two vector currents. It predicts the production cross-section dependence on the azimuthal angle ϕ and t$_1$ and t$_2$ – four-momentum transfer at the beam-fast and target-slow vertices respectively in a

form of $\dfrac{d\sigma}{dt_1 dt_2 d\phi} \sim G_E^{p\,2}(t_1) G_E^{p\,2}(t_2) F^2(t_1, t_2) A(t_1, t_2, \phi)$, where G$^P_E$(t) is the proton-Pomeron form-factor, F^2(t$_1$,t$_2$) is the Pomeron-Pomeron-meson form factor and A(t$_1$,t$_2$,ϕ) is a non-conserving vector current. The predictions for different J^{PC} states production are summarized in table 2.

0^{-+}	$A(t_1,t_2,\phi) = t_1 t_2 \sin^2(\phi)$ $F^2(t_1,t_2) = e^{-b_T(t_1+t_2)}$
1^{++}	$A(t_1,t_2,\phi) = (\sqrt{t_1}-\sqrt{t_2})^2 + 4\sqrt{t_1 t_2}\sin^2(\phi/2)$ $F^2(t_1,t_2) = e^{(-b_T t_1 + b_L t_j)}$
2^{-+}	$A(t_1,t_2,\phi) = (\sqrt{t_1}-\sqrt{t_2})^2 + 4\sqrt{t_1 t_2}\cos^2(\phi/2)$ $F^2(t_1,t_2) = e^{(-b_T t_1 + b_L t_j)}$
$0^{++} 2^{++}$	$F^2(t_1,t_2)\,A(t_1,t_2,\phi) =$ $t_1 t_2\,[e^{-b_L(t_1+t_2)/2} + \sqrt{t_1 t_2}/\mu^2\,e^{-b_T(t_1+t_2)/2}\cos(\phi)]^2$

Table2. Summary of the Close and Schuller model prediction for different J^{PC}

The model has only 3 free parameters, two if which, b_T and b_L, having meaning of the transversal and longitudinal Pomeron-Pomeron form-factors, can be fixed from the pseudo-scalar and vector meson production data. The experimental distributions and the fit results for 0^{-+} and 1^{++} systems are shown in figures 3 and 4 respectively.

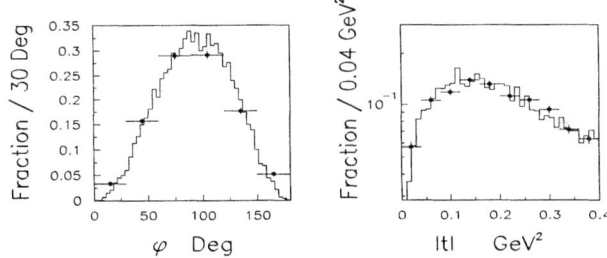

FIGURE 3. ϕ- and t-distributions for the $\eta'(958)$ production (dots), The curve is the fit results with the $b_T=0.5\text{GeV}^{-2}$

As it can be seen from the Table 2 formulas for the 1^{++} state, when $|t_2-t_1|$ is small $d\sigma/d\phi$ should be proportional to $\sin^2\phi/2$ while when $|t_2-t_1|$ is large $d\sigma/d\phi$ should be constant. The two bottom pictures of the figure 4 show that the expected trend is observed in the data.

The fit of the η' t-distribution to the exponential function (Table 2) give $b_T=0.5\text{GeV}^{-2}$. Using this value for the fit to the 1^{++} state t-distribution one can get $b_L=3\text{GeV}^{-2}$. The model prediction for the 2^{-+} state is now parameter free. The ϕ- and t-distributions for $\eta_2(1645)$-meson and the model prediction with b_L and b_T values mentioned above is shown in figure 5. The model fits to the data rather well.

451

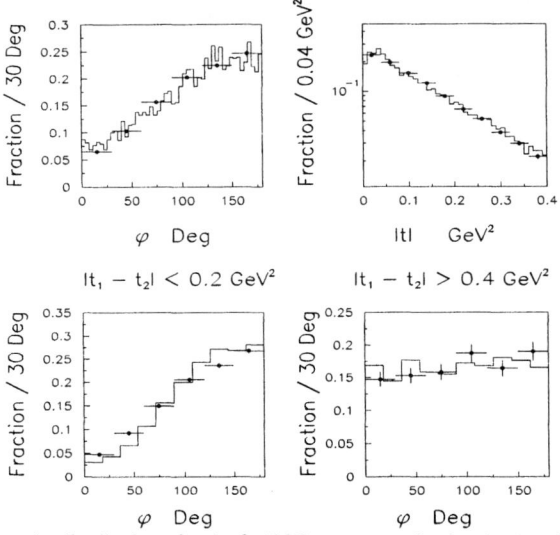

FIGURE 4. φ- and t-distributions for the $f_1(1285)$ meson production (top) and φ-distributions for different $|t_1-t_2|$ regions (bottom)

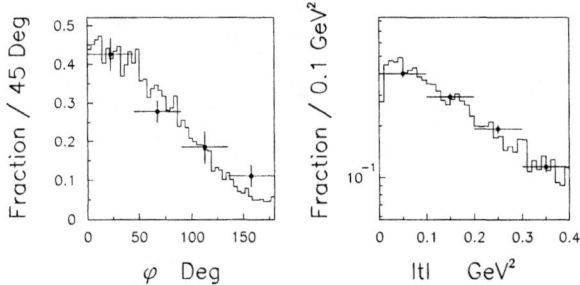

FIGURE 5. φ- and t-distributions for the $\eta_2(1645)$ production (dots), the curve is the fit to the model prediction (Table2) with the $b_L=0.3GeV^{-2}$ and $b_T=0.5GeV^{-2}$

The scalar and tensor meson production case is little more complicated. Two types of the φ-dependence where found experimentally: rising from small to large angles for $f_0(1370)$ and $f_2(1270)$ and falling down for $f_0(1500)$ and $f_2(1950)$ (see figure 6). These dependencies can be described by the model by varying μ^2 parameter. The fit sown in figure 6 give the following values:

Resonance	$f_0(1370)$	$f_0(1500)$	$f_2(1270)$	$f_2(1950)$
μ^2	-0.5	+0.7	-0.4	+0.7

The physical meaning of the parameter μ^2 is not clear. But it is interesting to mention that the sign and the value of this parameter is different for the glueball candidates $f_0(1500$ and $f_2(1950)$ and for the (most probably) "ordinary" $\theta\theta$-mesons $f_2(1270)$ and $f_0(1370)$.

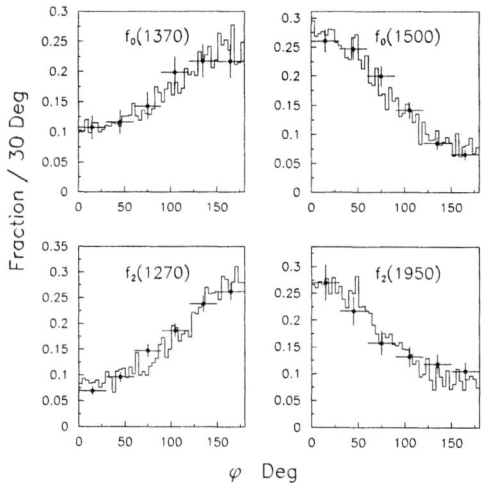

φ Deg

FIGURE 6. ϕ-distributions for the $f_0(1370)$, $f_0(1500)$, $f_2(1270)$ and $f_2(1950)$ mesons (dots). The curve is the fit to the model prediction (Table2) with the $b_L=0.3GeV^{-2}$ and $b_T=0.5GeV^{-2}$

Summary

The data coming from the CERN Ω-spectrometer Central Production experiment series:WA76, WA91, WA102, significantly update our understanding of the central production process. The "naïve" picture with the vacuum Pomerons do not fit to the experiments. The non-uniform azimuthal angle ϕ-distributions, where ϕ is defined as an angle between the momentum vectors of the outgoing protons, require the exchange particle spin grater that 0. The model of Close and Schuller with the exchange particle (Pomeron) transforming like a non-conserving vector current, fits well to the experimental data. One of the model parameters-μ^2 has different sign and values for the glueball candidates and $\theta\theta$-mesons and probably connected to the gluon content of the meson state.

References

1. T.A.Armstrong *et al.*, *Nucl. Instr. Meth.* **A274**, 165 (1989); F.Antinori *et al.*, *Nuove Cimento* **A107**, 1857 (1994).
2. D.Robson, *Nucl. Phys.* **B130**, 328 (1977); F.E.Close, *Rep. Prog. Phys.*, **51**, 833 (1988).
3. D.Barberis et al., *Phys. Lett.*, **B413**, 214 (1997); *Phys. Lett.*, **B413**, 225 (1997); *Phys. Lett.*, **B422**, 399 (1998); *Phys. Lett.*, **B427**, 398 (1998); *Phys. Lett.*, **B432**, 436 (1998); *Phys. Lett.*, **B436**, 204 (1998); *Phys. Lett.*, **B440**, 225 (1998); *Phys. Lett.*, **B446**, 342 (1999); *Phys. Lett.*, **B453**, 305 (1999); *Phys. Lett.*, **B453**, 316 (1999); *Phys. Lett.*, **B453**, 325 (1999); *Phys. Lett.*, **B462**, 462 (1999); *Phys. Lett.*, **B467**, 165 (1999); *Phys. Lett.*, **B471**, 429 (2000); *Phys. Lett.*, **B471**, 435 (2000); *Phys. Lett.*, **B471**, 440 (2000); *Phys. Lett.*, **B474**, 423 (2000); *Phys. Lett.*, **B479**, 59 (2000).
4. A.Kirk, *Phys. Lett.*, **B489**, 29 (2000).
5. P.Castoldi et al., *Phys. Lett.*, **B425**, 359 (1998); J.Ellis and D.Kharzeev, hep-ph/9811222, N.I.Kochelev, hep-ph/9902203; N.I.Kochelev, T.Morii and A.V.Vinnikov, hep-ph/9903279.
6. F.E.Close and G.Schuller, *Phys. Lett.*, **B464**, 279 (1999); F.E.Close, A.Kirk and G.Schuller, *Phys. Lett.*, **B477**, 13 (1999); F.E.Close and A.Kirk, *Phys. Lett.*, **B483**, 345 (2000);

SUMMARY

Summary of Photon 2000

Michael Krämer[a] and Stefan Söldner-Rembold[b]

[a] *Department of Physics and Astronomy, The University of Edinburgh, EH9 3JZ, Scotland*
[b] *CERN, CH-1211 Geneva 23, Switzerland, and Heisenberg Fellow*

Abstract. This summary of the International Conference on the Structure and Interactions of the Photon (Photon 2000) encompasses experimental and theoretical results on photon structure functions, inclusive and exclusive processes, photon diffraction and total cross sections as well as photon physics at a future Linear Collider.

I INTRODUCTION

The physics of photon interactions at high energies is becoming increasingly diverse. Photon-photon interactions at LEP and CLEO and photon-proton interactions at HERA are prolific testing grounds for Quantum Chromodynamics (QCD). Current studies in QCD are often motivated by the need to control background reactions and cross-section predictions for new physics searches. QCD, however, may also be thought of as a paradigm for a rich and complex quantum field theory, exhibiting features like asymptotic freedom and confinement. Insights into the various aspects of QCD will thus lead to new insights into the nature of quantum field theory. Photon 2000 covered the whole range of QCD studies in high-energy photon interactions, from photon structure functions and inclusive reactions to exclusive and diffractive processes.

In the future, many new opportunities for photon physics will open up at a future Linear Collider. Plans for a Photon Collider as part of the Linear Collider project were discussed, where high energy photon beams are obtained by scattering laser photons off electron beams. One of the most exciting possibilities of such a facility is to produce Higgs particles as resonances in photon-photon scattering, a powerful way of determining fundamental Higgs properties.

II STRUCTURE FUNCTION OF THE REAL PHOTON

Probing the structure of the photon in deep-inelastic electron-photon scattering is a classical QCD measurement [1]. The high energies accessible at LEP, the

CP571, *PHOTON 2000*, edited by A. J. Finch

application of new analysis techniques and the improvements in Monte Carlo generators have considerably increased the kinematic reach and the precision of the measurements.

At LEP the virtuality of the "probing" photon is usually denoted by $Q^2 = -q^2$ (the negative squared four-momentum of the photon) and the virtuality of the "probed" photon is $P^2 = -p^2 \approx 0$ GeV^2. Just like for the proton, the deep-inelastic scattering cross-section is written as

$$\frac{d^2\sigma_{e\gamma \to e+hadrons}}{dx dQ^2} = \frac{2\pi\alpha^2}{x\,Q^4}\left[\left(1 + (1-y)^2\right) F_2^\gamma(x, Q^2) - y^2 F_L^\gamma(x, Q^2)\right], \qquad (1)$$

where α is the fine structure constant, x and y are are the usual dimensionless variables of deep-inelastic scattering and $W^2 = (q + p)^2$ is the squared invariant mass of the hadronic final state. The scaling variable x is given by

$$x = \frac{Q^2}{Q^2 + W^2 + P^2}. \qquad (2)$$

The term proportional to $F_L^\gamma(x, Q^2)$ is small and is therefore usually neglected. The structure function $F_2^\gamma(x, Q^2)$ can be identified with the sum over the parton densities of the photon weighted by the square of the parton's charge. As a consequence deep-inelastic electron-photon scattering mainly probes the quark structure of the photon, gluons only enter through scaling violations.

The measurement of hadronic structure functions is difficult due to the necessity to obtain x from the reconstruction of the hadronic final state in the detector (Eq. 2) using regularized unfolding. Especially at low x where the hadron-like component of the structure function dominates, the hadronisation process is not well modelled by Monte Carlo generators leading to large systematic uncertainties. However, significant progress has recently been made in reducing the systematic uncertainties due to unfolding of detector effects and hadronisation corrections. ALEPH, L3 and OPAL have compared their combined data to the PHOJET and HERWIG generators [2]. An unbiased tune using informations from HERA has improved HERWIG significantly.

Furthermore new methods for regularised unfolding like the maximum entropy method or the singular value decomposition method have been used. ALEPH [3] and OPAL [4] have introduced two-dimensional unfolding and improved treatment of hadronic energy in the forward region. L3 is applying energy-momentum conservation using kinematic information from both hadrons and the electrons.

The uncertainty on the measurements at low x shown in Fig. 1a are therefore considerably reduced [4,5]. The hadron-like component dominates at low x and there could be a first indication of the low x rise of the photon structure function expected from QCD evolution. The F_2^γ data are compared to the GRV-LO [6], SaS-1D [7] and the WHIT1 [8] which all describe the data reasonably well within the uncertainties.

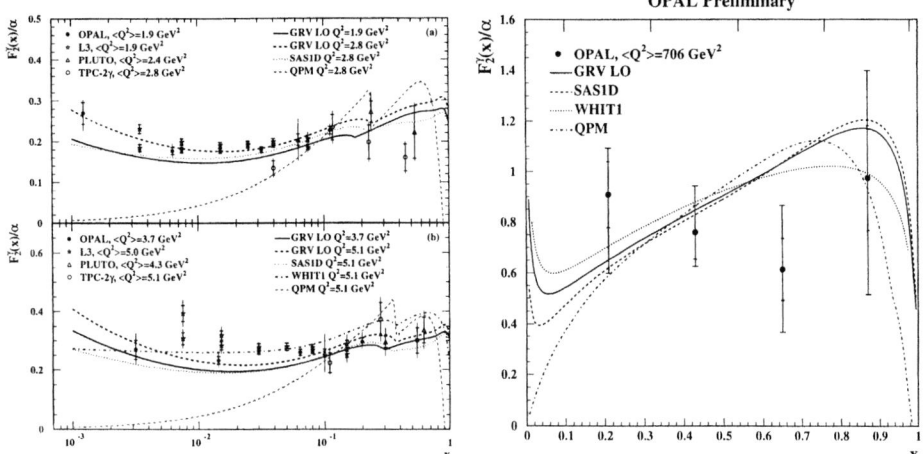

FIGURE 1. a) The measured hadronic structure function F_2^γ at low Q^2 compared to the GRV-LO [6], SaS-1D [7] and the WHIT1 [8] parametrisations and to the Quark Parton Model (QPM). In the case of L3 the values obtained using PHOJET and TWOGAM for unfolding are shown separately. b) preliminary high Q^2 measurement of F_2^γ presented by OPAL at this conference.

The only new high Q^2 measurement presented at this conference is shown in Fig. 1b at the currently highest measured $\langle Q^2 \rangle$ of about 706 GeV^2. The uncertainties are currently too large to distinguish between different parametrisations which all describe the data well in this kinematic region where the point-like contribution to F_2^γ is expected to be dominant.

It was pointed out at this conference that genuine QCD effects can be studied through the x dependence of the slope $dF_2^\gamma(x, Q^2)/dQ^2$ [10]. The Q^2 dependence of the structure function F_2^γ in bins of x is shown in Fig. 2 for all currently available measurements. The data are compared to the GRV-HO and the SaS-1D parametrisation, and to the sum of the asymptotic prediction [11] for 3 light flavours and the point-like part of the charm structure function taken from GRV. Positive scaling violation of the photon structure function is observed in all x ranges - different from the proton - due to the 'regular' QCD evolution at low x and due to the inhomogeneous term ($\gamma \to q\bar{q}$) at larger x. As expected, the asymptotic prediction fails to describe the data at low x, where the non-perturbative hadron-like contribution dominates, whereas all models give a reasonable description of the medium to high x, Q^2 region.

In Fig. 2a the charm threshold is clearly visible. Above the kinematic threshold for charm production, the c and u contribution to the point-like part of the photon structure function are of similar size. OPAL has measured the charm

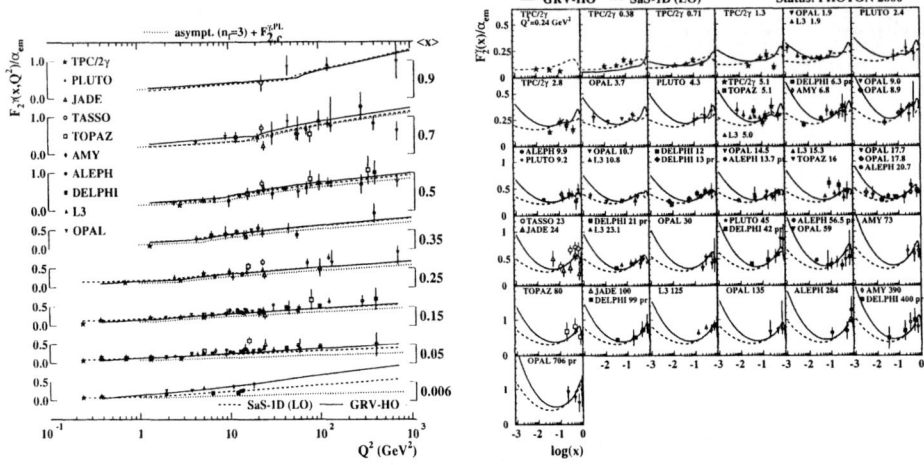

FIGURE 2. a) The Q^2 dependence of the hadronic structure function F_2^γ in bins of x compared to the GRV-HO, SaS-1D and the asymptotic prediction. b) the x dependence of F_2^γ in different Q^2 bins.

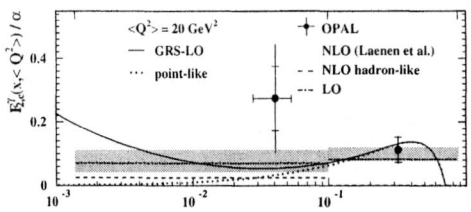

FIGURE 3. Charm structure function $F_{2,c}^\gamma$ of the photon as function of x for $\langle Q^2 \rangle = 20$ GeV^2.

structure function of the photon for the first time using D* decays [12]. The region $x > 0.1$ - which is dominated by the point-like component - is in good agreement with a NLO calculation [13]. In the region $x < 0.1$ the measurement suggests the existence of a hadron-like component with currently large errors. These uncertainties are expected to be significantly reduced in the future due to higher statistics and better MC modelling of charm production.

III INCLUSIVE PRODUCTION OF JETS AND HADRONS

A complementary approach to studying photon structure are measurements of inclusive cross-sections for the production of hadrons, prompt photons and jets in photon-photon and photon-proton interactions [15]. In all these processes the relevant scale is usually assumed to be the transverse momentum of the produced particles or jets.

New results on inclusive π^0 and K_s production in photon-photon colli-

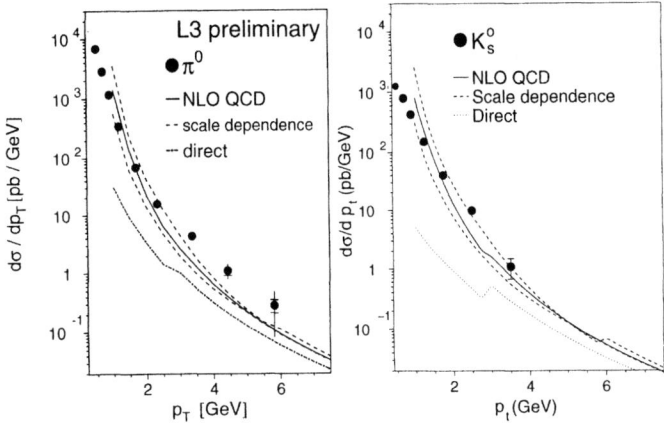

FIGURE 4. Differential cross-section $d\sigma/dp_T$ measured by L3 for the inclusive production of a) π^0 and b) K_s^0 mesons compared to NLO calculations.

sions as a function of the transverse momentum p_T of the hadron have been presented by L3 [14]. The results have also been compared to similar OPAL measurements of charged hadron and K_s production [16]. These measurements are important to test prediction of NLO perturbative QCD and to test the universality of fragmentation functions [17]. They also provide information about photon structure and they can also be used to improve Monte Carlo generators.

The additional direct component in photon-photon interactions leads to a harder p_T spectrum compared to purely hadronic interactions. This effect is clearly visible in the data (Fig. 4). The data are reasonably well described by the NLO calculations. As in the case of the photon structure function the charm contribution becomes important for transverse momenta p_T larger than the charm threshold.

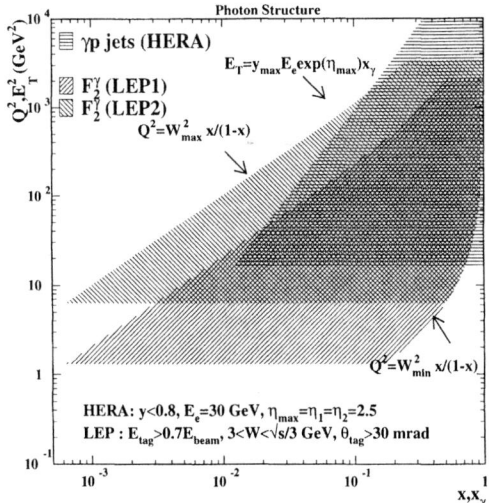

FIGURE 5. A comparison of the (Q^2, x) planes covered by LEP with the (E_T^2, x_γ) plane covered by HERA studying jet production.

Another process in which photon structure can be studied is the production of (di-)jets in photon-proton (HERA) or photon-photon (LEP) interactions. The interacting photons are almost real and the largest physical scale is the transverse energy of the jets. The variable x_γ which is related to the fraction of the photon's momentum participating in the hard interactions can be reconstructed from the pseudorapidities η^{jet} and transverse energies E_T^{jet} of the jets:

$$x_\gamma = \frac{E_T^{\text{jet1}} e^{-\eta^{\text{jet1}}} + E_T^{\text{jet2}} e^{-\eta^{\text{jet2}}}}{2 y E_e}, \qquad (3)$$

where $y E_e$ is the energy taken by the photon. At e^+e^- colliders the scattered electrons are not tagged to ensure that the photons are quasi-real. The two $x_\gamma^\pm$ values are therefore calculated from the hadronic final state using

$$x_\gamma^\pm = \frac{\displaystyle\sum_{\text{jets}=1,2} (E \pm p_z)}{\displaystyle\sum_{\text{hadrons}} (E \pm p_z)} \qquad (4)$$

where p_z is the longitudinal momentum component.

In leading order, x_γ is equivalent to x and we can relate the parton distributions probed in deep-inelastic electron-photon scattering and in jet production by $q_{\text{DIS}}(Q^2, x) \approx q_{\text{jet}}((E_T^{\text{jet}})^2, x_\gamma)$. In jet production gluon induced processes dominate the cross-section in most kinematic regions, i.e. different from deep-inelastic electron-photon scattering the results are directly sensitive to the gluon distribution in the photon.

In Fig. 5 the kinematic ranges accessible at LEP and HERA are compared. At HERA accessing the low x parton densities of the photon requires the reconstruction of jets at low E_T^{jet} and large η^{jet}. This is experimentally difficult. In addition, additional soft or hard interactions of the photon's and the proton's remnant can take place which need to be disentangled from the primary hard scattering process.

Inclusive jet cross-section in photon-photon interactions have previously been measured at TRISTAN [18] and by OPAL [19]. At this conference, new ALEPH [20] and OPAL measurements [21] have been presented which extend the kinematic range in x_γ and $(E_T^{\text{jet}})^2$ using the data taken at an e^+e^- centre-of-mass energy of 183 GeV (ALEPH) and 189-202 GeV (OPAL). Both experiments apply the KTCLUS jet finding algorithm. The data are corrected to the hadron level and compared to partonic NLO calculations [22]. Good agreement is found between data and Monte Carlo for the differential di-jet cross-section, $d\sigma/dE_T^{\text{jet}}$, as function of the the transverse energy E_T^{jet} apart from the lowest E_T^{jet} bin (Fig. 6a).

The hadronisation effects are much more visible in Fig. 6b where the x_γ distribution unfolded to the hadron level is compared to the NLO calculations [22] using the GRV parton densities. To separate the resolved and direct contribution, OPAL integrates over the regions $x_\gamma < 0.75$ and $x_\gamma > 0.75$. The deficit at low x_γ could indicate that the gluon density in the photon is underestimated but more systematic studies are needed.

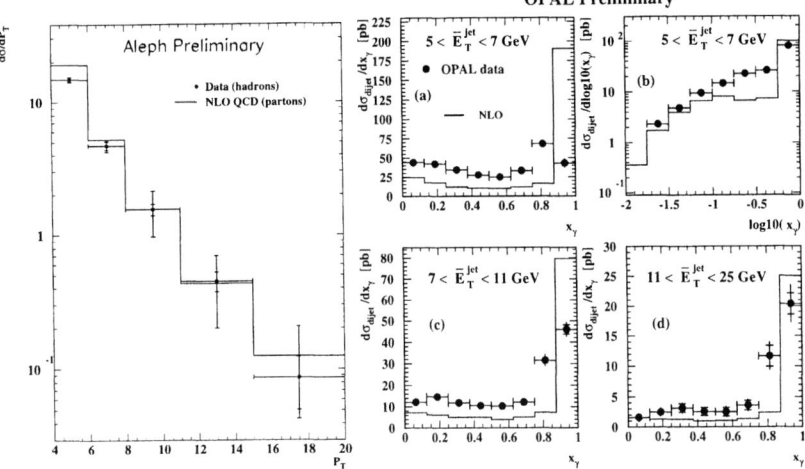

FIGURE 6. a) ALEPH: Differential di-jet cross-section, $d\sigma/dE_T^{jet}$; b) OPAL: The di-jet cross-section as a function of x_γ and $\log_{10} x_\gamma$ for different regions of mean transverse jet energy $(\bar{E}_T^{jet})^2$ of the two jets; In both cases the data are compared to NLO calculations by Klasen et al.

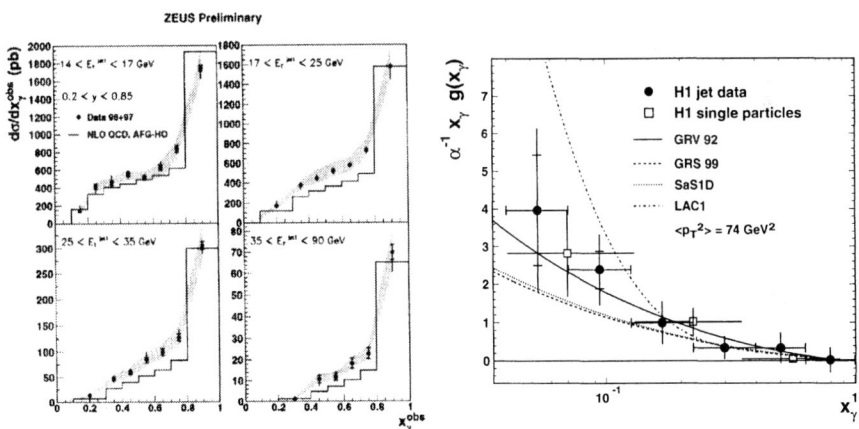

FIGURE 7. a) ZEUS: differential cross-section $d\sigma/dx_\gamma$ compared to a NLO QCD calculation using the AFG parton distributions. b) H1: the gluon distribution of the photon, $\alpha^{-1}x_\gamma g(x_\gamma)$ at an average scale p_T^2 of 74 GeV^2 (using di-jets) and 38 GeV^2 (using hadrons).

Fig. 7a shows a ZEUS measurement of the differential cross-section $d\sigma/dx_\gamma^{\text{obs}}$ for different E_T^{jet} bins using jets in the range $-1 < \eta_{\text{jet}} < 2$ and $Q^2 < 1 \text{ GeV}^2$ [23]. The NLO calculations using the AFG parton distributions [24] also lie systematically too low which could again indicate the need for more gluons in the parametrisations of the parton densities.

NLO calculations are performed at the parton level and contain no hadronisation effects and also no underlying event. In addition, scale uncertainties have to be taken into account. For the high E_T^{jet} region considered here, these effects are expected to be small enough so that the discrepancy between data and the NLO calculations can be attributed to inadequacies of the parametrisations of the parton densities.

The second approach to extract effective parton densities is shown in Fig. 7b. The effective parton densities are extracted from the di-jet data assuming a similar angular distribution for all resolved processes [26]. The effective parton density of the photon is given by

$$\tilde{q}(x_\gamma, p_T^2) \equiv \sum_{n_f} \left(q(x_\gamma, p_T^2) + \bar{q}(x_\gamma, p_T^2) \right) + \frac{9}{4} g(x_\gamma, p_T^2). \tag{5}$$

The LO quark density $q(x_\gamma, p_T^2) + \bar{q}(x_\gamma, p_T^2)$ is reasonable well constrained by e^+e^- data in this kinematic range and its contribution - shown as dashed line in Fig. 7 - is small. If the quark distribution is subtracted, a clear rise of the gluon distribution towards low x_γ can be observed.

IV PARTON DENSITIES OF THE VIRTUAL PHOTON

Until now we have studied the structure of (quasi-)real photons, i.e. $P^2 \approx 0 \text{ GeV}^2$ (LEP) or $Q^2 \approx 0 \text{ GeV}^2$ (HERA). The polarised and unpolarised structure functions of the virtual photon were also discussed at the conference [27]. In e^+e^- collisions the effective structure function of virtual photons can be measured if $Q^2 >> P^2 >> \Lambda_{\text{QCD}}^2$. This was first done by PLUTO [28]. For real photons only the cross-sections σ_{LT} and σ_{TT} contribute, where the indices refer to the longitudinal and transverse helicity states of the probe and target photon, respectively, i.e. $F_2^\gamma \simeq \sigma_{\text{LT}} + \sigma_{\text{TT}}$. For $P^2 >> 0$ other helicity states have to be taken into account, leading to the definition of the effective structure function $F_{\text{eff}}^\gamma \simeq \sigma_{\text{LT}} + \sigma_{\text{TT}} + \sigma_{\text{TL}} + \sigma_{\text{LL}}$ (interference terms are neglected here). This effective structure function has been measured by L3 and is shown in Fig. 8. We expect the non-perturbative part of the parton densities (VMD) at low x to decrease with increasing virtuality of the photon. Compared to the data as a function of P^2 in Fig. 8b, the QPM prediction therefore fails to describe the point at $P^2 = 0$. The shape of the P^2 dependence is consistent with the simple QPM ansatz but the errors are still large.

As in the case of the real photon, virtual photon structure can also be studied using jet production. At this conference ALEPH has presented a first LEP

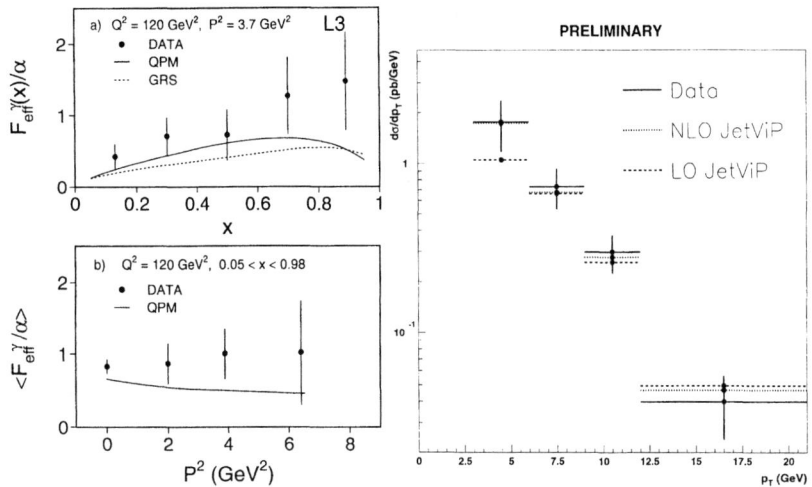

FIGURE 8. a) a) Effective structure function measured by L3 as function of x and P^2. b)

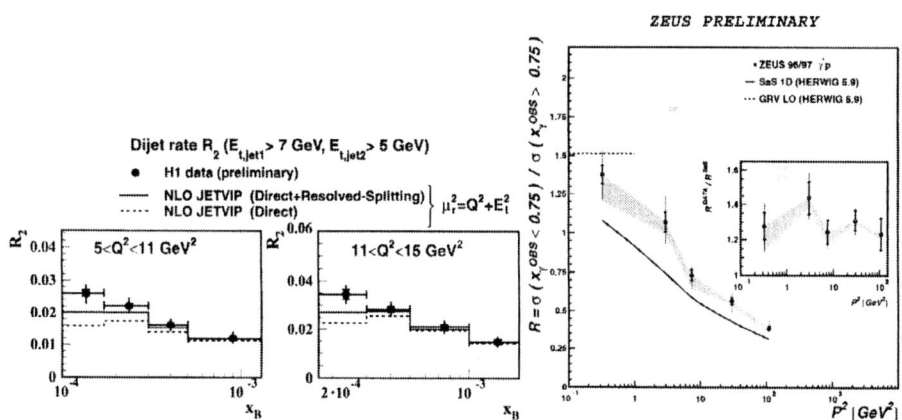

FIGURE 9. a) H1: di-jet rate as a function of x_{Bj} compared to JetViP with and without resolved contributions; b) ZEUS: the ratio of di-jet cross-section for $x_\gamma < 0.75$ and $x_\gamma > 0.75$ as a function of the photon virtuality P^2 (usually Q^2 in ep scattering) .

measurement of the di-jet cross-section for single-tagged events in the range $10 < Q^2 < 200$ GeV^2. The data in Fig. 8b are compared to the NLO calculation using JetViP [25]. The same calculation is used by H1 in Fig. 9a where di-jet rates as function of x_{Bj} are compared to a JetViP calculation with direct photon contributions only and after adding in a resolved contribution using SaS-1D parton distributions. Since $Q^2 < (E_{\mathrm{T}}^{\mathrm{jet}})^2$ the formalism of parton densities can be applied. Including the resolved contribution improves the data description. However, a big caveat has to added. The scale dependence of the NLO calculations is large leading to substantial uncertainties [29].

The decrease of the resolved photon contribution is demonstrated by ZEUS in Fig. 9b where the ratio of di-jet cross-section for $x_\gamma < 0.75$ and $x_\gamma > 0.75$ as a function of the photon virtuality P^2 (usually Q^2 in ep scattering) is plotted [30]. The jets are reconstructed in the γp frame and the minimum transverse energy for the two jets is $E_{\mathrm{T}}^{\mathrm{jet}1,2} > 7.5, 6.5$ GeV. The HERWIG prediction using the SaS-1D parametrisation [7] describes the P^2 dependence even though the normalisation is not in agreement with the data. This suggests that the SaS-1D model for the suppression of the resolved component describes the data.

The investigation of virtual photon structure has just started. Much more precise data are to be expected from LEP and HERA on the structure of the virtual photons in the next years.

V INCLUSIVE PRODUCTION OF CHARM AND BEAUTY

HERA has provided a wide spectrum of D* photoproduction measurements and first results on D$_\mathrm{s}$ mesons have been reported [31]. It is not guaranteed *a priori* that charm cross-sections can be reliably predicted in perturbation theory. The NLO calculations are indeed plagued by large uncertainties and, as shown in Fig. 10, they tend to underestimate the D* cross-section, in particular in the forward direction and at low x_γ where resolved photon processes contribute. Potentially large next-to-next-to-leading order corrections and higher-twist contributions may have to be included to improve the theoretical predictions. Other possible explanations for the discrepancies include non-perturbative string effects between the proton remnant and the charm quark or an enhancement of the low-x gluon component of the photon, which is currently not well constrained experimentally.

A similar picture has emerged in two-photon collisions at LEP [33–36]. The D* cross section is slightly underestimated by NLO theory in the experimentally visible region, in particular at low D* transverse momentum where resolved photon processes are prominent, see Fig. 11 (left). The L3 collaboration has presented a first measurement [34,37] of the differential $W_{\gamma\gamma}$ distribution for charm production. At larger $W_{\gamma\gamma}$ values this differential cross-section is completely dominated by the resolved channels, and can thus provide valuable information on the photonic gluon densities. From Fig. 11 (right) it is evident that the NLO calculation [38] reproduces

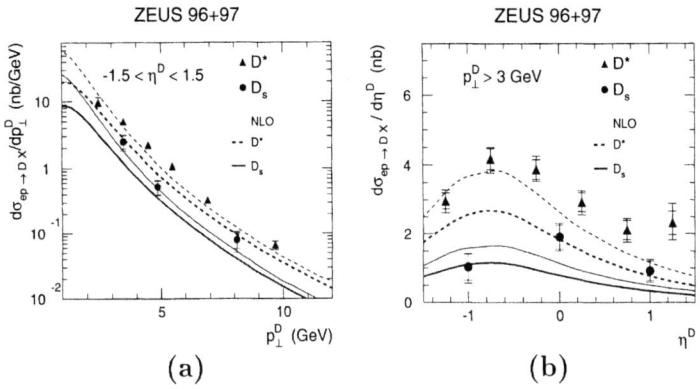

FIGURE 10. Differential cross-sections for the photoproduction reaction $ep \to DX$: (a) $d\sigma/dp_\perp^D$ and $d\sigma/d\eta^D$, where D stands for D* or D_s. Inner (outer) error bars show statistical (statistical and systematic added in quadrature) errors. The D_s (dots) and D* (triangles) data are compared with NLO predictions for D_s (full curves) and D* (dashed curves) with two parameter settings: $m_c = 1.5$ GeV, $\mu_R = m_\perp$ (thick curves) and $m_c = 1.2$ GeV, $\mu_R = 0.5m_\perp$ (thin curves). See [31] for further details.

the shape of this distribution rather well, but to agree in normalization a small charm mass of $m_c \approx 1.3$ GeV is required.

Due to the larger b quark mass, the theoretical predictions should be under better control for the beauty cross-section. Recent HERA and LEP measurements, however, show that the beauty photoproduction cross-section exceeds the expectation from NLO calculations. The results from the HERA collaborations [32] are displayed in Fig. 12, together with the NLO QCD prediction. The experimental results have been confirmed by a new and independent H1 analysis [32] based on a different data sample and a different tagging technique (microvertex-detection).

A large beauty photoproduction cross-section has also been observed recently in two-photon collisions at LEP. The L3 [34] and OPAL [33] measurements are compared to NLO QCD calculations in Fig. 13. Two predictions are presented, corresponding to different choices of the heavy quark mass renormalization scheme: the on-shell scheme (lower band) which has been used in previous calculations [38] and the $\overline{\text{MS}}$ scheme (upper band). Although formally of next-to-next-to-leading order, the difference between the two schemes is numerically significant, and the $\overline{\text{MS}}$ scheme in general gives a larger cross-section prediction. Still, for either renormalization scheme is the NLO calculation below the experimental data. Resolving this discrepancy remains as a challenge for the future.

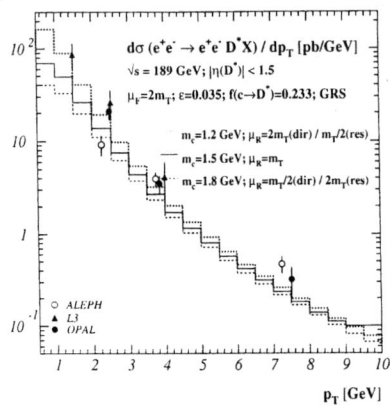 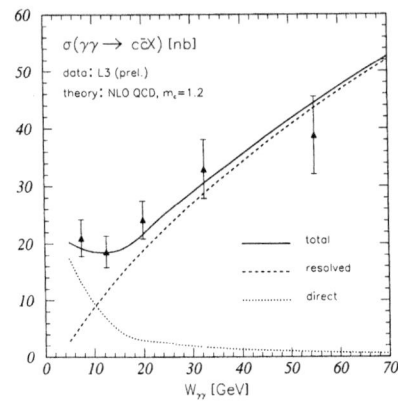

FIGURE 11. Left: comparison between the NLO QCD prediction and the OPAL/L3/ALEPH data [33–35] for the D* transverse momentum distribution. Right: comparison between the NLO QCD prediction ($m_c = 1.3$ GeV) and the L3 data [34] for the differential $W_{\gamma\gamma}$ distribution for inclusive charm production. See [38] for further details on the calculation.

VI   EXCLUSIVE PROCESSES

Exclusive reactions provide a powerful tool to study QCD at the amplitude level and to extract information on fundamental quantities like hadronic wavefunctions. For many exclusive processes involving large mass scales one can derive a factorization theorem that separates the short-distance physics of the parton reaction at the hard scale from the long-distance physics describing the binding of the partons inside the hadrons.

A   Hadron pairs

Two-photon exclusive processes such as $\gamma^{(*)}\gamma \to$ hadron pairs at large energies s and fixed scattering angle θ are important to test the validity of the leading-twist factorization theorems and to measure the shape of the hadron distribution amplitudes. The comparison of the leading-order perturbative QCD prediction [39] for the sum of $\pi^+\pi^-$ plus K^+K^- final states with recent CLEO data is shown in Fig. 16 (left). The data seem to support the validity of the perturbative description for energies above $W \gtrsim 2$ GeV. A recent calculation [40], however, which takes into account the partonic transverse momentum as well as Sudakov corrections predicts a perturbative cross-section significantly below the experimental results, Fig. 16 (right). The difference between the two theoretical analyses can be attributed to the phenomenological value for the pion form factor which enters the prediction in the collinear approximation, see [39,40] for further discussion.

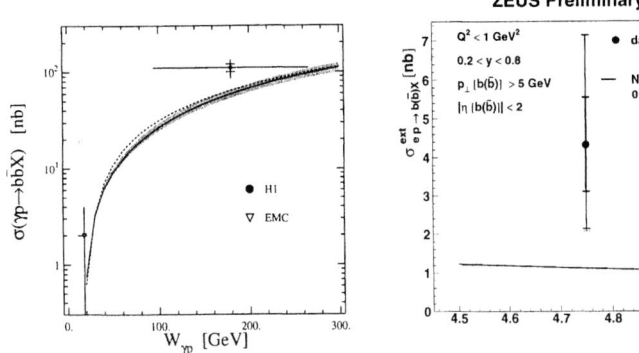

FIGURE 12. Left: total photoproduction cross-section for beauty production measured by H1 compared to the NLO QCD prediction. Right: cross section measured by ZEUS extrapolated to the parton level compared to the NLO QCD prediction. See [32] for further details.

Whether processes like $\gamma\gamma \to \pi^+\pi^-$ might be dominated by soft contributions up to large energies is thus still subject to debate.

B Charmonium resonances

Measurements of the two-photon width of charmonium resonances can be used to study QCD at scale $\sim m_c$. The two-photon width of a charmonium resonance R_c is usually measured in photon-photon interactions, $\gamma\gamma \to R$, or through the decay of the resonance produced in $\bar{p}p$ interactions, $\bar{p}p \to R \to \gamma\gamma$. The cross-section for the exclusive production of $\bar{p}p$ resonances in two-photon collisions is too small to be observed.

A compilation of measurements of the two-photon width of the η_c, the χ_{c2}, and the χ_{c0} (spin 1 resonances can not be produced in the collisions of two real photons) is shown in Fig. 14 [41]. The most striking feature confirmed at this conference by [41] is the discrepancy between the two-photon width of the χ_{c2} measured in photon-photon interactions and in $\bar{p}p$ interactions by E760 and E835. The origin of this discrepancy is still unclear but higher precision measurements using the full LEP/CLEO statistics could certainly help to clarify the situation.

C Glueball searches

The (non-)production of glueballs in two-photon interactions is considered an important tool to identify potential glueball states. In real photon-photon interactions only scalar and tensor resonances can be produced. There is no direct coupling of

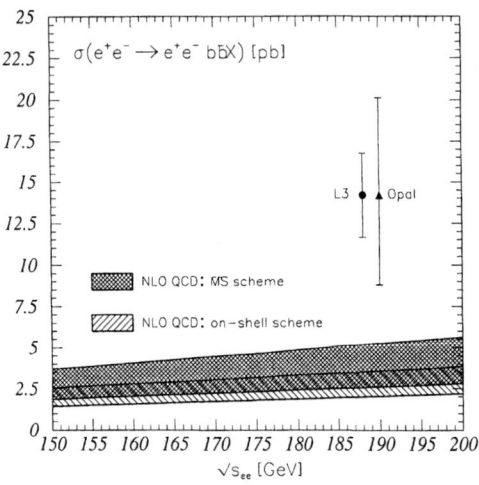

FIGURE 13. Total cross-section for beauty production in two-photon collisions at LEP. The NLO calculation is compared to the experimental data by L3 [34] and OPAL [33]. Two theoretical predictions are presented corresponding to different choices of the heavy quark mass renormalization scheme: the on-shell scheme (lower band) which has been used in previous calculations [38] and the $\overline{\text{MS}}$ scheme (upper band). The renormalization and factorization scales have been varied between 2 GeV and 15 GeV. For the $\overline{\text{MS}}$ heavy quark mass $\overline{m}_b(\overline{m}_b) = 4.25$ GeV is used and the on-shell mass is varied in the range 4.5 GeV $\leq m_b^{\text{os}} \leq 5$ GeV.

photons to gluons and production of scalar or tensor glueballs in two-photon interactions should therefore be highly suppressed. To measure this suppression the quantity stickiness was introduced which relates the coupling of the resonance to photons and the coupling to gluons using the two-photon width of the resonance and the width of the radiative decays of J/ψ mesons into the glueball candidate.

Already at Photon 99, Hans Paar has criticised the use of Stickiness since it is not normalised and has advocated to use Gluiness instead [42]. Stickiness is of order 10 for pure s$\bar{\text{s}}$ states due to the quark charge, whereas Gluiness is expected to be near 1 for any pure q$\bar{\text{q}}$ state, but experiments still predominately use Stickiness.

Several glueball candidates which have previously been observed in the gluon rich environment of radiative J/ψ decays have been searched for in $\gamma\gamma$ experiments and discussed at this conference: The scalar states $f_0(1500)$ and $f_0(1710)$ by ALEPH [43] and the pseudo-scalar $\eta(1440)$ by L3 [44]. In the case of the $f_0(1500)$ and the $f_0(1710)$, a relatively small lower limit on the stickiness of 1.4 and 0.3 was measured by ALEPH, mainly due to the small branching ratio in the $\pi^+\pi^-$ final state used here. For the $\eta(1440)$ L3 obtains a stickiness greater than 87 (at the 95% confidence level) in the $\eta\pi^+\pi^-$ channel which indicates a large gluonium content in this resonance.

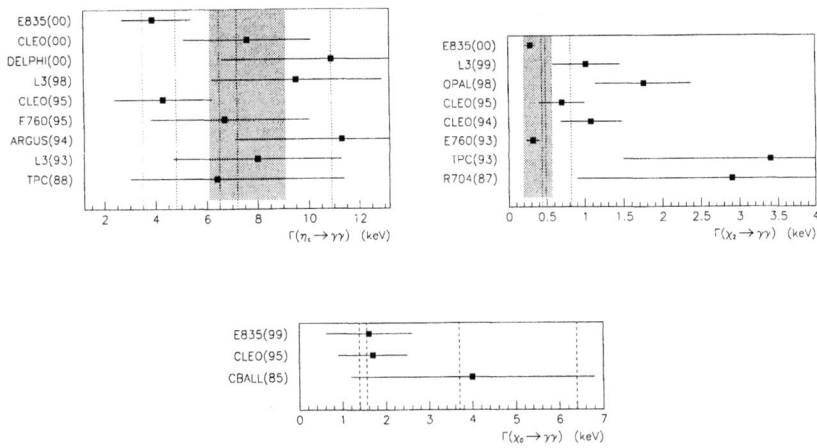

FIGURE 14. Compilation of the two-photon widths of the η_c, the χ_{c2}, and the χ_{c0} mesons [41].

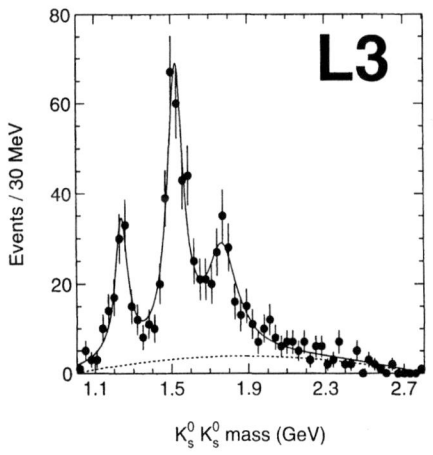

FIGURE 15. The $K_s K_s$ mass spectrum measured by L3.

The tensor candidate $f_j(2220)$ has been studied by CLEO [45] in the $\pi^+\pi^-$ and $K_S K_S$ final states and by L3 [46] in the $K_S K_S$ final state (Fig. 15). CLEO obtains a lower limit on the stickiness of 102 and L3 a lower limit of 73. This supports the hypothesis of the glueball nature of the $f_j(2220)$ but more confirmation of the existence of the state is needed.

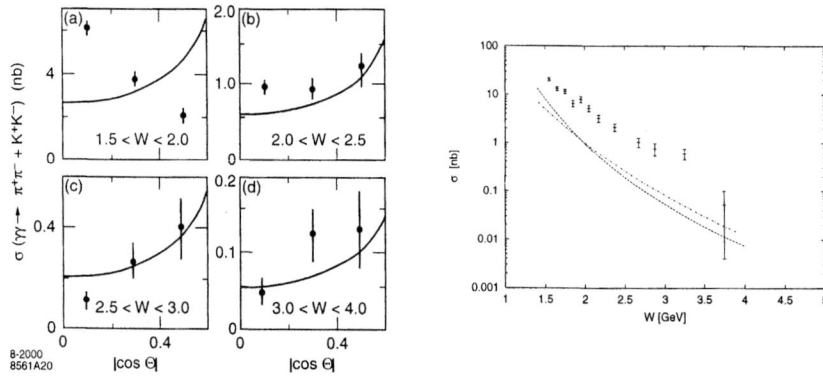

FIGURE 16. Comparison of two different perturbative predictions for $\gamma\gamma \to \pi^+\pi^- + K^+K^-$ with experimental data from CLEO. Left: the angular distribution compared to the prediction [39]. Right: The total cross-section compared to a calculation [40] which includes Sudakov corrections. See [39,40] for further details.

VII PHOTON DIFFRACTION AND TOTAL CROSS-SECTIONS

The behaviour of the cross-section in the limit of high energy and fixed momentum transfer is one of the outstanding open questions in the theory of strong interactions [47]. The BFKL equation adresses this issue in terms of the fundamental quanta, quarks and gluons, at least for special situations in which perturbation theory applies. Interesting responses to the challenge of the large higher-order corrections to the BFKL kernel have been discussed [48] but still more work is needed to arrive at a coherent theoretical picture.

The total hadronic cross-section for off-shell photons at e^+e^- colliders is considered to be the optimal test of BFKL dynamics. The process has been measured by the L3 [49] and OPAL [50] collaborations at LEP. BFKL effects as large as predicted by leading-order calculations are not confirmed by the data, see Fig. 17. First attempts to include the NLO corrections to the BFKL kernel seem in better agreement with the experimental results. However, for a consistent NLO prediction of the $\gamma^*\gamma^*$ cross-section the so far uncalculated NLO corrections to the photon impact factors need to be included.

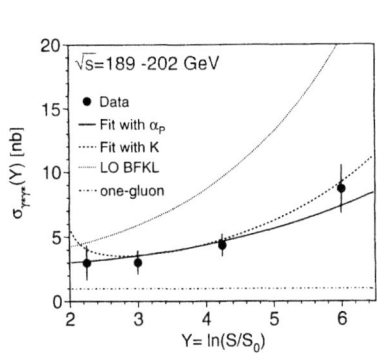

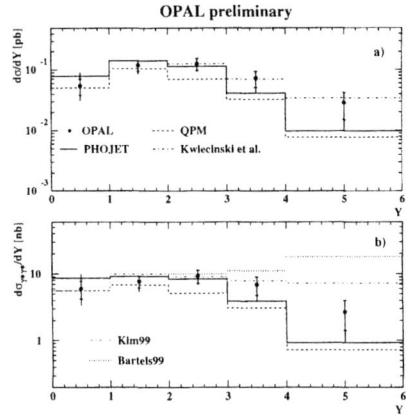

FIGURE 17. Cross-sections for double-tagged two-photon events $\gamma^*\gamma^* \to$ hadrons. Left: L3 measurement for $e^+e^- \to e^+e^-$ hadrons, after subtraction of the QPM contribution, at $\sqrt{s} \simeq 189 - 202$ GeV ($\langle Q^2 \rangle = 15$ GeV^2). The data are compared to the LO BFKL calculation (dotted line). Also shown are different fits to the data, see [49] for details. Right: OPAL measurement for $e^+e^- \to e^+e^-$ hadrons and $\gamma^*\gamma^* \to$ hadrons at $\langle Q^2 \rangle = 17$ GeV^2. The data are compared to the QPM prediction, the PHOJET Monte Carlo program and different BFKL calculations. See [50] for details.

VIII PHOTON PHYSICS AT A FUTURE LINEAR COLLIDER

Two-photon physics can be studied at a future linear e^+e^- collider through the scattering of bremsstrahlung photons. The bremsstrahlung spectrum, however, is peaked at low photon energies, so that the more interesting physics analyses at higher $\gamma\gamma$ energies are limited by statistics. Compton backscattering of laser light offers the prospect of intense beams of high-energy photons. At such a Photon Collider centre-of-mass energies of about 80% of the e^+e^- collider energy and $\gamma\gamma$ luminosities close to the e^+e^- luminosity can be achieved [51].

There has been a intense debate about how the physics case for the Photon Collider compares in strength with that for a linear collider in e^+e^- mode. The central physics motivation for a Photon Collider is to study the properties of a Standard Model Higgs boson or to explore the Higgs sector of theories beyond the Standard Model. Determining the $H\gamma\gamma$ coupling is particularly interesting because it is built up by loops of charged particles. If the mass of the loop particle is generated through the Higgs mechanism, the decoupling of the heavy particles is lifted. The $\gamma\gamma$ width thus reflects the spectrum of these heavy states with masses possibly beyond the direct reach of accelerator experiments. With an integrated luminosity of $\int \mathcal{L}_{\gamma\gamma} \approx 50$ fb $^{-1}$ the two-photon width of a light Higgs boson in the mass range $M_h \approx 120$ GeV can be measured with high statistical accuracy of about

2% [52].

A precise measurement of the $\gamma\gamma$ and $Z\gamma$ Higgs couplings can help to discriminate between the Standard Model and theories with extended Higgs sectors, like the Minimal Supersymmetric Model (MSSM) or the Two Higgs Doublet Model (2HDM) which contain five physical Higgs particles, see Fig. 18 [53,54].

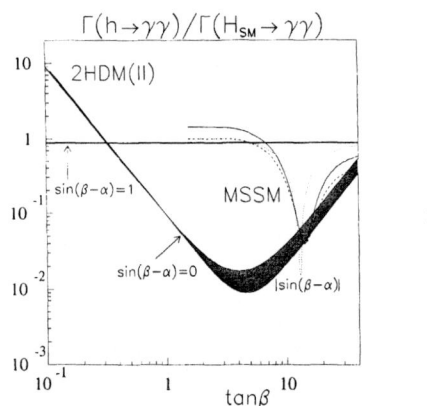 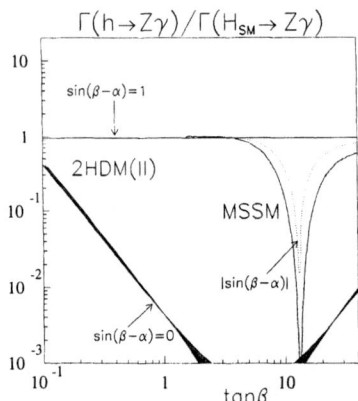

FIGURE 18. The Higgs boson $h \rightarrow \gamma\gamma$ and $h \rightarrow Z\gamma$ decay widths in the Minimal Supersymmetric Model (MSSM) and Two Higgs Doublet Model (2HDM) compared to the Standard Model prediction as functions of $\tan\beta$. The results correspond to a Higgs mass of M_h=100 GeV. In the 2HDM, the band for $\sin(\beta - \alpha) = 0$ corresponds to a mass of the charged Higgs boson ranging from 165 GeV to infinity. For the MSSM curves, the solid lines include effects of supersymmetric loop particles (default masses), whereas the dashed lines do not. The dotted curve describes the function $|\sin(\beta - \alpha)|$ as a function of $\tan\beta$ for fixed mass M_h=100 GeV in the MSSM. See [53,54] for further details.

To definitely establish the existence of an extended Higgs sector the additional heavy Higgs bosons have to be discovered. Detailed studies have shown that the heavy MSSM Higgs bosons H and A may escape detection at the LHC. At e^+e^- colliders these particles can only be found in associated production $e^+e^- \rightarrow HA$ with masses below approximately half the total e^+e^- energy. At Photon Colliders in comparison, Higgs particles can be produced as resonances, $\gamma\gamma \rightarrow H, A$, and the search for heavy Higgs bosons can be extended to masses not accessible elsewhere. The cross-sections for the resonant production of H and A decaying into $b\bar{b}$ or neutralino final states and the corresponding background cross sections are shown in Fig. 19 [55]. While for $b\bar{b}$ final states the background is strongly suppressed against the signal, detailed analyses of the final state topologies are needed to separate the neutralino decay channel from the background charginos above threshold. The negative parity of the pseudoscalar Higgs boson A could be established at a Photon Collider with polarized beams.

In addition to Higgs physics there is a substantial list of exciting topics for

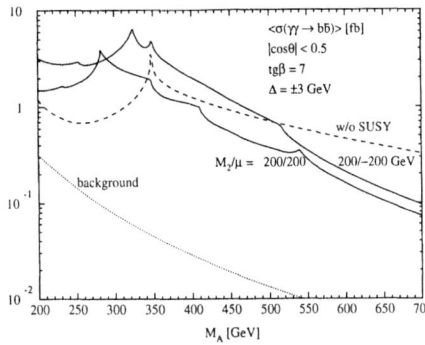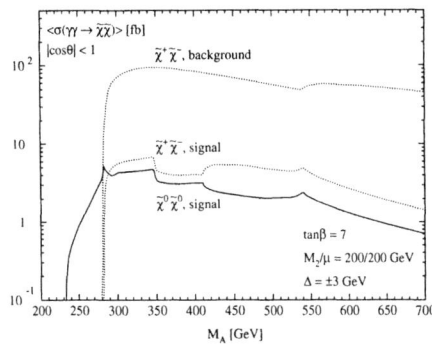

FIGURE 19. Cross-section for resonant heavy Higgs boson H, A production as a function of the Higgs boson mass M_A with final decay into $b\bar{b}$ pairs (left) and chargino and neutralino final states (right). See [55] for further details.

which a Photon Collider would extend and supplement studies done at a e^+e^- collider [51,54]. Anomalous couplings for example, or the effects of Kaluza-Klein excitations in theories with large extra dimensions, can be probed in W boson pair or top quark pair production $\gamma\gamma \rightarrow$ WW or $t\bar{t}$. The properties of leptoquarks or excited fermions can be studied in $e\gamma$ or polarized $\gamma\gamma$ collisions. And supersymmetric charged particles may be produced with potentially large rates in two-photon collisions or probed indirectly in $\gamma\gamma \rightarrow \gamma\gamma$ scattering.

Last but not least, two-photon interactions at a high-energy e^+e^-, $e\gamma$ or $\gamma\gamma$ colliders will allow a comprehensive program of QCD analyses, including detailed measurements of the photon structure and studies of the total $\gamma\gamma$ cross section [56].

The total cross-section of $\gamma\gamma$ interactions is not understood from first principles. Systematic comparisons between the $p\bar{p}/pp$, γp and $\gamma\gamma$ cross-sections over a wide range of energies could shed light on the underlying QCD dynamics. To discriminate between various phenomenological models the $\gamma\gamma$ cross-section need to be measured with a precision better than 10%. To achieve such a precision a Photon Collider may be necessary [56].

The $e\gamma$ mode of a Photon Collider has a great potential to measure the photon structure function $F_2^\gamma(x, Q^2)$ at high momentum transfer Q^2 and over a large range of Bjorken-x, as demonstrated in Fig. 20. While the analysis of electroweak neutral and charged current structure functions at high Q^2 will determine the up and down type quark content of the photon, the gluon content can be reconstructed from jet and charm production [57]. Polarised high-energy photon beams, in addition, offer a unique opportunity to study the polarised parton distributions of the photon, for which no experimental data are available to date. With a high-energy $\gamma\gamma$ and $e\gamma$ collider the structure of the photon could at last be measured with comparable precision to that of the proton.

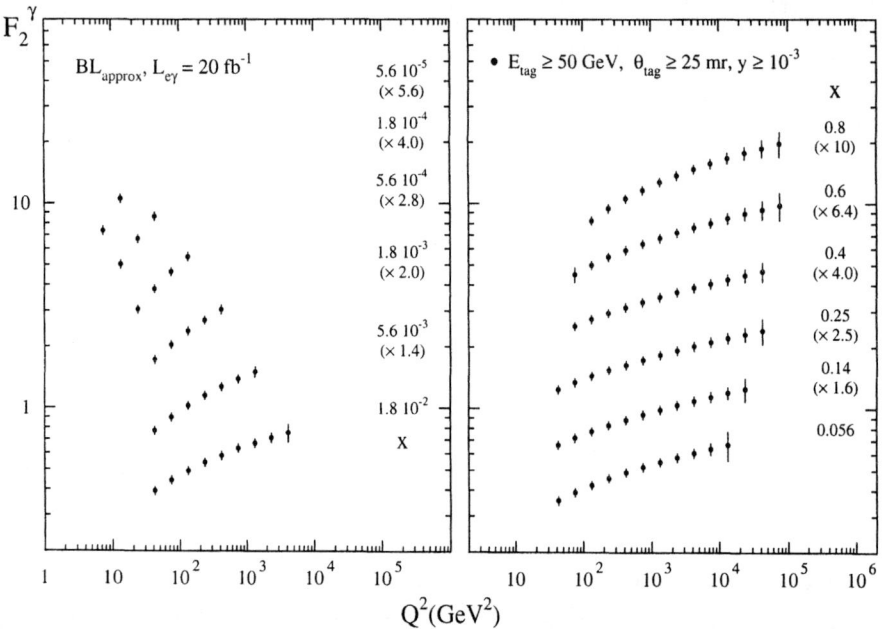

FIGURE 20. The kinematic coverage of F_2^γ measurements at a $e\gamma$ collider. The prospective data points are shown with statistical and (5%) systematic error, for 20 fb^{-1} photon collider luminosity. See [56] for further details.

IX   CONCLUSIONS

Many new theoretical and experimental results have been presented at this conference and only a biased selection of them could be discussed in this summary. New precise data from LEP and HERA on the structure of real and virtual photons calls for a common quantitative understanding of the LEP and HERA data on photon structure, especially in the form of new parton densities which make use of the new measurements. To accomplish this, we also still need a better understanding of the non-perturbative effects (hadronisation, underlying event) which still lead to systematic limitations in the interpretation of the data.

Also the classical field of two-photon interactions, resonance production, continues to produce interesting results: For example the difference between the two-photon width of χ_{c2} measured in $\bar{p}p$ and $\gamma\gamma$ interactions is not understood. Important contributions have been made to the search for glueball states.

ACKNOWLEDGEMENTS

We would like to thank Alex Finch and his colleagues for the perfect organisation of the conference. As summary speakers we unfortunately had not much time to enjoy Ambleside and the beautiful Lake District. Alex Finch and his team created a wonderful atmosphere for a successful and enjoyable conference.

REFERENCES

1. M. Krawczyk, these proceedings, hep-ph/0012179.
2. ALEPH, L3, OPAL, and the LEP two-photon group, CERN-EP-2000-109, hep-ex/0010041, subm. to Eur. Phys. J. C.
3. ALEPH Coll., ALEPH note 99-038.
4. E. Clay, these proceedings; OPAL Coll., G. Abbiendi et al., Eur. Phys. J. C18 (2000) 15.
5. L3 Coll., M. Acciarri et al., Phys. Lett. B 436 (1998) 403-416.
6. M. Glück, E. Reya and A. Vogt, Phys. Rev. **D45** (1992) 3986, M. Glück, E. Reya and A. Vogt, Phys. Rev. **D46** (1992) 1973.
7. G.A. Schuler and T. Sjöstrand, Z. Phys. C68 (1995) 607.
8. K. Hagiwara et al., Phys. Rev. D51 (1995) 3197.
9. R. Taylor, these proceedings, hep-ex/0010010.
10. J. Chyla, these proceedings, hep-ph/0010309.
11. E. Witten, Nucl. Phys. B120 (1977) 189; in the parametrisation of L.E. Gordon et al., Z. Phys. C56 (1992) 307.
12. R. Nisius, these proceedings, hep-ex/0010020. OPAL Coll., G. Abbiendi et al., Eur. Phys. J. C16 (2000) 579.
13. E. Laenen, S. Riemersma, J. Smith and W.L. van Neerven, Phys. Rev. D49 (1994) 5753; E. Laenen and S. Riemersma, Phys. Lett. B376 (1996) 169.
14. P. Achard, these proceedings, hep-ex/0010024.
15. M. Wing, these proceedings, hep-ex/0010042.
16. OPAL Coll., K. Ackerstaff et al., Eur. Phys. J. C6 (1999) 253.
17. B.A. Kniehl, G. Kramer, B. Pötter, hep-ph/0011155; J. Binnewies, B.A. Kniehl, G. Kramer, Phys. Rev. D53 (1996) 6110.
18. AMY Coll., B.J. Kim et al., Phys. Lett. B325 (1994) 248; TOPAZ Coll., H. Hayashii et al., Phys. Lett. B314 (1993) 149.
19. OPAL Coll., G. Abbiendi et al., Eur. Phys. J. C10 (1999) 547; OPAL Coll., K. Ackerstaff et al., Z. Phys. C73 (1997) 433.
20. P. Hodgson, M. Lehto, B. Pötter, these proceedings.
21. B. Surrow, Th. Wengler, these proceedings.
22. M. Klasen, T. Kleinwort, G. Kramer, Eur. Phys. J. Direct C1 (1998) 1, and private communications.
23. J. Terron, these proceedings.
24. P. Aurenche, M. Fontannaz, J.P. Guillet, Z. Phys. C64 (1994) 621.
25. B. Pötter, Comp. Phys. Comm. 119 (1999) 45.

26. S. Maxfield, Di-jet Cross-Sections in Photoproduction and Photon Structure, these proceedings; H1 Coll., C. Adloff et al., Phys. Lett. B483 (2000) 36; H1 Coll., C. Adloff et al., Eur. Phys. J. C10 (1993) 363.
27. K. Saeki, T. Uematsu, these proceedings, hep-ph/0010119.
28. PLUTO Coll., C. Berger et al., Phys. Lett. B142 (1984) 119.
29. S. Maxfield, Jets in DIS and the Virtual Photon Structure, these proceedings.
30. C. Glasman, these proceedings, hep-ex/0010017.
31. D. Hochman, these proceedings, hep-ex/0010050.
32. K. Daum, these proceedings.
33. A. Csilling, these proceedings, hep-ex/0010060.
34. S. Saremi, these proceedings.
35. U. Sieler, these proceedings.
36. A. Sokolov, these proceedings.
37. V. Andreev, private communication.
38. E. Laenen, these proceedings.
39. S. Brodsky, these proceedings, hep-ph/0010176.
40. C. Vogt, these proceedings, hep-ph/0010040.
41. W. Baldini, these proceedings.
42. H. Paar, Nucl. Phys. B (Proc. Suppl.) 82 (2000) 337.
43. R. Jones, these proceedings.
44. V. Andreev, these proceedings.
45. CLEO Coll., M.S. Alam et al., Phys. Rev. Lett. 81 (1998) 3328; CLEO Coll., R. Godang et al., Phys. Rev. Lett. 79 (1997) 3829.
46. S. Braccini, these proceedings, hep-ex/0010030.
47. G. Shaw, these proceedings, hep-ph/0011007.
48. D. Ross, these proceedings.
49. M. Wadhwa, these proceedings.
50. M. Przybycien, these proceedings.
51. V. Telnov, these proceedings, hep-ex/0010034.
52. G. Jikia, S. Söldner-Rembold, hep-ex/010105; M. Melles, W.J. Stirling, V.A. Khoze, Phys. Rev. D61 (2000) 054015.
53. M. Krawczyk, these proceedings.
54. I. Ginzburg, these proceedings
55. M. Mühlleitner in "Proceedings of the International Workshop on High Energy Photon Colliders", DESY, Hamburg, 2000 (hep-ph/0101177), M. Mühlleitner et al., hep-ph/0101083.
56. A. De Roeck, these proceedings, hep-ph/0101075.
57. P. Jankowski, these proceedings.

LIST OF PARTICIPANTS

Mr. Tetsuo Abe
ZEUS-Tokyo group
DESY F1J
Notkestrasse 85
22603 Hamburg
Germany
abe@mail.desy.de

Mr. Pablo Achard
24 quai E. Ansermet
1221 Genève 4
Switzerland
pablo.achard@cern.ch

Dr. Valeri Andreev
EP Division
CERN
CH-1211 Genève 23
Switzerland
Valeri.Andreev@cern.ch

Dr. Wander Baldini
Dipartimento di Fisica
Via paradiso 12
44100 Ferrara
Italy
baldini@fe.infn.it

Dr. Armin Böhrer
Dr. Arm Siegen University
D-57068 Siegen
Germany
armin.boehrer@cern.ch

Mr. Saverio Braccini
Univ. de Genève
DPNC
24 Quai Ernest-Ansermet
CH-1211 Genève 4
Switzerland
Saverio.Braccini@cern.ch

Prof. Stanley J. Brodsky
SLAC Theory Group
Stanford University
PO Box 4349
Stanford
CA 94309
USA
sjbth@slac.stanford.edu

Dr. Armen Bunyatyan
DESY-H1
Notkestrasse 85
22607 Hamburg
Germany
bunar@mail.desy.de

Dr. Jonathan Butterworth
Department of Physics and Astronomy
University College London
Gower St.
WC1E 6BT
London
United Kingdom
jmb@hep.ucl.ac.uk

Dr. Susan Cartwright
Dept of Physics and Astronomy
University of Sheffield
Hicks Building
Sheffield
S3 7RH
United Kingdom
susan.cartwright@cern.ch

Prof. Jiri Chýla
Institute of Physics
AS CR
Na Slovance 2
18221 Prague 8
Czech Republic
chyla@fzu.cz

Mr. David Clarke
Physics Dept.
Lancaster University
Lancaster
LA1 4YB
United Kingdom
david.clarke@cern.ch

Dr. Edmund Clay
44 Finstall Rd
Aston Fields
Bromsgrove
Worcs
B60 2EA
United Kingdom
ec@hep.ucl.ac.uk

Dr. Ákos Csilling
G06710
CERN
CH-1211 Genève 23
Switzerland
Akos.Csilling@cern.ch

Dr. Jaroslav Cvach
Na Slovance 2
CZ-182 21 Praha 8
Czech Republic
cvach@fzu.cz

Dr. Wilfrid Da Silva
Universités Paris 6 et 7
Tour 33 RdCh
BP200 4 Place Jussieu
75252 Paris 5
France
dasilva@in2p3.fr

Mrs. Dr. Karin Daum
DESY - F32
Notkestrasse 85
D-22603 Hamburg
Germany
daum@mail.desy.de

Dr. Albert De Roeck
CERN/EP
CH-1211 Geneva 23
Switzerland
deroeck@mail.cern.ch

Mr. Johannes Elmsheuser
Fakultät für Physik
Hermann-Herder Str. 3
79104 Freiburg
Germany
elmsheus@ruhpb.physik.uni-freiburg.de

Prof. Frederik Erné
NIKHEF
P.O. Box 41882
1009 DB Amsterdam
The Netherlands
z63@nikhef.nl

Dr. John Field
Departement de Physique Nucleaire et
Corpusculaire
Université de Genève
24 quai Ernest Ansermet
CH-1211 Genève 4
Switzerland
john.field@cern.ch

Dr. Alexander Finch
Department of Physics
University of Lancaster
Lancaster
LA1 4YB
United Kingdom
A.Finch@lancaster.ac.uk

Dr. Jeff Forshaw
Department of Physics and Astronomy
University of Manchester
Brunswick Street
Manchester
M13 9PL
United Kingdom
forshaw@mail.cern.ch

Mr. Christer Friberg
Dept. of Theoretical Physics
Lund University
Sölvegatan 14A
SE-223 62 LUND
Sweden
christer@thep.lu.se

Prof. Ilya Ginzburg
Institute of Mathematics SB RAS
Prosp.ac. Koptyug 4
630090 Novosibirsk
Russia
ginzburg@math.nsc.ru

Dr. Claudia Glasman
DESY F1
Notkestrasse 85
22603 Hamburg
Germany
claudia@mail.desy.de

Prof. Errol (Asher) Gotsman
School of Physics and Astronomy
Tel Aviv University
Tel Aviv 69978
Israel
gotsman@post.tau.ac.il

Dr. Thierry Gousset
SUBATECH B.P. 20722
F-44307 Nantes cedex 3
France
gousset@subatech.in2p3.fr

Miss. Claire Gwenlan
Dept of Physics and Astronomy
UCL
Gower Street
London
WC1E 6BT
United Kingom
cg@hep.ucl.ac.uk

Mr. Daniel Haas
E00410
CERN
CH-1211 Geneva 23
Switzerland
Daniel.Haas@cern.ch

Mr. Johannes Hess
Johannes Hess Universität Siegen
ALEPH-Gruppe
Emmy-Noether Campus
Walter-Flex-Str. 3
57068 Siegen
Germany
hess@aleph.physik.uni-siegen.de

Dr. Don Hochman
Particle Physics Dept.
Weizmann Institute of Science
Rehovot
76100 Israel
fhhochmn@wicc.weizmann.ac.il

Mr. Paul Hodgson
Department of Physics and Astronomy
Hicks building
Hounsfield Rd
University of Sheffield
S3 7RH
United Kingdom
p.hodgson@shef.ac.uk

Mr. Pawel Jankowski
ul. Hoza 69 00-681 Warsaw
Poland
pjank@fuw.edu.pl

Mrs. Carrie Johnson
17 Kestrel Grove
Bournville
Birmingham
West Midlands
B30 1TQ
United Kingdom
clj@hep.ph.bham.ac.uk

Dr. Roger Jones
Lancaster University
Lancaster
LA1 4YB
United Kingdom
Roger.Jones@cern.ch

Dr. Sergey Kananov
School of Physics
Tel Aviv University
69978 Ramat Aviv
Israel
sergey@zufo.tau.ac.il

Dr. Frederic Kapusta
LPNHE
Universités Paris 6 et 7
Tour 33 RdCh
BP200 4 Place Jussieu
75252 Paris 5
France
kapusta@in2p3.fr

Prof. Uri Karshon
Particle Physics Department
Weizmann Institute of Science
Rehovot 76100
Israel
uri.karshon@weizmann.ac.il

Prof. Maria Novella Kienzle-Focacci
DPNC Physics Institute
Geneva University
24 quai E. Ansermet
Genève
CH-1211 Genève 4
Switzerland
maria.kienzle@cern.ch

Ms. Katarzyna Klimek
Institute of Nuclear Physics
ul. Kawiory 26 A 30-055
Krakow
Poland
kklimek@mail.desy.de

Dr. Michael Krämer
University of Edinburgh
Edinburgh
EH9 3JZ
United Kingdom
Michael.Kraemer@ed.ac.uk

Professor. Maria Krawczyk
Institute of Theoretical Physics
Warsaw University
ul. Hoza 69
00681 Warsaw
Poland
krawczyk@fuw.edu.pl

Dr. Eric Laenen
Kruislaan 409
1098 SJ Amsterdam
The Netherlands
Eric.Laenen@nikhef.nl

Dr. Benno List
Paul-Scherrer-Institut
WLGA E19
CH-5232 Villigen-PSI
Switzerland
Benno.List@cern.ch

Dr. Anna Macchiolo
EP Division Cern
CH-1211 Geneva 23
Switzerland
Anna.Macchiolo@cern.ch

Dr. Hanna Mahlke-Krüger
Cornell University
Ithaca
NY 14853-5001
USA
mahlke@mail.desy.de

Prof. Shigeo Matsumoto
Department of Physics
Chuo Univ.
1-13-27 Kasuga Bunkyo-ku
Tokyo 112-8551
Japan
matumoto@phys.chuo-u.ac.jp

Dr. Stephen Maxfield
Department of Physics
University of Liverpool
Oxford Street
Liverpool
L69 7ZE
United Kingdom
sjm@hep.ph.liv.ac.uk

Prof. David J. Miller
Physics and Astronomy
UCL
Gower Street
London WC1E 6BT
United Kingdom
dave.miller@cern.ch

Dr. Richard Nisius
CERN
Bat 28 R-024
CH-1211 Genève 23
Switzerland
Richard.Nisius@cern.ch

Dr. Saro Ong
IPN Orsay/Université Paris Sud
F-91406 Orsay Cedex
France
ong@ipno.in2p3.fr

Prof. Hans P. Paar
Physics Department
0319 9500 Gilman Drive
La Jolla
CA 92093
USA
hpaar@ucsd.edu

Dr. Giulia Pancheri
Via Enrico Fermi 40
I00044 Frascati (Rome)
Italy
pancheri@lnf.infn.it

Prof. Michael Pennington
Physics Department
University of Durham
Durham
DH1 3LE
United Kingdom
m.r.pennington@durham.ac.uk

Dr. Björn Pötter
Foehringer Ring 6
D-80807 Munich
Germany
poetter@mppmu.mpg.de

Mr. Gerrit Prange
Raum A 008
FB Physik
Emmy-Noether-Campus
57068 Siegen
Germany
gerrit.prange@cern.ch

Dr. Mariusz Przybycien
Institute of Nuclear Physics
Kawiory 26A
30055 Krakow
Poland
Mariusz.Przybycien@cern.ch

Prof. Douglas Ross
Department of Physics and Astronomy
University of Southampton
Southampton
SO17 1BJ
United Kingdom
dar@phys.soton.ac.uk

Mr. Sepehr Saremi
4415 Saugus Ave.
Apt. 201 Sherman Oaks
CA 91403-4060
USA
Sepehr.Saremi@cern.ch

Dr. Alexander Savin
DESY-F1
Notkestrasse 85
22607 Hamburg
Germany
savin@mail.desy.de

Prof. Valeri Serbo
Physics Department
Novosibirsk State University
Pirogova 2
Novosibirsk 630090
RUSSIA
serbo@math.nsc.ru

Dr. Graham Shaw
Department of Physics and Astronomy
University of Manchester
Manchester
M13 9PL
United Kingdom
graham.shaw@man.ac.uk

Mr. Uwe Sieler
Fachbereich Physik
Emmy-Noether-Campus
Gebäude A
Universität-Gesamthochschule Siegen
D - 57068 Siegen
Germany
uwe.sieler@cern.ch

Mr. Alexander Singovski
CERN/PE
CH-1211 Geneva 23
Switzerland
Alexander.Singovski@cern.ch

Dr. Anatoli Sokolov
EP 13-1-020
CERN
CH-1211 Genève 23
Switzerland
Anatoli.Sokolov@cern.ch

Dr. Stefan Söldner-Rembold
CERN-EP
CH-1211 Geneva 23
Switzerland
stefan.soldner-rembold@cern.ch

Mr. Rainer Stamen
IIHE CP 230
Université Libre de Bruxelles
Boulevard du Triomphe
B 1050
Bruxelles
stamen@physik.uni-dortmund.de

Dr. Bernd Surrow
CERN EP-Division/OPAL
CH-1211 Geneva 23
Switzerland
bernd.surrow@cern.ch

Mr. Russell Taylor
CERN
G06710
CH-1211 Genève 23
Switzerland
Russell.Taylor@cern.ch

Prof. Valery Telnov
DESY - MPY
Notkestrasse 85
D-22603 Hamburg
Germany
telnov@mail.desy.de

Dr. Juan Terron
DESY F1
Notkestrasse 85
22603 Hamburg
Germany
terron@mail.desy.de

Dr. Sharka Todorova-Nova
CERN
CH-1211 Geneva 23
Switzerland
todorovova@cern.ch

Prof. Tsuneo Uematsu
Yoshida
Sakyo-ku
Kyoto 606-8501
Japan
uematsu@phys.h.kyoto-u.ac.jp

Dr. Pierre Van Mechelen
University of Antwerpen (UIA)
Physics Department
Universiteitsplein 1
2610 Antwerpen
Belgium
Pierre.VanMechelen@uia.ua.ac.be

Mr. Carsten Vogt
Universität Wuppertal
Fachbereich Physik
Gaussstrasse. 20
42097 Wuppertal
Germany
cvogt@theorie.physik.uni-wuppertal.de

Dr. Maneesh Wadhwa
EP Division
CERN
CH-1211 Geneva 23
Switzerland
Maneesh.Wadhwa@cern.ch

Dr. Thorsten Wengler
Bat 28/2-009
CERN
CH-1211 Geneva 23
Switzerland
thorsten.wengler@cern.ch

Dr. Mike Whalley
Physics Department
Science Laboratories
South Road
Durham City
DH1 3LE
United Kingdom
M.R.Whalley@durham.ac.uk

Dr. Matthew Wing
ZEUS
DESY
Notkestrasse 85
22607 Hamburg
Germany
wing@mail.desy.de

Miss. Angela Wyatt
Dept. of Physics and Astronomy
University of Manchester
Manchester
M13 9PL
United Kingdom
wyatt@mail.desy.de

Author Index

A

Abe, T., 305
Achard, P., 172
Andreev, V. P., 380

B

Badełek, B., 18
Baldini, W., 358
Braccini, S., 366
Brodsky, S. J., 315
Bunyatyan, A., 231

C

Carimalo, C., 430
Chapkin, M., 252
Chýla, J., 49, 55
Clay, E. W., 31
Csilling, Á., 276

D

Da Silva, W., 430
Daum, K., 283
De Roeck, A., 416
Diehl, M., 333

F

Friberg, C., 197
Frixione, S., 239

G

Ginzburg, I. F., 326, 409
Glasman, C., 203, 209
Godbole, R. M., 131
Gotsman, E., 76
Gousset, T., 333
Grau, A., 131

H

Haas, D., 339
Hochman, D., 292
Hodgson, P., 185

J

Jankowski, P., 423
Johnson, C. L., 102
Jones, R. W. L., 374

K

Kananov, S., 394
Kapusta, F., 430
Kienzle-Focacci, M. N., 154
Klimek, K., 125
Krämer, M., 239, 457
Krawczyk, M., 3, 18
Kweiciński, J., 18

L

Laenen, E., 239
Lehto, M., 185
Levin, E., 76
List, B., 108

M

Macchiolo, A., 439
Mahlke-Krüger, H., 88
Maor, U., 76
Maxfield, S., 191, 216

N

Naftali, E., 76
Nisius, R., 37

O

Obraztsov, V., 252

P

Paar, H. P., 351
Pancheri, G., 131
Pennington, M. R., 388
Pire, B., 333
Pötter, B., 185
Przybycień, M., 140

R

Ross, D., 82

S

Saremi, S., 266
Sasaki, K., 37
Savin, A. A., 299
Schiller, A., 326
Serbo, V. G., 326
Shaw, G., 63
Sieler, U., 245
Singovski, A., 448

Sokolov, A., 252
Söldner-Rembold, S., 457
Stamen, R., 117
Staśto, A. M., 18
Surrow, B., 179

T

Taylor, R. J., 25
Telnov, V., 403
Terrón, J., 225

U

Uematsu, T., 37

V

Vogt, C., 345

W

Wadhwa, M., 140
Wengler, T., 179
Wing, M., 163
Wyatt, A., 96